Current Plant Science:
Novel Concepts and Innovative Strategies: Chittaranjan Kole

1. Next Generation Crop Improvement for Agricultural Sustainability and Food Security
   Belaghihalli N Gnanesh and *Chittaranjan Kole (Editors), 2026*
2. Next Generation Crop Production for Agricultural Sustainability and Food Security
   Belaghihalli N Gnanesh, Megha N Parajulee and *Chittaranjan Kole (Editors), 2026*
3. Next Generation Crop Protection for Agricultural Sustainability and Food Security
   Belaghihalli N Gnanesh, Megha N Parajulee, Ankush M Raut and *Chittaranjan Kole (Editors), 2026*

Current Plant Science: Novel Concepts and Innovative Strategies: Chittaranjan Kole

Next Generation Crop Protection for Agricultural Sustainability and Food Security

Editors:

Belaghihalli N Gnanesh
Indian Council of Agricultural Research
JSS Krishi Vigyan Kendra, Suttur, Mysore, India

Megha N Parajulee
Regents Fellow Professor
Texas A&M University AgriLife Research and Extension Center
Texas, USA

Ankush M Raut
Associate Professor, Department of Entomology
School of Agriculture, Lovely Professional University
Jalandhar, India

Chittaranjan Kole
Prof. Chittaranjan Kole Foundation for Science & Society
Kolkata, India

CRC Press
Taylor & Francis Group
Boca Raton London New York

CRC Press is an imprint of the
Taylor & Francis Group, an **informa** business
A SCIENCE PUBLISHERS BOOK

Cover page designed by Prof. Chittaranjan Kole

First edition published 2026
by CRC Press
2385 NW Executive Center Drive, Suite 320, Boca Raton FL 33431

and by CRC Press
4 Park Square, Milton Park, Abingdon, Oxon, OX14 4RN

CRC Press is an imprint of Taylor & Francis Group, LLC

Library of Congress Cataloging-in-Publication Data (applied for)

ISBN: 978-1-041-00428-8 (hbk)
ISBN: 978-1-041-01463-8 (pbk)
ISBN: 978-1-003-61486-9 (ebk)

DOI: 10.1201/9781003614869

Typeset in Times New Roman
by Prime Publishing Services

Message

Climate change and extreme weather, emerging pests and diseases, and environmental degradation are confronting agriculture with unprecedented challenges in the 21st century. Ensuring food and nutritional security for a growing global population demands innovative and sustainable solutions in crop protection. Conventional methods alone are no longer sufficient to safeguard crops, ecosystems, and human health.

The book ***Next Generation Crop Protection for Agricultural Sustainability and Food Security*** provides valuable insights into crop protection and sustainability. The fifteen chapters compiled in this volume provide a comprehensive overview of multidisciplinary strategies that not only safeguard crops but also promote environmental sustainability and long-term agricultural resilience.

The book emphasises the integration of artificial intelligence and machine learning in agriculture and highlights novel approaches in crop protection. It presents sustainable and futuristic pest management strategies, such as nanotechnology for agrochemical delivery and the use of digital tools for effective pest control. Additionally, it underscores the need to understand the effects of climate change on pest behaviour to adapt agricultural practices accordingly. Overall, this compendium advocates for a transformative, interdisciplinary approach to developing resilient agricultural systems. This book would certainly serve the urgent needs of farmers. We wish the editors and authors all the very best in their future endeavours.

His Holiness Sri Shivarathri Deshikendra Mahaswamiji

Chancellor of the JSS Academy of Higher Education and Research (JSS AHER), Mysuru and JSS University, Noida

Preface

The growing demand for sustainable and environmentally safe crop production practices has led to significant scientific breakthroughs and innovative field approaches. This book, ***Sustainable Crop Protection: Emerging Innovations and Integrated Approaches***, brings together comprehensive insights from multiple disciplines, offering readers a holistic perspective on sustainable crop health management. The following chapters provide an organised framework of emerging concepts, technological advancements, and practical strategies.

Chapter 1 covers biotechnological interventions, including transgenics, CRISPR, and molecular breeding for sustainable and resilient crop production. **Chapter 2** presents an advanced exploration of omics technologies—genomics, transcriptomics, proteomics, and metabolomics—for sustainable insect pest management, offering novel molecular insights.

Chapter 3 examines how genomic selection and predictive breeding tools can accelerate crop improvement and protection strategies for long-term food security, and **Chapter 4** provides an in-depth discussion on RNA interference technology, its application in pest and disease management, and its potential as an eco-friendly alternative to chemical pesticides. Nanotechnology-based innovations are presented in **Chapter 5** with a focus on controlled delivery systems that improve the efficiency of agrochemicals and biopesticides while minimizing environmental risks.

Chapter 6 explores digital and precision agriculture tools for pest detection and management, emphasising AI-driven decision-making and site-specific interventions, and **Chapter 7** addresses how climate variability influences pest population dynamics, with insights into adaptive management strategies under changing climatic scenarios.

Chapter 8 provides a detailed account of nematode diversity, their ecological roles, and sustainable management practices across cropping systems and **Chapter 9** highlights the potential of algae-derived compounds as bio-based alternatives for pest and disease control. A focused discussion on wheat diseases is presented in **Chapter 10** with strategies to safeguard one of the world's most important staple crops under climate stress.

Chapter 11 emphasizes the ecological role of beneficial arthropods in maintaining agroforestry systems and their potential in natural pest regulation. Case-specific study in **Chapter 12** addresses nematode problems in tuberose cultivation and provides sustainable solutions tailored to this high-value ornamental crop. An in-depth analysis of the cryptic *Bemisia tabaci* species complexes is presented **in Chapter 13**, highlighting advanced strategies for its management. **Chapter 14** underscores food safety concerns surrounding Ochratoxin A and reviews sustainable approaches for mitigating its risks in food and feed systems. The closing **Chapter 15** highlights the central role of farmers in driving innovation and adoption of sustainable practices, emphasising participatory research and community-based initiatives.

This book collectively addresses the urgent need for holistic, environmentally responsible, and scientifically grounded crop protection methods. By blending molecular tools, ecological processes, and participatory techniques, it aims to support practitioners, researchers, and policymakers in building resilient agricultural systems.

We thank all contributors for their excellent scholarly efforts and hope this work will inspire sustainable agricultural practices for future generations. The editors acknowledge the help from all the staff of Taylor & Francis Group, LLC / CRC Press at all stages.

Mysuru, Karnataka, India — **Belaghihalli N Gnanesh**
Lubbock, Texas, USA — **Megha N Parajulee**
Phagwara, Punjab, India — **Ankush M Raut**
Kolkata, West Bengal, India — **Chittaranjan Kole**

Contents

List of Contributors

A. Najitha Banu
Department of Zoology, School of Bioengineering and Bioscience, Lovely Professional University, Phagwara, Punjab, India. Email: najirila2010@gmail.com

Abha Chauhan
New York State Institute for Basic Research in Developmental Disabilities, Staten Island, New York, USA. Email: abhachauhan1@gmail.com

Ankush M Raut
Department of Entomology, School of Agriculture, Lovely Professional University, Phagwara, Punjab, India. Email: ankushento@gmail.com

Aradhana Panda
Division of Entomology, Faculty of Agriculture, SKUAST Kashmir, Wadura Sopore 193 201, J&K, India. Email: aradhanapanda97@gmail.com

Arshdeep Singh
Department of Agronomy, School of Agriculture, Lovely Professional University, Phagwara 144 411, India. Email: harrydeep628@gmail.com

Arunkumar Nagarathinam
Kumaraguru Institute of Agriculture, affiliated to Tamil Nadu Agriculture University, Erode, Tamil Nadu 638315, India. Email: drarunmicrobiologylab@gmail.com

Balagere Narayanaswamy Balaji
Central Silk Board, Bengaluru, Karnataka 560068, India. Email: balajibvn123@gmail.com

Barnik Debnath
ICAR-Central Institute for Cotton Research, Nagpur-440 010, Maharashtra, India. Email: barnikdebnath4@gmail.com

Basanagouda Gonal
Central Sericultural Research and Training Institute, Central Silk Board, Galander, Pampore-192 121, J&K, India. Email: basanagoudagonal@gmail.com

Basavaraj N. Hadimani
Division of Entomology, ICAR-Indian Agricultural Research Institute, New Delhi 110012, India. Email: hadimanibasavaraj6@gmail.com

Belaghihalli N Gnanesh
Genetics and Plant Breeding, ICAR JSS - Krishi Vigyan Kendra, Mysuru-571 129, Karnataka, India Email: gnaneshbn@gmail.com

Bhumika H. S.
Department of Biochemistry, Adichunchanagiri School of Natural Sciences, Adichunchanagiri University, B.G. Nagara, Mandya-571448, India. Email: bhumikasiddesh5727@gmail.com or Center for Research and Innovation, Adichunchanagiri University, B.G. Nagara, Mandya-571448, India.

Chinnannan Karthik
Department of Biology, West Virginia State University Institute, WV, 25112-1000, USA. Email: kar7612bt@gmail.com

Chitta Ranjan Satpathi
Department of Agricultural Entomology, Bidhan Chandra Krishi Vishwavidyalaya, Mohanpur, Nadia, West Bengal, India. Email: csatpathi2003@yahoo.co.in

Danishta Aziz
Division of Entomology, Faculty of Agriculture, SKUAST Kashmir, Wadura Sopore 193 201, J&K, India. Email: daniaziz9974@gmail.com

Devina Seram
Department of Entomology, School of Agriculture, Lovely Professional University, Phagwara 144411, Punjab, India. Email: devnah@gmail.com

Diego Jarquin
Assistant Professor, Integration & Application of AI & Omics in Plant Breeding, Department of Agronomy, University of Florida, Gainesville, Florida, USA. Email: jhernandezjarqui@ufl.edu

Dinesha Ramadas
Department of Biochemistry, Adichunchanagiri School of Natural Sciences, Adichunchanagiri University, B.G. Nagara, Mandya-571448, India. Email: drdinesha@acu.ac.in

Eajaz Ahmad Dar
Krishi Vigyan Kendra Ganderbal, 190 006, SKUAST Kashmir, J&K, India. Email: darajaz9@gmail.com

Elango Kolandhaisamy
Kumaraguru Institute of Agriculture, affiliated to Tamil Nadu Agriculture University, Erode, Tamil Nadu 638315, India. Email: elaento@gmail.com

G. R. Vishwas Gowda
Department of Plant Pathology, University of Agricultural Sciences, Dharwad 580005. Email: vishwasgowda450@gmail.com

Gowrisri Nagaraj
Kumaraguru Institute of Agriculture, affiliated to Tamil Nadu Agriculture University, Erode, Tamil Nadu 638315, India. Email: gowrisrin@gmail.com

Habib Ali
Department of Agricultural Engineering, Khwaja Fareed University of Engineering and Information Technology, Rahim Yar Khan, 64200, Pakistan. Email: habib_ali1417@yahoo.com

Haobijam James Watt
Department of Agricultural Economics and Extension, School of Agriculture, Lovely Professional University, Phagwara 144411, Punjab, India. Email: haobijamjameswatt@gmail.com

Ishtiyaq Ahad
Division of Entomology, Faculty of Agriculture, SKUAST Kashmir, Wadura Sopore 193 201, J&K, India. Email: drishtiyaqraina@skuastkashmir.ac.in

Iza Fatima
College of Agriculture Food and Wine, The University of Adelaide, Urbrrae, South Australia, Australia. Email: ifatima@okstate.edu

Jaba Jagdish
International Crops Research Institute for the Semi-Arid Tropics, Hyderabad, Telangana, 502324, India. Email: Jagdish.Jaba@icrisat.org

Julian Garcia Abadillo
Graduate Researcher, Integration & Application of AI & Omics in Plant Breeding, Department of Agronomy, University of Florida, Gainesville, Florida, USA. Email: garciaabadillo.j@ufl.edu

Kalavathi K. Kambali
ICAR-KVK, Dharwad, University of Agricultural Sciences, Dharwad 580005. Email: kambalikk@uasd.in

Kalenahalli Yogendra
International Crops Research Institute for the Semi-Arid Tropics, Hyderabad, Telangana, 502324, India. Email: yogendra.kalenahalli@icrisat.org

Kiran Kumar Kommu
ICAR-Central Citrus Research Institute, Nagpur 440033, Maharashtra, India. Email: kiranag0734@gmail.com

Kolanchi Prakash
Department of Entomology, Tamil Nadu Agricultural University, Coimbatore, Tamil Nadu 641003, India. Email: prakash280899@gmail.com

Mahantesha B.N. Naika
Centre for Biotechnology Research, Department of BCI, College of Horticulture, Bagalkot 587 104, University of Horticultural Sciences, Bagalkot, Karnataka, India. Email: mahantesha.bnn@uhsbagalkot.edu.in

Mahesh Math
Sampoorna International Institute of Agri Science & Horticultural Technology, Maddur 571 433, Karnataka, India. Email: maheshento@gmail.com

Marimuthu Murugan
Department of Entomology, Tamil Nadu Agricultural University, Coimbatore, Tamil Nadu, 641003, India. Email: muruganmarimuthu@tnau.ac.in

Meenakshi Rana
School of Agriculture, Lovely Professional University, Phagwara 144 411, Punjab, India. Email: meenakshi.20557@lpu.co.in

Megha N Parajulee
Cotton Entomology Program, Texas A&M AgriLife Research and Extension Center 1102 East Drew Street, Lubbock, Texas 79403, USA. Email: megha.parajulee@ttu.edu / m-parajulee@tamu.edu

Mohd Salim Mir
Division of Agronomy, Faculty of Agriculture, SKUAST Kashmir, Wadura Sopore 193 201, J&K, India. Email: mirsalim@skuastkashmir.ac.in

Murugesan Naveen
Department of Entomology, Tamil Nadu Agricultural University, Coimbatore, Tamil Nadu, 641003, India. Email: naveenm1696@gmail.com

Onkarappa Dhanyakumar
International Crops Research Institute for the Semi-Arid Tropics, Hyderabad, Telangana 502324, India. or
Department of Entomology, Tamil Nadu Agricultural University, Coimbatore, Tamil Nadu 641003, India. Email: dhanyakumarento@gmail.com

Pandareesh M. D.
Department of Biochemistry, Adichunchanagiri School of Natural Sciences, Adichunchanagiri University, B.G. Nagara, Mandya-571448, India. Email: pandareesh@acu.ac.in or
Center for Research and Innovation, Adichunchanagiri University, B.G. Nagara, Mandya-571448, India.

Pawan Saini
Central Sericultural Research and Training Institute, Central Silk Board Galander, Pampore 192 121, J&K, India. Email: pawansaini-coapbg@pau.edu

Prasakha Dutta
Department of Agricultural Entomology, Bidhan Chandra Krishi Vishwavidyalaya, Mohanpur, Nadia, West Bengal, India. Email: shakhabina@gmail.com

Prashanth Babu H.
Division of Genetics, ICAR-Indian Agricultural Research Institute, New Delhi 110012, India. Email: prashanthbabuh@gmail.com

Priyakharini Pragash
Kumaraguru Institute of Agriculture, affiliated to Tamil Nadu Agriculture University, Erode, Tamil Nadu 638315, India. Email: priyapragash169@gmail.com

Priyank Hanuman Mhatre
ICAR-Central Potato Research Institute, Regional Station, Ooty 643004, The Nilgiris, Tamil Nadu, India. Email: priyank.mhatre-icar@nic.in

Radha Sivarajan Sajeevan
Department of Plant Protection Biology, Swedish University of Agricultural Sciences, SE-23422, Lomma, Sweden. Email: sajeevan.radha.sivarajan@slu.se

Raju Sapkota
Texas A&M University, Department of Entomology, Texas A&M AgriLife Research and Extension Center, Lubbock, Texas 79403, USA. E-mail: raju.sapkota@ag.tamu.edu

Ravi V. Mural
Assistant Professor of Quantitative Genetics and Genomics, Department of Agronomy, Horticulture & Plant Science, South Dakota State University, Brookings, South Dakota, USA. Email: ravi.mural@sdstate.edu

Ravindra Kumar
ICAR-Indian Institute of Wheat and Barley Research, Karnal 132001, Haryana, India. Email: ravindrakumarbhu@gmail.com

Revanayya Gothe
International Crops Research Institute for the Semi-Arid Tropics, Hyderabad, Telangana, 502324, India.

Sahana K.P.
Sampoorna International Institute of Agri Science & Horticultural Technology, Maddur 571 433, Karnataka, India. Email: sahanasagar94@gmail.com

Samy Sayed
Department of Economic Entomology and Pesticides, Faculty of Agriculture, Cairo University, GIZA 12613, Egypt. Email: samy_mahmoud@hotmail.com

Satish N Chavan
ICAR Indian Institute of Rice Research, Hyderabad 500030, Telangana, India. Email: satishagrico@gmail.com

Seweta Srivastava
School of Agriculture, Lovely Professional University, Phagwara 144 411, Punjab, India. Email: seweta.21896@lpu.co.in

Shanmugasundaram Pavithran
Department of Entomology, Tamil Nadu Agricultural University, Coimbatore, Tamil Nadu, 641003, India. Email: sspavithranmsd@gmail.com

Shimpy Sarkar
Department of Entomology, School of Agriculture, Lovely Professional University, Phagwara 144 411, India. Email: shimpy610@gmail.com

Shyam Sundar S.
Department of Oral and Maxillofacial Surgery, JSS Dental College and Hospital, JSS Academy of Higher Education and Research, SS Nagar, Mysuru-570015, India. Email: shyam.surgeon@gmail.com

Sirisha Tadigiri
ICAR-Directorate of Floriculture Research, Regional Station, Rajahmundry 533125, Andhra Pradesh, India. Email: tadigiri.sirisha3@gmail.com

Srijan Ambati
Assistant Professor, Genetics & Plant Breeding, Professor Jayashankar Telangana Agricultural University, Warangal, Telangana, India. Email: srijan.ambati1@gmail.com

Suchitra Rakesh
Assistant Professor, Department of Microbiology, Central University of Tamil Nadu, Thiruvarur, India. Email: suchitrar@cutn.ac.in

Summira Rafiq
Division of Entomology, Faculty of Agriculture, SKUAST Kashmir, Wadura Sopore 193 201, J&K, India. Email: summirarafiq5@gmail.com

Sunil Kumaraswamy
Department of Biology, Oklahoma State University, Stillwater, OK 74078, United States. Email: sunil.lnu@okstate.edu

Sunil Shivananjappa
School of Biological Sciences, University of Western Australia, Perth, WA, Australia. Email: Sunil.shivananjappa@research.uwa.edu.au

Vishnutej Ellur
Regulatory Scientist/Technical Expert, Syngenta Seeds LLC, Durham, North Carolina, USA. Email: vishnutej.ellur@outlook.com

Zahoor Ahmad Shah
Division of Agri. Extension, Faculty of Agriculture, SKUAST Kashmir, Wadura Sopore 193 201, J&K, India. Email: s.zahoor37@gmail.com

Abbreviations

@	at the rate
&	And
%	Percentage
°C	Degree Celsius
<	Less Than
>	Greater Than
16S rRNA	16S ribosomal RNA
4-OH-OTA	4-(R)-Hydroxy Ochratoxin A
AChE	Acetylcholinesterase
AD	Alzheimer's Disease
ADHD	Deficit Hyperactivity Disorder
ADP	Adenosine Diphosphate
AEI	Augmented Expected Improvement
AFS	Agroforestry systems
Ag	Silver
AgNP	Silver Nanoparticle
AGO	Argonaute
AhR	Aryl Hydrocarbon Receptor
AI	Artificial Intelligence
AI	Active Ingredient
ALP	Alkaline Phosphatase
AMT	Agrobacterium-Mediated Transformation
ANN	Artificial Neural Networks
ANS	Autonomic Nervous System
APSIM	Agricultural Production Systems sIMulator
ARD	Agricultural Research and Development
ASAL	Allium Sativum Leaf Agglutinin
ASK-1	Apoptosis Signal-Regulated Kinase 1
ATP	Adenosine triphosphate
AUDPC	Area Under the Disease Progress Curve
BAPC	Branched amphiphilic peptide capsule
BayesCπ	Bayesian parametric model Cπ
BayesR	Bayesian parametric model R
BEN	Balkan Endemic Nephropathy
BHC	Benzene Hexachloride
BLAST	Basic Local Alignment Search Tool
BLUP	Best Linear Unbiased Prediction
BMD	Bone Mineral Density
Bt	*Bacillus thuringiensis*
C:N	Carbon: Nitrogen
C2F	Coarse-to-Fine

C_3	3-carbon molecule
C_4	4-carbon molecule
CA	Conservation Agriculture
CABI	Centre for Agriculture and Bioscience International
CARTs	Classification and regression trees
Cas9	CRISPR-associated Protein 9
CBO	Community-Based Organizations
CBPM	Community-based Pest Management
CCPs	Critical Control Points
CDNB	1-chloro-2,4-dinitrobenzene
CEA	Colocasia Esculenta Tuber Agglutinin
CFNet	Channel Fusion Network
CHiDO	Characterization and Integration of Driven Omics
CIV	Chilo Iridescent Virus
cm	Centimetre
CMSA	Community-Managed Sustainable Agriculture
CMV	Cucumber mosaic virus
CNN	Convolutional Neural Network
CO_2	Carbon dioxide
COI	Cytochrome Oxidase Subunit I
CP	Coat protein
CPO	Crop Protection Online
CPPI	Chickpea Protease Inhibitor
CPRD	Collaborative Participatory Research and Development
CRISPR	Clustered Regularly Interspaced Short Palindromic Repeats
CRISPRa	CRISPR Activation
CRISPRi	CRISPR Interference
crRNA	CRISPR RNA
CuO	Copper Oxide
cv.	Cultivar
CYP450	Cytochrome P450
DAP	*Days After Planting*
DAPG	2,4-diacetylphloroglucinol
DCR	Dicer
DDT	Dichlorodiphenyltrichloroethane
DIMBOA	2,4-Dihydroxy-7-methoxy-1,4-benzoxazin-3-one
DL	Deep Learning
DNA	Deoxyribonucleic acid
DON	Deoxynivalenol
DSBs	Double-Strand Breaks
dsRNA	Double-Stranded Ribonucleic Acid
DSSAT	Decision Support System for Agrotechnology Transfer
e.g.	For Example
EC	Electrical Conductivity
ECFP	Enhanced Cyan Fluorescent Protein
ECO_2	Equivalent Carbon dioxide
EFAO	Ecological Farmers Association of Ontario
EFSA	The European Food Safety Authority
ENM	Engineered Nanomaterial
eNOS	Endothelial Nitric Oxide Synthase

EPA	Environmental Protection Agency
EPA	Extension Planning Areas
EPF	Entomopathogenic fungi
EPN	Entomopathogenic Nematode
ER	Endoplasmic Reticulum
ERA	Ecological Risk Assessments
eRNAi	Environmental RNAi
et al.,	And others
EU	European Union
EWAS	Epigenome-Wide Association Studies
FAO	Food and Agriculture Organization
FCR	Fusarium crown rot
FFS	Farmers' Field School
FHB	Fusarium head blight
FIPA	Farmer-led Initiatives and Participatory Approach
FLI	Farmer-led Innovations
FLRP	Farmer–led Research Program
FMNR	Farmer Managed Natural Regeneration
FO	Farmer organizations
FPAR	Farmer-led Participatory Action Research
FPR	Farmer participated projects
FRN	Farmer Research Networks
FT–ICR–MS	Fourier Transform Ion Cyclotron Resonance Mass Spectrometry
G-BLUP	Genomic Best Linear Unbiased Prediction
G×E	Genotype-by-Environment Interaction
GAP	Good Agricultural Practice
GC-MS	Gas Chromatography Mass Spectrometry
GEBVs	Genomic Estimated Breeding Values
GGT	Gamma-Glutamyl Transpeptidase
GHGs	Greenhouse Gases
GIS	Geographic Information System
GMOs	Genetically Modified Organisms
GMPs	Good Manufacturing Practices
GNA	Galanthus Nivalis Lectin
GNSS	Global Navigation Satellite System
GP	Genomic Prediction
GPS	Global Positioning System
GRP-75	Glucose-Regulated Protein 75
GS	Genomic Selection
GSPs	Good Storage Practices
GST	Glutathione S-transferases
HACCP	Hazard Analysis and Critical Control Points
HDR	Homology-Directed Repair
HIGS	Host-Induced Gene Silencing
HIPV	Herbivory induced plant volatiles
HK-2	Human Proximal Tubular Epithelial Cells
HM	humidity model
HPLC	High Performance Liquid Chromatography
HSP	Heat Shock Proteins
ICRISAT	International Crops Research Institute for the Semi-Arid Tropics

IFA	International Fertilizer Association
IgE	Immunoglobulin E
IL-6	Interleukin-6
INM	Integrated Nematode Management
iNOS	Inducible Nitric Oxide Synthase
IoT	Internet of Things
IPCC	Intergovernmental Panel on Climate Change
IPM	Integrated Pest Management
IPPM	Integration of Production and Pest Management
IRM	Insecticide Resistance Management
IRRI	International Rice Research Institute
ISR	Induced Systemic Resistance
ISTK	Iridoviral Serine/Threonine Kinase
ITK	Indigenous Traditional Knowledge
ITS1	nuclear ribosomal intergenic spacer I
J2	Second-stage juvenile
JECFA	Joint FAO/WHO Expert Committee on Food Additives
k-NN	k-Nearest Neighbors
KB	Karnal bunt
kdr	knock down resistance
KEGG	Kyoto Encyclopedia of Genes and Genomes
LAI	Leaf Area Index
LAMP	Loop-mediated Isothermal Amplification
LASSO	Least Absolute Shrinkage and Selection Operator
LCMS	Liquid Chromatography–Mass Spectrometry
LCST	Lower Critical Solution Temperature
LD	Linkage Disequilibrium
LDH	Layered double hydroxide
LDH	Lactate Dehydrogenase Activity
LED	Light Emitting Diode
LL	Lentil Lectin
LR	leaf rust
LSKNet	Large Selective Kernel Network
m/s	Metre per second
MAE	Mean Absolute Error
MAS	Marker Assisted Selection
MBT	Mother and Baby Trial
MEAM1	Middle East Asia Minor 1
METs	Multi-Environment Trials
MFO	Mixed Function Oxidase
mg/kg	Milligram/Kilogram
ML	Machine Learning
MOF	Metal-Organic Framework
MOH	Ministry of Health
MoT	*Magnaporthe oryzae* pathotype *Triticum*
MP	Movement protein
MPE	Mean Percent Error
MRL	Maximum Residue Limit
mRNA	Messenger Ribonucleic Acid
MS	Multiple Sclerosis

MSCs	Mesenchymal Stem Cells
MSE	Mean Square Error
MT	Minimum tillage
mtCOI	Mitochondrial cytochrome oxidase gene subunit-1
NAL	National Agricultural Library
NBTs	New Breeding Techniques
NCBI	National Center for Biotechnology Information
NDVI	Normalized Difference Vegetation Index
NDWI	Normalized Difference Water Index
NF	Nematophagous fungi
ng/g	Nanogram Per Gram
ng/kg	Nanogram/Kilogram
ng/L	Nanogram/Liter
ng/ml	Nanogram/Millilitre
NGO	Non-governmental organizations
NGS	Next-Generation Sequencing
NGS	next-generation sequencing
NGTs	New Genomic Techniques
NHEJ	Non-homologous End Joining
NHEJ	Non-Homologous End Joining
nm	Nanometer
NMR	Nuclear Magnetic Resonance Spectroscopy
NN	Neural Networks
NO	Nitric Oxide
NP	Nanoparticle
npylr1	Neuropeptide Y-Like Receptors 1 Gene
Nrf2	Nuclear Factor Erythroid 2-Related Factor 2
NT	No tillage
NTOs	Non-target Organisms
NWE	Nadaraya-Watson Estimator
ODM	Oligonucleotide-Directed Mutagenesis
OGTR	Australia's Office of the Gene Technology Regulator
OP	organophosphate
OTA	Ochratoxin A
OTs	Ochratoxins
PA	Precision Agriculture
PAI-2	Plasminogen Activator Inhibitor 2
PAM	Protospacer Adjacent Motif
PAR	Participatory Action Research
PAZ	Piwi/Argonaute/Zwille domain
PBDM	Physiologically Based Demographic Models
PBL	Planetary Boundary Layer
PCN	Potato cyst nematodes
PCWDE	Plant Cell Wall–Degrading Enzymes
PD	Parkinson's Disease
PDIFTP	Participatory Domestication of Indigenous Fruit Trees Program
PEI	Polyethylenimine
PGPM	Plant growth-promoting microorganisms
PGPR	Plant growth-promoting rhizobacteria
PIPs	Plant-Incorporated Protectants

pka	Acid Dissociation Constant
PLGA	Poly Lactic-Co-Glycolic Acid
PNIPAM	Poly(N-Isopropylacrylamide)
PPA	Pinellia Pedatisecta Agglutinin
ppb	Parts Per Billion
PPE	Personal Protective Equipment
ppm	Parts Per Million
PPM	Precision Pest Management
ppm	Parts per million
PPN	Plant-Parasitic Nematodes
PR	Pathogenesis Related
PRISE	The Pest Risk Information Service
PTWI	Provisional Tolerable Weekly Intake
PVX	Potato virus X
QoIs	Quinone outside Inhibitors
QSMAS	Quesungual Slash and Mulch Agroforestry System
QTL	Quantitative trait loci
R gene	Resistant gene
RAAS	Renin-Angiotensin-Aldosterone System
RAPD-PCR	Random Amplified Polymorphic DNA Polymerase Chain Reaction
RAS	Renin-Angiotensin System
RdRP	RNA-dependent RNA Polymerase
REACH	Registration, Evaluation, Authorization and Restriction of Chemicals
REP	Red Edge Position
RF	Random Forest
RFLP	Restriction Fragment Length Polymorphism
RH	Relative Humidity
RILs	Recombinant Inbred Lines
RISC	RNA-Induced Silencing Complex
RKHS	Reproducing Kernel Hilbert Space
RKN	Root-Knot Nematode
RMSE	Root Mean Square Error
RNA	Ribonucleic Acid
RNAi	Ribonucleic Acid interference
ROS	Reactive Oxygen Species
RR-BLUP	Ridge Regression Best Linear Unbiased Prediction
RVDs	Repeat Variable Di-residues
S gene	Susceptibility gene
SA	Sustainable Agriculture
SADIE	Spatial Analysis by Distance Indices
SAR	Systemic Acquired Resistance
SCAR	Sequence Characterized Amplified Region
SCN	Soybean cyst nematode
SDHI	Succinate Dehydrogenase Inhibitor
SDM	Species Distribution Model
SDN	Site-Directed Nucleases
sgRNA	Single-guide RNA
SIGS	Spray-Induced Gene Silencing
SiO_2	Silicon dioxide
siRNA	Small Interfering Ribonucleic Acid

SIT	Sterile Insect Technique
SLNs	Solid Lipid Nanoparticles
SMR	Symbiont-mediated RNAi
SNP	Single nucleotide polymorphism
sp	Species
spp.	Species (plural form)
SR	Stripe rust
SSD	Super Saturated Design
SSMZ	Site-Specific Management Zones
SSPM	Site-Specific Pest Management
STB	Septoria tritici blotch
SVM	Support Vector Machine
TAD	Take-all decline
TALEN	Transcription Activator-Like Effectors
TF	Tissue Factor
TiO_2	Titanium Dioxide
TMV	Tobacco mosaic virus
TP	Training Population
tracrRNA	Trans-activating CRISPR RNA
TSWV	Tomato spotted wilt virus
TYLCV	Tomato yellow leaf curl virus
USA	United States of America
UGT	UDP-glycosyltransferase
USAID	United States Agency for International Development
USDA	United States Department of Agriculture
UV	Ultraviolet
VIPs	Vegetative Insecticidal Proteins
VRA	variable rate application
VRS	Variable Rate Seeding
VRT	Variable Rate Technologies
W/m^2	Watt per metre square
WGS	Whole genome sequencing
XIoU	Extended Intersection over Union
YOLO-DPD	You Only Look Once - Diseases and Pests Detection
ZFN	Zinc Finger Nucleases
ZFs	Zinc Fingers
ZnO	Zinc Oxide
µg/kg	Microgram/Kilogram

CHAPTER 1

Novel Biotechnological Strategies for Safeguarding Crops and Promoting Sustainable Agriculture

Sunil Kumaraswamy[1*], Onkarappa Dhanyakumar[2,3], Balagere Narayanaswamy Balaji[4], Basavaraj N. Hadimani[5], Kolanchi Prakash[3] and Belaghihalli N Gnanesh[6]

1. Introduction

Global food security is increasingly threatened by complex challenges, and innovative biotechnological strategies have a potential role in safeguarding crops and promoting sustainable agriculture (Schmidhuber & Tubiello, 2007). The world's population is expected to reach nearly 10 billion by 2050, placing unprecedented pressure on agricultural systems to produce more food with fewer resources (Wheeler & Von Braun, 2013). Traditional agricultural practices, while essential, are often insufficient to address these multifaceted challenges. Crop resistance breeding utilizing host plant resistance traits from plants is crucial for cost-effective and long-term integrated pest management (Kumaraswamy et al., 2024). However, these conventional methods of crop protection are time consuming and less precise. Moreover, current pest management is heavily reliant on chemical inputs that may have adverse environmental impacts. Therefore, the agricultural sector is increasingly turning to biotechnology as a means to enhance crop resilience, productivity, and sustainability (Dara, 2019; Nawaz et al., 2019).

Genetic engineering and plant biotechnology in pest management techniques have reached new horizons attaining numerous achievements. Various molecular techniques involving genetic mechanisms have been reported, and such studies have been reproduced to conduct *in vitro* experiments. The microbial interventions represent another frontier in agricultural biotechnology (Kamthan et al., 2016). *Bacillus thuringiensis* (*Bt*) gene encodes a protein that is toxic to certain insect pests but is harmless to humans, animals, and beneficial insects. When introduced into crop plants through genetic engineering, the *Bt* gene enables these plants to produce the *Bt* toxin which targets

[1] Department of Biology, Oklahoma State University, Stillwater, OK 74078, United States.
[2] International Crops Research Institute for the Semi-Arid Tropics, Hyderabad, Telangana 502324, India.
[3] Department of Entomology, Tamil Nadu Agricultural University, Coimbatore, Tamil Nadu 641003, India.
[4] Central Silk Board, Bengaluru, Karnataka 560068, India.
[5] Division of Entomology, ICAR-Indian Agricultural Research Institute, New Delhi 110012, India.
[6] Genetics and Plant Breeding, ICAR-JSS Krishi Vigyan Kendra, Mysuru 571129, Karnataka, India.
* Corresponding author: sunil.lnu@okstate.edu

DOI: 10.1201/9781003614869-1

the gut cells of specific pests causing them to stop feeding and eventually die (Khan et al., 2012). Over years of evolution, plants have developed a wide array of protein-based strategies to defend themselves against a variety of biotic stressors, including insects, fungi, and bacteria (Gimenez et al., 2018; Jha & Mohamed, 2022). Moreover, plant proteins and secondary metabolites play an important role in the natural defense mechanisms of plants against insect pests (Kumaraswamy et al., 2025). These proteins can act as inhibitory enzymes, toxins, or signaling molecules that disrupt the life cycles of insect pests (Hussain et al., 2011; Jha & Mohamed, 2022)

A wide range of advanced tools and techniques has been used in a new era of agricultural innovation with advanced biotechnological tools and techniques. Genetic modification (GM) has been one of the earliest and most well known applications, allowing for the development of crops with enhanced traits such as pest resistance and improved nutritional profiles. However, the advent of more precise and versatile technologies like RNA interference (RNAi), CRISPR-Cas9 gene editing has further revolutionized the field. RNAi can silence specific genes associated with pests and diseases, providing a novel, environmentally friendly alternative to chemical pesticides (Adeyinka et al., 2020; Mehlhorn et al., 2021). CRISPR-Cas9 enables targeted modifications to plant genomes with unparalleled accuracy, offering the potential to not only improve crop resilience to environmental stresses but also to address specific challenges against insect pests (Perkin et al., 2016; Singh et al., 2022) .

In this chapter, we explain the continued development and adoption of biotechnological innovations which are essential for overcoming the challenges faced by modern agriculture. The integration of these novel biotechnological strategies into agricultural systems holds the promise of transforming the future of farming by reducing reliance on chemical inputs, increasing crop resilience to climate change, and improving the efficiency of resource use. This helps to conserve biodiversity of natural enemies and ensures farmers can meet the growing demand for food in a changing world.

2. Genetic Transformation

Genetic transformation is a groundbreaking technology that allows for the direct alteration of an organism's genetic material. The genome modification and editing methods previously discussed, often known as new breeding techniques (NBTs), which are transforming the rapid creation of innovative crop traits (Schaart et al., 2016). Unlike traditional breeding methods, which rely on selecting and crossing plants with desirable traits over multiple generations, genetic engineering allows to introduce, remove, or modify specific genes with high precision. This technology has wide range applications in plant breeding and has a pivotal role in crop improvement that are more resistant/tolerant to pests, diseases, and environmental stresses, as well as enhanced nutritional content and yield (Robinson, 2001). By enabling the transfer of genes across species barriers, genetic engineering opens up possibilities that were previously unimaginable, transforming the way we approach the development of new crop varieties (Lal, 2020). New breeding approaches modify a plant's genes to develop a variety that is superior to existing cultivars/varieties. Sometimes, this involves "wider crosses" where species or even genera that are not closely related are crossed. However, these crosses often require advanced techniques like genetic transformation (Tarafdar et al., 2014).

Genetic transformation refers to the integration of foreign genetic material into a specific genotype through genetic engineering (Alves et al., 1999). The inserted gene is called a transgene and organisms developed through successful gene transfer are known as transgenics (Babaoglu et al., 2000). Recently, researchers combined molecular biology, tissue culture, and gene transfer techniques to create a powerful tool for introducing traits of interest into plants. This approach allows the insertion of genes from animals, microorganisms, or different plant species into a recipient plant's genome in a controlled manner, bypassing fertilization (Anjanappa & Gruissem, 2021).

2.1 Plant Transformation Methods

Various biotechnical methods have been used to introduce foreign genes into plants and alter their biochemical and physiological properties. A common approach in all these methods is that the transforming DNA first needs to enter the plant cell by penetrating both the plant cell wall and the plasma membrane. It must then reach the nucleus and integrate into the resident chromosomes. Gene transfer is carried out, for most species, using explant competent regeneration to develop complete, fertile plants. Thus, the development of tissue culture technology is frequently seen as a combination of art and science. Gene transfer technology has become a routine in working with several plant species, yet the limiting step in others is not the transformation itself but rather the lack of efficient regeneration protocols (Peña, 2004). The most widely used and successful method of plant transformation is the "*Agrobacterium*-mediated plant transformation" (AMT) (Kronfoth, 2018). Other methods for plant transformation are direct transformation methods such as "Microinjection" (De la Pena et al., 1987), "Polyethylene glycol-mediated transfer" (Uchimiya et al., 1986), "Electroporation" (Fromm et al., 1985), and "Gene gun methodology" (Sanford, 1988).

2.2 Agrobacterium-mediated Transformation

AMT is a highly complex and well-developed method used to introduce foreign genes into a wide range of host plants (Gelvin, 2003). This technique is commonly employed for generating transgenic plants because of its advantages over other transformation methods, including stable integration of transgenes and cost-effectiveness (Rahman et al., 2023). Agrobacterium tumefaciens is a rod-shaped, gram-negative, soil-borne bacterium recognized as a natural genetic engineer and has been known as a plant pathogen since the early 20th century (Gelvin, 2003). Since the early 1980s, the ability of Agrobacterium to transfer DNA to plant cells has been utilized in plant genetic engineering (Gelvin, 2003). During this process, the bacterium transfers a segment of its plasmid DNA, called T-DNA, into host plant cells. This DNA can integrate into the genetic material of the plant cell nucleus, resulting in stable transformation where the introduced genes are inherited by future generations (Hwang et al., 2017). Alternatively, the bacterial DNA may remain transient in the nucleus without genome integration, replicating alongside the plant DNA and expressing genes to produce specific proteins (Hwang et al., 2017).

AMT provides several advantages compared to other gene transfer techniques such as particle bombardment or electroporation (Rahman et al., 2023). These advantages include the transfer of DNA fragments with defined ends and minimal rearrangement, the capacity to transfer relatively large DNA segments, and the integration of a low number of gene copies into plant chromosomes. This often results in fewer transgene copies with reduced rearrangements, which is beneficial for regulatory approval and consistent gene expression since multiple transgene insertions can cause gene silencing. AMT has played a key role in enhancing various crop species for multiple traits (Rahman et al., 2023). Arabidopsis, a widely studied model plant with extensive genetic and genomic resources, is particularly suitable for transformation by *A. tumefaciens* for both stable and transient gene expression, although transformation efficiency varies among different accessions (Hwang et al., 2017; Opabode, 2006; Tzfira & Citovsky, 2006). Applications include developing disease-resistant apples (for scab and fire blight), insect resistance, altering rootstock characteristics, and improving fruit quality through genetic modification. AMT has also been used to increase stress tolerance and modify floral traits in ornamental plants (Ziemienowicz, 2013). The transient expression of genes delivered by *Agrobacterium* serves as a valuable tool to study gene function and promoter activity.

2.3 Gene Gun (Biolistics Method)

The gene gun, also known as biolistics or particle bombardment, is a physical method of gene transfer that allows for the direct delivery of foreign DNA into plant cells, tissues, or even organelles like chloroplasts. This technique involves coating microscopic particles (usually gold or tungsten) with DNA and then propelling them at high speeds into plant cells using a burst of helium gas. The DNA-coated particles penetrate the cell wall and membrane, enabling the DNA to enter the cell's nucleus or other target sites, where it can integrate into the plant genome. Unlike AMT, which relies on biological processes, the gene gun is a versatile tool that can be used across a wide range of plant species (Ozyigit & Yucebilgili Kurtoglu, 2020).

The biolistic gene gun has been commercially available and used in both transient gene expression and transgenic plant production (Finer et al., 2000). Biolistics has proven especially useful in transforming monocotyledonous plants like rice, maize, and wheat, which are often difficult to transform using *Agrobacterium*-based methods. Additionally, the gene gun allows for the direct transformation of organelles such as chloroplasts, offering the possibility of high level gene expression and reducing the risk of gene flow to other plants through pollen. It is an efficient and widely utilized gene transfer method with reproducible results. While the technique is powerful, it also has some limitations, including the potential for cell damage and the randomness of DNA integration, which can lead to variable expression levels and potential disruptions of endogenous genes. Despite these challenges, the gene gun remains a valuable tool in plant genetic engineering, particularly for species that are otherwise difficult to transform (Akbar et al., 2017; Plackett et al., 2014).

2.4 Current Global Regulatory Landscape for Plant Genome Editing

The regulatory status of genome editing differs widely among countries and regions. Some countries exempt certain genome edited products from genetically modified organism (GMO) regulations, while others enforce stricter controls (Sprink et al., 2022; Menz et al., 2020; Smyth, 2019). This changing landscape reflects the fast progress in genome editing technologies and the varied approaches to their regulation (Schmidt et al., 2020; Sprink et al., 2022). Many nations are currently updating their regulatory frameworks to accommodate these emerging technologies (Sprink et al., 2022). In the United States, regulation of genome edited crops follows the Coordinated Framework for Biotechnology (Wolt et al., 2015). Since 2018, the USDA (The United States Department of Agriculture) has generally chosen not to regulate genome editing technologies on a case-by-case basis, especially for plants developed with site-directed nucleases (SDN-1 or SDN-2) once foreign genes are removed (Schmidt et al., 2020). US biosafety laws focus on risks like the presence of T-DNA from pathogens or toxicity, allowing genome edited crops without recombinant DNA or novel safety concerns to avoid regulatory review (Schmidt et al., 2020; Wolt & Wolf, 2018). Conversely, the European Union regulates all genome edited organisms as GMOs under strict rules, as ruled by the Court of Justice of the EU in 2018 (Schmidt et al., 2020; Wasmer, 2019; Zimny & Eriksson, 2020). This decision has been criticized for hindering innovation, prompting ongoing discussions to amend EU legislation (Schmidt et al., 2020). Meanwhile, China leads in agricultural gene editing research but lacks a comprehensive regulatory framework, relying instead on fragmented biosafety measures (Schmidt et al., 2020; Li et al., 2019).

In 2020, India released a draft document and guidelines through the Department of Biotechnology under the Ministry of Science and Technology, proposing a tiered regulatory approach for genome edited plants (Schmidt et al., 2020; Menz et al., 2020). This system categorizes products based on the type and extent of genome editing. Group 1 includes plants with one or a few base pair edits or deletions, typically created by SDN-1 or ODM (Oligonucleotide-Directed Mutagenesis) techniques, which are considered low risk. Group 2 covers plants with a few or several base pair edits made using SDN-2 and a template, although the distinction between "a few" and 'several' base pairs is not clearly defined. For these two groups, risk assessment focuses on confirming the targeted edits, excluding significant off-target effects, and verifying trait efficacy to ensure equivalence with reference varieties, usually evaluated case by case. Group 3 consists of plants with substantial DNA changes or insertion of foreign DNA and is subject to rigorous risk assessments like traditional transgenic plants (Schmidt et al., 2020). As of 2022, India's policy exempts products derived from SDN-1 and SDN-2 techniques, which are free of exogenous DNA, from certain regulatory provisions, while SDN-3 products remain regulated under existing biosafety rules (Prasad & Chimata, 2023).

In other countries, regulatory approaches vary considerably. Canada bases regulation on the concept of novelty, exempting edited plants similar to natural variants with a safe breeding history (Schmidt et al., 2020). New Zealand treats all genome edited events as GMOs (Schmidt et al., 2020). Australia's Office of the Gene Technology Regulator (OGTR) excludes SDN-1 applications from regulation from 2019 (Schmidt et al., 2020). Japan and Israel do not regulate SDN-1 and SDN-2 products lacking foreign DNA, with Japan providing transparency through pre-submission consultations and notifications (Schmidt et al., 2020; Kondo & Taguchi, 2022). Several African countries, including South Africa, Kenya, and Nigeria, are developing policies to regulate genome editing (Schmidt et al., 2020). Meanwhile, Switzerland, Norway, and the UK are considering legislation to streamline approval processes, with Norway proposing exemptions and expedited assessments for certain genome edits in 2018 (Schmidt et al., 2020).

Globally, the regulatory landscape is evolving, generally trending towards exempting some genome editing techniques from traditional GMO regulations (Schmidt et al., 2020; Sprink et al., 2022). Argentina was the first in 2015 to exempt crops without foreign DNA from biosafety rules, followed by Chile, Brazil, and Colombia, which regulate new products as GMOs only if they contain a novel combination of genetic material (Schmidt et al., 2020). Paraguay and Uruguay also intend to adopt similar frameworks, reflecting growing recognition of the need for balanced regulation to foster innovation while ensuring safety (Schmidt et al., 2020).

3. New Genomic Techniques

Insect pests pose a significant threat to global agricultural productivity, causing substantial crop losses and compromising food security. Traditional pest control methods, including chemical pesticides, have limitations such as environmental impact, pest resistance, and effect on non-target organisms (NTOs). Advances in genomic techniques offer promising new strategies for managing biotic stress caused by insect pests. Use of zinc finger nucleases (ZFN), Transcription activator-like effectors (TALEN), CRISPR-Cas9 and other gene editing technologies and transgenic approaches have emerged as powerful tools to engineer crop plants with enhanced resistance/tolerance to insects (Bhuyan et al., 2023). CRISPR-Cas9 allows precise modifications in crop genomes to enhance pest resistance by knocking out susceptibility genes or introducing novel resistance traits. RNAi technology, on the other hand, enables the silencing of target resistant genes in pests, leading to their death or reduced virulence. Moreover, the development of GM crops expressing pest-specific toxins or repellents has shown efficacy in controlling pest populations.

3.1 Zinc Finger Nucleases

ZFNs were first discovered in 1985 on gene transcription in *Xenopus laevis*, commonly known as the African clawed frog. This research was instrumental in uncovering the role of zinc fingers (ZFs) in gene regulation. The study specifically examined thc transcription of a 5S RNA sequence, which was found to bind with a transcription factor. This binding was facilitated by the presence of ZFs, a type of protein motif that interacts with DNA sequences (Miller et al., 1985). Moreover, ZFs are characterized by their ability to bind through a specific domain composed of two cysteine and two histidine residues, known as the Cys2-His2 domain. This domain is a highly conserved structure in many proteins and is crucial for the protein's ability to recognize and bind to specific DNA sequences (Beerli & Barbas III, 2002; Choo & Klug, 1994). The Cys2-His2 structure is composed of approximately 30 amino acids, which adopt a distinctive beta-beta-alpha (bba) configuration. This configuration is critical for the structural integrity of the ZF and its ability to interact with DNA (Pavletich & Pabo, 1991). Importantly, each ZF domain is capable of binding to a specific sequence of 3-4 base pairs (bp) of DNA. The specificity of this interaction allows each ZF to recognize a single subunit of a ZFN. When several ZFs are linked together, they can collectively bind to longer DNA sequences, typically ranging from 9 to 18 base pairs in length. This multivalent binding capacity enables ZFNs to target specific sites within the genome with high precision, making them powerful tools for gene editing (Cathomen & Joung, 2008).

ZFNs, a type of engineered DNA-binding protein, offer a promising genomic tool for enhancing crop resistance to insect pests and directly targeting pest populations. ZFNs function by creating double-strand breaks (DSBs) at specific DNA sequences, allowing for targeted gene disruption or modification. This technology can be applied to disrupt genes in insect pests that are critical for their survival or reproduction, effectively reducing pest populations. Additionally, ZFNs can be used to introduce resistance traits into crops, making them less susceptible to pest attacks. The precision of ZFNs allows for minimal off-target effects, making them a potentially safer alternative to other genetic modification techniques. A ZFN is a heterodimer in which every subunit comprises a zinc finger domain and a FokI endonuclease domain (Urnov et al., 2010). Genome editing by ZFNs has been demonstrated in plants, including rice and *Arabidopsis* (Ainley et al., 2013; Gallego-Bartolomé et al., 2019). ZFNs are effectual genome editing elements; however, they were not extensively adopted because of the complexity in nature of the contact between ZFs and DNA. Other limitations include the inherent difficulty in designing, interest dependent specificity, and difficulty in authenticating such proteins for a particular DNA locus of context (Sander et al., 2011).

3.1.1 Mechanism of the ZFNs

ZFNs have a modular structure that allows them to recognize specific DNA sequences by assembling Cys2-His2 ZF domains into a customized array. Each ZF domain is linked to the FokI enzyme, which serves as a nonspecific nuclease domain (Bibikova et al., 2003). The natural FokI enzyme possesses independent binding and cleavage activities. Consequently, ZF domains are designed to bind to the FokI nuclease, enabling it to cleave DNA at targeted sites (Bibikova et al., 2003; Kim et al., 1996). By engineering novel ZFs, a wide range of DNA sequences can be specifically targeted. For precise DNA cleavage, the FokI enzyme requires dimerization (Smith et al., 2000). However, due to the relatively weak dimerization interface, ZFNs are typically designed in pairs, with each pair recognizing two specific sequences flanking the cleavage site. ZFNs bind to opposite sides of the target site, allowing the FokI domains to dimerize and cleave the DNA within 5-7 base pair (bp) spacers, resulting in DSBs with 5' overhangs (Bibikova et al., 2001; Shimizu et al., 2009). The ability of ZFNs to induce DSBs at virtually any location within a genome makes them a powerful tool for genome editing in a wide range of organisms (Gaj et al., 2013; Urnov et al., 2010) (Figure 1).

3.1.2 Genetic Engineering Using ZFN in Insects

A strategy combining the site-specific DNA binding capabilities of ZFNs with transposases has been developed to enhance genome targeting efficiency. The gene-targeting tool ZFN has been successfully applied in *Drosophila melanogaster*. Specifically, the genomic loci *ry* and *bw* were targeted for gene editing targets and injected into *D. melanogaster* eggs, resulting in 89% and 67% mutation rates in females and males respectively (Beumer et al., 2006). The disruption of the *BmBLOS2* gene was targeted by injecting mRNA encoding ZFNs directly into *Bombyx mori* eggs leading to change its phenotypic character. The effectiveness of the ZFN system has also been demonstrated in *B. mori* (Takasu et al., 2010). Additionally, ZFNs were applied to target the *orco* gene in mosquitoes in *Aedes aegypti* and mutants that exhibited reduced odor-evoked responses (DeGennaro et al., 2013). Similarly, targeting the neuropeptide Y-like receptors 1 (*npylr*1) gene in *A. aegypti* using ZFNs and an HR vector did not inhibit host-seeking behaviors suggesting that this peptide might regulate host-seeking behavior in *A. aegypti* (Liesch et al., 2013). Fusing the *Mos1* and *piggyBac* transposases significantly increased the efficiency of plasmid-based transposition assays in *A. aegypti* (Maragathavally et al., 2006). Similarly, *Gal4-piggyBac* transposases have been used for transposition in *D. melanogaster* and *B. mori* (Wang et al., 2010).

3.2 Transcription Activator-like Effectors

TALENs are innovative engineered enzymes that have significantly advanced genetic engineering tools providing an alternative to ZFNs, It is derived from the plant pathogens of *Xanthomonas* group and introduced a distinct platform for engineering programmable DNA-binding proteins (Bogdanove & Voytas, 2011). TALENs are dimeric transcription nucleases or factors built from arrays of 33-35 amino acid modules, each one of which is targeted to a single nucleotide (Moscou & Bogdanove, 2009). The nucleotide specificity of each domain is determined by hypervariable amino acids, known as repeat variable di-residues (RVDs). RVDs are crucial for recognizing specific nucleotides: guanine (G), cytosine (C), thymine (T), and adenine (A) (Boch et al., 2009). Understanding the role of RVDs led to the development of a new programmable nuclease, where domains of TALE repeats are fused to the *FokI* nuclease domain, resulting in the creation of TALENs (Li et al., 2011). TALENs induce DSBs at specific genomic locations, allowing for targeted gene knockout or knock-in, which are then repaired by the cell's natural mechanisms. This repair process can result in the insertion, deletion, or alteration of genetic material. The modular nature of TALENs makes them highly effective tools for gene modification across a wide range of organisms (Budhagatapalli et al., 2015).

3.2.1 Mechanism of the TALENs

TALENs are similar to ZFNs, with the primary difference being their ability to target additional loci within genomes. While ZFNs use ZF DNA-binding domains, TALENs utilize transcription activator-like effector (TALE) proteins as their DNA recognition modules. Both TALENs and ZFNs employ the FokI restriction enzyme as their DNA cleavage module. Although TALENs offer greater design flexibility compared to ZFNs, the genes encoding TALENs are approximately three times longer than those for ZFNs (Moscou & Bogdanove, 2009). The TALE arrays are assembled and fused with the FokI nuclease, and when the FokI cleavage domains dimerize, they precisely cut the genome at the targeted site (Cermak et al., 2011). TALENs are designed as pairs that target neighboring sequences. When they bind to either side of the recognition site, the *FokI* domains dimerize and cleave the DNA. TALENs use a short linker and a spacer of 16 to 19 base pairs (bp) between the binding domains (Reyon et al., 2012). This straightforward design has facilitated the use of TALENs for inducing DSBs in gene editing across various species (Figure 1).

3.2.2 Genetic Engineering Using TALEN in Insects

TALEN has been utilized for knockout studies in *D. melanogaster* (Kondo & Ueda, 2013). *trh* mutants in the G1 generation in *D. melanogaster* eggs resulting in 30-100% by using mRNA mixtures with TALEN pairs A and B were separately and TALEN pair B provided a 17% mutation rate in G0 generation and observed 6-15% mutation rate in the G1 generation (Kondo & Ueda, 2013). TALENs targeting the *kynurenine 3-monooxygenase (kmo)* gene in *Aedes aegypti* and procedures and detection of mutations in the *kmo* gene in *Ae. aegypti* (Aryan et al., 2014).

3.3 CRISPR-Cas9 in Insect Pest Management

CRISPR-Cas9 mediated genome editing represents one of the most transformative advancements in molecular biology and genetics, offering unprecedented precision in the manipulation of genetic material (Cui et al., 2017). This editing is originally derived from a bacterial immune defense mechanism, and has rapidly transformed into a powerful tool for precisely altering the genomes of diverse organisms. This technology operates by harnessing the Cas9 enzyme, which is directed to specific DNA sequences by a customizable RNA molecule. Upon recognizing a target sequence, facilitated by the recognition of protospacer adjacent motifs (PAMs) in the foreign DNA, the CRISPR-associated nucleases (Cas) are guided by CRISPR RNAs (crRNAs) to introduce targeted double-stranded breaks (DSBs) in the DNA (Cui et al., 2017; Gantz & Akbari, 2018). The ensuing DNA breaks are then repaired by the cell's natural mechanisms, primarily through non-homologous end joining (NHEJ) or homology-directed repair (HDR) pathways. This process can result in insertions or deletions at the site of the break, allowing for precise genetic modifications. The most widely utilized CRISPR-Cas9 system for genome editing employs the *Streptococcus pyogenes* Cas9 nuclease, which specifically recognizes the PAM sequence NGG. This CRISPR-Cas system provides an invaluable tool for targeted genome editing, offering vast potential across various fields (Jinek et al., 2012).

3.3.1 Mechanism of CRISPR-Cas9 System in Genome Editing

The CRISPR-Cas9 system, derived from a bacterial immune mechanism, has been extensively studied, with detailed mechanisms covered in various reviews (Hsu et al., 2014; Sander & Joung, 2014). Once introduced into target cells, the Cas9-gRNA complex recognizes the PAM sequence (NGG for Cas9) and binds to the target DNA by pairing crRNA with the DNA sequence. The HNH and RuvC domains of Cas9 then create a double-stranded break (DSB) near the PAM sequence. Cas9-gRNA complex undergoes significant conformational changes to facilitate DNA binding and cleavage (Jiang et al., 2015).

The DSBs are primarily repaired through NHEJ or HDR. NHEJ often introduces indel mutations, leading to potential gene disruption, while HDR, which uses a homologous repair template, allows for precise gene editing, such as knockout (KO) or knock-in (KI) modifications. Mutated forms of Cas9, known as ' dead' or dCas9 bind DNA without cutting enabling applications like transcriptional regulation, epigenetic modifications and genomic labelling (Doudna & Charpentier, 2014). Additionally employing a double-nicking approach with dCas9 on-target specificity can be greatly enhanced and reduce off-target effects. These innovations have broadened the scope of CRISPR-Cas9 beyond simple genome editing enabling more complex genetic modifications and functional studies (Doudna & Charpentier, 2014; Ochiai et al., 2015) (Figure 1).

3.3.2 Genetic Engineering Using CRISPR-Cas9 in Insects

The CRISPR-Cas9 system is a powerful and flexible tool that has transformed genetic engineering in many organisms, including insects, by allowing precise and efficient gene editing (Gilles et al., 2015; Petersen, 2017; Wang et al., 2016). This technology provides a platform for genetic manipulation across a wide range of insect species, supporting advances in basic research, disease vector control, and pest management (Gilles et al., 2015). CRISPR-Cas9 has driven significant progress in modifying the genetics of various insects, from model species like *Drosophila melanogaster* and *Tribolium castaneum* to important pests and vectors such as *Aedes aegypti, Bombyx mori, Spodoptera frugiperda*, and *Culex quinquefasciatus* (Gilles et al., 2015; Salum et al., 2024).

The CRISPR-Cas9 system has demonstrated remarkable efficiency and versatility across various insect models. In *Drosophila* S2 cells, targeting the yellow and white genes with single guide RNAs (sgRNAs) achieved up to 88% mutagenesis efficiency via homologous recombination, coupled with the development of high-resolution melting analysis for detecting indels in the injected generation, facilitating large-scale genetic screening (Gratz et al., 2013). In *Drosophila melanogaster*, two stable transgenic strains were created: one expressing Cas9 driven by the germline-specific nanos promoter, and the other expressing sgRNA under the U6 promoter, enabling targeted genome editing (Kondo & Ueda, 2013). The first application of CRISPR-Cas9 in mosquitoes targeted the enhanced cyan fluorescent protein (ECFP) gene in *Aedes aegypti*, where two designed sgRNAs yielded a 5.5% knockout rate, confirming the system's feasibility in this vector species (Dong et al., 2015). Further advances include multiplex targeting of six genes involved in key biological processes in *Aedes aegypti*, such as *kmo*, *loqs*, and *nix*, which produced edited mutants and expanded the toolkit for functional genetics in mosquitoes (Basu et al., 2015). Transgenic *Aedes aegypti* strains expressing Cas9 were developed via piggyBac-mediated transformation, enabling rapid disruption of genes related to vision (*kh*), flight (*white*), and blood feeding (*yellow*), thereby enhancing the efficiency and throughput of genome editing in this species (Li et al., 2017). In lepidopterans, the CRISPR-Cas9 system was applied to the silkworm *B. mori* by targeting the *BmBLOS2* gene, which affects lysosome-related organelles. Injection of sgRNAs and Cas9 mRNA into preblastodermal embryos resulted in a 95% mutation rate, evident from changes in embryo translucency, and co-injection led to a large 3.5 kb deletion, showcasing the system's capacity for inducing precise, large genomic modifications (Wang et al., 2013).

Technology plays a vital role in improving insect resistance in key crops such as cotton and potato (Ahmed et al., 2024). A high-throughput CRISPR-Cas9 mutagenesis library was created for cotton, focusing on genes associated with insect resistance, achieving genome editing efficiencies as high as 97.29% (Sun et al., 2023). Overexpression of the *GhMLP423* gene, discovered through this library, boosted insect resistance in cotton by triggering the plant's systemic acquired resistance (SAR) (Sun et al., 2023). Insect resistance threatens the long-term success of *Bt* transgene technologies. Plants use jasmonic acid (JA) signaling, regulated by *JAZ* repressors, to balance defense and growth. Mutations in JAZ proteins enhance resistance to pests by increasing defense metabolites and trichomes, which also affect cotton fiber. Since 2017, resistance to *Bt* traits like *Vip3A* in bollworms has grown, raising concerns for U.S. cotton. Targeting *JAZ* genes through CRISPR-Cas9 offers a promising strategy to boost internal herbivore resistance, despite challenges from genetic redundancy and fitness trade-offs (Rock et al., 2024).

4. Plant-incorporated Protectants

PIPs are pesticidal substances with defensive or protective properties against insect pests produced by genetically modified plants, typically through the introduction of genes from other organisms. These protectants are expressed directly within the tissues of genetically modified plants (GMPs) to defend crops against insect pests (Barton et al., 1987; Baum et al., 2007; Mao et al., 2007; Schellenberger et al., 2016) and viruses (Fitch et al., 1992; Scorza et al., 2001). When insect pests feed on these crops, they ingest the PIPs along with the plant tissue, which then disrupts their development and induces mortality (Schuler et al., 1998; Zhang et al., 2017). The development of transgenic plants producing PIPs has significantly advanced insect resistance in agriculture, exemplified by the introduction of *Bt* toxins in crops like cotton, maize, sweet potato, and tomato (Chang et al., 2017; Zhong et al., 2019; Katta et al., 2020; Soliman et al., 2017).

4.1 Insecticidal Proteins

4.1.1 Microbial Derived (Bt) Proteins

Entomopathogenic bacteria, *Bt*, have been used for insect control for over a century, targeting pests through their digestive systems. *Bt* cotton, first commercialized in 1996 in the U.S., has become a crucial tool for pest management globally (James, 2014; Mendelsohn et al., 2003). GM crops expressing PIPs were first reported in 1987, with *Bt* crops containing genes from *B. thuringiensis* that produce insecticidal crystal (*Cry*) proteins (*Bt* toxins) (Barton et al., 1987; Fischhoff et al., 1987; Vaeck et al., 1987). These proteins, typically 63–68 kDa (Mendelsohn et al., 2003), provide effective control against major cotton pests like tobacco budworm (*Heliothis virescens*), bollworm (*Helicoverpa zea*), and pink bollworm (*Pectinophora gossypiella*). The first *Bt* cotton, Bollgard, expressed *Cry1Ac*, while later versions like Bollgard II (*Cry1Ac+Cry2Ab*) and WideStrike (*Cry1Ac+Cry1F*), which express multiple *Cry* proteins, offered enhanced pest control (Gore et al., 2001; Chitkowski et al., 2003; Tindall et al., 2009). In India, *Bt* cotton was first commercialized in

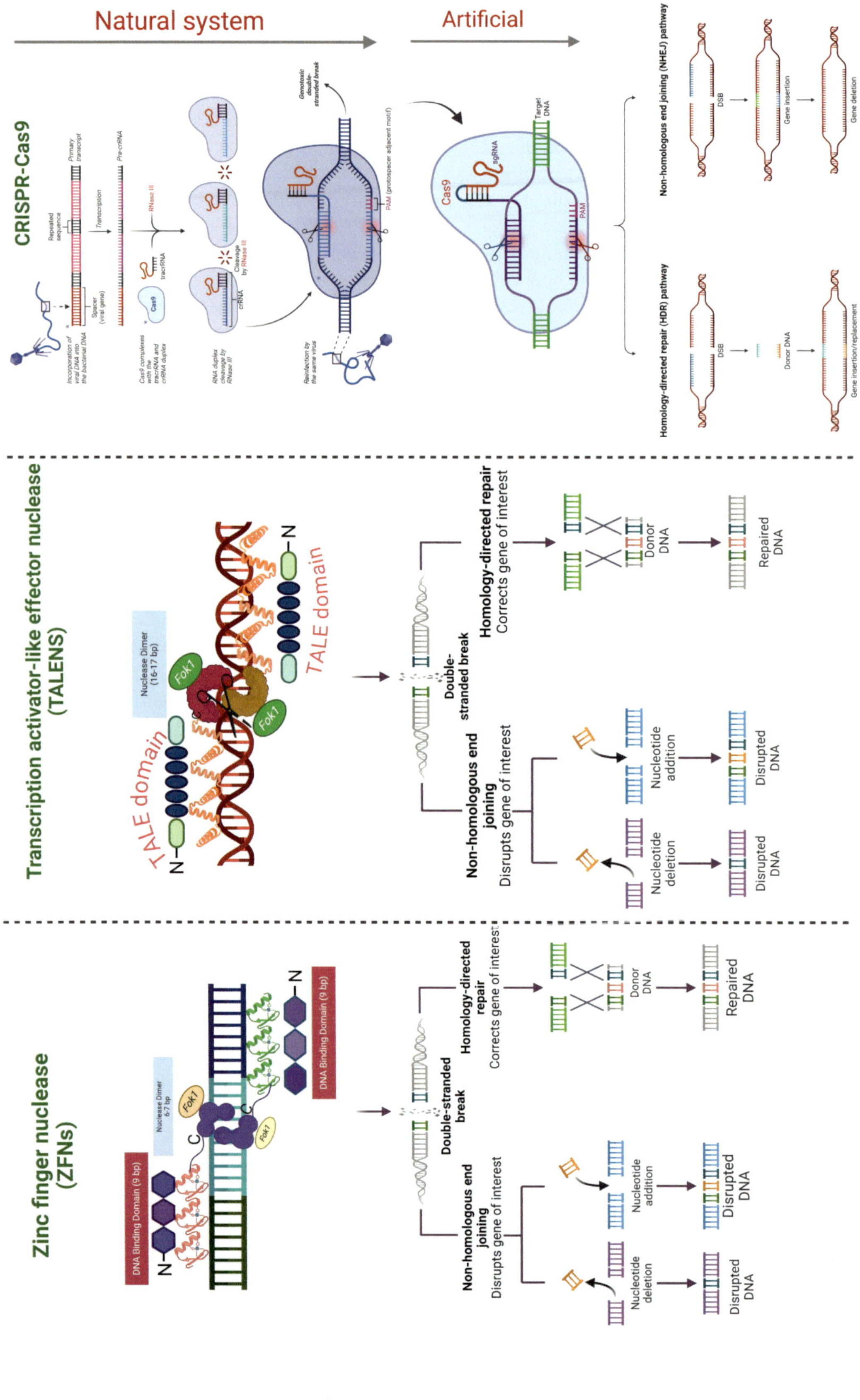

Fig. 1 Overview on the mechanism of ZFN, TALEN and CRISPR-Cas9 based genome editing system.

2002; with the second-generation Bollgard II approved in 2006 it now dominates 95% of the cotton-growing area.

Another example is *Bt* maize which provides protection against Lepidopteran and Coleopteran pests (Boucias & Pendland, 1998; Palma et al., 2014; Van Frankenhuyzen, 2009; 2013). Likewise, in Bangladesh, four insect resistant *Bt* brinjal varieties were commercialized in 2014 (Koch et al., 2015). Transgenic soybean expressing *Cry1Ac+Cry1Ab* was approved for commercial use in Latin America (Koch et al., 2015). Similarly, synthetic *cry1Ab* gene in transgenic tomato exhibited 100% larval mortality against tomato leafminer *Phthorimaea absoluta* (Soliman et al., 2017). Another study showed that modified *Cry1Ac* expression in tomatoes resulted in 38-100% larval mortality against *P. absoluta* (Selale et al., 2017). Rice lines expressing *Cry2A* conferred 80% mortality against rice leaf folders (*Cnaphalocrocis medinalis*) (Gunasekara et al., 2017). In another study, transgenic pigeonpea lines with *cry1Ac* and *cry2Aa* genes caused mortality in gram pod borer (*Helicoverpa armigera*) (Ghosh et al., 2017). Likewise, *Cry2Aa* gene expression in pigeon pea led to larval mortality of *H. armigera* (Singh et al., 2018). Sweet potato expressing *Cry1Aa* gene showed resistance against tobacco cutworm (*Spodoptera litura*) (Zhong et al., 2019). Furthermore, soybean expressing *Cry8* like genes provided resistance against black chafer (*Holotrichia parallela*) (Qin et al., 2019). *Cry* proteins are highly attractive for pest management due to their selective insecticidal activity, providing protection against insect feeding damage (Van Frankenhuyzen, 2013). Different classes of *Cry* proteins have structure-dependent toxicities specific to certain insect orders (Crickmore et al., 2016; Xu et al., 2014). Additionally, *Cry1* targets Lepidoptera, while *Cry3* targets Coleoptera (Clark et al., 2005).

Some *Cry* proteins also show significant toxicity against hemipteran pests like Lygus plant bugs and aphids (Baum et al., 2012; Porcar et al., 2009). Although *Cry* proteins typically have a narrow insecticidal spectrum, cross-order activity is increasingly observed (Van Frankenhuyzen, 2009). *Cry2Aa* targets insects from Lepidoptera and Diptera (mosquitoes) (De Maagd et al., 2001; Van Frankenhuyzen, 2009; Widner & Whiteley, 1989), while *Cry3Aa* is effective against Coleoptera, Hemiptera, and Hymenoptera (Van Frankenhuyzen, 2009). *Cry51Aa2* shows increased activity against hemipteran pests like Lygus (Gowda et al., 2020). *Cry51Aa1* and *Cry51Aa2* are also active against Coleoptera (Xu et al., 2015). Native *Cry51Aa2* protein is effective against hemipteran species *Lygus lineolaris*, *Lygus hesperus*, and the coleopteran Colorado potato beetle (*Leptinotarsa decemlineata*) but not Lepidoptera (Baum et al., 2012). *Bt* crops also feature glyphosate resistance (Que et al., 2010). GM crop adoption is widespread in the Americas, Asia, Australia, and parts of Africa, but limited in Europe.

4.1.2 Mode of Action of Bt Proteins

Bt toxins must be ingested, overcome digestive and physical barriers, and reach the target site to become effective. In the insect midgut, *Cry* proteins interact with specific receptors on epithelial cells, inserting into the membrane to form transmembrane pores. These pores cause cell lysis and pest mortality (Bravo et al., 2007) (Figure 2). The specificity of *Bt* proteins arises from their solubility, stability, proteolytic activation, and receptor binding on the insect midgut epithelium (Pigott & Ellar, 2007). All steps, ingestion, activation, and receptor binding are necessary for activity and contribute to high specificity (Van Frankenhuyzen, 2013). *Bt* produces insecticidal crystal proteins during sporulation, which are activated by midgut proteinases, bind to specific receptors, create pores, and kill the insect pests (Whalon & Wingerd, 2003).

4.2 Vegetative Insecticidal Proteins

B. thuringiensis produces vegetative insecticidal proteins (VIPs) like *Vip1*, *Vip2*, and *Vip3* types during its vegetative stage (Crickmore et al., 1998). These proteins are inactive until activated by enzymes in the insect midgut, causing cell swelling and disruption (Baranek et al., 2015). *Vip3*

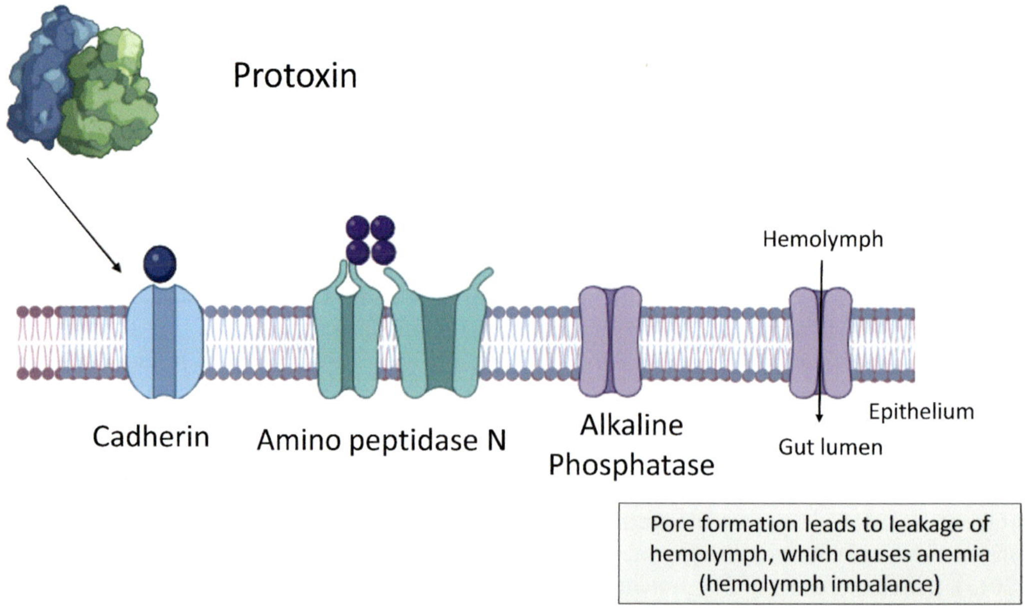

Fig. 2 Schematic representation of mode of action of *Bt* toxins.

proteins, such as *Vip3Aa19*, are used in crops like cotton and maize, with *Vip3Aa19* first introduced in Bollgard cotton (*COT102*) in 2008 to protect against major pests like *H. virescens*, *H. zea*, and *P. gossypiella*. Commercial *Bt*-crops use a combination of *Vip3* and *Cry* proteins to manage insect resistance through gene pyramiding. This approach involves incorporating multiple genes producing toxins with different target sites of action to minimize resistance development (Hernández-Rodríguez et al., 2009; Jouzani et al., 2017). For example, cotton lines expressing *VIP3A* alone or in combination with *Cry1Ab* show strong resistance to *H. zea* and *H. virescens* (Bommireddy et al., 2011). Additionally, a chimeric protein, *Vip3AcAaa*, demonstrates broader insect resistance, making it a promising option for pest management (Chen et al., 2018). *Vip3Aa20*, introduced in corn as *MIR162* and commercialized in 2009, was later combined with *Cry* genes (*Cry1Ab* + *Vip3Aa20*, *Cry1Ab*, *Vip3Aa20*, and *mCry3A*). Additionally, corn events expressing *Vip3Aa19* showed high toxicity to black cutworm (*Agrotis ipsilon*) (Yan et al., 2020), while sugarcane lines with *Vip3A* exhibited high mortality rates against stem borer (*Chilo infuscatellus*) (Riaz et al., 2020). *Vip3Ba1*, from Australian *Bt* isolates, was used in cowpea for legume pod borer (*Maruca vitrata*) protection (Bett et al., 2017). Tobacco plants transformed with a mutant *Vip3BR* protein achieved full protection against African cotton leafworm (*Spodoptera littoralis*), *A. ipsilon* and *H. armigera* (Gayen et al., 2015).

4.3 Lectins

Plant lectins are carbohydrate-binding proteins that target specific sugars on cell membranes. Transgenic rice expressing Allium sativum leaf agglutinin (ASAL) and Galanthus nivalis lectin (GNA) exhibited resistance to major sap-sucking pests like brown planthopper (*Nilaparvata lugens*), white-backed planthopper (*Sogatella furcifera*) and green leafhopper (*Nephotettix nigropictus*). Additionally, rice with mannose-specific GNA under a phloem-specific promoter showed partial resistance to *N. lugens* (Bharathi et al., 2011), while Arabidopsis expressing a recombinant fusion protein (*Hv1a/GNA*) proved effective against peach potato and grain aphids (Nakasu et al., 2014).

The GNA gene in potatoes conferred aphid resistance compared to non-transgenic plants (Mi et al., 2017). Transformed *Brassica juncea* with lentil lectin (LL) and chickpea protease inhibitor (CPPI) showed enhanced resistance to aphids (Rani et al., 2017b). Cotton lines expressing *Sclerotium rolfsii* lectin reduced aphid populations by 69% and killed *S. litura* in 96 hours (Vanti et al., 2018). Pinellia pedatisecta agglutinin (PPA) conferred resistance to wheat aphid in BE104 (Duan et al., 2018). Transgenic mustard expressing *Colocasia esculenta* tuber agglutinin (CEA) showed 70-81% higher insect mortality than controls, proving CEA as an effective, non-allergenic insecticidal protein (Das et al., 2018). Transgenic tomato with lectin genes (*rvl1 and srl1*) had a 4.59% transformation efficiency and better resistance to root knot nematode Meloidogyne incognita (Bhagat et al., 2019). The *Arcelin 1* gene from *Phaseolus vulgaris* conferred significant resistance against bruchid pests in cowpea, reducing egg counts, emerging insects, and grain loss (Grazziotin et al., 2020).

4.4 Fusion Proteins

Fusion proteins combining different insecticidal domains are more effective against pests. For instance, transgenic rice with *Cry2AX1* showed improved resistance to lepidopteran pests (Chakraborty et al., 2016), while *Cry2Aa* enhanced aphid mortality (Chougule et al., 2013) and *Cry51Aa2* targeted lygus species in cotton (Gowda et al., 2016). A fusion protein of lectin and protease inhibitors in *B. juncea* conferred resistance to aphids (Rani et al., 2017a). *Bt* maize with *Cry1Ab/cry2Aj* was resistant to lepidopteran pests like *S. exigua* and *Harmonia axyridis* (Chang et al., 2017) and transgenic cotton with *Cry1Be+cry1Fa* exhibited resistance to *S. litura* and *O. nubilalis* (Meade et al., 2017). Various studies have developed crops with enhanced pest resistance by expressing fusion proteins. For example, rice expressing *Cry1Ac+ASAL* showed increased resistance to yellow stem borer, leaf folder, and brown plant hopper (Boddupally et al., 2018), while *Cry1Ab+vip3A* conferred resistance to Asiatic rice borer and rice leaf folder (Xu et al., 2018). Rice expressing *cry2Aa+cry1Ca* targeted *Chilo suppressalis* (Qiu et al., 2019), and sugarcane with *cry1Ab-cry1Ac* showed resistance to shoot borer (Koerniati et al., 2020). Combining *Cry* and *VIP* genes offers a sustainable strategy for managing various insect pests.

4.5 Iridoviral Serine/Threonine Kinase System

The Iridoviral Serine/Threonine Kinase (ISTK) system, derived from the Chilo iridescent virus (CIV), is a promising alternative for insect pest management due to its novel mechanism that helps combat insecticide resistance and targets a broad range of pests (Ganapathy et al., 2021). Iridoviruses like CIV are large double-stranded DNA viruses infecting various insects without affecting mammals. ISTK, a gene product of CIV, produces a 48-kDa polypeptide and a 35-kDa cleavage product called iridoptin, both exhibiting insecticidal activity by inducing apoptotic blebbing and host protein shutoff in insect cells (Ganapathy et al., 2021; Chitnis et al., 2011). Iridoptin has demonstrated significant efficacy, causing 50% mortality in cotton aphids and 100% mortality in green peach aphids. ISTK is notable as the first iridovirus-encoded kinase known to induce apoptosis in insect cells, indicating a novel mode of action. GM crops, especially those expressing insecticidal *Bt δ-endotoxin* genes, have been widely used against lepidopteran and coleopteran pests. However, extensive and prolonged use of *Bt* toxins has resulted in the development of pest resistance. Additionally, phloem-feeding insects such as aphids are not affected by *Bt* crops, underscoring the necessity to discover new polypeptide toxins with different modes of action (Ganapathy et al., 2021).

4.6 RNA Interference (RNAi) Approach

A new class of GM crops expressing double-stranded RNA (dsRNA) PIPs has been developed to utilize RNAi for gene silencing. It was discovered in 1998, is a process that suppresses gene expression by targeting specific RNA sequences (Fire et al., 1998). These dsRNA molecules interfere with gene expression in pests, providing crop protection by disrupting essential proteins. RNAi crops first emerged in 2007, designed to target proteins in pest insects (Baum et al., 2007; Mao et al., 2007). Typically, insecticidal dsRNAs are around 150-320 nucleotides long (Huvenne & Smagghe, 2010). In June 2017, the U.S. Environmental Protection Agency (U.S. EPA) approved the first RNAi crop targeting corn rootworm (*Diabrotica* spp.), which also expresses a *Cry* protein and is glyphosate-resistant (U.S. EPA, 2017; Fishilevich et al., 2016; Ramaseshadri et al., 2013).

Application of dsRNAs in maize induced RNAi-mediated gene knockdown, increasing insect mortality (Li et al., 2015). In rice, knockdown of APN1 and APN2 genes reduced susceptibility to *C. suppressalis* and conferred resistance to *Bt* varieties (Qiu et al., 2017). Silencing dvvgr and dvbol genes in maize enhanced resistance to western corn rootworm (Niu et al., 2017). Soybean expressing dsRNA targeting *SpbP0* showed reduced larval mortality even at T3 generation (Meng et al., 2017). Combining *Bt* toxin with RNAi in cotton targeted the *JHMT* gene, affecting *H. armigera* (Ni et al., 2017). Transgenic potatoes with hairpin RNAi targeting EcR gene showed increased resistance to Colorado potato beetle, with 15-80% insect mortality (Hussain et al., 2019). Kola et al. (2019) developed rice lines with *Acetylcholinesterase* gene knockdown, showing resistance to yellow stem borer. Yan et al. (2020) used dsRNA/nano carriers to target specific genes in soybeans, leading to high soybean aphid mortality. Mamta et al. (2016), and Mamta and Rajam (2017) demonstrated that silencing the chitinase gene (*HaCHI*) in tobacco and tomato via HI-RNAi resulted in insect resistance, causing larval, pupal, and adult mortality. These studies highlight RNAi as an effective method for developing insect-resistant plants.

4.6.1 Mode of Action of dsRNA

After ingestion, dsRNA PIPs are transported into target cells within the pest insect. Inside the cell, the dsRNA is enzymatically cleaved into small interfering RNA molecules (siRNA, approximately 20 nucleotides long). These siRNAs guide the degradation of complementary mRNA by the insect's endogenous RNAi machinery. This degradation prevents the translation of mRNA into essential proteins, leading to sublethal effects such as reduced growth or even pest mortality (Figure 3). The first approved dsRNA PIP disrupts the synthesis of the *Snf7* protein, a crucial vacuolar sorting protein (Ramaseshadri et al., 2013). Specifically, dsRNA targeting an essential gene in the insect pest is processed by the RNAase III enzyme Dicer into siRNAs, which then guide the Argonaute protein within the RNA-induced silencing complex (RISC) to degrade complementary mRNA (Scott et al., 2013).

4.7 Ecological Impacts of PIPs

Unlike synthetic pesticides, the extent and timing of PIPs release from GM crops are less predictable, as PIPs are produced and released within the plant. PIPs can enter the environment through root exudates (Saxena et al., 1999) or pollen (Losey et al., 1999; Pleasants et al., 2001; Sears et al., 2001). Major PIP release occurs from decaying tissue left in the field after harvest (Palm et al., 1996; Zwahlen et al., 2003). Cry protein release is variable, with concentrations differing significantly between tissue types and crop strains, ranging from low in maize seeds to high in leaves and roots (U.S. EPA, 2015). Estimates suggest that one hectare of soil may receive between 3.6–5,400 g of Cry proteins (Clark et al., 2005). For dsRNA PIPs, while specific release pathways are similar to Cry proteins, dsRNA tissue concentrations are much lower, ranging from 0.13 ng-dsRNA/g in pollen to 85 ng-dsRNA/g in whole plant tissue (U.S. EPA, 2015).

Compared to *Cry* proteins, dsRNA released into environmental systems is expected to be in smaller amounts. However, lower dsRNA concentrations in soil don't necessarily eliminate potential ecological effects. For instance, dsRNA sequences have been shown to inhibit insect growth and cause mortality at ~10,000-fold lower concentrations than *Cry3Bb1* (Levine et al., 2015). Quantifying dsRNA in RNAi crops and environmental compartments can help determine

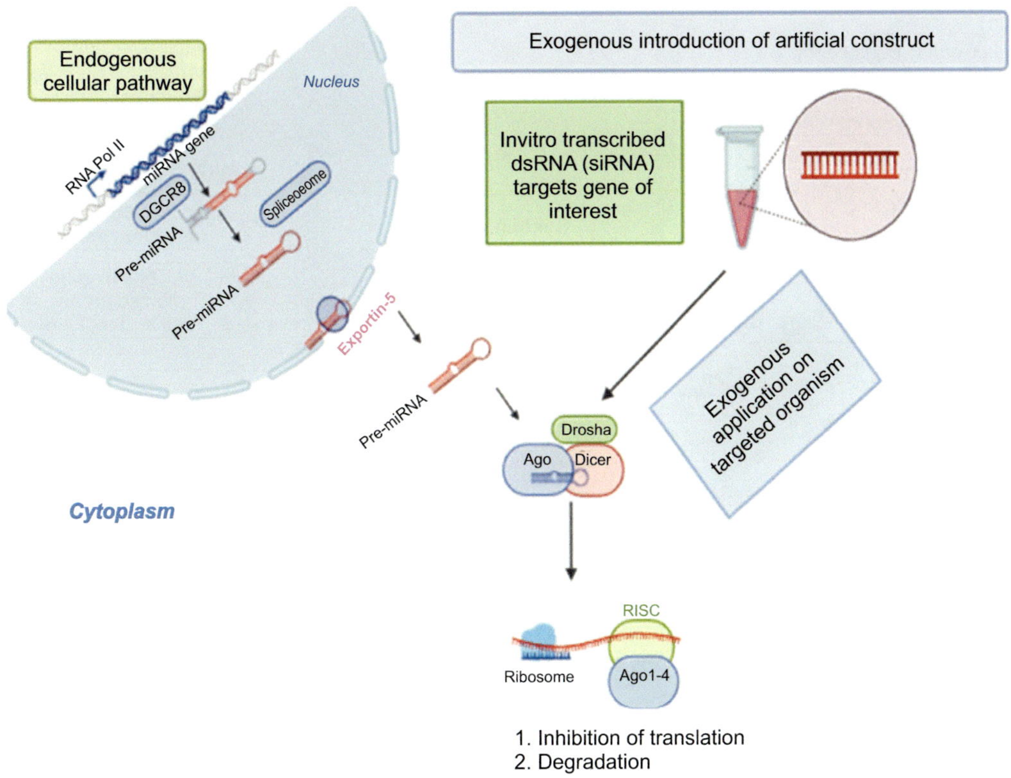

Fig. 3 Schematic representation of mechanism of dsRNA function.

how dsRNA is released during and after the growing season. Additionally, the impact of PIPs on NTOs should be considered. While *Cry* proteins are highly toxic to specific insect orders, they are generally less toxic to distantly related taxa (Rukarwa et al., 2013; Van Frankenhuyzen, 2009; Van Frankenhuyzen, 2013). Widespread cultivation of *Bt* crops has not shown adverse impacts on NTOs (Betz et al., 2000). Characterizing the activity spectrum of PIPs is crucial for selecting and testing NTOs in ecological risk assessments (ERA) (Bachman et al., 2013; Romeis et al., 2008).

4.8 Legal Requirements for PIPs

ERA of PIPs evaluates the likelihood of adverse effects from their use and environmental release, considering both hazard and exposure. With the advent of RNAi crops expressing dsRNA PIPs, the U.S. EPA convened Scientific Advisory Panels in 2014 (U.S. EPA, 2014) and 2016 (U.S. EPA, 2016) to address ERA and related knowledge gaps (Auer & Frederick, 2009; Lundgren & Duan, 2013; Roberts et al., 2015). The European Food Safety Authority (EFSA) also assessed the ecological risks of RNAi crops and considered potential human consumption risks (European Food Safety Authority, 2014).

5. Sustainable Agriculture through Biotechnological Strategies

Sustainable agriculture is an approach to farming that focuses on producing long-term crops and livestock without harming the environment. It involves using techniques that protect public health, human communities, and animal welfare. One of the most promising areas in sustainable agriculture is the application of biotechnological strategies, which can enhance crop resilience and conserve biodiversity (Glick, 2015) (Figure 4).

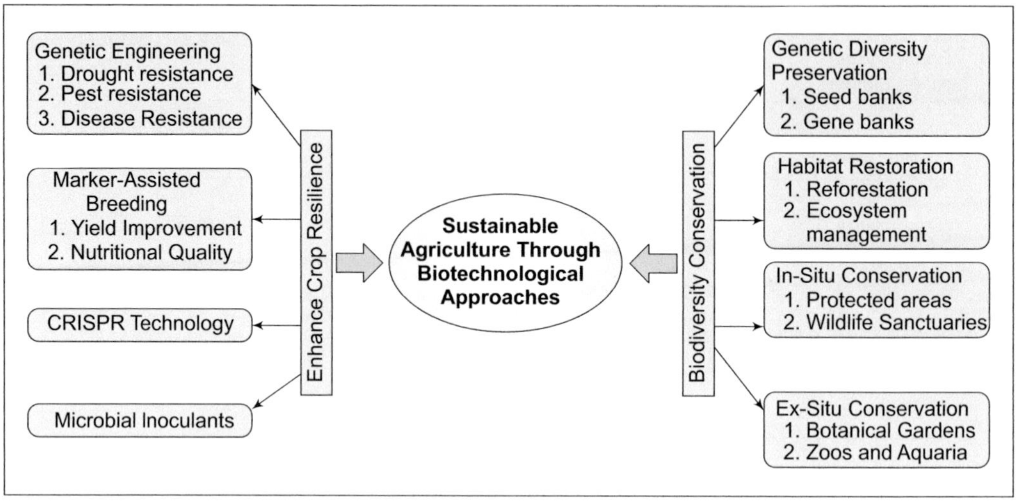

Fig. 4 Overview of biotechnological approaches in sustainable agriculture.

5.1 Biotechnological Approaches to Enhance Crop Resilience

Crop resilience, the ability of crops to withstand and recover from adverse conditions such as drought, salinity, pests, and diseases, is essential for ensuring food security amid climate change and a growing global population. Biotechnological approaches offer various methods to improve crop resilience. Genetic engineering has significantly enhanced crop diversity, with transgenic crops exhibiting pest resistance, drought tolerance, and enhanced nutritional content, reducing reliance on chemical inputs and promoting environmental sustainability (Brookes & Barfoot, 2020). For example, *Bt* cotton, engineered to express *Bt* toxin, shows enhanced pest resistance, increasing yields and reducing pesticide use (ISAAA, 2017). Similarly, glyphosate-resistant soybean, herbicide-tolerant crop, has facilitated more effective weed management (Duke, 2018). The advent of CRISPR-Cas9 technology has revolutionized crop improvement by enabling precise genome editing, allowing targeted modifications to enhance stress tolerance and disease resistance. Recent studies have successfully used CRISPR-Cas9 to confer drought tolerance in rice by targeting the *OsPYL9* gene, which plays a crucial role in abscisic acid signaling (Mishra et al., 2018). Additionally, CRISPR-based editing has improved disease resistance in crops like wheat and tomato by targeting susceptibility genes (Elsharawy & Refat, 2023).

In areawide programs addressing food and nutritional security, gene-edited crops have shown significant promise. A notable example is the deployment of genetically edited rice in Asia, where the adoption of salt-tolerant varieties has improved yields in coastal and saline-prone regions, thereby enhancing food security in these vulnerable areas (Reddy et al., 2017). Similarly, the introduction of biofortified crops, such as zinc-enriched wheat in India and vitamin A-enriched cassava in Africa, has directly addressed nutritional deficiencies among local populations, contributing to both food security and improved health outcomes (Bouis & Saltzman, 2017). These programs illustrate the potential of gene-edited crops to not only improve resilience to environmental stresses but also to directly enhance the nutritional quality of food crops, offering a comprehensive solution to food security challenges.

Molecular breeding, which combines traditional breeding methods with molecular biology tools, has also significantly contributed to crop resilience. Marker-assisted selection (MAS) and genomic selection (GS) are prominent techniques in this domain. MAS has been effectively used to introgress drought-resistant genes into maize, resulting in varieties that perform better under water-limited conditions (Ragot et al., 2007). GS utilizes genome-wide markers to predict the performance of breeding lines, accelerating the development of resilient crops (Meuwissen et al., 2001). Additionally, transgenic approaches and epigenetic modifications enhance crop resilience by creating transgenic plants that express certain proteins or enzymes to tolerate extreme environmental conditions. For instance, inserting genes encoding antifreeze proteins from cold-water fish into plants improves frost resistance (Glick, 2015). Epigenetic modifications, such as DNA methylation and histone modifications, regulate gene expression in response to environmental stresses, improving tolerance to drought and salinity in crops like rice and barley (Kumar, 2018). This approach offers a reversible and non-transgenic method to enhance resilience. Furthermore, microbiome engineering, which involves introducing beneficial microbes or enhancing native microbial communities, has shown promise in improving crop resilience. Studies have demonstrated that inoculating crops with drought-tolerant endophytes enhances their tolerance to water stress (Naylor et al., 2017). Additionally, microbiome engineering can improve nutrient uptake and disease resistance in crops (Mendes et al., 2013).

5.2 Biotechnological Strategies for Biodiversity Conservation

Biodiversity conservation is critical for the sustainability and resilience of agricultural systems. The loss of biodiversity negatively impacts ecosystem services, agricultural productivity, and food security (Cardinale et al., 2012). Biotechnology offers innovative tools and strategies to conserve biodiversity while promoting sustainable agriculture. It plays a crucial role in the conservation of genetic resources and restoration of ecosystems. Through techniques like cryopreservation and tissue culture, biotechnology ensures the long-term storage of genetic material, thereby preserving plant and animal species threatened by habitat loss or climate change (Engelmann, 2011). Tissue culture also enables the rapid propagation of endangered species and helps restore degraded ecosystems (Reed et al., 2011). Additionally, biotechnological methods such as bioremediation use genetically engineered microorganisms or plants to clean up polluted or degraded environments, effectively restoring these habitats (Rafeeq et al., 2023). In Spain, the cryopreservation of olive, *Olea europaea* germplasm has been a successful biotechnological strategy for conserving this economically and culturally significant species. By cryopreserving embryonic axes and somatic embryos, researchers have secured genetic material from diverse olive varieties, including rare and endangered ones. This initiative has not only conserved the genetic diversity of olive trees but also provided a valuable resource for breeding programs aimed at improving disease resistance and climate adaptability (Bradaï & Sánchez-Romero, 2021).

The field of biotechnology also significantly contributes to *ex-situ* conservation and genetic rescue. Techniques like tissue culture and micropropagation are essential for conserving plant species outside their natural habitats, which is especially useful for rare and endangered species at risk from habitat destruction or climate change (Engelmann, 2015). Furthermore, in vitro conservation techniques and germplasm banks maintain genetic diversity for future breeding and research (Engelmann, 2015). Genetic rescue efforts, facilitated by biotechnology, enhance the genetic diversity of small, isolated populations by introducing genes from other populations or species, reducing inbreeding risks and improving adaptability to changing environments (Jaganathan et al., 2018).

Biotechnology also addresses challenges in agriculture and biodiversity through synthetic biology, soil microbiome management, pollinator conservation, and the development of climate-resilient crops. Synthetic biology creates crops with novel traits, such as improved photosynthetic efficiency and biomass production under stress conditions (Schwander et al., 2016), and synthetic microbes that promote soil health or provide habitats for other species (Glick, 2015). Soil microbiome management employs biofertilizers and biopesticides to enhance soil fertility and pest management, promoting a balanced soil ecosystem and reducing dependence on synthetic agrochemicals (Singh et al., 2011). For pollinator conservation, biotechnology develops pollinator-friendly crops with enhanced nutritional value to support pollinator health and diversity (Klein et al., 2007). In the

Philippines, the International Rice Research Institute (IRRI) has developed climate-resilient rice varieties using advanced biotechnological methods. These varieties are designed to withstand extreme weather conditions such as flooding and drought, which are becoming more frequent due to climate change. The introduction of these resilient rice varieties has improved food security and contributed to the conservation of rice biodiversity by maintaining a wide range of genetic traits within cultivated species (Stuecker et al., 2018). Moreover, biotechnological innovations produce climate-resilient crops to mitigate climate change impacts, ensuring food security and conserving biodiversity (Wassmann et al., 2009).

6. Critical Gaps, Challenges, and Priorities

Biotechnological advancements often require expensive tools for developing crop protection technologies. Bridging the gap between cutting edge research and field level application is crucial, especially in resource-limited environments. Cost-effective and scalable solutions that are accessible to farmers are essential for widespread adoption. Regulatory frameworks for biotechnological interventions, like GM and CRISPR, differ significantly across countries, hindering global implementation. A balance must be struck between promoting innovation and ensuring safety, transparency, and public acceptance. Harmonizing global regulatory standards is important to address ethical concerns, including environmental impacts and food safety. Pest resistant crops face the threat of pests evolving resistance over time. Developing long-term solutions that adapt to emerging diseases and climate change remains a challenge. Integrating biotechnology with holistic approaches, such as crop diversification and biological control, enhances resilience. This integration promotes sustainable agricultural practices.

There is limited understanding of the long-term ecological effects of GMPs and gene edited crops on biodiversity and ecosystems. It's crucial to ensure biotechnological interventions do not disrupt ecological balance, such as harming NTOs or degrading soil health. Conducting thorough environmental assessments and developing mitigation strategies are essential to protect biodiversity and ecosystem services in agriculture. Many biotechnological advancements lack sufficient input from farmers and local communities, leading to a disconnect between innovations and real needs. Biotechnological solutions must align with the socio-economic realities of farmers, including cultural practices and market demands. Involving farmers and local communities in research and development ensures more practical and sustainable outcomes. Some biotechnological innovations rely on external inputs like specific fertilizers or pesticides, which may conflict with sustainability goals. Reducing dependency on these inputs is necessary to promote sustainable agricultural practices. Low-input, resource efficient technologies should be encouraged to enhance resilience and prevent environmental degradation. Addressing these gaps is key to ensuring that biotechnological innovations are effective, sustainable, and equitable.

7. Conclusion

Novel biotechnological strategies represent a transformative approach for safeguarding crops and promoting sustainable agriculture in the face of increasing global challenges such as climate change, pest outbreaks, and resource scarcity. By leveraging advances in genetic transformation, new genomic techniques, molecular biology, and precision breeding, it is possible to develop crop varieties with enhanced resistance to biotic and abiotic stresses, improved nutritional profiles, and greater resilience to environmental fluctuations. These innovations not only help to protect crop yields and ensure food security but also reduce the dependency on chemical inputs, thereby supporting eco-friendly farming practices. As biotechnology continues to evolve, interdisciplinary collaborations and policy frameworks must be strengthened to ensure responsible deployment, public acceptance, and equitable access to these technologies, ultimately contributing to a more sustainable and resilient agricultural future.

References

Adeyinka, O.S., Riaz, S., Toufiq, N., Yousaf, I., Bhatti, M.U., Batcho, A., Olajide, A.A., Nasir, I.A. and Tabassum, B. (2020). Advances in exogenous RNA delivery techniques for RNAi-mediated pest control. *Molecular Biology Reports*, *47*, 6309–6319.

Ahmed, A.I., Khan, A.I., Negm, M.a.M., Iqbal, R., Azhar, M.T., Khan, S.H. and Rana, I.A. (2024). Enhancing cotton resilience to challenging climates through genetic modifications. *Journal of Cotton Research*, *7*(1). https://doi.org/10.1186/s42397-024-00171-4

Ainley, W.M., Sastry-Dent, L., Welter, M.E., Murray, M.G., Zeitler, B., Amora, R., Corbin, D.R., Miles, R.R., Arnold, N.L., Strange, T.L., Simpson, M.A., Cao, Z., Carroll, C., Pawelczak, K.S., Blue, R., West, K., Rowland, L.M., Perkins, D., Samuel, P., ... Petolino, J.F. (2013). Trait stacking via targeted genome editing. *Plant Biotechnology Journal*, *11*(9), 1126–1134. https://doi.org/10.1111/pbi.12107

Akbar, S., Tahir, M., Wang, M.-B. and Liu, Q. (2017). Expression Analysis of Hairpin RNA Carrying Sugarcane mosaic virus (SCMV) Derived Sequences and Transgenic Resistance Development in a Model Rice Plant. *Bio.Med. Research International*, *2017*, 1–10. https://doi.org/10.1155/2017/1646140

Alves, A.C., Quecini, V.M. and Vieira, M.L.C. (1999). Plant transformation: Advances and perspectives. *Scientia Agricola*, *56*, 1–8.

Anjanappa, R.B. and Gruissem, W. (2021). Current progress and challenges in crop genetic transformation. *Journal of Plant Physiology*, *261*, 153411.

Aryan, A., Myles, K.M. and Adelman, Z.N. (2014). Targeted genome editing in Aedes aegypti using TALENs. *Methods*, *69*(1), 38–45.

Auer, C. and Frederick, R. (2009). Crop improvement using small RNAs: Applications and predictive ecological risk assessments. *Trends in Biotechnology*, *27*(11), 644–651. https://doi.org/10.1016/j.ti*Bt*ech.2009.08.005

Babaoglu, M., Davey, M.R. and Power, J.B. (2000). Genetic engineering of grain legumes: Key transformation events. *AgBiotechNet*, *2*, 1–8.

Bachman, P.M., Bolognesi, R., Moar, W.J., Mueller, G.M., Paradise, M.S., Ramaseshadri, P., Tan, J., Uffman, J.P., Warren, J., Wiggins, B.E. and Levine, S.L. (2013). Characterization of the spectrum of insecticidal activity of a double-stranded RNA with targeted activity against Western Corn Rootworm (Diabrotica virgifera virgifera LeConte). *Transgenic Research*, *22*(6), 1207–1222. https://doi.org/10.1007/s11248-013-9716-5

Baranek, J., Kaznowski, A., Konecka, E. and Naimov, S. (2015). Activity of vegetative insecticidal proteins Vip3Aa58 and Vip3Aa59 of Bacillus thuringiensis against lepidopteran pests. *Journal of Invertebrate Pathology*, *130*, 72–81. https://doi.org/10.1016/j.jip.2015.06.006

Barton, K.A., Whiteley, H.R. and Yang, N.S. (1987). Bacillus thuringiensis §-Endotoxin Expressed in Transgenic Nicotiana tabacum Provides Resistance to Lepidopteran Insects. *Plant Physiology*, *85*(4), 1103–1109. https://doi.org/10.1104/pp.85.4.1103

Basu, S., Aryan, A., Overcash, J.M., Samuel, G.H., Anderson, M.A.E., Dahlem, T.J., Myles, K.M. and Adelman, Z.N. (2015). Silencing of end-joining repair for efficient site-specific gene insertion after TALEN/CRISPR mutagenesis in Aedes aegypti. *Proceedings of the National Academy of Sciences*, *112*(13), 4038–4043. https://doi.org/10.1073/pnas.1502370112

Baum, J.A., Bogaert, T., Clinton, W., Heck, G.R., Feldmann, P., Ilagan, O., Johnson, S., Plaetinck, G., Munyikwa, T., Pleau, M., Vaughn, T. and Roberts, J. (2007). Control of coleopteran insect pests through RNA interference. *Nature Biotechnology*, *25*(11), 1322–1326. https://doi.org/10.1038/n*Bt*1359

Baum, J.A., Sukuru, U.R., Penn, S.R., Meyer, S.E., Subbarao, S., Shi, X., Flasinski, S., Heck, G.R., Brown, R.S. and Clark, T.L. (2012). Cotton Plants Expressing a Hemipteran-Active Bacillus thuringiensis Crystal Protein Impact the Development and Survival of Lygus hesperus (Hemiptera: Miridae) Nymphs. *Journal of Economic Entomology*, *105*(2), 616–624. https://doi.org/10.1603/ec11207

Beerli, R.R. and Barbas III, C.F. (2002). Engineering polydactyl zinc-finger transcription factors. *Nature Biotechnology*, *20*(2), 135–141.

Bett, B., Gollasch, S., Moore, A., James, W., Armstrong, J., Walsh, T., Harding, R. and Higgins, T.J. (2017). Transgenic cowpeas (Vigna unguiculata L. Walp) expressing Bacillus thuringiensis Vip3Ba protein are protected against the Maruca pod borer (Maruca vitrata). *Plant Cell Tiss. Org. Cult.*, *131*, 335–345.

Betz, F.S., Hammond, B.G. and Fuchs, R.L. (2000). Safety and Advantages of Bacillus thuringiensis-Protected Plants to Control Insect Pests. *Regulatory Toxicology and Pharmacology*, *32*(2), 156–173. https://doi.org/10.1006/rtph.2000.1426.

Beumer, K., Bhattacharyya, G., Bibikova, M., Trautman, J.K. and Carroll, D. (2006). Efficient gene targeting in Drosophila with zinc-finger nucleases. *Genetics*, *172*(4), 2391–2403.

Bhagat, Y.S., Bhat, R.S., Kolekar, R.M., Patil, A.C.,, Lingaraju, S., Patil, R.V. and Udikeri, S.S. (2019). Remusatia vivipara lectin and Sclerotium rolfsii lectin interfere with the development and gall formation activity of Meloidogyne incognita in transgenic tomato. *Transgenic Res.*, *28*, 299–315.

Bharathi, Y., Kumar, S.V., Pasalu, I.C., Balachandran, S.M., Reddy, V.D. and Rao, K.V. (2011). Pyramided rice lines harbouring Allium sativum (asal) and Galanthus nivalis (gna) lectin genes impart enhanced resistance against major sap sucking pests. *J. Biotechnol.*, 152, 63–71.

Bhuyan, S.J., Kumar, M., Devde, P.R., Rai, A.C., Mishra, A.K., Singh, P.K. and Siddique, K.H.M. (2023). Progress in gene editing tools, implications, and success in plants: A review. *Frontiers in Genome Editing*, *5*. https://doi.org/10.3389/fgeed.2023.1272678

Bibikova, M., Beumer, K., Trautman, J.K. and Carroll, D. (2003). Enhancing Gene Targeting with Designed Zinc Finger Nucleases. *Science*, *300*(5620), 764. https://doi.org/10.1126/science.1079512

Bibikova, M., Carroll, D., Segal, D.J., Trautman, J.K., Smith, J., Kim, Y.-G. and Chandrasegaran, S. (2001). Stimulation of Homologous Recombination through Targeted Cleavage by Chimeric Nucleases. *Molecular and Cellular Biology*, *21*(1), 289–297. https://doi.org/10.1128/MCB.21.1.289-297.2001

Boch, J., Scholze, H., Schornack, S., Landgraf, A., Hahn, S., Kay, S., Lahaye, T., Nickstadt, A. and Bonas, U. (2009). Breaking the Code of DNA Binding Specificity of TAL-Type III Effectors. *Science*, *326*(5959), 1509–1512. https://doi.org/10.1126/science.1178811

Boddupally, D., Tamirisa, S., Gundra, S.R., Vudem, D.R. and Khareedu, V.R. (2018). Expression of hybrid fusion protein (Cry1Ac: ASAL) in transgenic rice plants imparts resistance against multiple insect pests. *Sci. Rep.*, *8*, 1–10.

Bogdanove, A.J. and Voytas, D. F. (2011). TAL Effectors: Customizable Proteins for DNA Targeting. *Science*, *333*(6051), 1843–1846. https://doi.org/10.1126/science.1204094

Bommireddy, P.L., Leonard, B.R., Temple, J., Price, P., Emfinger, K., Cook, D. and Hardke, J.T. (2011). Field performance and seasonal efficacy profiles of transgenic cotton lines expressing Vip3A and VipCot against Helicoverpa zea (Boddie) and Heliothis virescens (F.). *J. Cotton Sci.*, *15*, 251–259.

Boucias, D.G. and Pendland, J.C. (1998). Bacillus thuringiensis: Producer of Potent Insecticidal Toxins. In: *Springer eBooks*, pp. 217–257. https://doi.org/10.1007/978-1-4615-4915-4_7

Bouis, H.E. and Saltzman, A. (2017). Improving nutrition through biofortification: A review of evidence from HarvestPlus, 2003 through 2016. *Global Food Security*, *12*, 49–58.

Bradaï, F. and Sánchez-Romero, C. (2021). Effect of cryopreservation on the *ex vitro* establishment of olive plants regenerated via somatic embryogenesis. *Plants*, *10*(2), 396.

Bravo, A., Gill, S.S. and Soberón, M. (2007). Mode of action of Bacillus thuringiensis Cry and Cyt toxins and their potential for insect control. *Toxicon*, *49*(4), 423–435. https://doi.org/10.1016/j.toxicon.2006.11.022

Brookes, G. and Barfoot, P (2020). *GM crops: Global Socio-economic and Environmental Impacts 1996–2018*. PG Econ. Ltd., UK, pp. 1–213.

Budhagatapalli, N., Rutten, T., Gurushidze, M., Kumlehn, J. and Hensel, G. (2015). Targeted modification of gene function exploiting homology-directed repair of TALEN-mediated double-strand breaks in barley. *G3: Genes, Genomes, Genetics*, *5*(9), 1857–1863.

Cardinale, B.J., Duffy, J.E., Gonzalez, A., Hooper, D.U., Perrings, C., Venail, P., Narwani, A., Mace, G.M., Tilman, D., Wardle, D.A. and Kinzig, A.P. (2012). Biodiversity loss and its impact on humanity. *Nature*, *486*(7401), 59–67.

Cathomen, T. and Joung, J.K. (2008). Zinc-finger nucleases: The next generation emerges. *Molecular Therapy*, *16*(7), 1200–1207.

Cermak, T., Doyle, E.L., Christian, M., Wang, L., Zhang, Y., Schmidt, C., Baller, J.A., Somia, N.V., Bogdanove, A.J. and Voytas, D.F. (2011). Efficient design and assembly of custom TALEN and other TAL effector-based constructs for DNA targeting. *Nucleic Acids Research*, *39*(12), e82–e82.

Chakraborty, M., Reddy, P.S., Mustafa, G., Rajesh, G., Narasu, V.L., Udayasuriyan, V. and Rana, D. (2016). Transgenic rice expressing the cry2AX1 gene confers resistance to multiple lepidopteran pests. *Transgenic Res.*, *25*, 665–678.

Chang, X., Lu, Z., Shen, Z., Peng, Y. and Ye, G. (2017). Bitrophic and tritrophic effects of transgenic cry1Ab/cry2Aj maize on the beneficial, nontarget Harmonia axyridis (Coleoptera: Coccinellidae). *Envir. Entomol.*, *46*, 1171–1176.

Chen, W., Liu, C., Lu, G., Cheng, H., Shen, Z. and Wu, K. (2018). Effects of Vip3AcAa? Cry1Ac cotton on midgut tissue in Helicoverpa armigera (Lepidoptera: Noctuidae). *J. Insect Sci.*, *18*, 13.

Chitkowski, R.L., S.G. Turnipseed, M.J. Sullivan and W.C. Bridges, Jr. (2003). Field and laboratory evaluations of transgenic cottons expressing one or two Bacillus thuringiensis var. kurstaki Berliner proteins for management of Noctuid (Lepidoptera) pests. *J. Econ. Entomol.*, *96*, 755–762.

Chitnis, N.S., Paul, E.R., Lawrence, P.K., Henderson, C.W., Ganapathy, S., Taylor, P.V., Virdi, K.S., D'Costa, S.M., May, A.R. and Bilimoria, S.L. (2011). A Virion-Associated protein kinase induces apoptosis. *Journal of Virology*, *85*(24), 13144–13152. https://doi.org/10.1128/jvi.05294-11

Choo, Y. and Klug, A. (1994). Selection of DNA binding sites for zinc fingers using rationally randomized DNA reveals coded interactions. *Proceedings of the National Academy of Sciences*, *91*(23), 11168–11172. https://doi.org/10.1073/pnas.91.23.11168

Chougule, N., P., Li, H., Liu, S., Linz, L.B., Narva, K.E., Meade, T. and Bonning, B.C. (2013). Retargeting of the Bacillus thuringiensis toxin Cyt2Aa against hemipteran insect pests. *Proc. Natl. Acad. Sci. USA*, *110*, 8465–8470.

Clark, B.W., Phillips, T.A. and Coats, J.R. (2005). Environmental Fate and Effects of Bacillus thuringiensis (*Bt*) Proteins from Transgenic Crops: A Review. *Journal of Agricultural and Food Chemistry*, *53*(12), 4643–4653. https://doi.org/10.1021/jf040442k

Crickmore, N., Zeigler, D.R., Feitelson, J., Schnepf, E., Van Rie, J., Lereclus, D., et al. (1998). Revision of the nomenclature for the Bacillus thuringiensis pesticidal crystal proteins. *Microbiol. Mol. Biol. Rev.*, *62*, 807–813.

Crickmore, N., Baum, J., Bravo, A., Lereclus, D., Narva, K., Sampson, K., Schnepf, E., M, S., Ziegler, D.R., (2016). Bacillus thuringiensis toxin nomenclature. http://www.*Bt*nomenclature.info/.

Cui, Y., Sun, J.-L. and Yu, L. (2017). Application of the CRISPR gene-editing technique in insect functional genome studies: A review. *Entomologia Experimentalis et Applicata*, *162*(2), 124–132. https://doi.org/10.1111/eea.12530

Dara, S.K. (2019). The new integrated pest management paradigm for the modern age. *Journal of Integrated Pest Management*, *10*(1), 12.

Das, A., Ghosh, P. and Das, S. (2018) Expression of Colocasia esculenta tuber agglutinin in Indian mustard provides resistance against Lipaphis erysimi and the expressed protein is non-allergenic. *Plant Cell Rep.*, *37*, 849–863.

De la Pena, A., Lörz, H. and Schell, J. (1987). Transgenic rye plants o*Bt*ained by injecting DNA into young floral tillers. *Nature*, *325*(6101), 274–276.

De Maagd, R.A., Bravo, A. and Crickmore, N. (2001). How Bacillus thuringiensis has evolved specific toxins to colonize the insect world. *Trends in Genetics*, *17*(4), 193–199. https://doi.org/10.1016/s0168-9525(01)02237-5

DeGennaro, M., McBride, C.S., Seeholzer, L., Nakagawa, T., Dennis, E.J., Goldman, C., Jasinskiene, N., James, A.A. and Vosshall, L.B. (2013). Orco mutant mosquitoes lose strong preference for humans and are not repelled by volatile DEET. *Nature*, *498*(7455), 487–491.

Dong, S., Lin, J., Held, N.L., Clem, R.J., Passarelli, A.L. and Franz, A.W. (2015). Heritable CRISPR/Cas9-mediated genome editing in the yellow fever mosquito, Aedes aegypti. *PloS One*, *10*(3), e0122353.

Doudna, J.A. and Charpentier, E. (2014). The new frontier of genome engineering with CRISPR-Cas9. *Science*, *346*(6213), Article 6213. https://doi.org/10.1126/science.1258096

Duan, X., Hou, Q., Liu, G., Pang, X., Niu, Z., Wang, X., Zhang, Y., Li, B. and Liang, R. (2018). Expression of Pinellia pedatisecta lectin gene in transgenic wheat enhances resistance to wheat aphids. *Molecules*, *23*, 748.

Duke, S.O. (2018). The history and current status of glyphosate. *Pest Management Science*, *74*(5), 1027–1034.

Elsharawy, H. and Refat, M. (2023). CRISPR/Cas9 genome editing in wheat: Enhancing quality and productivity for global food security: A review. *Functional and Integrative Genomics*, *23*(3), 265.

Engelmann, F. (2011). Use of biotechnologies for the conservation of plant biodiversity. *In Vitro Cellular and Developmental Biology-Plant*, *47*, 5–16.

European Food Safety Authority. (2014). Event Report: International Scientific Workshop *Risk Assessment Considerations for RNAi-Based GM Plants* : Brussels. Belgium.

Finer, J.J., Finer, K.R. and Ponappa, T. (2000). Particle Bombardment Mediated Transformation. In: J. Hammond, P. McGarvey and V. Yusibov (Eds.), *Plant Biotechnology*, *240*, 59–80. Springer Berlin Heidelberg. https://doi.org/10.1007/978-3-642-60234-4_3

Fire, A., Xu, S., Montgomery, M.K., Kostas, S.A., Driver, S.E. and Mello, C.C. (1998). Potent and specific genetic interference by double-stranded RNA in Caenorhabditis elegans. *Nature*, *391*(6669), 806–811. https://doi.org/10.1038/35888

Fischhoff, D.A., Bowdish, K.S., Perlak, F.J., Marrone, P.G., McCormick, S.M., Niedermeyer, J.G., Dean, D.A., Kusano-Kretzmer, K., Mayer, E.J., Rochester, D.E., et al. (1987). Insect Tolerant Transgenic Tomato Plants. *Nat. Biotechnol.*, 5, (8), 807–813.

Fishilevich, E., Vélez, A.M., Storer, N.P., Li, H., Bowling, A.J., Rangasamy, M., Worden, S.E., Narva, K.E. and Siegfried, B.D. (2016). RNAi as a management tool for the western corn rootworm, Diabrotica virgifera virgifera. *Pest Management Science*, *72*(9), 1652–1663. https://doi.org/10.1002/ps.4324

Fitch, M.M.M., Manshardt, R.M., Gonsalves, D., Slightom, J.L. and Sanford, J.C. (1992). Virus Resistant Papaya Plants Derived from Tissues Bombarded with the Coat Protein Gene of Papaya Ringspot Virus. *Nature Biotechnology*, *10*(11), 1466–1472. https://doi.org/10.1038/n*Bt*1192-1466

Fromm, M., Taylor, L.P. and Walbot, V. (1985). Expression of genes transferred into monocot and dicot plant cells by electroporation. *Proceedings of the National Academy of Sciences*, *82*(17), 5824–5828. https://doi.org/10.1073/pnas.82.17.5824

Gaj, T., Gersbach, C.A. and Barbas, C.F. (2013). ZFN, TALEN, and CRISPR/Cas-based methods for genome engineering. *Trends in Biotechnology*, *31*(7), 397–405.

Gallego-Bartolomé, J., Liu, W., Kuo, P.H., Feng, S., Ghoshal, B., Gardiner, J., Zhao, J. M.-C., Park, S.Y., Chory, J. and Jacobsen, S.E. (2019). Co-targeting RNA polymerases IV and V promotes efficient *de novo* DNA methylation in Arabidopsis. *Cell*, *176*(5), 1068–1082.

Ganapathy, S., Parajulee, M.N., Francisco, M.S., Zhang, H. and Bilimoria, S.L. (2021). Novel-iridoviral kinase induces mortality and reduces performance of green peach aphids (Myzus persicae) in transgenic Arabidopsis plants. *Plant Biotechnology Reports*, *15*(1), 13–25. https://doi.org/10.1007/s11816-020-00659-w

Romeis, J., Bartsch, D., Bigler, F., Candolfi, M.P., Gielkens, M.M.C., Hartley, S.E., Hellmich, R.L., Huesing, J.E., Jepson, P.C., Layton, R., Quemada, H., Raybould, A., Rose, R.I., Schiemann, J., Sears, M.K., Shelton, A.M., Sweet, J., Vaituzis, Z. and Wolt, J.D. (2008). Assessment of risk of insect-resistant transgenic crops to nontarget arthropods. *Nature Biotechnology*, *26*(2), 203–208. https://doi.org/10.1038/n*Bt*1381

Rukarwa, R.J., Mukasa, S.B., Odongo, B., Ssemakula, G. and Ghislain, M. (2013). Identification of relevant non-target organisms exposed to weevil-resistant *Bt* sweetpotato in Uganda. *3 Biotech*, *4*(3), 217–226. https://doi.org/10.1007/s13205-013-0153-1

Salum, Y.M., Yin, A., Zaheer, U., Liu, Y., Guo, Y. and He, W. (2024). CRISPR/Cas9-Based Genome Editing of Fall Armyworm (Spodoptera frugiperda): Progress and Prospects. *Biomolecules*, *14*(9), 1074. https://doi.org/10.3390/biom14091074

Sander, J.D. and Joung, J.K. (2014). CRISPR-Cas systems for editing, regulating, and targeting genomes. *Nature Biotechnology*, *32*(4), 347–355.

Sander, J.D., Dahlborg, E.J., Goodwin, M.J., Cade, L., Zhang, F., Cifuentes, D., Curtin, S.J., Blackburn, J.S., Thibodeau-Beganny, S. and Qi, Y. (2011). Selection-free zinc-finger-nuclease engineering by context-dependent assembly (CoDA). *Nature Methods*, *8*(1), 67–69.

Sanford, J.C. (1988). The biolistic process. *Trends in Biotechnology*, *6*(12), 299–302.

Saxena, D., Flores, S. and Stotzky, G. (1999). Insecticidal toxin in root exudates from *Bt* corn. *Nature*, *402*(6761), 480. https://doi.org/10.1038/44997

Schaart, J.G., van de Wiel, C.C., Lotz, L.A. and Smulders, M.J. (2016). Opportunities for products of new plant breeding techniques. *Trends in Plant Science*, *21*(5), 438–449.

Schellenberger, U., Oral, J., Rosen, B.A., Wei, J.Z., Zhu, G., Xie, W., McDonald, M.J., Cerf, D.C., Diehn, S.H., Crane, V.C., Sandahl, G.A., Zhao, J.Z., Nowatzki, T.M., Sethi, A., Liu, L., Pan, Z., Wang, Y., Lu, A.L., Wu, G. and Liu, L. (2016). A selective insecticidal protein from Pseudomonas for controlling corn rootworms. *Science*, *354*(6312), 634–637. https://doi.org/10.1126/science.aaf6056

Schmidhuber, J. and Tubiello, F.N. (2007). Global food security under climate change. *Proceedings of the National Academy of Sciences*, *104*(50), 19703–19708. https://doi.org/10.1073/pnas.0701976104

Schmidt, S.M., Belisle, M. and Frommer, W.B. (2020). The evolving landscape around genome editing in agriculture. *EMBO Reports*, *21*(6). https://doi.org/10.15252/embr.202050680

Schuler, T.H., Poppy, G.M., Kerry, B.R. and Denholm, I. (1998). Insect-resistant transgenic plants. *Trends in Biotechnology*, *16*(4), 168–175. https://doi.org/10.1016/s0167-7799(97)01171-2.

Schwander, T., Schada von Borzyskowski, L., Burgener, S., Cortina, N.S. and Erb, T.J. (2016). A synthetic pathway for the fixation of carbon dioxide *in vitro*. *Science*, *354*(6314), 900–904.

Scorza, R., Callahan, A., Levy, L., Damsteegt, V., Webb, K. and Ravelonandro, M. (2001). Post-transcriptional gene silencing in plum pox virus resistant transgenic European plum containing the plum pox potyvirus coat protein gene. *Transgenic Research*, *10*(3), 201–209. https://doi.org/10.1023/a:1016644823203

Scott, J.G., Michel, K., Bartholomay, L.C., Siegfried, B.D., Hunter, W.B., Smagghe, G., Zhu, K.Y. and Douglas, A.E. (2013). Towards the elements of successful insect RNAi. *J. Insect Physiol.*, *59*, 1212–1221.

Sears, M.K., Hellmich, R.L., Stanley-Horn, D.E., Oberhauser, K.S., Pleasants, J.M., Mattila, H.R., Siegfried, B.D. and Dively, G.P. (2001). Impact of *Bt* corn pollen on monarch butterfly populations: A risk assessment. *Proceedings of the National Academy of Sciences*, *98*(21), 11937–11942. https://doi.org/10.1073/pnas.211329998

Selale, H., Dağlı, F., Mutlu, N., Doğanlar, S. and Frary, A. (2017). Cry1Ac-mediated resistance to tomato leaf miner (Tuta absoluta) in tomato. *Plant Cell Tiss. Org. Cult.*, *131*, 65–73.

Shimizu, Y., Bhakta, M.S. and Segal, D.J. (2009). Restricted spacer tolerance of a zinc finger nuclease with a six amino acid linker. *Bioorganic and Medicinal Chemistry Letters*, *19*(14), 3970–3972.

Singh, S., Kumar, N.R., Maniraj, R., Lakshmikanth, R., Rao, K.Y., Muralimohan, N., Arulprakash, T., Karthik, K., Shashibhushan, N.B., Vinutha, T., Pattanayak, D., Kumar, P.A. and Sreevathsa, R. (2018). Expression of Cry2Aa, a Bacillus thuringiensis insecticidal protein in transgenic pigeon pea confers resistance to gram pod borer, Helicoverpa armigera. *Sci. Rep.*, *8*, 1–2.

Singh, J.S., Pandey, V.C. and Singh, D.P. (2011). Efficient soil microorganisms: A new dimension for sustainable agriculture and environmental development. *Agriculture, Ecosystems, and Environment*, *140*(3–4), 339–353.

Singh, S., Rahangdale, S., Pandita, S., Saxena, G., Upadhyay, S.K., Mishra, G. and Verma, P.C. (2022). CRISPR/Cas9 for insect pests management: A comprehensive review of advances and applications. *Agriculture*, *12*(11), 1896.

Smith, J., Bibikova, M., Whitby, F.G., Reddy, A.R., Chandrasegaran, S. and Carroll, D. (2000). Requirements for double-strand cleavage by chimeric restriction enzymes with zinc finger DNA-recognition domains. *Nucleic Acids Research*, *28*(17), 3361–3369.

Smyth, S.J. (2019). Global status of the regulation of genome editing technologies. *CABI Reviews*, 1–6. https://doi.org/10.1079/pavsnnr201914021

Soliman, H.I., Abo-El-Hasan, F.M., El-Seedy, A.S. and Mabrouk, Y.M. (2017). Agrobacterium-mediated transformation of tomato (Lycopersicon esculentum mill.) using a synthetic cry1ab gene for enhanced resistance against Tuta absoluta (Meyrick). *J. Microbiol. Biotech. Food Sci.*, *7*(1), 67–74.

Sprink, T., Wilhelm, R. and Hartung, F. (2022). Genome editing around the globe: An update on policies and perceptions. *Plant Physiology, 190*(3), 1579–1587. https://doi.org/10.1093/plphys/kiac359

Stuecker, M.F., Tigchelaar, M. and Kantar, M.B. (2018). Climate variability impacts on rice production in the Philippines. *PloS One*, *13*(8), e0201426.

Sun, L., Alariqi, M., Wang, Y., Wang, Q., Xu, Z., Zafar, M.N., Yang, G., Jia, R., Hussain, A., Chen, Y., Ding, X., Zhou, J., Wang, G., Wang, F., Li, J., Zou, J., Zhu, X., Yu, L., Sun, Y., . . . and Jin, S. (2023). Construction of host plant Insect-Resistance mutant library by High-Throughput CRISPR/CAS9 system and identification of a Broad-Spectrum insect Resistance gene. *Advanced Science*, *11*(4). https://doi.org/10.1002/advs.202306157

Takasu, Y., Kobayashi, I., Beumer, K., Uchino, K., Sezutsu, H., Sajwan, S., Carroll, D., Tamura, T. and Zurovec, M. (2010). Targeted mutagenesis in the silkworm Bombyx mori using zinc finger nuclease mRNA injection. *Insect Biochemistry and Molecular Biology*, *40*(10), 759–765.

Tarafdar, A., Kamle, M., Prakash, A. and Padaria, J.C. (2014). Transgenic plants: Issues and future prospects. *Biotechnol. Adv.*, *2*, 1–47.

Tindall, K.V., M.W. Siebert, B.R. Leonard, J. All and F.J. Haile. (2009). Efficacy of Cry1Ac:Cry1F proteins in cotton leaf tissue against fall armyworm, beet armyworm, and soybean looper (Lepidoptera: Noctuidae). *J. Econ. Entomol.*, *102*, 1497–1505.

Tzfira, T. and Citovsky, V. (2006). Agrobacterium-mediated genetic transformation of plants: Biology and biotechnology. *Current Opinion in Biotechnology*, *17*(2), 147–154. https://doi.org/10.1016/j.copbio.2006.01.009

U.S. Environmental Protection Agency. (2017). EPA Registers Innovative Tool to Control Corn Rootworm. https://www.epa.gov/newsreleases/epa-registers-innovative-tool-control-corn-rootworm.

U.S. Environmental Protection Agency. (2014). A Set of Scientific Issues Being Considered by The Environmental Protection Agency Regarding: RNAi Technology as a Pesticide: Problem Formulation for Human Health and Ecological Risk Assessment, SAP Minutes No. 2014–02: Arlington, VA.

U.S. Environmental Protection Agency. (2015). Environmental Risk Assessment for a FIFRA Section 3 Limited Seed Increase Registration of DvSnf7 Double Stranded RNA (dsRNA) and Cry3Bb1 Bacillus Thuringiensis Insecticidal Protein as Expressed in MON 87411 Maize: Washington, D.C.

U.S. Environmental Protection Agency. (2016). Transmission of Meeting Minutes of the September 27–28, 2016 FIFRA SAP Meeting Held to Consider and Review Scientific Issues Associated with "RNAi Technology: Human Health and Ecological Risk Assessments for SmartStax PRO", SAP Minutes No. 2016–02: Arlington, VA.

Uchimiya, H., Fushimi, T., Hashimoto, H., Harada, H., Syono, K. and Sugawara, Y. (1986). Expression of a foreign gene in callus derived from DNA-treated protoplasts of rice (Oryza sativa L.). *Molecular and General Genetics MGG*, *204*(2), 204–207. https://doi.org/10.1007/BF00425499

Urnov, F.D., Rebar, E.J., Holmes, M.C., Zhang, H.S. and Gregory, P.D. (2010). Genome editing with engineered zinc finger nucleases. *Nature Reviews Genetics*, *11*(9), 636–646.

Vaeck, M., Reynaerts, A., Höfte, H., Jansens, S., de Beuckeleer, M., Dean, C., Zabeau, M., Montagu, M.V. and Leemans, J. (1987). Transgenic plants protected from insect attack. *Nature*, *328* (6125), 33–37.

Van Frankenhuyzen, K. (2009). Insecticidal activity of Bacillus thuringiensis crystal proteins. *Journal of Invertebrate Pathology*, *101*(1), 1–16. https://doi.org/10.1016/j.jip.2009.02.009

Van Frankenhuyzen, K. (2013). Cross-order and cross-phylum activity of Bacillus thuringiensis pesticidal proteins. *Journal of Invertebrate Pathology*, *114*(1), 76–85. https://doi.org/10.1016/j.jip.2013.05.010

Vanti, G.L., Katageri, I.S.,, Inamdar, S.R., Hiremathada, V. and Swamy, B.M. (2018). Potent insect gut binding lectin from Genetic engineering of crops for insect resistance Page 11 of 12 114 Sclerotium rolfsii impart resistance to sucking and chewing type insects in cotton. *J. Biotechnol.*, *278*, 20–27.

Walters, F.S. and English, L.H. (1995). Toxicity of Bacillus thuringiensisδ-endotoxins toward the potato aphid in an artificial diet bioassay. *Entomologia Experimentalis Et Applicata*, *77*(2), 211–216. https://doi.org/10.1111/j.1570-7458.1995.tb02003.x

Wang, H., La Russa, M. and Qi, L.S. (2016). CRISPR/CAS9 in genome editing and beyond. *Annual Review of Biochemistry*, *85*(1), 227–264. https://doi.org/10.1146/annurev-biochem-060815-014607

Wang, N., Jiang, C., Jiang, M., Zhang, C. and Cheng, J. (2010). Using chimeric piggyBac transposase to achieve directed interplasmid transposition in silkworm Bombyx mori and fruit fly Drosophila cells. *Journal of Zhejiang University SCIENCE B*, *11*(9), 728–734. https://doi.org/10.1631/jzus.B1000139

Wang, Y., Li, Z., Xu, J., Zeng, B., Ling, L., You, L., Chen, Y., Huang, Y. and Tan, A. (2013). The CRISPR/Cas system mediates efficient genome engineering in Bombyx mori. *Cell Research*, *23*(12), 1414–1416.

Wasmer, M. (2019). Roads Forward for European GMO Policy Uncertainties in Wake of ECJ Judgment Have to be Mitigated by Regulatory Reform. *Frontiers in Bioengineering and Biotechnology*, *7*. https://doi.org/10.3389/fbioe.2019.00132

Wassmann, R., Jagadish, S.V.K., Sumfleth, K., Pathak, H., Howell, G., Ismail, A., Serraj, R., Redona, E., Singh, R.K. and Heuer, S (2009). Regional vulnerability of climate change impacts on Asian rice production and scope for adaptation. *Advances in Agronomy*, *102*, 91–133.

Whalon, M.E. and Wingerd, B.A. (2003). *Bt*: mode of action and use. Archives of Insect Biochemistry and Physiology. *Entol. Soc. America*, *54*, 200–211.

Wheeler, T. and Von Braun, J. (2013). Climate Change Impacts on Global Food Security. *Science*, *341*(6145), 508–513. https://doi.org/10.1126/science.1239402

Widner, W.R. and Whiteley, H.R. (1989). Two highly related insecticidal crystal proteins of Bacillus thuringiensis subsp. kurstaki possess different host range specificities. *Journal of Bacteriology*, *171*(2), 965–974. https://doi.org/10.1128/jb.171.2.965-974.1989

Wolt, J.D. and Wolf, C. (2018). Policy and governance perspectives for regulation of genome edited crops in the United States. *Frontiers in Plant Science*, *9*. https://doi.org/10.3389/fpls.2018.01606

Wolt, J.D., Wang, K. and Yang, B. (2015). The regulatory status of genome-edited crops. *Plant Biotechnology Journal*, *14*(2), 510–518. https://doi.org/10.1111/pbi.12444

Xu, C., Cheng, J., Lin, H., Lin, C., Gao, J. and Shen, Z. (2018). Characterization of transgenic rice expressing fusion protein Cry1Ab/Vip3A for insect resistance. *Sci. Rep.*, *8*, 1–8.

Xu, C., Chinte, U., Chen, L., Yao, Q., Meng, Y., Zhou, D., Bi, L.J., Rose, J., Adang, M.J., Wang, B.C., Yu, Z. and Sun, M. (2015). Crystal structure of Cry51Aa1: A potential novel insecticidal aerolysin-type β-pore-forming toxin from Bacillus thuringiensis. *Biochemical and Biophysical Research Communications*, *462*(3), 184–189. https://doi.org/10.1016/j.bbrc.2015.04.068

Xu, C., Wang, B.C., Yu, Z. and Sun, M. (2014). Structural Insights into Bacillus thuringiensis Cry, Cyt, and Parasporin Toxins. *Toxins*, *6*(9), 2732–2770. https://doi.org/10.3390/toxins6092732

Yan, S., Qian, J., Cai, C., Ma, Z., Li, J., Yin, M., Ren, B. and Shen, J. (2020). Spray method application of transdermal dsRNA delivery system for efficient gene silencing and pest control on soybean aphid Aphis glycines. *J. Pest Sci.*, *93*, 449–459.

Zhang, J., Khan, S.A., Heckel, D.G. and Bock, R. (2017). Next-Generation Insect-Resistant Plants: RNAi-Mediated Crop Protection. *Trends in Biotechnology*, *35*(9), 871–882. https://doi.org/10.1016/j.ti*Bt*ech.2017.04.009

Zhong, Y., Ahmed, S., Deng, G., Fan, W., Zhang, P. and Wang, H. (2019). Improved insect resistance against Spodoptera litura in transgenic sweet potato by overexpressing Cry1Aa toxin. *Plant Cell Rep.*, *38*, 1439–1448.

Ziemienowicz, A. (2013). Agrobacterium-mediated plant transformation: Factors, applications, and recent advances. *Biocatalysis and Agricultural Biotechnology*, *3*(4), 95–102. https://doi.org/10.1016/j.bcab.2013.10.004

Zimny, T. and Eriksson, D. (2020). Exclusion or exemption from risk regulation? *EMBO Reports*, *21*(12). https://doi.org/10.15252/embr.202051061

Zwahlen, C., Hilbeck, A., Gugerli, P. and Nentwig, W. (2003). Degradation of the Cry1Ab protein within transgenic Bacillus thuringiensis corn tissue in the field. *Molecular Ecology*, *12*(3), 765–775. https://doi.org/10.1046/j.1365-294x.2003.01767.x

Chapter 2

Multi-omics Approach for Sustainable Insect Management Practices

Onkarappa Dhanyakumar[1,2], Murugesan Naveen[3], Shanmugasundaram Pavithran[4], Revanayya Gothe[5], Marimuthu Murugan[6], Jaba Jagdish[7] and Kalenahalli Yogendra[8*]

1. Introduction

Climate stands among the most formidable challenges to global food and nutritional security. As the planet's climate continues to shift, the stability of food systems is increasingly threatened, leading to serious consequences for both the quantity and quality of food production. Shifts in temperature, altered rainfall patterns, and an escalation in extreme weather events are negatively impacting agriculture, causing a decline in crop yields (Schmidhuber & Tubiello, 2007). Plants, being immobile, are constantly exposed to challenging environmental factors and biotic invasions in their natural habitat. The interaction of biotic factors with plants can either be beneficial or deleterious having impacts on population dynamics and evolutionary relationship between plants and insects, and also in cycling of nutrients in the ecosystem. (Nawaz et al., 2023). Among the biotic stress, Insects play an important role in ecosystems and cause significant damage to agricultural production. Plants have evolved a variety of morphological, biochemical, and molecular-level defense survival tactics in response to insect pests (Jaba et al., 2017; War et al., 2017). Farmers primarily rely on chemical-based control methods for insect pest management. The excessive use of these synthetic pesticides has sparked concerns due to their environmental impact, potential harm to non-target species, and the development of resistance against insecticides. To address these challenges, the scope is to explore multi-omics approaches as a pathway toward more sustainable insect management practices.

Toxins and defense proteins produced by plants target the insect's physiological functions (Mithöfer & Boland, 2012). Plants employ both direct and indirect defenses that involve complex biochemical and molecular systems against herbivores (Gatehouse, 2002; War et al., 2017). Multi-omics is a unified strategy that integrates various omics disciplines, such as genomics, transcriptomics, proteomics, and metabolomics, to study biological systems comprehensively. In insect management, multi-omics provides a detailed understanding of insects at the molecular and cellular levels,

[1,5,7,8] International Crops Research Institute for the Semi-Arid Tropics, Hyderabad, Telangana, 502324, India.
[1,3,4,6] Department of Entomology, Tamil Nadu Agricultural University, Coimbatore, Tamil Nadu, 641003, India.
[5] Department of Plant Breeding and Genetics, Punjab Agricultural University, Ludhiana, Punjab, 141004, India.
* Corresponding author: yogendra.kalenahalli@icrisat.org

DOI: 10.1201/9781003614869-2

offering insights into their behavior and environmental interactions (Li et al., 2019; Sheikh et al., 2020). The application of multi-omics in insect management can potentially revolutionize the field by examining the genetic makeup of insects and pinpointing genes that contribute to insecticide resistance or adaptability to changing environments (Sai Reddy et al., 2022; Yang et al., 2021). Transcriptomics allows for the analysis of gene expression patterns, providing clues about insect development and responses to stressors. Proteomics and metabolomics provide an increased understanding of the biochemical pathways and physiological processes in insects, which can be targeted to disrupt their life cycles or deter their activity (Jiang et al., 2022; S. Wang et al., 2023).

The strategic integration of multi-omics approaches enables the development of sustainable and environmentally sound alternatives to conventional chemical insecticides. It can direct the development of biological control agents, like predators, parasitoids, entomopathogenic bacteria, entomopathogenic bacteria, entomopathogenic nematodes, and provide guidance for the manufacture of genetically modified organisms (GMOs) for the purpose of controlling certain insect populations (Harith-Fadzilah et al., 2021). Additionally, by providing insightful information on plant-insect interactions, multi-omics approaches can create crop varieties that show resistance to insects, reducing the need for chemical management measures (Zogli et al., 2020). This chapter explores the potential application of a multi-omics approach to sustainable insect management practices. We explore different omics disciplines, their roles in insect research, and their potential to drive innovative, environmentally friendly solutions. A comprehensive utilization of multi-omics approach has the potential to develop more sustainable and effective insect management strategies, with benefits extending to agriculture, ecosystems, and human health.

2. Understanding Plant-insect Interactions through Multi-omics

2.1 Genomics

More than 20 years ago, the first insect genome was sequenced. Since then, sequencing technologies and insect genome assemblies have come a long way. Now, more than 600 nuclear genomes of insects have been sequenced and are available in the GenBank database (Sayers et al., 2020). However, this only makes up around 0.06% of the roughly 1 million described insect species (Stork, 2018).

Insect species with assembled genomes are represented across 18 orders, including Archaeognatha, Blattodea, Collembola, Diptera, Ephemeroptera, Odonata, Hemiptera, Orthoptera, Lepidoptera, Coleoptera, Phasmatodea, Phthiraptera, Plecoptera, Siphonaptera, Strepsiptera, Thysanoptera, Hymenoptera, and Trichoptera. The order with the most assembled genomes is Diptera, followed by Lepidoptera, Hymenoptera, and Hemiptera. Currently, 5,694 submissions of insect genomes are accessible through National Center for Biotechnology Information (NCBI). Due to the challenges faced in annotating the genome sequences, only 526 insect genome submissions (~10%) were annotated despite the diversity among assembled genomes (Li et al., 2019). The genome sequencing of insects began in 2000 with *Drosophila melanogaster* 2000 (Adams et al., 2000) and was facilitated by the relatively small genome sizes of many well-studied organisms (Grbić et al., 2011; Gregory, 2005).

The i5K project, or the 5,000 Insect Genome Project, launched in 2011 (i5K Consortium, 2013: https://i5k.github.io/publications), is an ambitious effort to sequence and analyze the genomes of 5,000 insect and arthropod species within a five-year timeframe. Genome sequencing offers critical insights for overcoming pesticide resistance in disease-vectoring insects while protecting beneficial pollinators like bees. The i5K project, which is open and volunteers to participate in, highlights the accomplishments of collaboration among scientists who have a deep interest in researching arthropods in addition to the information that can be obtained from genomic and genetic analysis. To support the genome hosting needs of the i5K initiative, the National Agricultural Library (NAL) has established the i5K Workspace@NAL (http://i5k.nal.usda.gov/). All i5k members can share their genome project details with the i5k Workspace. Once submitted, the newly generated data will be accessible via organism-specific pages, genome browsers, and BLAST search platforms (Poelchau et al., 2015). The 10K genome project was founded in 2009 by a consortium of biologists and genome scientists with the goal of sequencing and comprehensively analyzing the complete genomes of 10,000 vertebrate species (Koepfli et al., 2015).

The taxonomy and systematics of *Bemisia tabaci* have sparked debate among scientists for over a century (Wang et al., 2025). However, advances in molecular and phylogenomic tools, together with affordable high-throughput DNA sequencing and sophisticated bioinformatics, have greatly improved the ability to define and classify cryptic species in this globally important group, addressing long-standing taxonomic challenges and presenting viable solutions (Brown et al., 2023). Whole genome sequencing (WGS) data provide critical insights into phylogenetic relationships, evolutionary trajectories, extinct and cryptic taxa, as well as the genetic diversity present in ancestral populations of a species (Theissinger et al., 2023). *B. tabaci*, initially described in 1899, has now proliferated to around ~40 cryptic species (Wang et al., 2024). Whole genome phylogenomic analysis of *B. tabaci* suggests a more than commonly used 3.5–4.0% difference in the COI gene is needed to differentiate cryptic species within the *B. tabaci* group. However, it can be used to differentiate among subspecies within this group. A 1% difference in the nuclear genome is a more accurate indicator of the geographical and evolutionary divisions between the cryptic species of *B. tabaci* (Brown et al., 2023). Similarly, WGS of *B. tabaci* (MEAM1 strain) revealed the presence of 142 genes that were horizontally transmitted from bacteria or fungi, which are not found in other insects. These genes are involved in hopanoid/sterol biosynthesis and xenobiotic detoxification, potentially underpinning distinctive biological traits such as polyphagy and resistance to insecticides (Chen et al., 2016).

2.2 Transcriptomics

Understanding the molecular basis of insecticide resistance is crucial for effectively managing resistance in agricultural pests. In certain insects, the molecular mechanisms behind the insecticide resistance have been elucidated, notably involving the overexpression of detoxification enzymes such as cytochrome P450s (CYP), esterases, and glutathione S-transferases (GST) (Hsu et al., 2012). Additionally, insect whole genome sequencing of insects has uncovered many genes that may be responsible for the detoxification of chemical insecticides, notably those belonging to the cytochrome P450, esterase, and glutathione S-transferase families (Kalajdzic et al., 2012). Populations of *Helicoverpa armigera* have been documented to exhibit resistance levels exceeding 100-fold against organophosphates (OPs) and over 1000-fold against pyrethroids (Farnsworth et al., 2010). Overexpression of a single CYP gene, CYP6P3, was responsible for pyrethroid resistance in *Anopheles gambiae* mosquitoes (Müller et al., 2008). Overexpression of *CYP6CM1* in *B. tabaci* was responsible for metabolizing imidacloprid, leading to neonicotinoid resistance (Karunker et al., 2009). Similarly, elevated expression of CYP6G1 has been observed in *Drosophila* strains resistant to imidacloprid and DDT (Daborn et al., 2002). Transcriptomic studies across multiple insect developmental stages can help identify numerous genes involved in normal growth and development, which may serve as targets for pest management. Transcriptome sequencing of *Spodoptera exigua* revealed 9 candidate genes for RNAi-based pest control, namely, ADP-ribosylation factor 1 (arf1), ADP-ribosylation factor 2 (arf2), chitinase1, chitinase7, tubulin1, tubulin2, plasma glutamate carboxypeptidase (PGCP), helicase, and ATPase (Li et al., 2013) Genes related to the juvenile hormone and ecdysteroid biosynthetic pathways can also be targeted for gene silencing, reducing larval body weight, growth, and development (Camargo et al., 2015). Injection of dsRNA targeting the Ace1 gene in *Rhopalosiphum padi* and *Sitobion avenae* significantly altered AChE activity, resulting in heightened susceptibility to pirimicarb and malathion (Xiao et al., 2015). Transcriptomic comparison between imidacloprid-resistant and susceptible strains of *Aphis gossypii* revealed that 12 UDP-glycosyltransferase (UGT) genes were differentially expressed, of which 9 UGT genes were up-regulated. Knockdown of these upregulated UGT genes by using RNAi significantly increases the sensitivity of imidacloprid to resistant strains of *A. gossypii* (Chen et al., 2019).

Transcriptome sequencing of *Frankliniella occidentalis* revealed notable differences in gene expression between plants infected and uninfected with Tomato spotted wilt virus (TSWV). TSWV-infected individuals exhibited elevated expression of mucin, phospholipase C, tight junction protein ZO-1, collagen, laminin, and vinculin, which aid in the virus's infection, replication, and spread within the vector, as well as its transmission to plants. Notably, mucin aids in trapping a more significant number of virus particles within the mucus lining of midgut and associated tissues. In TSWV-infected thrips, gene expression analysis showed approximately 37 genes were upregulated while about 96 genes were downregulated, alterations that could disrupt vital metabolic pathways within the insect (Zhang et al., 2013).

2.3 Proteomics

Proteomics is recognized as a high-throughput approach for large-scale protein profiling, with a primary focus on elucidating protein structures and functions. Advances in mass spectrometry-based proteomics have allowed researchers to reliably, accurately, and quantitatively analyze the proteome, from individual cells and tissues to the entire body of an organism (Guo et al., 2025; Mayorga-Martino et al., 2025). This overview examines the advancements in proteomics research across various insect pests and highlights its potential applications in their management.

About 4 differentially expressed UDP-glycosyltransferase genes (UGT344A16, UGT344B4, UGT344C7, and UGT344N4) were discovered through proteomic analysis of imidacloprid-resistant and susceptible strains of *A. gossypii*. Knockdown of UGT344C7 and UGT344B4 by using dsRNA increased the susceptibility of the imidacloprid of *A. gossypii* (Chen et al., 2019). Moreover, this approach is instrumental in elucidating plant–insect interactions, a key step in formulating effective strategies for controlling insect pests. Proteomics can identify proteins in plant defense responses to insects fed on a particular genotype/plant. Tomato plants respond to aphid attacks by elevating oxidative stress levels and producing detoxification proteins (Coppola et al., 2013). Comparative analyses reveal that pest-infested plants exhibit altered proteomes compared to healthy ones, aiding in the identification of key defense-related proteins that can be targeted to enhance rice resistance against the small brown planthopper (*Laodelphax striatellus*) (Dong et al., 2017). Proteomics also helps understand complex interactions involving plants, insects, and their endosymbionts (Kaul et al., 2016; Renoz et al., 2017).

Proteomic analysis also helps select potential target genes for RNAi without causing unintended effects on non-target organisms, including predators, parasitoids, and other beneficial insect species. The proteomic analysis of *B. tabaci* identified 453 proteins, including 246 non-redundant proteins, that may serve to locate unique nucleotide sequences within the insect genome. These sequences help identify target genes for pest management through RNAi (Mishra et al., 2016).

Numerous proteins have been detected at the interface of virus acquisition and transmission by insect vectors through proteomic approaches, offering insights into potential targets for insect–plant–vector management (Liu et al., 2015; Mittapelly & Rajarapu et al., 2020). Cell walls of plants mainly comprise biopolymer cellulose, a rich energy source for leaf-feeding organisms. Phytophagous beetles belonging to *Chrysomeloidea* and *Curculionoidea* possess endogenous genes in their gut that encode plant cell wall–degrading enzymes (PCWDEs), identified through transcriptomic and proteomic analyses. Through proteomic approaches, 13 proteins degraded polysaccharides of plant cell walls, mainly belonging to xylanases, β-1,4-glucanases or cellulases, and polygalacturonases or pectinases families (Kirsch et al., 2012). Identifying and characterizing highly resistant plant proteins, along with their corresponding targets in the insect digestive system, offer valuable insights into insect–host plant interactions. Applications of multi-omics in sustainable insect management play a vital role, and understanding insect–plant interaction at a deeper level helps find the candidate metabolite/gene for further validation and application.

2.4 Metabolomics

Metabolomics, a high-throughput technique for characterizing all small-molecule metabolites, offers profound insights into biological processes by detecting and identifying a broad spectrum of molecules. This approach uncovers the metabolic dynamics and adaptive responses of living organisms to both genetic and environmental variations (Idle & Gonzalez, 2007; Palli et al., 2012). It encompasses untargeted metabolomics for detecting both known and novel metabolites, targeted metabolomics for precise quantification, and metabolic profiling for examining crude extracts and classifying the full spectrum of metabolites. The primary techniques employed in metabolomics include gas chromatography–mass spectrometry (GC–MS), liquid chromatography–mass spectrometry (LC–MS), Fourier transform ion cyclotron resonance mass spectrometry (FT–ICR–MS), and nuclear magnetic resonance (NMR) spectroscopy. While GC-MS and LC-MS are particularly adept at identifying and quantifying known metabolites, NMR excels in identifying unknown ones (Palli et al., 2012). In addition to its broad applications, metabolomics is crucial in enhancing host plant resistance and developing biopesticides, addressing the global challenge of rapidly spreading pests in agriculture and horticulture.

In plant systems, metabolomics serves as a powerful comparative approach for identifying bioactive compounds involved in defense against insect herbivory. By comparing the metabolomes of resistant and susceptible cultivars, researchers can identify the compounds responsible for resistance. *Barbarea vulgaris* exhibits natural resistance or tolerance to pollen beetles (*Brassicogethes aeneus*) owing to the presence of secondary metabolites such as glucobarbarin and saponins in its green flower buds (Austel et al., 2021). In cereals, metabolomics has uncovered defense mechanisms against pathogens and pests. Among these, feeding Western corn rootworm (*Diabrotica virgifera*) on maize roots triggers the production of DIMBOA. This antifeedant is further elevated by *Spodoptera littoralis* feeding (Erb et al., 2009). Moreover, comparisons between resistant wild and susceptible cultivated tomato varieties revealed that the resistant types contained acyl sugars, recognized by their anti-herbivory properties (Mirnezhad et al., 2010). NMR-based metabolomic analysis identified compounds linked to thrips resistance, including the flavonoid luteolin, the phenylpropanoid sinapic acid, and β-alanine. Additionally, monomeric and dimeric acyclic diterpene glycosides (capsianosides) in pepper were associated with resistance, and plant metabolites such as pyrrolizidine alkaloids (Jacobine, Jaconine), Flavanoid (Kaempferol-glucoside), Phenylpropanoid (Chlorogenic acid), Phenylpropanoid (Feruloyl quinic acid) and acyl sugars were involved in host-plant resistance mainly against *F. occidentalis* (Macel et al., 2019). These findings underscore the value of metabolomics in elucidating plant–insect interactions, demonstrating how plant metabolites can influence the host range of insect pests (Sanchez-Arcos et al., 2019). Research on legumes infested by host-specific and non-specific races of the pea aphid (*Acyrthosiphon pisum*) revealed distinct metabolite profiles between host-specific and non-host-specific interactions. While secondary metabolites often contribute significantly to plant defense against insect herbivores (Mazid et al., 2011). Metabolomics thus holds significant potential for discovering active compounds that can enhance direct defense mechanisms in plant resistance against insects (Figure 1).

3. Integration of Multi-Omics in Sustainable Insect Management

Understanding the cellular components and complex behaviors of insects is crucial for unraveling the mechanisms underlying insect pest management. Integrating multiple omics approaches is therefore essential to comprehensively assess an organism's responses under specific environmental and biological conditions. Historically, the integration of metabolomics with genomics, transcriptomics, and proteomics has offered a holistic understanding of biological functions, bridging genomic data with phenotypic responses to environmental interactions (Weckwerth, 2011). The integration of various NGS techniques has transformed crop improvement. While single-omics approaches have limitations in sensitivity and specificity, combining multiple-omics technologies addresses these

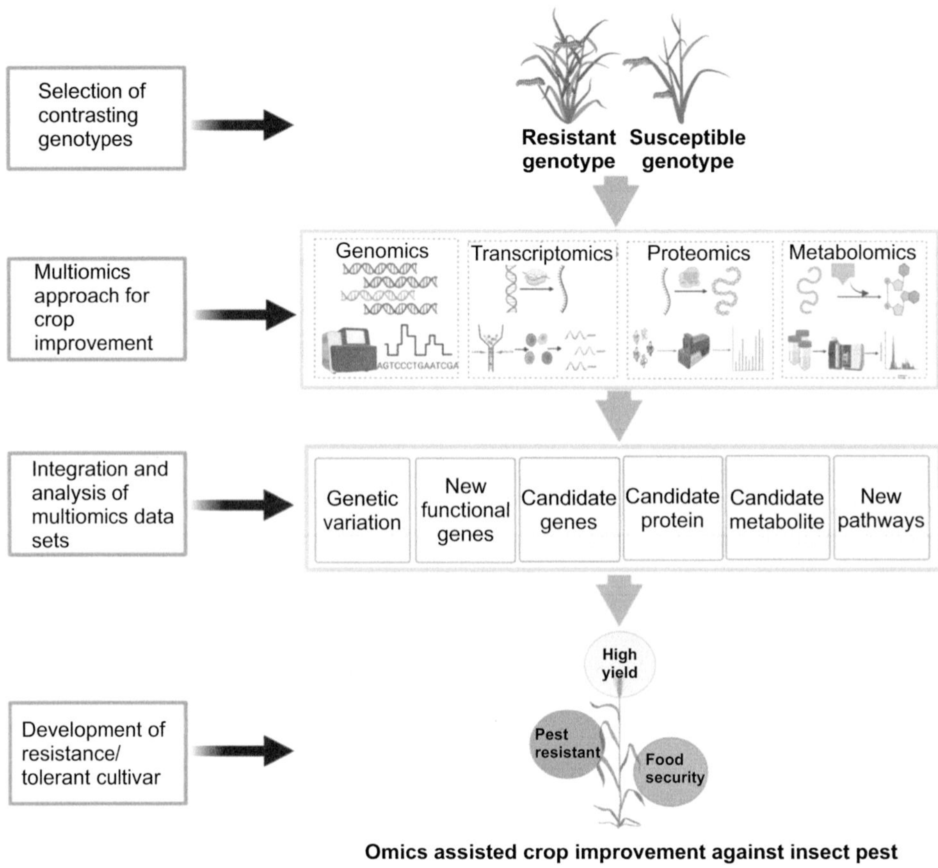

Fig. 1 Overview of the multi-omics approach for crop improvement against insect pest management.

challenges. Integrating multi-omics data has become a standard practice for investigating genes that control critical traits. Comprehensive omics datasets are crucial for thoroughly identifying genes, proteins, biological pathways, and the roles of biomolecules (Bisht et al., 2023). High-throughput omics technologies such as genomics, transcriptomics, metabolomics, and proteomics have been widely utilized in research focused on insect–plant interactions. The abundance of large-scale omics datasets has facilitated the emergence of integrated multi-omics approaches, enabling the elucidation of biological complexity across scales from molecular mechanisms at the cellular level to interactions within ecological communities. Given the substantial molecular and ecosystem-level information available, insect–plant interactions serve as an ideal model for these investigations (Singh et al., 2023; Zogli et al., 2020).

Herbivores have developed diverse feeding techniques to obtain nutrients from their host plants, affecting plants at various levels, from molecular to higher trophic levels (Heil, 2008). Additionally, organisms associated with insects, including parasites and endosymbionts, play a crucial role in shaping plant–insect interactions (Douglas, 2013). Plants have developed diverse defense mechanisms against herbivory, encompassing morphological, biochemical, and molecular adaptations. The multi-omics approach targets genes, metabolites, and biological pathways to devise insect pest management strategies within plant–insect interaction frameworks, taking into account the influence of both biotic and abiotic factors on insect population dynamics (Kiechle et al., 2004; Vemuri & Aristidou, 2005). The integration of molecular and ecological approaches, facilitated by the accessibility of omics data, deepens our understanding of insect–plant interactions across scales, from community dynamics to gene-level processes (Baldwin 2001; Stam et al. 2014). High-throughput analyses of insect genomes, epigenomes, and metagenomes enable the identification of plant responses to herbivory, connecting genetic modifications to observable phenotypic changes.

Integrated metabolomic and transcriptomic reveal how tomato and brinjal respond differently to *Tuta absoluta* attacks, and changes were observed in salicylic acid. Jasmonic acid (JA) levels decrease in tomatoes, but increase in brinjal after infestation. Transcriptomic analysis identified 1,072 and 2,834 differentially expressed genes in tomato and brinjal, respectively, with KEGG analysis highlighting the activation of genes such as PR1b1, NPR1, NPR3, MAPKs, and ANP1 in response to infestation (Chen et al., 2021). High-throughput quantitative proteomics, a pivotal modern technology, facilitates the detailed characterization of proteomes and their dynamic changes throughout plant development and in response to stress. Moreover, the physiological factors affecting *Plutella xylostella* larvae feeding on herbivore-susceptible and resistant *Arabidopsis thaliana* lines (Collins et al., 2010). Similarly, comparative proteomic analyses have identified potato plant responses to mechanical damage and as well as herbivory by the Colorado potato beetle and potato aphid (Duceppe et al., 2012). Plants produce secondary metabolites such as alkaloids, terpenoids, cyanogenic glycosides, glucosinolates, and phenolics that function either as signaling molecules or as direct defensive agents, contributing significantly to the evolution of plant chemical defenses against herbivore attacks (Zebelo & Maffei, 2012). For example, glucoiberin presence in *Brassica oleracea* negatively correlates with herbivore abundance (Poelman et al., 2010). Variations in cytochrome P450s producing glucosinolate compounds in *Boechera stricta* provide resistance to cabbage looper (Brukhin et al., 2019). Metabolic engineering in *A. thaliana* phloem has modified carbon allocation, leading to enhanced resistance to green peach aphid feeding (Cao et al., 2013). Metabolomic studies involving *B. oleracea* and *Pieris rapae* demonstrated an induction of phenylpropanoids in both plant tissues and insect bodies during feeding (Jansen et al., 2009).

Plants produce volatile organic compounds crucial to pests' host selection process. Analyzing these volatiles through a volatile-omics approach can help design environmentally sustainable pest management strategies. Recent advancements in omics technologies have driven the development of high-throughput, high-resolution phenomics tools that are crucial for assessing phenotypic traits influenced by genetic variations and environmental conditions (Fiorani & Schurr, 2013; Furbank & Tester, 2011).

4. Application of Multi-omics Approach through Gene Editing and Genetic Engineering

Genetic modification of insects is an innovative field of biotechnology that involves altering the insect genomes to achieve desired traits or characteristics. Such genetically modified insects are being engineered to reduce their native populations, which pose significant threats to food security and public health (Devos et al., 2022; Harvey-Samuel et al., 2017; Knipling et al., 1968).

The origin of insect genetic modification traces back to the 1930s with the development of the Sterile Insect Technique (SIT) pioneered by E.F. Knipling, which marked a significant milestone in biological pest control. SIT involves mass-rearing of insects, sterilizing them using gamma irradiation (Cobalt-60), and releasing them into the natural environments, where they mate with wild populations but produce no offspring, leading to a gradual reduction in the insect population (Klassen et al., 2021). This strategy was effectively adopted in the 1950s to manage the screw-worm fly, *Cochliomyia hominivorax,* across the southern United States, Mexico, Central America, and Panama (Klassen et al., 2021; Lindquist et al., 1992; Vargas-Terán et al., 2021). The technique was later applied to other pests, including confinement of the Mediterranean fruit fly, *Ceratitis capitata* in Guatemala and Mexico, South Africa, Jordan and Israel (Mumford, 2004; Vreysen et al., 2013); suppression of the cotton pink bollworm, *Pectinophora gossypiella*, in the southern United States and northern Mexico and regulation of the codling moth, *Cydia pomonella* in the Okanagan Valley of British Colombia, Canada (Bloem et al., 2005).

Over the decades, the field evolved with the introduction of molecular techniques, such as transposable elements, between the 1980s and 1990s, allowing foreign genes to be inserted into insect genomes. At the dawn of the 21st century, advanced techniques such as RNA interference (RNAi) were introduced, initially demonstrated in *Caenorhabditis elegans* (Fire et al., 1998) and gene drive systems, enabling precise gene silencing and population control. Advanced gene editing approaches like CRISPR/Cas9 have revolutionized genetic modification by offering a highly efficient, accurate, and versatile tool for editing insect genomes, significantly enhancing our ability to modify insects for pest control, disease vector management, and biological research.

4.1 RNA Interference

RNA interference (RNAi) has undergone extensive investigation for insect functional genomics and pest management since it is an effective method for controlling gene expression. RNA interference, commonly known as RNAi, is a natural biological process in most eukaryotic organisms. In addition to protecting against invading viruses, it controls the stability and translation of mRNA (Rodrigues et al., 2021). The systemic nature of RNAi has been extensively investigated in Coleoptera, Hemiptera, Lepidoptera, and Orthoptera, with transgenerational silencing from injected females to their eggs has also been observed. However, insects lack RNA-dependent RNA polymerase (RdRP) in their genomes, and the movement of dsRNA between tissues is still not completely understood.

4.1.1 Mechanism of Action

Once inside the cell, the dsRNA is identified and cleaved by the enzyme Dicer, which processes the long dsRNA molecules into smaller fragments called small interfering RNAs (siRNAs), typically 21–25 nucleotides in length. These siRNA fragments remain double-stranded after Dicer processing. Furthermore, siRNA duplexes become integrated into a multi-protein assembly known as the RNA-induced silencing complex (RISC). Within RISC, siRNA duplex is separated into single strands: the guide strand, complementary to the target mRNA remains bound to the complex, whereas the passenger strand is typically degraded. Then, the guide strand within the RISC complex directs it to the complementary target mRNA through sequence complementarity. The RISC complex effectively silences the gene by cleaving and degrading the mRNA upon attachment, preventing its translation into protein (Zhu & Palli, 2020) (Figure 2).

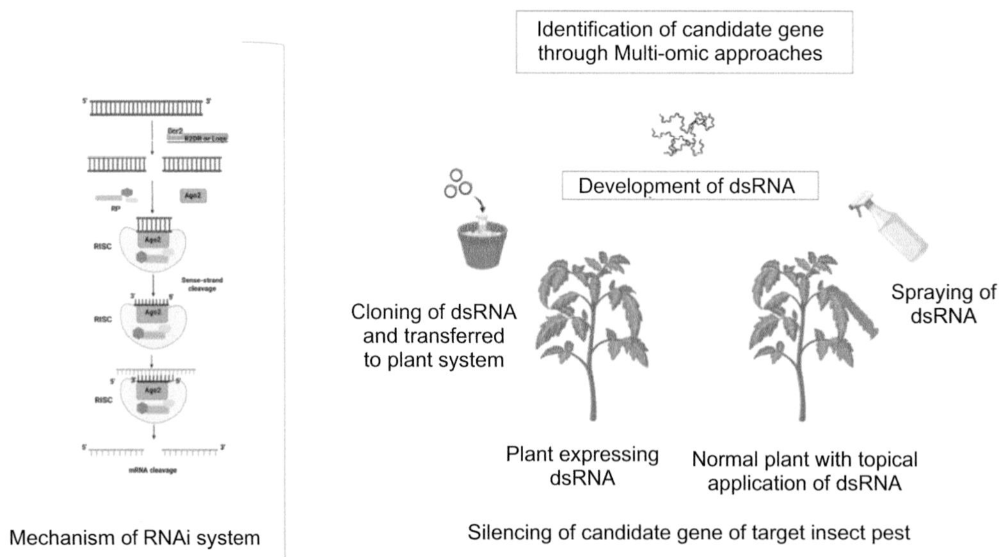

Fig. 2 Mechanism of RNAi technology and its application in insect pest management.

4.2 CRISPR/Cas9

The origin of CRISPR/Cas9 traces back to the late 1980s, when clustered, regularly interspaced short palindromic repeats (CRISPR) were first identified in the DNA of *Escherichia coli.* Still, their function remained a mystery for many years. It was only in the early 2000s that researchers uncovered that CRISPR sequences constitute a bacterial adaptive immune system, defending against viral infections by bacteriophages. This groundbreaking revelation was made by multiple research groups, including the teams led by Francisco Mojica, who identified the CRISPR sequences in archaea, and Philippe Horvath, who linked these sequences to bacterial immunity (Horvath & Barrangou, 2010; Wiedenheft et al., 2012). However, the eukaryotic genome lacks the CRISPR sequences. In 2012, the field of genome editing was revolutionized by the work of Jennifer Doudna and Emmanuelle Charpentier, who demonstrated how the CRISPR-associated protein 9 (Cas9) could be repurposed as an efficient tool for cutting DNA at specific locations. This discovery was a pivotal moment in genetics, providing scientists with a highly versatile and efficient method for editing genes (Doudna & Charpentier, 2014).

4.2.1 Components of the CRISPR/Cas9 System

The CRISPR/Cas9 system comprises a set of critical components that operate collaboratively for precise genetic modification. At its core, the system includes CRISPR RNA (crRNA), a short RNA molecule of approximately 21 nucleotides complementary to the target DNA sequence. The crRNA guides the Cas9 enzyme to the specific location in the genome where the cut will be made. Trans-activating CRISPR RNA (tracrRNA) binds to crRNA to form a complex that is essential for guiding the Cas9 enzyme to the specific target DNA sequence. In laboratory settings, this tracrRNA-crRNA duplex is frequently engineered into a single-guide RNA (sgRNA) to simplify the system. The Cas9 enzyme, an endonuclease, cleaves DNA strands at the targeted site, producing double-strand breaks (DSBs) that are crucial for facilitating the gene editing process.

4.2.2 Mechanism of Action

The CRISPR/Cas9 system functions through a sequence of well-coordinated steps: Targeting and binding, DNA cleavage, and DNA repair. The sgRNA directs the Cas9 enzyme to a specific genomic DNA sequence that is complementary to the crRNA segment of the sgRNA. This target sequence must be positioned next to a protospacer adjacent motif (PAM), which is typically 'NGG' in the event of Cas9 from *Streptococcus pyogenes*. The PAM is essential as it enables the Cas9 enzyme to differentiate the target DNA from the CRISPR array within the bacterial genome, thereby preventing self-targeting and ensuring specificity. Upon binding of the sgRNA to the complementary DNA sequence, the Cas9 enzyme undergoes a conformational shift that activates its nuclease domains. The enzyme cleaves both strands of DNA, generating a DSB at the target site. The cell repairs this DSB primarily through two mechanisms: non-homologous end joining (NHEJ) and homology-directed repair (HDR). NHEJ is an error-prone pathway that frequently introduces insertions or deletions of nucleotides, resulting in mutations commonly referred to as 'indels'. These mutations can disrupt the target gene's function, effectively "knocking out" the gene. In contrast, HDR utilizes a homologous DNA template to accurately mend the break, enabling the precise insertion of desired genetic sequences and the generation of 'knock-in' mutations (Cannon & Kiem, 2021; Deltcheva et al., 2011; Knott & Doudna, 2018) (Figure 3).

4.2.3 Advancements and Modifications to CRISPR/Cas9

Since its discovery, the CRISPR/Cas9 system has undergone various modifications aimed at improving its accuracy, efficiency, and flexibility. These enhancements have broadened its range of applications while also increasing its safety.

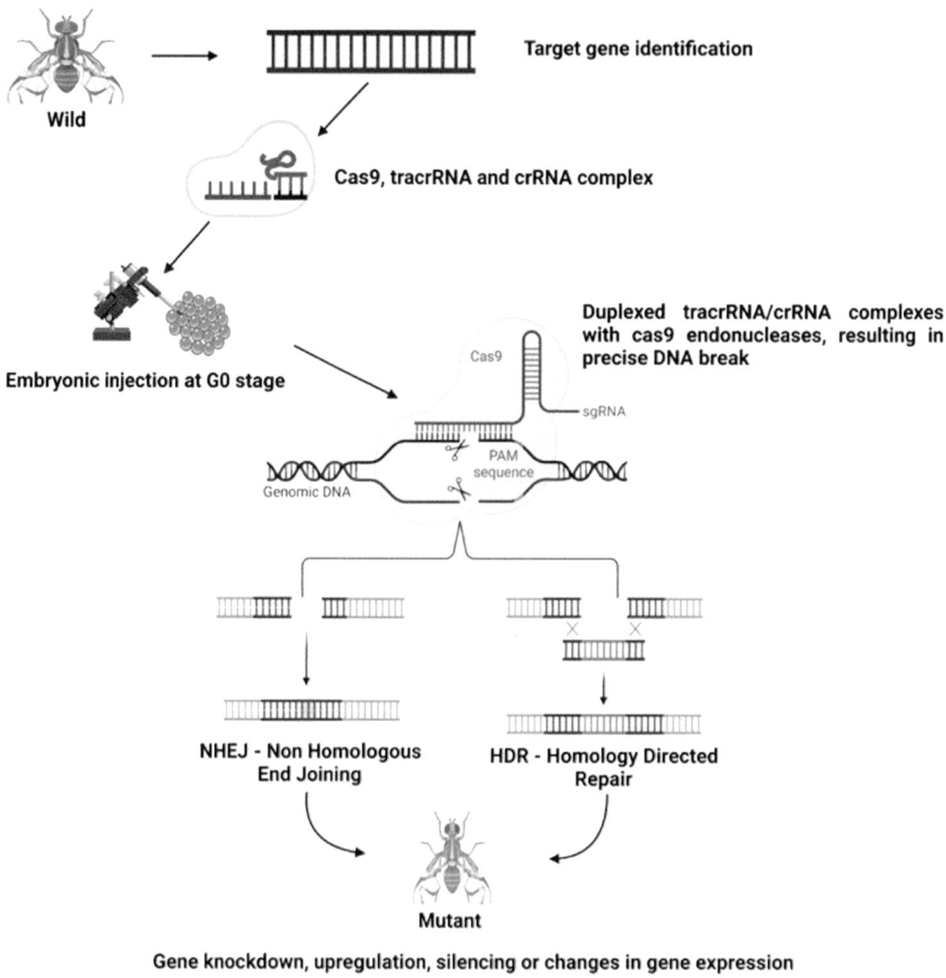

Fig. 3 Mechanism of CRISPR/cas9 genome editing in an insect system.

- **Base Editing:** The development of base editors, which allow the precise transformation of a one nucleotide to another without causing a double-strand break, represents a significant advancement in CRISPR technology. Base editors consist of a chemically inactive Cas9 (dCas9) linked to a deaminase enzyme that can change specific DNA bases, such as converting cytosine to thymine. This technique enables more accurate genetic modifications while minimizing the chances of inducing indels and off-target effects.
- **Prime Editing:** Prime editing is a newer technique that expands the capabilities of CRISPR/Cas9 by allowing for the precise insertion, deletion, or replacement of DNA sequences without requiring a donor template. Prime editors are composed of a Cas9 nickase linked to reverse transcriptase enzyme, enabling the direct incorporation of a specified DNA sequence from an RNA template into the genome. This approach provides enhanced flexibility and precision compared to traditional CRISPR/Cas9 editing and is less prone to errors and off-target effects.
- **CRISPR Interference (CRISPRi) and CRISPR Activation (CRISPRa):** CRISPRi and CRISPRa are techniques that use a modified, catalytically inactive Cas9 (dCas9) to repress or activate gene expression without cutting the DNA. In CRISPRi, dCas9 is connected to a transcriptional repressor domain that attaches itself to the target gene's promoter region and prevents transcription. Conversely, in CRISPRa, dCas9 is fused to a transcriptional activator

domain that promotes increased gene expression. These techniques allow researchers to study gene function and regulation without permanently altering the DNA sequence.

- **Anti-CRISPR Proteins:** To further improve the safety and control of CRISPR/Cas9 editing, researchers have discovered natural proteins, known as anti-CRISPR proteins, that can inhibit the activity of the Cas9 enzyme. These proteins can prevent off-target effects and provide temporal control over gene editing, allowing for more precise and controlled genetic modifications.

5. Future Aspects and Strategic Applications of Omics in IPM

The integration of multi-omics: genomics, transcriptomics, proteomics, and metabolomics offers transformative potential to push Integrated Pest Management (IPM) strategies into a new era of precision and sustainability. Despite their availability for several years, the full-scale application of these technologies in managing insect resistance has been slow, hindered by fragmented interdisciplinary collaboration, limited infrastructure for large-scale data integration, challenges in translating molecular findings into practical solutions, and insufficient capacity-building among stakeholders. Addressing these barriers calls for robust platforms to consolidate and interpret multi-omics datasets, strong partnerships bridging molecular science with field-level IPM, and greater investment in advanced bioinformatics and machine learning tools capable of providing actionable insights. Looking ahead, the next generation of IPM could employ real-time omics-based surveillance, predictive biomarker-guided interventions, and targeted disruption of resistance mechanisms, thus enabling rapid, site-specific decision-making. This chapter underscores how strategic integration of omics, combined with interdisciplinary collaboration and technology adoption, can drive a paradigm shift from reactive pest control to proactive, knowledge-driven, and ecologically balanced resistance management.

6. Conclusion

The multi-omics approach provides a promising pathway for sustainable insect management practices by exploring plant secondary metabolomics. Integrating various omics technologies, including genomics, transcriptomics, proteomics, and metabolomics, enables a holistic understanding of the intricate biological networks that regulate plant–insect interactions. Plant secondary metabolites including alkaloids, terpenoids, and phenolics, are crucial components of plant defense strategies against herbivorous insects. However, traditional single-omics studies often fail to identify the full range of metabolites involved or how their biosynthetic pathways are regulated. The multi-omics approach overcomes these limitations by enabling the simultaneous examination of multiple biological layers, from gene expression to metabolic pathways and protein interactions, thus providing a holistic view of how plants deploy chemical defenses. This comprehensive analysis is essential for identifying specific genes and regulatory networks responsible for producing bioactive compounds that deter insect feeding or impair insect development. By leveraging multi-omics data, breeders and researchers can identify key metabolites and the genes controlling their synthesis, allowing for the development of crop varieties with improved pest resistance. RNAi technology allows for the targeted silencing of essential insect genes, thereby reducing pest populations without harming non-target species, thus offering an environmentally friendly alternative to chemical pesticides. In parallel, CRISPR-based gene editing offers precise tools for modifying plant and insect genomes, enhancing plant resistance traits, and disrupting insect survival pathways. This approach also facilitates the identification of novel compounds that help control insect populations and decrease dependence on chemical pesticides, which often pose risks to ecosystems and human health. Ultimately, the multi-omics strategy supports sustainable agriculture by enhancing natural plant defenses, minimizing the use of synthetic chemicals, and promoting ecological balance within pest management systems.

References

Adams, M.D., Celniker, S.E., Holt, R.A., Evans, C.A., Gocayne, J.D., Amanatides, P.G., Scherer, S.E., Li, P.W., Hoskins, R.A., Galle, R.F., George, R.A., Lewis, S.E., Richards, S., Ashburner, M., Henderson, S.N., Sutton, G.G., Wortman, J.R., Yandell, M.D., Zhang, Q., …and Venter, J.C. (2000). The Genome Sequence of *Drosophila melanogaster*. *Science*, *287*(5461), 2185–2195.

Austel, N., Böttcher, C. and Meiners, T. (2021). Chemical defence in Brassicaceae against pollen beetles revealed by metabolomics and flower bud manipulation approaches. *Plant, Cell, and Environment*, *44*(2), 519–534.

Baldwin, I.T. (2001). An ecologically motivated analysis of plant-herbivore interactions in native tobacco. *Plant Physiology, 127*(4), 1449–1458.

Bisht, A., Saini, D.K., Kaur, B., Batra, R., Kaur, S., Kaur, I., Jindal, S., Malik, P., Sandhu, P.K., Kaur, A., Gill, B.S., Wani, S.H., Kaur, B., Mir, R.R., Sandhu, K.S. and Siddique, K.H.M. (2023). Multi-omics assisted breeding for biotic stress resistance in soybean. *Molecular Biology Reports*, *50*(4), 3787–3814.

Bloem, S., Hight, S.D., Carpenter, J.E. and Bloem, K.A. (2005). Development of the most effective trap to monitor the presence of the cactus moth *Cactoblastis cactorum* (Lepidoptera: Pyralidae). *Florida Entomologist*, *88*(3), 300–306.

Brown, J.K., Paredes-Montero, J.R. and Stocks, I.C. (2023). The *Bemisia tabaci* cryptic (sibling) species group: Imperative for a taxonomic reassessment. *Current Opinion in Insect Science*, *57*, 101032.

Brukhin, V., Osadtchiy, J.V., Florez-Rueda, A.M., Smetanin, D., Bakin, E., Nobre, M. S. and Grossniklaus, U. (2019). The *Boechera* genus as a resource for apomixis research. *Frontiers in Plant Science*, *10*, 392.

Camargo, R.D.A., Herai, R.H., Santos, L.N., Bento, F.M.M., Lima, J.E., Marques-Souza, H. and Figueira, A. (2015). De novo transcriptome assembly and analysis to identify potential gene targets for RNAi-mediated control of the tomato leafminer (*Tuta absoluta*). *BMC Genomics*, *16*(1), 635.

Cannon, P.M. and Kiem, H.P. (2021). The genome-editing decade. *Molecular Therapy*, *29*(11), 3093–3094.

Cao, T., Lahiri, I., Singh, V., Louis, J., Shah, J. and Ayre, B.G. (2013). Metabolic engineering of raffinose-family oligosaccharides in the phloem reveals alterations in carbon partitioning and enhances resistance to green peach aphid. *Frontiers in Plant Science*, *4*, 263.

Chen, L., Li, X., He, T., Li, P., Liu, Y., Zhou, S., Wu, Q., Chen, T., Lu, Y. and Hou, Y. (2021). Comparative biochemical and transcriptome analyses in tomato and eggplant reveal their differential responses to *Tuta absoluta* infestation. *Genomics*, *113*(4), 2108–2121.

Chen, W., Hasegawa, D.K., Kaur, N., Kliot, A., Pinheiro, P.V., Luan, J., Stensmyr, M.C., Zheng, Y., Liu, W., Sun, H., Xu, Y., Luo, Y., Kruse, A., Yang, X., Kontsedalov, S., Lebedev, G., Fisher, T.W., Nelson, D.R., Hunter, W.B., … and Fei, Z. (2016). The draft genome of whitefly *Bemisia tabaci* MEAM1, a global crop pest, provides novel insights into virus transmission, host adaptation, and insecticide resistance. *BMC Biology*, *14*(1), 110.

Chen, X., Xia, J., Shang, Q., Song, D. and Gao, X. (2019). UDP-glucosyltransferases potentially contribute to imidacloprid resistance in *Aphis gossypii* glover based on transcriptomic and proteomic analyses. *Pesticide Biochemistry and Physiology*, *159*, 98–106.

Collins, R.M., Afzal, M., Ward, D.A., Prescott, M.C., Sait, S.M., Rees, H.H. and Tomsett, A.B. (2010). Differential proteomic analysis of *Arabidopsis thaliana* genotypes exhibiting resistance or susceptibility to the insect herbivore, *Plutella xylostella*. *PLoS One*, *5*(4), e10103.

Coppola, V., Coppola, M., Rocco, M., Digilio, M.C., D'Ambrosio, C., Renzone, G., Martinelli, R., Scaloni, A., Pennacchio, F., Rao, R. and Corrado, G. (2013). Transcriptomic and proteomic analysis of a compatible tomato-aphid interaction reveals a predominant salicylic acid-dependent plant response. *BMC Genomics*, *14*(1), 515.

Daborn, P.J., Yen, J.L., Bogwitz, M.R., Le Goff, G., Feil, E., Jeffers, S., Tijet, N., Perry, T., Heckel, D., Batterham, P., Feyereisen, R., Wilson, T.G. and Ffrench-Constant, R.H. (2002). A Single P450 Allele Associated with Insecticide Resistance in *Drosophila*. *Science*, *297*(5590), 2253–2256.

Deltcheva, E., Chylinski, K., Sharma, C.M., Gonzales, K., Chao, Y., Pirzada, Z.A., Eckert, M.R., Vogel, J. and Charpentier, E. (2011). CRISPR RNA maturation by trans-encoded small RNA and host factor RNase III. *Nature*, *471*(7340), 602–607.

Devos, Y., Mumford, J.D., Bonsall, M.B., Camargo, A.M., Firbank, L.G., Glandorf, D.C.M., Nogué, F., Paraskevopoulos, K. and Wimmer, E.A. (2022). Potential use of gene drive modified insects against disease vectors, agricultural pests, and invasive species poses new challenges for risk assessment. *Critical Reviews in Biotechnology*, *42*(2), 254–270.

Dong, Y., Fang, X., Yang, Y., Xue, G.P., Chen, X., Zhang, W., Wang, X., Yu, C., Zhou, J. and Mei, Q. (2017). Comparative proteomic analysis of susceptible and resistant rice plants during early infestation by small brown planthopper. *Frontiers in Plant Science*, *8*, 1744.

Doudna, J.A. and Charpentier, E. (2014). The new frontier of genome engineering with CRISPR-Cas9. *Science*, *346*(6213), 6213.

Douglas, A.E. (2013). Microbial Brokers of Insect-Plant Interactions Revisited. *Journal of Chemical Ecology*, *39*(7), 952–961.

Duceppe, M.O., Cloutier, C. and Michaud, D. (2012). Wounding, insect chewing and phloem sap feeding differentially alter the leaf proteome of potato, *Solanum tuberosum* L. *Proteome Science*, *10*(1), 73.

Erb, M., Flors, V., Karlen, D., De Lange, E., Planchamp, C., D'Alessandro, M., Turlings, T.C.J. and Ton, J. (2009). Signal signature of aboveground-induced resistance upon belowground herbivory in maize. *The Plant Journal*, *59*(2), 292–302.

Farnsworth, C.A., Teese, M.G., Yuan, G., Li, Y., Scott, C., Zhang, X., Wu, Y., Russell, R.J. and Oakeshott, J.G. (2010). Esterase-based metabolic resistance to insecticides in heliothine and spodopteran pests. *Journal of Pesticide Science*, *35*(3), 275–289.

Fiorani, F. and Schurr, U. (2013). Future Scenarios for Plant Phenotyping. *Annual Review of Plant Biology*, *64*(1), 267–291.

Fire, A., Xu, S., Montgomery, M.K., Kostas, S.A., Driver, S.E. and Mello, C.C. (1998). Potent and specific genetic interference by double-stranded RNA in Caenorhabditis elegans. *Nature*, *391*(6669), 806–811.

Furbank, R.T. and Tester, M. (2011). Phenomics: Technologies to relieve the phenotyping bottleneck. *Trends in Plant Science*, *16*(12), 635–644.

Gatehouse, J.A. (2002). Plant resistance towards insect herbivores: A dynamic interaction. *New Phytologist*, *156*(2), 145–169.

Grbić, M., Van Leeuwen, T., Clark, R.M., Rombauts, S., Rouzé, P., Grbić, V., Osborne, E.J., Dermauw, W., Thi Ngoc, P.C. and Ortego, F. (2011). The genome of *Tetranychus urticae* reveals herbivorous pest adaptations. *Nature*, *479*(7374), 487–492.

Gregory, T.R. (2005). Genome size evolution in animals. *The Evolution of the Genome*, Elsevier, pp. 3–87.

Guo, T., Steen, J.A. and Mann, M. (2025). Mass-spectrometry-based proteomics: From single cells to clinical applications. *Nature*, *638*(8052), 901–911.

Harith-Fadzilah, N., Abd Ghani, I. and Hassan, M. (2021). Omics-based approach in characterizing mechanisms of entomopathogenic fungi pathogenicity: A case example of *Beauveria bassiana. Journal of King Saud University - Science*, *33*(2), 101332.

Harvey-Samuel, T., Ant, T. and Alphey, L. (2017). Towards the genetic control of invasive species. *Biological Invasions*, *19*(6), 1683–1703.

Heil, M. (2008). Indirect defence via tritrophic interactions. *New Phytologist*, *178*(1), 41–61.

Horvath, P. and Barrangou, R. (2010). CRISPR/Cas, the Immune System of Bacteria and Archaea. *Science*, *327*(5962), 167–170.

Hsu, J.C., Chien, T.Y., Hu, C.C., Chen, M.J. M., Wu, W.J., Feng, H.T., Haymer, D.S. and Chen, C.Y. (2012). Discovery of genes related to insecticide resistance in Bactrocera dorsalis by functional genomic analysis of a *de novo* assembled transcriptome, *PLoS One*, 7(8): e40950.

Idle, J.R. and Gonzalez, F.J. (2007). Metabolomics. *Cell Metabolism*, *6*(5), 348–351.

Jaba, J., Agnihotri, M. and Chakravarty, S. (2017). Screening for host plant resistance to *Helicoverpa armigera* (Hubner) in chickpea using novel techniques. *Legume Research*, *40*(5), 5.

Jansen, J.J., Allwood, J.W., Marsden-Edwards, E., Van Der Putten, W.H., Goodacre, R. and Van Dam, N.M. (2009). Metabolomic analysis of the interaction between plants and herbivores. *Metabolomics*, *5*(1), 150–161.

Jiang, D., Lin, R., Tan, M., Yan, J. and Yan, S. (2022). The mycorrhizal-induced growth promotion and insect resistance reduction in *Populus alba* x *P. berolinensis* seedlings: A multi-omics study. *Tree Physiology*, *42*(5), 1059–1069.

Kalajdzic, P., Oehler, S., Reczko, M., Pavlidi, N., Vontas, J., Hatzigeorgiou, A.G. and Savakis, C. (2012). Use of mutagenesis, genetic mapping and next generation transcriptomics to investigate insecticide resistance mechanisms. *PLoS One*, *7*(6), e40296.

Karunker, I., Morou, E., Nikou, D., Nauen, R., Sertchook, R., Stevenson, B.J., Paine, M.J., Morin, S. and Vontas, J. (2009). Structural model and functional characterization of the *Bemisia tabaci* CYP6CM1vQ, a cytochrome P450 associated with high levels of imidacloprid resistance. *Insect Biochemistry and Molecular Biology*, *39*(10), 697–706.

Kaul, S., Sharma, T. and Dhar, M.K. (2016). "Omics" tools for better understanding the plant–endophyte interactions. *Frontiers in Plant Science*, *7*, 955.

Kiechle, F.L., Zhang, X. and Holland-Staley, C.A. (2004). The omics era and its impact. *Archives of Pathology and Laboratory Medicine*, *128*(12), 1337–1345.

Kirsch, R., Wielsch, N., Vogel, H., Svatoš, A., Heckel, D.G. and Pauchet, Y. (2012). Combining proteomics and transcriptome sequencing to identify active plant-cell-wall-degrading enzymes in a leaf beetle. *BMC Genomics*, *13*(1), 587.

Klassen, W., Curtis, C.F. and Hendrichs, J. (2021). History of the sterile insect technique. In: Dyck, V.A., Hendrichs, J. & Robinson, A.S. (Eds.). *Sterile Insect Technique*. CRC Press, pp. 1–44.

Knipling, E.F., Laven, H., Craig, G.B., Pal, R., Kitzmiller, J.B., Smith, C.N. and Brown, A.W.A. (1968). Genetic control of insects of public health importance. *Bulletin of the World Health Organization*, *38*(3), 421.

Knott, G.J. and Doudna, J.A. (2018). CRISPR-Cas guides the future of genetic engineering. *Science*, *361*(6405), 866–869.

Koepfli, K.P., Paten, B., the Genome 10K Community of Scientists and O'Brien, S. J. (2015). The Genome 10K Project: A Way Forward. *Annual Review of Animal Biosciences*, *3*(1), 57–111.

Li, F., Zhao, X., Li, M., He, K., Huang, C., Zhou, Y., Li, Z. and Walters, J.R. (2019). Insect genomes: Progress and challenges. *Insect Molecular Biology*, *28*(6), 739–758.

Li, H., Jiang, W., Zhang, Z., Xing, Y. and Li, F. (2013). Transcriptome analysis and screening for potential target genes for RNAi-mediated pest control of the Beet armyworm, *Spodoptera exigua*. PLoS ONE *8*(6): e65931.

Lindquist, D.A., Abusowa, M. and Hall, M.J.R. (1992). The New World screwworm fly in Libya: A review of its introduction and eradication. *Medical and Veterinary Entomology*, *6*(1), 2–8.

Liu, W., Gray, S., Huo, Y., Li, L., Wei, T. and Wang, X. (2015). Proteomic analysis of interaction between a plant virus and its vector insect reveals new functions of hemipteran cuticular protein. *Molecular and Cellular Proteomics*, *14*(8), 2229–2242.

Macel, M., Visschers, I.G.S., Peters, J.L., Kappers, I.F., De Vos, R.C.H. and Van Dam, N.M. (2019). Metabolomics of Thrips Resistance in Pepper (*Capsicum* spp.) Reveals Monomer and Dimer Acyclic Diterpene Glycosides as Potential Chemical Defenses. *Journal of Chemical Ecology*, *45*(5–6), 490–501.

Mayorga-Martino, V., Mansurova, M., Calla-Quispe, E. and Ibáñez, A.J., (2025). Unlocking the Secrets of Insects: The Role of Mass Spectrometry to Understand the Life of Insects. *Mass Spectrometry Reviews*, *44*(4), 543–565.

Mazid, M., Khan, T.A. and Mohammad, F. (2011). Role of secondary metabolites in defense mechanisms of plants. *Biology and Medicine*, *3*, 232–249.

Mirnezhad, M., Romero-González, R.R., Leiss, K.A., Choi, Y.H., Verpoorte, R. and Klinkhamer, P.G.L. (2010). Metabolomic analysis of host plant resistance to thrips in wild and cultivated tomatoes. *Phytochemical Analysis*, *21*(1), 110–117.

Mishra, M., Saurabh, S., Maurya, R., Mudawal, A., Parmar, D. and Singh, P.K. (2016). Proteome analysis of *Bemisia tabaci* suggests specific targets for RNAi mediated control. *Journal of Proteomics*, *132*, 93–102.

Mithöfer, A. and Boland, W. (2012). Plant Defense Against Herbivores: Chemical Aspects. *Annual Review of Plant Biology*, *63*(1), 431–450.

Mittapelly, P. and Rajarapu, S.P. (2020). Applications of proteomic tools to study insect vector–plant virus interactions. *Life*, *10*(8), 143.

Müller, P., Warr, E., Stevenson, B.J., Pignatelli, P.M., Morgan, J.C., Steven, A., Yawson, A.E., Mitchell, S.N., Ranson, H. and Hemingway, J. (2008). Field-caught permethrin-resistant *Anopheles gambiae* overexpress CYP6P3, a P450 that metabolises pyrethroids. *PLoS Genetics*, *4*(11), e1000286.

Mumford, J.D. (2004). Economic analysis of area-wide fruit fly management. In; *Proceedings of the 6th International Symposium Fruit Flies of Economic Importance*. International Atomic Energy Agency (IAEA), pp. 189–193.

Nawaz, M., Sun, J., Shabbir, S., Khattak, W.A., Ren, G., Nie, X., Bo, Y., Javed, Q., Du, D. and Sonne, C. (2023). A review of plants strategies to resist biotic and abiotic environmental stressors. *Science of the Total Environment*, 165832.

Palli, S.R., Bai, H. and Wigginton, J. (2012). Insect genomics. In: Gilbert, L.I. (Eds.) *Insect Molecular Biology and Biochemistry*. Elsevier, pp. 1–29.

Poelchau, M., Childers, C., Moore, G., Tsavatapalli, V., Evans, J., Lee, C.Y., Lin, H., Lin, J.W. and Hackett, K. (2015). The i5k Workspace@NAL: Enabling genomic data access, visualization, and curation of arthropod genomes. *Nucleic Acids Research*, *43*(D1), D714–D719.

Poelman, E.H., Van Loon, J.J.A., Van Dam, N.M., Vet, L.E.M. and Dicke, M. (2010). Herbivore-induced plant responses in *Brassica oleracea* prevail over effects of constitutive resistance and result in enhanced herbivore attack. *Ecological Entomology*, *35*(2), 240–247.

Renoz, F., Champagne, A., Degand, H., Faber, A.M., Morsomme, P., Foray, V. and Hance, T. (2017). Toward a better understanding of the mechanisms of symbiosis: A comprehensive proteome map of a nascent insect symbiont. *PeerJ*, *5*, e3291.

Rodrigues, T.B., Mishra, S.K., Sridharan, K., Barnes, E.R., Alyokhin, A., Tuttle, R., Kokulapalan, W., Garby, D., Skizim, N.J. and Tang, Y. (2021). First sprayable double-stranded RNA-based biopesticide product targets proteasome subunit beta type-5 in Colorado potato beetle (*Leptinotarsa decemlineata*). *Frontiers in Plant Science*, *12*, 728652.

Sai Reddy, M.S., Karthik, S., Raju, B.J. and Yashaswini, G. (2022). Multi-omics Approaches in Insect-Plant Interactions. In: Tanda, A.S. (Eds.), *Molecular Advances in Insect Resistance of Field Crops*. Springer International Publishing, pp. 335–368.

Sanchez-Arcos, C., Kai, M., Svatoš, A., Gershenzon, J. and Kunert, G. (2019). Untargeted metabolomics approach reveals differences in host plant chemistry before and after infestation with different pea aphid host races. *Frontiers in Plant Science*, *10*, 188.

Sayers, E.W., Cavanaugh, M., Clark, K., Ostell, J., Pruitt, K.D. and Karsch-Mizrachi, I. (2020). GenBank. *Nucleic Acids Research*, *48*(D1), D84–D86.

Schmidhuber, J. and Tubiello, F.N. (2007). Global food security under climate change. *Proceedings of the National Academy of Sciences*, *104*(50), 19703–19708.

Sheikh, N., Barman, D. and Bhattacharjee, K. (2020). Abiotic and Biotic Stress Research in Plants: A Gizmatic Approach of Modern Omics Technologies. In: Roychowdhury, R., Choudhury, S., Hasanuzzaman, M. and Srivastava S. (Eds.), *Sustainable Agriculture in the Era of Climate Change*. Springer International Publishing, pp. 413–439.

Singh, J., Kumar, D., Chauhan, S., Dhillon, H.K., Kumar, S., Kumar, V. and Kapoor, R. (2023). Role of omics approaches in vegetable breeding for insect pest resistance. *SN Applied Sciences*, *5*(12), 364.

Stam, J.M., Kroes, A., Li, Y., Gols, R., Van Loon, J.J.A., Poelman, E.H. and Dicke, M. (2014). Plant interactions with multiple insect herbivores: From community to genes. *Annual Review of Plant Biology, 65*(1), 689–713.

Stork, N.E. (2018). How Many Species of Insects and Other Terrestrial Arthropods Are There on Earth? *Annual Review of Entomology*, *63*(1), 31–45.

Theissinger, K., Fernandes, C., Formenti, G., Bista, I., Berg, P.R., Bleidorn, C., Bombarely, A., Crottini, A., Gallo, G.R. and Godoy, J.A. (2023). How genomics can help biodiversity conservation. *Trends in Genetics, 39*(7), 545–559. https://www.cell.com/trends/genetics/fulltext/S0168-9525(23)00020-3?s=09.

Vargas-Terán, M., Spradbery, J.P., Hofmann, H.C. and Tweddle, N.E. (2021). Impact of screwworm eradication programmes using the sterile insect technique. In: *Sterile Insect Technique*. CRC Press, pp. 949–978.

Vemuri, G.N. and Aristidou, A.A. (2005). Metabolic Engineering in the omics Era: Elucidating and Modulating Regulatory Networks. *Microbiology and Molecular Biology Reviews*, *69*(2), 197–216.

Vreysen, M.J., Seck, M.T., Sall, B. and Bouyer, J. (2013). Tsetse flies: Their biology and control using area-wide integrated pest management approaches. *Journal of Invertebrate Pathology*, *112*, S15–S25.

Wang, H., Geng, S., Liu, S., Li, Z., Cameron, S., Lei, T., Xu, W., Liu, Q., Zuo, S., Omongo, C.A. & Maruthi, M.N. (2025). Unraveling the cryptic *Bemisia tabaci* species complex: Global phylogenomic analysis reveals evolutionary relationships and biogeographic patterns. *Insect Science*, *32*(1).

Wang, H., Lei, T., Wang, X., Cameron, S., Navas-Castillo, J., Liu, Y., Maruthi, M.N., Omongo, C.A., Delatte, H., Lee, K., Krause-Sakate, R., Ng, J., Seal, S., Fiallo-Olivé, E., Bushley, K., Colvin, J. and Liu, S. (2024). A comprehensive framework for the delimitation of species within the *Bemisia tabaci* cryptic complex, a global pest-species group. *Insect Science*, *32*(1), 321–342.

Wang, S., Yang, H., Hu, Y., Zhang, C. and Fan, D. (2023). Multi-omics reveals the effect of population density on the phenotype, transcriptome, and metabolome of *Mythimna separata. Insects*, *14*(1), 68.

War, A.R., Murugesan, S., Boddepalli, V.N., Srinivasan, R. and Nair, R.M. (2017). Mechanism of Resistance in Mungbean [*Vigna radiata* (L.) R. Wilczek var. radiata] to bruchids, *Callosobruchus* spp. (Coleoptera: Bruchidae). *Frontiers in Plant Science*, *8*.

Weckwerth, W. (2011). Green systems biology: From single genomes, proteomes and metabolomes to ecosystems research and biotechnology. *Journal of Proteomics*, *75*(1), 284–305.

Wiedenheft, B., Sternberg, S.H. and Doudna, J.A. (2012). RNA-guided genetic silencing systems in bacteria and archaea. *Nature*, *482*(7385), 331–338.

Xiao, D., Lu, Y., Shang, Q., Song, D. and Gao, X. (2015). Gene silencing of two acetylcholinesterases reveals their cholinergic and non-cholinergic functions in *Rhopalosiphum padi* and *Sitobion avenae. Pest Management Science*, *71*(4), 523–530.

Yang, Y., Saand, M.A., Huang, L., Abdelaal, W.B., Zhang, J., Wu, Y., Li, J., Sirohi, M.H. and Wang, F. (2021). Applications of multi-omics technologies for crop improvement. *Frontiers in Plant Science*, *12*, 563953.

Zebelo, S.A. and Maffei, M.E. (2012). Signal Transduction in Plant–Insect Interactions: From Membrane Potential Variations to Metabolomics. In: Volkov A.G. (Eds.), *Plant Electrophysiology*. Springer Berlin Heidelberg , pp. 143–172.

Zhang, Z., Zhang, P., Li, W., Zhang, J., Huang, F., Yang, J., Bei, Y. and Lu, Y. (2013). De novo transcriptome sequencing in *Frankliniella occidentalis* to identify genes involved in plant virus transmission and insecticide resistance. *Genomics*, *101*(5), 296–305.

Zhu, K.Y. and Palli, S.R. (2020). Mechanisms, Applications, and Challenges of Insect RNA Interference. *Annual Review of Entomology*, *65*(1), 293–311.

Zogli, P., Pingault, L., Grover, S. and Louis, J. (2020). Ento (o) mics: The intersection of 'omic' approaches to decipher plant defense against sap-sucking insect pests. *Current Opinion in Plant Biology*, *56*, 153 161.

CHAPTER 3

Emerging Approaches to Sustainable Agriculture and Food Security
Genomic Selection, Prediction, and Machine Learning

Vishnutej Ellur[1*], Srijan Ambati[2], Diego Jarquin[3], Julian Garcia Abadillo[4] and Ravi V. Mural[5]

1. Introduction

Plant breeding strategies have evolved considerably over time. Initially, efforts concentrated primarily on improving yield. Subsequently, breeding shifted toward incorporating resistance traits to further enhance productivity. More recently, advancements such as speed breeding have emerged, significantly shortening seed-to-seed cycles and accelerating breeding programs (Kumar & Walia, 2024). The rise of molecular breeding, particularly DNA marker technologies, has revolutionized traditional methods that depended heavily on phenotypic observations or pedigree records. This progress has led to a transition from marker-assisted selection (MAS) toward the more comprehensive genomic selection (GS) approach.

MAS typically focuses on a limited number of markers linked to major genomic regions or quantitative trait loci (QTLs) with substantial effects, excluding markers weakly correlated or unrelated to traits. However, for complex traits influenced by numerous small-effect genes, identifying QTLs has proven to be a challenging task. Consistent detection of the same QTL across various genetic backgrounds and environmental conditions is difficult due to confounding non-genetic factors (Zhu et al., 2008). Furthermore, linkage analyses in biparental populations suffer from limited recombination, weakening the ability to detect marker-trait associations. To overcome these obstacles, association mapping was introduced.

1 Regulatory Scientist/Technical Expert, Syngenta Seeds LLC, Durham, North Carolina, USA.
2 Assistant Professor, Genetics & Plant Breeding, Professor Jayashankar Telangana Agricultural University, Warangal, Telangana, India.
3 Assistant Professor, Integration & Application of AI & Omics in Plant Breeding, Department of Agronomy, University of Florida, Gainesville, Florida, USA.
4 Graduate Researcher, Integration & Application of AI & Omics in Plant Breeding, Department of Agronomy, University of Florida, Gainesville, Florida, USA.
5 Assistant Professor of Quantitative Genetics and Genomics, Department of Agronomy, Horticulture & Plant Science, South Dakota State University, Brookings, South Dakota, USA.
* Corresponding author: vishnutej.ellur@outlook.com

DOI: 10.1201/9781003614869-3

Association mapping leverages fine-scale recombination events within chromosome regions to detect trait-marker linkages in diverse, non-biparental populations (Rashid et al., 2018). Nonetheless, this method struggles to identify rare variants affecting economically important traits. Although genes may be localized precisely, the predictive accuracy of these selected regions often falls short compared to whole-genome approaches. Advances in high-density single nucleotide polymorphism (SNP) arrays, which now contain tens of thousands of markers, alongside reduced costs and the ubiquity of SNPs as genetic variants, have popularized their use for association studies and advanced statistical modeling (Crossa et al., 2017).

Machine Learning (ML) represents a transformative tool in agriculture, complementing GS by capturing intricate interactions among genotypes, phenotypes, and environments. ML algorithms analyze high-dimensional data to uncover patterns and enhance predictive accuracy. They facilitate identification of genetic markers linked to resistance traits and improve the prediction of plant resistance by integrating genomic, phenotypic, and environmental datasets (Crossa et al., 2024).

Historically, plant breeding involved backcrossing to transfer a known major gene of economic importance. In the absence of such genes in germplasm, early breeders used random mutagenesis (mutation breeding) or wild hybridization to introduce novel resistance traits. Later, transgenic methods enabled the insertion of foreign genes. More recently, genome editing techniques like CRISPR/Cas9 and CRISPR/Cas12 have provided precise targeted genetic modifications (Zhou et al., 2023). From a modeling perspective, GS has become a pivotal tool in modern breeding by integrating genome-wide marker data with phenotypes in statistical frameworks to improve trait prediction. This chapter delves into the deployment of GS and genomic prediction (GP) in breeding programs, highlighting the potential of ML to augment their effectiveness.

2. Mapping Genomic Information to Phenotypes: MAS and GS

Both MAS and GS are approaches to solve a supervised ML task involving a first step of training, in which a training population (TP) containing both the predictors (genomic markers) and target (phenotypes) variables are observed, and a second step of predicting, in which a new set of predictors in given to the model to predict their unknown targets. Figure 1 presents a schematic overview of the parameters and factors affecting genomic selection.

Although marker-trait association is the fundamental principle underlying both MAS and GS, their applications and methods for estimating breeding values differ. While GS estimates the effects of all markers, assuming they are linked to genes that influence the trait, it selects favorable individuals based on linked markers (Spindel et al., 2015). MAS focuses on evaluating the relevance of specific gene effects on traits (Meuwissen et al., 2016) it combines all the marker effects to determine Genomic Estimated Breeding Values (GEBVs), capturing the total genetic variance of each individual (Arruda et al., 2016). Furthermore, GS utilizes genotype data from the entire genome, while MAS targets specific regions (Shamshad & Sharma, 2018).

Given that the model will learn from the TP, optimizing it is key for building/developing GS-based models that can reliably predict the performance of untested individuals. GS involves three key components: selecting a suitable reference population, choosing an appropriate genotyping method, and developing a predictive statistical model. Selecting a reference population for GS requires identifying a diverse group of individuals or lines for which genotypic data and phenotypic data are available and represents the target breeding population and capture a wide range of genetic variation to ensure accurate predictions and maximize the effectiveness of GS models (Berro et al., 2019). The genome-wide and phenotypic data should also ensure precise and consistent trait measurements (Yang et al., 2020). A suitable genotyping method must enable high-throughput analysis of genetic markers spread throughout the genome (Meuwissen et al., 2016). Additionally, a reliable/robust model is essential to accurately capture the relationship between genetic markers and the target phenotypes (Crossa et al., 2017). Below are some strategies for optimizing training sets depending on the specific circumstances of the problem being solved:

2.1 Untargeted vs Targeted Optimization

Untargeted optimization involves a generalized approach to improving the accuracy of GPs without focusing on a specific target population. Key aspects include selecting a random TP, utilizing all available genetic markers for prediction regardless of their relevance to specific traits, and applying general-purpose models. Although this approach may be less precise for specific traits, it offers broad applicability and captures overall genetic variance. It is particularly beneficial in early-stage breeding programs, also when knowledge of trait architecture is limited, and for maintaining maximum genomic variability/ diversity among selected breeding lines and populations (Fernández-González et al., 2023).

Targeted optimization can be applied when the test set or target population is already known, i.e., the genomic information for the unphenotyped individuals is available prior to training the GS model. By leveraging the genomic information in the said test set, it is possible to tailor the TP that will be used to train optimal models for the specific prediction task. Such a targeted approach maximizes genetic gains while minimizing costs and time (Alemu et al., 2024; Fernández-González et al., 2023). Targeted optimization methods outperform untargeted methods, especially in low heritability scenarios. A training population comprising approximately 50–55% of the total dataset is typically enough to achieve high prediction accuracy (Fernández-González et al., 2023).

2.2 Diversity and Similarity

Optimizing the training set requires careful evaluation and balance of genetic diversity, similarity or representativeness to the testing population, and phenotypic variation (Rutkoski et al., 2016; Sørensen et al., 2023). High similarity could limit the model's ability to generalize, while too much diversity can decrease prediction accuracy for specific populations. Additionally, population size is also a key factor in optimizing TPs (Sørensen et al., 2023). Strategies such as prediction core, which emphasizes genetic similarity and diversity, have shown to enhance prediction abilities across various phenotypes in wheat and oat populations (Sørensen et al., 2022).

Lorenz et al. (2012) investigated different strategies for developing populations to assess resistance to Fusarium head blight and reduce deoxynivalenol (DON) accumulation in grain. They applied a cross-validation technique that partitioned training and validation sets based on both genotype and environment, while utilizing genome-wide marker data to represent genomic relationships among individuals. This approach ensured sufficient linkage disequilibrium between markers and QTLs, enhancing prediction accuracy—even for candidates with more distant genetic relationships to the training population. Nonetheless, certain limitations were noted, such as the reliance on relatively high marker densities in simulations. Additionally, incorporating germplasm from distinct breeding programs or subpopulations into the selection pool slightly reduced prediction accuracy. In contrast, Persa et al. (2023) adopted a different method from the traditional random sampling in soybean populations, known as the Super Saturated Design (SSD). This technique leverages genomic data to reflect the genetic diversity among recombinant inbred lines (RILs). SSD selects a fixed number of genotypes from a larger pool, aiming to either maximize (SSD.max) or minimize (SSD.min) genetic diversity within the training set. When training set size was held constant, both SSD.max and SSD.min approaches led to a reduction in prediction accuracy compared to random sampling.

Several reports have shown that optimizing TPs can significantly reduce phenotyping costs for complex traits, as demonstrated in wheat for the gliadin-to-glutenin ratio (Lemeunier et al., 2022). However, it is key to address potential challenges, such as high computational demands and the need for diverse genetic backgrounds to ensure reliable predictions across various environments. Some further research in TP optimization includes the use of alternative optimization criteria: while the accuracy between predicted GEBVs and observed phenotypic values is considered the standard to establish GS model performance, alternatives such as the Augmented Expected Improvement

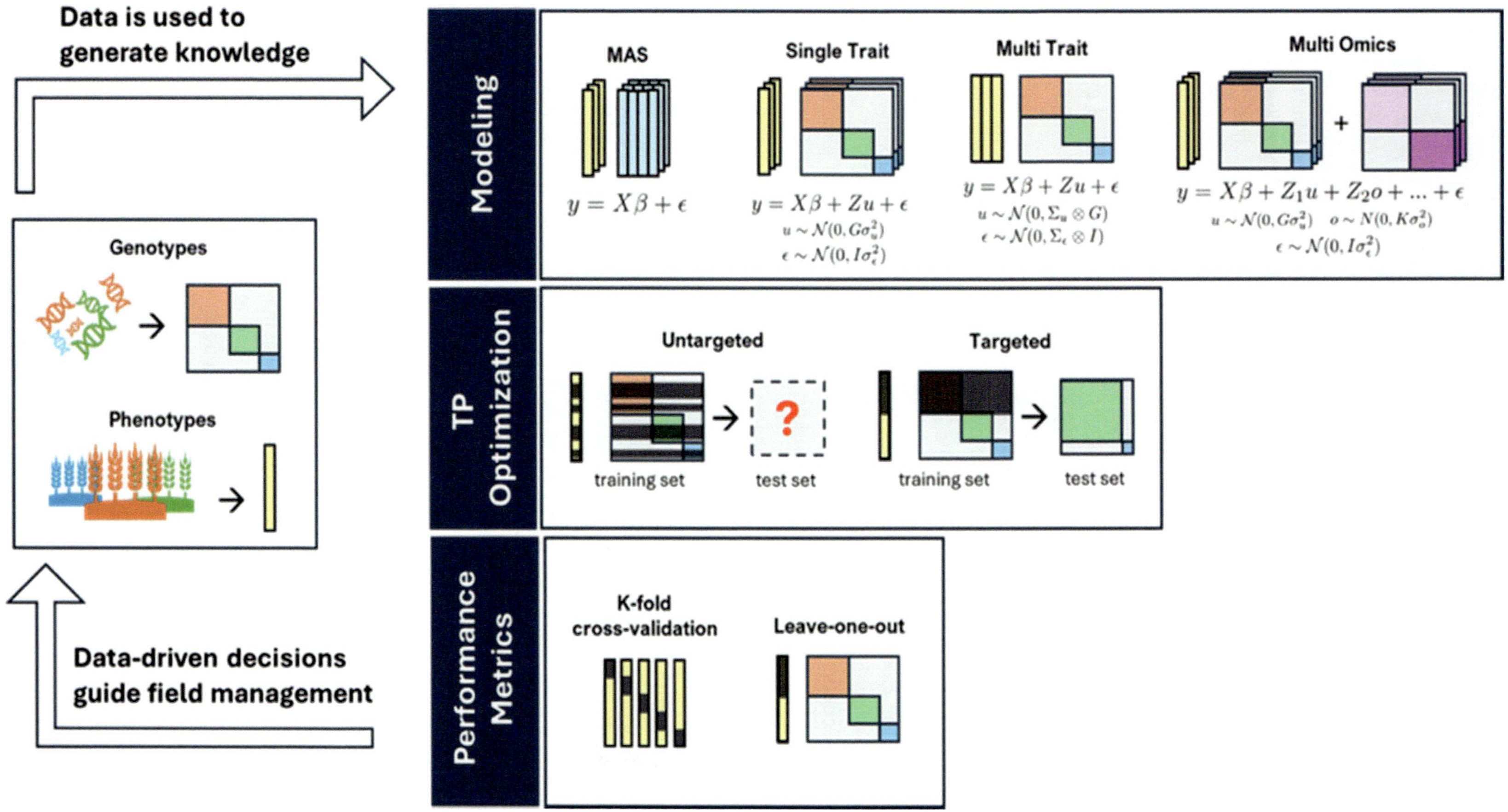

Fig. 1 Conceptual Framework Highlighting Key Parameters and Influencing Factors in Genomic Selection

(AEI), which was implemented using Bayesian optimization methods in Tu & Liao (2024) have been shown to enhance correlations among estimated and true breeding values with a competitive performance in real datasets.

2.3 Advantages of Genomic Selection

1. **Accelerated Breeding Cycles:** GS significantly reduces the time required for breeding by enabling rapid identification of high-performing genotypes (Veerendra Kumar et al., 2024). Shortened breeding cycles and significantly increased yield were demonstrated in soybean where haplotype-based GS improved prediction accuracy by 7% (Yoosefzadeh-Najafabadi et al., 2022).
2. **Increased Genetic Gain:** GS accurately predicts an individual's genetic potential, enabling the rapid identification of superior genotypes. This accelerates genetic progress and increases overall genetic gain, particularly when applied during the initial/early phases of a breeding program (Medina et al., 2021). For instance, in soybean breeding, optimal cross prediction methods were developed to identify the best parental combinations, enhancing genetic gain and breeding efficiency (Miller et al., 2023).
3. **Reduced Phenotyping Costs:** GS reduces the costs associated with extensive phenotyping by relying solely on genotype data and by also shortening the breeding cycle duration in comparison to that of traditional breeding methods (Budhlakoti et al., 2022; Heffner et al., 2010).
4. **Selection for Complex Traits:** GS models are more robust than other methods when selecting for low-heritability and complex traits (Budhlakoti et al., 2022).
5. **Broader Genetic Diversity:** GS can incorporate a wider range of genetic variation into breeding programs using genome-wide markers that lead to more robust and adaptable varieties (Spindel et al., 2016).

2.4 Improvements, Challenges, and Future Directions

1. **Integration with MAS:** Combining GS with MAS allows for more precise selection of traits, improving the overall effectiveness of breeding programs (Verma et al., 2024).
2. **Enhanced Prediction Models:** Advanced statistical approaches, such as Bayesian methods and ML techniques, have increased the accuracy of breeding value predictions, especially for complex traits (Ramadoss et al., 2024). Furthermore, multivariate genomic selection models have demonstrated improved prediction accuracy for correlated traits, resulting in enhanced overall yield performance in crops like soft red winter wheat (Boyles et al., 2023).
3. **Data Limitations:** GS requires high-quality data; however, limitations such as small or unrepresentative TPs, limited genome-wide marker coverage, inaccurate or inconsistent phenotypic data, uneven trait distribution, and insufficient environmental data can reduce prediction accuracy.

Recent advancements in optimal cross-prediction methods are transforming modern breeding strategies. These methods employ genomic data to identify and predict the most promising parental line combinations, enabling breeders to design crosses that maximize genetic gain. Cross-prediction in GS uses phenotypic data and genomic data from one of the TPs sets to assess or estimate the performance of another population with related genetic makeup (Jarquin et al., 2014; Pérez-Rodríguez et al., 2015). By simulating potential offspring outcomes, breeders can prioritize crosses that enhance critical traits, including higher yield, disease and insect, and improved adaptation to environmental stresses. This strategy shortens the breeding cycle while making resource use more efficient, ensuring that breeding efforts focus on achieving greater genetic and phenotypic improvements (Miller et al., 2023). GS potential lies in its ability to harness genomic data to make more accurate predictions about genetic merit, helping breeders focus on the most promising individuals and combinations. Consequently, this enables breeders to accelerate the development of cultivars with favorable characteristics.

Table 1 List of Topmost Implemented Methods in Genomic Selection.

Prediction Methods	*Application*	*Reference*
Best Linear Unbiased Prediction (BLUP)	It estimates genetic values by considering both fixed and random effects present in the dataset. Fixed effects refer to consistent factors across the entire population—such as location, year, or management practices—while random effects capture the genetic variation among individuals that contributes to differences in phenotype. BLUP is especially effective for evaluating genetic values of complex traits governed by numerous small-effect genes.	Badu-Apraku et al. (2019)
Genomic BLUP (G-BLUP) and Ridge Regression Best Linear Unbiased Prediction (RR-BLUP)	G-BLUP and RR-BLUP are considered equivalent approaches, with G-BLUP incorporating genomic data using a genomic relationship matrix. It operates under the assumption that all markers have equal contribution to the trait being analyzed. In plant breeding, G-BLUP is widely applied to estimate genetic merit using genome-wide SNP information. Whereas RR-BLUP is a variant of G-BLUP where ridge regression is applied to the BLUP framework to handle multicollinearity (correlations among predictors) in genomic data. RR-BLUP effectively shrinks the marker effects to provide better predictions. Commonly used in plant breeding when many small-effect markers are expected to influence a trait.	Clark & van der Werf (2013); Ben Hassen et al. (2017); Sinha et al. (2023); Ogutu et al. (2012); Van Raden (2008)
LASSO (Least Absolute Shrinkage and Selection Operator)	Ability to handle large datasets with many predictors, such as those encountered when dealing with genetic markers like SNPs. Not all genetic markers contribute equally to a target trait (e.g., yield, disease resistance). LASSO aids in selecting a subset of genetic markers that best predict the target trait, effectively serving as a feature selection method.	Tibshirani (1996); Xu et al. (2018); Rio et al. (2018)
Elastic Net	It is a linear regression approach that integrates both L1 (lasso) and L2 (ridge) regularization techniques. This method is particularly effective for managing correlated markers and is well-suited for datasets with many predictors. Applied in situations where markers are expected to be correlated, such as in genomic regions with linkage disequilibrium.	Zhang et al. (2022)
Bayes A	Assumes that the effects of genetic markers are normally distributed.	Rio et al. (2018)
Bayes B	Assumes some genetic markers have no effect.	Ben Hassen et al. (2017); Rio et al. (2018)
Bayes Cπ	Introduces a probability that a marker has a non-zero effect.	Rio et al. (2018)
Bayesian LASSO	Uses L1 regularization, as in LASSO regression, to reduce some marker effects to zero. This approach is particularly advantageous when the number of markers is greater than the number of observations.	Grenier et al. (2015)

Contd.

Table 1 *Contd.*

Prediction Methods	*Application*	*Reference*
Kernel-based methods such as Reproducing Kernel Hilbert Space (RKHS)	A non-parametric method that employs kernel functions to capture complex associations between genetic markers and phenotypic traits. RKHS is especially useful for modeling non linear effects and interactions (epistasis) among markers, which are often overlooked by conventional linear models.	Paulsen & Raghupathi (2016); Ben Hassen et al. (2017); Rio et al., (2018); Sinha et al. (2023)
Random Forests	A machine learning technique that constructs an ensemble of decision trees to predict traits from genomic data. Each tree is trained on a randomly selected subset of markers, and the overall prediction is derived by averaging the outputs of all trees. This method effectively captures non linear patterns and interactions among genetic markers, making it well-suited for complex traits.	Breiman (2001); Nicolaou & Constandinou (2016); Sinha et al. (2023)
Support Vector Machines (SVMs)	A supervised learning algorithm that identifies the optimal hyperplane to separate data points in high-dimensional space. In genomic selection, SVMs can classify individuals based on genetic marker profiles and are particularly useful for binary traits, such as the presence or absence of disease, or for distinguishing between distinct genetic classes.	Xu et al. (2018)
Artificial Neural Networks (ANN)	Computational models inspired by the structure and function of the human brain, capable of learning intricate patterns within data. ANNs are used to model complex, non linear relationships between genetic markers and traits and are increasingly applied in genomic prediction, especially where interactions among markers play a significant role.	Rosado et al. (2020); Coelho de Sousa et al. (2022)

3. Advances in Genomic Selection Models

The concept of GS emerged in the early 2000s, with Bernardo (1994) being the first to propose the concept or the idea. At that time, genomic data were limited, and Restriction Fragment Length Polymorphism (RFLP) markers were used to predict hybrid crosses in maize. Later, Meuwissen et al. (2001) highlighted the issue of high dimensionality, which occurs when the number of genomic variables exceeds the available data points for model training. To overcome this challenge, they introduced Bayesian models capable of accurately estimating the total breeding value of individuals, even with sparse phenotypic information. This innovation led to the widespread adoption of genomic selection in plant breeding programs.

Identifying quantitative traits markers in MAS is challenging because all effects must be identified and accounted for. In contrast, the foundation of GS is to estimate all markers concurrently, regardless of their individual significance, to capture the full spectrum of genetic effects, genomic selection has the potential to accelerate genetic gains and shorten breeding cycles for complex traits. As noted earlier, the accuracy of GS models relies on multiple factors that can be integrated into the model, making the choice of an appropriate model essential (Bhat et al., 2016; Meuwissen et al., 2001).

GS models play a crucial role in speeding up genetic improvement and lowering breeding costs by predicting the genetic potential, or breeding values, of individuals (Meuwissen et al., 2001). A diverse set of models are available for GS, allowing breeders to choose based on the complexity of genetic architecture, computational resources, and prediction accuracy needs (Spindel & McCouch, 2016). GS models are categorized into parametric models (including penalized and Bayesian approaches), non-parametric models, semi-parametric models, and hybrid models.

Parametric models evolved as an improvement on the over-parameterized simple linear model (Meuwissen et al., 2001; Lorenz et al., 2009). Advantage of parametric design is that because all of the selection rules are hard coded into the design, a variable may be changed to see the effects (Howard et al., 2014; Jeon et al, 2023). Penalized parametric models in genomic selection are statistical approaches that apply regularization techniques to handle high-dimensional genomic datasets, where the number of markers often surpasses the number of individuals. These models aim to predict breeding values or phenotypes while reducing the risks of overfitting and multicollinearity. Prominent penalized models include RR-BLUP, Lasso Regression, Elastic Net, SCAD (Smoothly Clipped Absolute Deviation), and MCP (Minimax Concave Penalty) (Gianola et al., 2009; Fan & Li, 2001; Heslot et al., 2012; Meuwissen et al., 2001; Zhang et al., 2010).

Bayesian parametric models utilize Bayesian statistical frameworks to estimate genetic effects and predict breeding values by incorporating prior distributions on marker effects, which are then updated with observed data to generate posterior distributions. These models are effective in analyzing complex, high-dimensional genetic data and can integrate pedigree information. Examples of Bayesian models include BayesA, BayesB, BayesCπ, Bayesian LASSO, BayesR, and Bayesian Ridge Regression (Meuwissen et al., 2001; Habier et al., 2009; de los Campos et al., 2009; Erbe et al., 2012; Tibshirani, 1996).

Both parametric and nonparametric methods predict phenotypes by analyzing historical datasets that include genotypic and phenotypic information (Howard et al., 2014). Nonparametric models do not rely on strict assumptions about the underlying genetic architecture or the precise relationships between genotypes and phenotypes. Being data-driven, these models can accommodate diverse genetic architectures and are capable of capturing complex, non-linear interactions such as dominance and epistasis. Recent research indicates that parametric methods often struggle to accurately predict phenotypes for traits governed by epistatic genetic architectures (Howard et al., 2014), whereas nonparametric approaches perform well in such cases. Examples of nonparametric models include the Nadaraya-Watson estimator (NWE), k-Nearest Neighbors (k-NN), random forests (RF), deep learning (DL) models, gradient boosting machines, and Gaussian kernel methods (Gianola & van Kaam, 2008; Heslot et al., 2012; Montesinos-López et al., 2019; Long et al., 2011; Crossa et al., 2010; Pérez & de los Campos, 2014; Azodi et al., 2019).

Semi-parametric models in GS combine parametric with non-parametric components to capture additive genetic effects alongside more complex patterns like dominance, epistasis, and genotype-by-environment interactions. As they use both linear and non linear components, they provide greater flexibility compared to purely parametric methods. They are very well-suitable for traits influenced by non linear genetic architectures. Some examples include RKHS (Reproducing Kernel Hilbert Space), extended GBLUP, Bayesian kernel methods, neural networks (NN) and, support vector machine (SVM) regression (Crossa et al., 2010; de los Campos et al., 2010; Gianola & van Kaam 2008; Pérez & de los Campos, 2014).

Recent advances in genomic profiling and phenotyping have significantly accelerated the adoption of genomic selection (GS) in plant breeding (Crossa et al., 2017; Varshney, 2021). The growing complexity of data generated from high-throughput genotyping, phenotyping, and multi-omics platforms calls for the integration of ML approaches to enhance predictive accuracy (Varshney, 2021). The pioneering application of ML in plant breeding was reported by Heffner et al. (2009), who employed random forest algorithms to capture non-additive genetic effects. They compared these results with traditional approaches such as Bayesian Lasso and Bayesian shrinkage regression, demonstrating the potential of ML to improve GS model performance. Since then, ML has increasingly attracted attention in breeding research. For instance, Ornella et al. (2012) showcased some of the earliest uses of ML techniques for genomic prediction in plants, emphasizing their value in enhancing prediction models for breeding programs. In this chapter, we provide an overview of the applications of ML and DL models in genomic selection and prediction, highlighting their role in protecting commercial crops and promoting sustainable agriculture and food security.

4. Use of GS and ML for Disease Resistance in Plant Breeding

Detecting and diagnosing diseases and pests is key in agriculture due to the food security threats. Early detection enables farmers to implement preventive measures, stopping diseases from spreading to other parts of the plant. Traditional diagnoses are often inaccurate, labor-intensive, and unreliable. It's crucial to develop an efficient method for accurate classification to tackle these issues. Table 1 showcases additional advancements in this field.

Once disease is detected, it is typically phenotyped using incidence (percentage of affected plants) and severity (level of damage on individual plants) (Mesterhazy, 1999). To capture phenotypic and genetic variation, repeated measurements can be taken over time, but integrating these into a single phenotypic value is challenging and strategies such as Area Under the Disease Progress Curve (AUDPC) are commonly utilized (Jeger & Viljanen-Rollinson, 2001; Simko & Piepho, 2012). Garcia-Abadillo et al. (2023) found that alternative methods, such as time to reach 50% of maximum severity or using the most variable measurement, were more predictive using GS for Fusarium resistance in winter wheat, especially in scenarios with unbalanced measurements.

Cassava is a clonally propagated food crop, and the breeding strategy relies on the phenotypic characterization of clonally propagated mature plants. Typically, selection cycles last three to six years before evaluating interesting genotypes for variety release. Genomic selection has recently been applied in cassava breeding to reduce breeding cycle duration. Wolfe et al. (2017) evaluated the accuracy of seven different prediction models across seven traits using three prediction scenarios: within-population cross-validation, cross-population prediction, and cross-generation prediction. Among the models tested, the ML-based Random Forest (RF) method was compared to various Bayesian approaches. RF regression, which utilizes marker data, is known for its ability to capture epistatic interactions and accurately estimate total genetic values. In this study, RF demonstrated the highest prediction accuracy for cassava mosaic disease severity at a single site across multiple breeding programs. Although these findings were preliminary, Wolfe et al. (2017) highlighted the potential advantages of the RF approach in cassava genomic selection.

Detecting pests and diseases in plants is a difficult task that often leads to ineffective control strategies and lower yields. AI technology could enable breeders to precisely identify and categorize plant diseases, making it easier to select and develop disease-resistant plant varieties. These resistant crops support sustainable agriculture by reducing the reliance on chemical pesticides and boosting yields and food security. Therefore, advancements in AI-driven plant disease detection have significant implications for breeding efforts aimed at creating high-yielding resilient varieties (Wang et al., 2023).

Ullah et al. (2017) applied SVM classification models to detect thrips, a major pest affecting strawberry greenhouses. One of the primary challenges was accurately identifying thrips within images of the crop canopy. The researchers developed a novel image-processing method to locate these pests on strawberry plants. The SVM model, utilizing different kernel functions, was employed to classify parasites and detect thrips. Key features included the ratio of major to minor diameter as a regional index, along with color indices based on Hue, Saturation, and Intensity. Model performance was evaluated using metrics such as mean square error (MSE), root mean square error (RMSE), mean absolute error (MAE), and mean percent error (MPE). The results showed excellent classification accuracy, with a MPE below 2.25%.

5. Leveraging Artificial Intelligence Tools for Phenomics Data Generation and Genomic Model Training

Artificial Intelligence (AI) has revolutionized many aspects of today's science. Recently, AI has started gaining popularity in GS. GS relies on the integration of phenomics data with genomic markers data to predict breeding values. AI-driven technologies can be leveraged to enhance this process. AI

Table 2 Machine Learning Algorithms and Models Used in Plant Breeding.

Model	*Crop*	*Application*	*References*
DeepPestNet framework	Tea	Pest identification	Ullah et al. (2022)
Support Vector Machine	Strawberry	Pest identification	Brahimi et al. (2017)
Convolutional Neural Networks	Rice	Diseases and pest identification	Rahman et al. (2020)
Convolutional Neural Networks	Any plant	Diseases identification	Barbedo (2019)
Neural Network	Rice	Diseases identification	Bhagawati et al. (2015)
Convolutional Neural Networks	Tomato	Diseases identification	Brahimi et al. (2017)
Convolutional Neural Networks	Olive	Diseases identification	Cruz et al. (2017)
Convolutional Neural Networks	Maize	Diseases identification	DeChan at al. (2017)
Convolutional Neural Networks	25 different plants	Diseases identification	Ferentinos (2018)
Convolutional Neural Network, Fully Convolutional Network, Single Shot Multibox Detector	Tomato	Diseases and pest identification	Fuentes et al. (2017)
Deep Convolutional Neural Network	Rice	Diseases and pest identification	Li et al. (2020)
CFNet- VoV-GCSP -LSKNet-YOLOv8s	Cotton	Diseases and pest identification	Li et al. (2024)
Deep Convolutional Neural Network	Tomato	Diseases and pest identification	Liu & Wang (2021)
Convolutional Neural Network	Citrus	Pest identification	Khanramaki et al.)2021)

technologies and tools provide accurate data and high-throughput measurements of complex traits. Using AI, phenomics, and genomics, breeders can select superior genotypes efficiently.

GS approaches rely on extensive and diverse biological and environmental datasets to enhance predictions of phenotypic responses to environmental factors and genomic signals/pulses. González et al. (2024) developed the Characterization and Integration of Driven Omics (CHiDO) platform, a user-friendly, no-code tool designed to seamlessly combine large-scale omics datasets for effective interaction modeling. Future developments of CHiDO will include capabilities for choosing among various AI/ML algorithms. Ongoing assessments are exploring the effectiveness of deep learning and Random Forest methods in capturing genotype-by-environment (G×E) interactions (Crossa et al., 2019). Overall, CHiDO marks a significant step forward in computational solutions for agricultural research.

Li et al. (2024) advanced the YOLO (You Only Look Once) framework of Jiang et al. (2022), a widely used system in agriculture for its speed, multi-scale adaptability, automation, and efficient data handling. Their upgraded architecture, termed CFNet-VoV-GCSP-LSKNet-YOLOv8s, incorporates enhancements to the CFNet module, C2F modules in the YOLOv8s head, the LSKNet attention mechanism, and the XIoU loss function. These refinements improve multi-scale object feature integration, optimize accuracy, lower computational demands, and accelerate model convergence. Additionally, the model employs a multi-channel scale attention mechanism designed to better capture features associated with plant diseases. Results demonstrate that CFNet-VoV-GCSP-LSKNet-YOLOv8s can accurately detect cotton pests and diseases in complex environments, with exceptional performance in identifying cotton leaf spot disease, making it a strong tool for pest and disease surveillance and management.

6. Single Trait vs Multi Traits Selection

Recently, efforts have been made to enhance the accuracy of GS models for both single-trait and multi-trait approaches, such as multivariate analyses. Single-trait models are designed to predict the value of one continuous or categorical phenotype in a test dataset. In contrast, multi-trait models are developed to predict two or more traits at the same time (Montesinos-López et al., 2021). The transition from single-trait to multi-trait linear mixed models, which estimate and utilize trait correlations to compute BLUPs of genetic value, is well established (Henderson & Quaas, 1976). The genetic correlation between traits forms the foundation for the advantages of multivariate analysis in GS. The higher the absolute genetic correlation between traits, the greater the benefit of using multivariate analysis (Montesinos-López et al., 2021). Parameter estimates become more precise when accounting for genetic and residual correlations between traits and environments. Additionally, multi-trait models enhance indirect selection by increasing the accuracy of genetic correlation estimates between traits.

In plant breeding, gene-level and genome-wide association mapping are extensively applied to identify natural genetic variants that influence both quantitative and qualitative traits. Numerous association panels have been developed, each comprising diverse plant genotypes from different crop species. These panels allow researchers to investigate a broad spectrum of traits across multiple environments, offering opportunities to examine pleiotropy—where a single genetic locus influences multiple phenotypes—and genotype-by-environment interactions, in which an allele's effect on a phenotype varies depending on environmental conditions (Mural et al., 2022).

Recent advances in statistical approaches for multi-trait selection have produced methods that are broadly applicable, computationally feasible, and capable of analyzing genomic datasets containing numerous effects across varied conditions. Unlike simpler models that assume shared or condition-specific effects, these methods account for arbitrary correlations in effect sizes, thereby increasing statistical power, improving effect estimates, and enabling more nuanced quantification of effect-size heterogeneity (Urbut et al., 2019).

In GS, statistical models are essential for distinguishing true associations from spurious signals caused by population structure and linkage disequilibrium (LD). Compared with single-trait models, multi-trait genomic models generally achieve higher prediction accuracy because they incorporate genetic correlations, which enhances predictive ability (Guo et al., 2014; Huang et al., 2019; Merrick et al., 2022). By leveraging correlations both among genotypes and between traits, multi-trait models capture complex inter-trait relationships more effectively. The extent of genetic correlation between traits is central to the benefits of multivariate analysis: the stronger the correlation, the greater the advantage. Such models also yield more precise parameter estimates by considering both genetic and residual correlations across traits and environments, and they improve indirect selection by refining genetic correlation estimates (Runcie et al., 2021). These advantages are especially pronounced for traits with low heritability or sparse phenotypic data.

Further, multi-locus models now allow the simultaneous testing of multiple markers, expanding the scope of genetic analysis (Guo et al., 2014; Huang et al., 2019; Merrick et al., 2022; Liu et al., 2016). Recommendations derived from multi-trait analyses are often more robust than those from single-trait approaches, particularly when traits exhibit strong correlations (Sampaio et al., 2023). However, multi-trait models have certain drawbacks: they are computationally more demanding, may require consensus across trait-specific results (Toomey et al., 2020; Xu et al., 2021), and show no clear advantage over single-trait models when phenotypic data are missing for some traits (Persa et al., 2020).

7. Applications of Genomic Selection, Prediction, and Machine Learning in Sustainable Agriculture

Sustainable agriculture increasingly benefits from cutting-edge genomic and computational tools that accelerate breeding and improve crop management. Techniques such as GS, GP, and ML are transforming how we address challenges related to yield, disease resistance, and environmental adaptability.

One notable example involves the use of ML algorithms to forecast crop yields. Jeong and colleagues (2016) demonstrated that RF models outperform traditional linear regressions in predicting yields of key crops like wheat, maize, and potato across diverse regions. Their results highlighted the robustness and adaptability of ML in handling complex agricultural data, though prediction accuracy may diminish under extreme or novel conditions.

In precision agriculture, integrating diverse data sources has proven vital. Pantazi et al. (2015) combined soil sensor outputs and satellite-derived vegetation indices to predict wheat yield variability within fields. Among several ML approaches tested, supervised Kohonen Networks achieved the highest accuracy, particularly in identifying low-yield zones. This underscores ML's ability to capture intricate soil-crop interactions, enhancing resource-efficient farming.

Advancements in DL have also improved genomic prediction accuracy. Montesinos-López et al. (2023) compared deep NNs with traditional GP models using multi-environment wheat datasets. Deep learning showed superior performance in certain scenarios, especially when predicting traits under varying irrigation and drought conditions. This illustrates the promise of integrating diverse genomic and environmental data to refine breeding decisions.

Beyond breeding, ML supports sustainable crop protection. Cassava, a vital carbohydrate crop, faces viral diseases threatening food security in sub-Saharan Africa. Ramcharan et al. (2017) applied deep convolutional neural networks trained on field images to detect multiple cassava diseases and pest damages with over 90% accuracy. The deployment of such models via mobile applications offers smallholder farmers accessible, affordable tools for early disease detection, enabling timely interventions.

Genomic selection has also accelerated the breeding of disease-resistant wheat varieties, particularly against rust diseases. By combining genotypic and phenotypic data, breeders can predict resistance levels without extensive field trials, speeding up the release of resilient cultivars and reducing reliance on fungicides.

Additionally, ML models have been instrumental in forecasting pest outbreaks such as fall armyworm infestations in maize. By analyzing environmental and historical pest data, these models provide early warnings, allowing farmers to apply targeted control measures, thereby reducing crop losses and minimizing pesticide use. This technology is actively used in several African countries grappling with this invasive pest.

Together, these applications of GS, GP, and ML illustrate how integrating genomics, and computational intelligence supports sustainable agriculture by enhancing yield prediction, accelerating breeding for stress resistance, and improving real-time crop protection strategies.

8. Conclusion and Future Perspective

GS can be viewed as an advanced form of MAS in plant breeding, where the focus shifts from targeting specific QTLs to utilizing genome-wide SNP marker profiles in combination with phenotype data. This approach enables the rapid and accurate identification of superior crop lines. Modern GS achieves even greater genetic gains by incorporating ML-based predictive modeling into MAS, which shortens the crop breeding cycle. These advancements are crucial for addressing the increasing challenges posed by climate change, resource limitations, and expanding world population. By harnessing high-throughput genotyping, sophisticated computational models, and large-scale phenotypic data, GS provides a sustainable approach to protecting commercial crops and ensuring long-term food security (Alemu et al., 2024; Budhlakoti et al., 2022; Sinha et al., 2023).

Despite more than two decades of research in GS, there remains significant room for improvement in this field. Improving the imputation of missing genotypic data is required for developing more accurate methods, which will enhance dataset quality and increase the predictive power of models. Challenges such as complex population structures, low marker density, and computational complexity/demands still pose difficulties in imputation. To potentially address these issues, promising approaches include the use of advanced statistical models (parametric) and AI methods (ML, DL), population-specific imputation strategies, and the integration of other relevant biological datasets or data types (Money et al., 2015; Li et al., 2009; Wang et al., 2022).

Genotype-by-environment (G × E) interactions are a key consideration in GS as they reflect how genotypes vary in performance across different environmental conditions. In plant breeding, addressing G × E is essential for making reliable predictions. Nonetheless, these interactions introduce complexity into the selection process, making it more challenging. As a result, successful GS strategies must incorporate methods to account for G × E effects. Several key approaches can enhance statistical models for G × E interactions, leading to more informed breeding decisions, broader environmental representation, and improved crop adaptation to diverse and changing environments. These include developing advanced statistical models that better capture G × E effects and conducting multi-environment trials (METs) to gather more accurate data. Incorporating environmental covariates, such as climate (Jarquin et al., 2014) and soil (Vieira et al., 2022) information, into prediction models is crucial, as is integrating G×E models with environmental GP models to simulate performance under various conditions. Leveraging ML and AI can help detect complex, non linear G × E interactions that traditional models may overlook, thereby improving prediction accuracy. Additionally, dynamic and adaptive models (crop growth models) that adjust over time with new data will enhance long-term predictions (Technow et al., 2015). Cross-validation methods tailored to G × E interactions can enhance the robustness of predictive models. Incorporating temporal environmental information alongside high-throughput phenotyping technologies can further enrich the data used for analysis. Moreover, improved modeling of complex genetic architectures, including gene-gene interactions, will lead to more accurate representation of G × E effects. Collectively, the discussed approaches strengthen the ability to incorporate G × E interactions into GS, resulting in more resilient and productive crop varieties (Bomma et al., 2024; Fradgley et al., 2022; Miao et al., 2024).

An emerging direction for enhancing GS involves integrating multi-omics data—including epigenomics, genomics, metabolomics, phenomics, proteomics, and transcriptomics—which can strengthen the robustness and practical utility of prediction models (González et al., 2024). Incorporating epigenetic information and haplotype data into GS frameworks holds significant promise for advancing breeding programs. However, challenges remain, such as obtaining high-quality epigenetic datasets across diverse populations and environments, managing computational demands, and interpreting complex epigenomic signals. Key approaches include the integration of epigenetic markers like DNA methylation and histone modifications, while considering their dynamic changes in response to environmental stimuli and developmental phases. Epigenome-Wide Association Studies (EWAS) offer a valuable tool to pinpoint epigenetic variants associated with complex traits, and combining multiple omics layers—such as transcriptomics and metabolomics—can yield a more holistic view of biological mechanisms. The use of haplotypes, representing linked groups of genetic variants, alongside improved phasing and imputation methods, is critical for capturing intricate genetic variation. Additionally, accounting for population-specific haplotype patterns and enhancing the efficiency and affordability of epigenetic data collection will further refine prediction models. Collectively, these improvements will deepen our understanding of trait heritability, gene regulation, and overall crop performance, ultimately driving more effective breeding strategies (Fallet et al., 2023; Kakoulidou et al., 2021; Rozek et al., 2014).

In conclusion, GS tools have demonstrated accelerated genetic gains in crop breeding programs. However, to maximize its efficiency, it is crucial to integrate all factors that influence models' performance including the implementation of appropriate cross-validation techniques specifically designed for the problem to solve (e.g. prediction of new traits, environments, genetic backgrounds or treatments and management practices) (Muvunyi et al., 2022; Montesinos-López et al., 2018; Montesinos-López et al., 2021), regularly updating and optimizing training datasets by incorporating new markers and phenotypic information with generations, and carefully interpreting the results in conjugation with domain-experts, ensuring ensuring the full potential of this data-driven, predictive approach (Alemu et al., 2024; Budhlakoti et al., 2022).

References

Alemu, A., Åstrand, J., Montesinos-López, O.A., Isidro y Sánchez, J., Fernández-Gónzalez, J., Tadesse, W., Vetukuri, R.R., Carlsson, A.S., Ceplitis, A., Crossa, J., Ortiz, R. and Chawade, A. (2024). Genomic selection in plant breeding: Key factors shaping two decades of progress. *Molecular Plant*, *17*(4), 552–578. https://doi.org/10.1016/j.molp.2024.03.007.

Arruda, M.P., Lipka, A.E., Brown, P.J., Krill, A.M., Thurber, C., Brown-Guedira, G. and Kolb, F.L. (2016). Comparing genomic selection and marker-assisted selection for Fusarium head blight resistance in wheat (Triticum aestivum L.). *Molecular Breeding*, *36*(7). https://doi.org/10.1007/s11032-016-0508-5.

Azodi, C.B., Bolger, E., McCarren, A., Roantree, M., de los Campos, G. and Shiu, S.H. (2019). Benchmarking machine learning methods for genomic prediction of complex traits. *G3: Genes|Genomes|Genetics*, *9*(11), 3691–3702. https://doi.org/10.1534/g3.119.400498.

Badu-Apraku, B., Talabi, A.O., Fakorede, M.A.B., Fasanmade, Y., Gedil, M., Magorokosho, C. and Asiedu, R. (2019). Yield gains and associated changes in an early yellow bi-parental maize population following genomic selection for Striga resistance and drought tolerance. *BMC Plant Biology*, *19*(1), 129. https://doi.org/10.1186/s12870-019-1740-z.

Barbedo, J.G.A. (2019). Plant disease identification from individual lesions and spots using deep learning. *Biosystems Engineering*, *180*, 96–107. https://doi.org/10.1016/j.biosystemseng.2019.02.002.

Ben Hassen, M., Cao, T.V., Bartholomé, J., Orasen, G., Colombi, C., Rakotomalala, J. and Ahmadi, N. (2017). Rice diversity panel provides accurate genomic predictions for complex traits in the progenies of biparental crosses involving members of the panel. *Theoretical and Applied Genetics*, *131*(2), 417–435. https://doi.org/10.1007/s00122-017-3011-4.

Berro, I., Lado, B., Nalin, R.S., Quincke, M. and Gutiérrez, L. (2019). Training population optimization for genomic selection. *The Plant Genome*, *12*(3), 190028. https://doi.org/10.3835/plantgenome2019.04.0028.

Bhagawati, R., Bhagawati, K., Singh, A.K.K., Nongthombam, R., Sarmah, R. and Bhagawati, G. (2015). Artificial neural network assisted weather-based plant disease forecasting system. *International Journal on Recent and Innovation Trends in Computing and Communication*, *3*(6), 4168–4173. https://doi.org/10.17762/ijritcc.v3i6.4612.

Bhat, J.A., Ali, S., Salgotra, R.K., Mir, Z.A., Dutta, S., Jadon, V. and Prabhu, K.V. (2016). Genomic selection in the era of next generation sequencing for complex traits in plant breeding. *Frontiers in Genetics*, *7*, 221. https://doi.org/10.3389/fgene.2016.00221.

Bomma, N., Shruthi, H.B., Soregaon, C.D., Gaddameedi, A., Suma, K., Pranati, J. and Gangashetty, P.I. (2024). Multi-environment testing for G×E interactions and identification of high-yielding, stable, medium-duration pigeonpea genotypes employing AMMI, GGE biplot, and YREM analyses. *Frontiers in Plant Science*, *15*, 1396826. https://doi.org/10.3389/fpls.2024.1396826.

Boyles, R.E., Ballén-Taborda, C., Brown-Guedira, G., Costa, J., Cowger, C., DeWitt, N. and Winn, Z.J. (2023). Approaching 25 years of progress towards Fusarium head blight resistance in southern soft red winter wheat (Triticum aestivum L.). *Plant Breeding*. https://doi.org/10.1111/pbr.13137

Brahimi, M., Boukhalfa, K. and Moussaoui, A. (2017). Deep learning for tomato diseases: Classification and symptoms visualization. *Applied Artificial Intelligence*, *31*(4), 299–315. https://doi.org/10.1080/08839514.2017.1315516.

Breiman, L. (2001). Random forests. *Machine Learning*, *45*(1), 5–32. https://doi.org/10.1023/a:1010933404324.

Budhlakoti, N., Kushwaha, A.K., Rai, A., Chaturvedi, K.K., Kumar, A., Pradhan, A.K., Kumar, U., Kumar, R.R., Juliana, P., Mishra, D.C. and Kumar, S. (2022). Genomic selection: A tool for accelerating the efficiency of molecular breeding for development of climate-resilient crops. *Frontiers in Genetics*, *13*, 832153. https://doi.org/10.3389/fgene.2022.832153.

Canella Vieira, C., Persa, R., Chen, P. and Jarquin, D. (2022). Incorporation of soil-derived covariates in progeny testing and line selection to enhance genomic prediction accuracy in soybean breeding. *Frontiers in Genetics*, *13*, 905824. https://doi.org/10.3389/fgene.2022.905824.

Clark, S.A. and van der Werf, J. (2013). Genomic best linear unbiased prediction (gBLUP) for the estimation of genomic breeding values. In: *Methods in Molecular Biology*. Springer, pp. 321–330. https://doi.org/10.1007/978-1-62703-447-0_13.

Coelho de Sousa, I., Nascimento, M., de Castro Sant'anna, I., Teixeira Caixeta, E., Ferreira Azevedo, C., Damião Cruz, C. and Vergara Lopes Serão, N. (2022). Marker effects and heritability estimates using additive-dominance genomic architectures artificial neural networks in Coffea canephora. *PLoS One*, *17*(1), e0262055. https://doi.org/10.1371/journal.pone.0262055.

Costa-Neto, G., Crespo-Herrera, L., Fradgley, N., Gardner, K., Bentley, A.R., Dreisigacker, S. and Crossa, J. (2023). Envirome-wide associations enhance multi-year genome-based prediction of historical wheat breeding data. *G3: Genes|Genomes|Genetics*, *13*(2). https://doi.org/10.1093/g3journal/jkac313.

Crossa, J., de los Campos, G., Pérez, P., Gianola, D., Burgueño, J., Araus, J.L. and Reynolds, M.P. (2010). Prediction of genetic values of quantitative traits in plant breeding using pedigree and molecular markers. *Genetics*, *186*(2), 713–724. https://doi.org/10.1534/genetics.110.118521.

Crossa, J., Montesinos-López, O.A., Costa-Neto, G., Vitale, P., Martini, J.W., Runcie, D. and Ortiz, R. (2024). Machine learning algorithms translate big data into predictive breeding accuracy. *Trends in Plant Science*, *30*(2), 167–184. https://doi.org/10.1016/j.tplants.2024.09.011.

Crossa, J., Pérez-Rodríguez, P., Cuevas, J., Montesinos-López, O., Jarquín, D., de los Campos, G. and Varshney, R.K. (2017). Genomic selection in plant breeding: Methods, models, and perspectives. *Trends in Plant Science*, *22*(11), 961–975. https://doi.org/10.1016/j.tplants.2017.08.011.

Cruz, A.C., Luvisi, A., De Bellis, L. and Ampatzidis, Y. (2017). X-FIDO: An effective application for detecting Olive quick decline syndrome with deep learning and data fusion. *Frontiers in Plant Science*, *8*. https://doi.org/10.3389/fpls.2017.01741.

de los Campos, G., Gianola, D., Rosa, G.J., Weigel, K.A. and Crossa, J. (2010). Semi-parametric genomic-enabled prediction of genetic values using reproducing kernel Hilbert spaces methods. *Genetics Research*, *92*(4), 295–308. https://doi.org/10.1017/S0016672310000285.

de los Campos, G., Naya, H., Gianola, D., Crossa, J., Legarra, A., Manfredi, E., Weigel, K. and Cotes, J.M. (2009). Predicting quantitative traits with regression models for dense molecular markers and pedigree. *Genetics*, *182*(1), 375–385.

DeChant, C., Wiesner-Hanks, T., Chen, S., Stewart, E.L., Yosinski, J., Gore, M.A. and Lipson, H. (2017). Automated identification of northern leaf blight-infected maize plants from field imagery using deep learning. *Phytopathology*, *107*(11), 1426–1432. https://doi.org/10.1094/phyto-11-16-0417-r.

Erbe, M., Hayes, B.J., Matukumalli, L.K., Goswami, S., Bowman, P.J., Reich, C.M., Mason, B.A. and Goddard, M.E. (2012). Improving accuracy of genomic predictions within and between dairy cattle breeds with imputed high-density single nucleotide polymorphism panels. *Journal of Dairy Science*, *95*(7), 4114–4129.

Fallet, M., Blanc, M., Di Criscio, M., Antczak, P., Engwall, M., Guerrero Bosagna, C., … and Keiter, S.H. (2023). Present and future challenges for the investigation of transgenerational epigenetic inheritance. *Environment International*, *172*, 107776. https://doi.org/10.1016/j.envint.2023.107776.

Fan, J. and Li, R. (2001). Variable selection via nonconcave penalized likelihood and its oracle properties. *Journal of the American Statistical Association*, *96*(456), 1348–1360.

Ferentinos, K.P. (2018). Deep learning models for plant disease detection and diagnosis. *Computers and Electronics in Agriculture*, *145*, 311–318. https://doi.org/10.1016/j.compag.2018.01.009.

Fernández-González, J., Akdemir, D. and Isidro y Sánchez, J. (2023). A comparison of methods for training population optimization in genomic selection. *Theoretical and Applied Genetics*, *136*(30). https://doi.org/10.1007/s00122-023-04265-6.

Fradgley, N.S., Gardner, K., Kerton, M., Swarbeck, S.M. and Bentley, A.R. (2022). Trade-offs in the genetic control of functional and nutritional quality traits in UK winter wheat. *Heredity*. https://doi.org/10.1038/s41437-022-00503-7

Fuentes, A., Yoon, S., Kim, S. and Park, D. (2017). A robust deep-learning-based detector for real-time tomato plant diseases and pests recognition. *Sensors*, *17*(9), 2022. https://doi.org/10.3390/s17092022.

Garcia-Abadillo, J., Morales, L., Buerstmayr, H., Michel, S., Lillemo, M., Holzapfel, J., ... and Isidro-Sánchez, J. (2023). Alternative scoring methods of Fusarium head blight resistance for genomic assisted breeding. *Frontiers in Plant Science*, *13*, 1057914.

Gianola, D. and van Kaam, J.B. (2008). Reproducing kernel Hilbert spaces regression methods for genomic-assisted prediction of quantitative traits. *Genetics*, *178*(4), 2289–2303. https://doi.org/10.1534/genetics.107.084285.

Gianola, D., Hill, W.G., Manfredi, E. and Fernando, R. (2009). Additive genetic variability and the Bayesian alphabet. *Genetics*, *183*(1), 347–363. https://doi.org/10.1534/genetics.109.103952.

González, F., García-Abadillo, J. and Jarquín, D. (2024). Introducing CHiDO: A no-code genomic prediction software implementation for the characterization and integration of driven omics. *The Plant Genome*, *18*(1). https://doi.org/10.1002/tpg2.20519.

Grenier, C., Cao, T.-V., Ospina, Y., Quintero, C., Châtel, M. H., Tohme, J., … and Ahmadi, N. (2015). Accuracy of genomic selection in a rice synthetic population developed for recurrent selection breeding. *PLoS One*, *10*(8), e0136594. https://doi.org/10.1371/journal.pone.0136594

Guo, G., Zhao, F., Wang, Y., Zhang, Y., Du, L. and Su, G. (2014). Comparison of single-trait and multiple-trait genomic prediction models. *BMC Genetics*, *15*, 30. https://doi.org/10.1186/1471-2156-15-30

Habier, D., Fernando, R.L. and Dekkers, J.C. (2009). The impact of genetic relationship information on genome-assisted breeding values. *Genetics*, *177*(4), 2389–2397.

Heffner, E.L., Lorenz, A.J., Jannink, L. and Sorrells, M.E. (2010). Plant breeding with genomic selection: Gain per unit time and cost. *Crop Science*, *50*(5), 1681–1690. https://doi.org/10.2135/cropsci2009.11.0662

Heffner, E.L., Sorrells, M.E. and Jannink, J.-L. (2009). Genomic selection for crop improvement. *Crop Science*, *49*(1), 1–12. https://doi.org/10.2135/cropsci2008.08.0512.

Henderson, C.R. and Quaas, R.L. (1976). Multiple trait evaluation using relatives' records. *Journal of Animal Science*, *43*(6), 1188–1197. https://doi.org/10.2527/jas1976.4361188x.

Heslot, N., Yang, H.-P., Sorrells, M.E. and Jannink, J.-L. (2012). Genomic selection in plant breeding: A comparison of models. *Crop Science*, *52*(1), 146–160. https://doi.org/10.2135/cropsci2011.06.0297.

Howard, R., Carriquiry, A.L. and Beavis, W.D. (2014). Parametric and nonparametric statistical methods for genomic selection of traits with additive and epistatic genetic architectures. *G3: Genes|Genomes|Genetics*, *4*(6), 1027–1046. https://doi.org/10.1534/g3.114.010298.

Huang, M., Liu, X., Zhou, Y., Summers, R.M. and Zhang, Z. (2019). BLINK: A package for the next level of genome-wide association studies with both individuals and markers in the millions. *GigaScience*, *8*(2). https://doi.org/10.1093/gigascience/giy154.

Jiang, P., Ergu, D., Liu, F., Cai, Y. and Ma, B. (2022). A review of YOLO algorithm developments. *Procedia Computer Science*, 199, 1066–1073. https://doi.org/10.1016/j.procs.2022.01.135

Jarquín, D., Crossa, J., Lacaze, X., Du Cheyron, P., Daucourt, J., Lorgeou, J., Piraux, F., Guerreiro, L., Pérez, P., Calus, M., Burgueño, J. and de los Campos, G. (2014). A reaction norm model for genomic selection using high-dimensional genomic and environmental data. *Theoretical and Applied Genetics*, *127*(3), 595–607. https://doi.org/10.1007/s00122-013-2243-1.

Jeger, M.J. and Viljanen-Rollinson, S.L.H. (2001). The use of the area under the disease-progress curve (AUDPC) to assess quantitative disease resistance in crop cultivars. *Theoretical and Applied Genetics*, *102*, 32–40.

Jeon, D., Kang, Y., Lee, S., Choi, S., Sung, Y., Lee, T.-H. and Kim, C. (2023). Digitalizing breeding in plants: A new trend of next-generation breeding based on genomic prediction. *Frontiers in Plant Science*, *14*, 1092584. https://doi.org/10.3389/fpls.2023.1092584.

Jeong, J.H., Resop, J.P., Mueller, N.D., Fleisher, D.H., Yun, K., Butler, E.E., Timlin, D.J., Shim, K.-M., Gerber, J.S., Reddy, V.R. and Kim, S.-H. (2016). Random forests for global and regional crop yield predictions. *PLoS One*, *11*(6), e0156571. https://doi.org/10.1371/journal.pone.0156571.

Kakoulidou, I., Avramidou, E.V., Baránek, M., Brunel-Muguet, S., Farrona, S., Johannes, F., ... and Maury, S. (2021). Epigenetics for crop improvement in times of global change. *Biology*, *10*(8), 766. https://doi.org/10.3390/biology10080766.

Kumar, K.P. and Walia, P. (2024). Speed breeding in cereal crops: Accelerating genetic improvement for rapid agricultural advancement. *Journal of Experimental Agriculture International*, *46*(6), 465–477. https://doi.org/10.9734/jeai/2024/v46i62498.

Lemeunier, P., Paux, E., Babi, S., Auzanneau, J., Goudemand-Dugué, E., Ravel, C. and Rincent, R. (2022). Training population optimization for genomic selection improves the predictive ability of a costly measure in bread wheat, the gliadin to glutenin ratio. *Euphytica*, *218*(8). https://doi.org/10.1007/s10681-022-03062-4.

Li, D., Wang, R., Xie, C., Liu, L., Zhang, J., Li, R., ... and Liu, W. (2020). A recognition method for rice plant diseases and pests video detection based on deep convolutional neural network. *Sensors*, *20*(3), 578. https://doi.org/10.3390/s20030578.

Li, R., He, Y., Li, Y., Qin, W., Abbas, A., Ji, R., ... and Yang, J. (2024). Identification of cotton pest and disease based on CFNet-VoV-GCSP-LSKNet-YOLOv8s: A new era of precision agriculture. *Frontiers in Plant Science*, *15*, 1348402. https://doi.org/10.3389/fpls.2024.1348402.

Li, Y., Willer, C., Sanna, S. and Abecasis, G. (2009). Genotype imputation. *Annual Review of Genomics and Human Genetics*, *10*, 387–406. https://doi.org/10.1146/annurev.genom.9.081307.164242.

Liu, J. and Wang, X. (2021). Plant diseases and pests detection based on deep learning: A review. *Plant Methods*, *17*(1), 22. https://doi.org/10.1186/s13007-021-00722-9.

Liu, X., Huang, M., Fan, B., Buckler, E.S. and Zhang, Z. (2016). Iterative usage of fixed and random effect models for powerful and efficient genome-wide association studies. *PLoS Genetics*, *12*(2), e1005767. https://doi.org/10.1371/journal.pgen.1005767.

Long, N., Gianola, D., Rosa, G.J.M. and Weigel, K.A. (2011). Machine learning classification procedure for selecting SNPs in genomic selection: Application to early mortality in broilers. *Journal of Animal Breeding and Genetics*, *128*(2), 109–119. https://doi.org/10.1111/j.1439-0388.2010.00904.x.

Lorenz, A.J., Smith, K.P. and Jannink, J.L. (2011). Potential and optimization of genomic selection for crop improvement. *Crop Science*, *51*(2), 491–507.

Medina, C.A., Kaur, H., Ray, I. and Yu, X. (2021). Strategies to increase prediction accuracy in genomic selection of complex traits in alfalfa (Medicago sativa L.). *Cells*, *10*(12), 3372. https://doi.org/10.3390/cells10123372.

Merrick, L.F. and Carter, A.H. (2021). Comparison of genomic selection models for exploring predictive ability of complex traits in breeding programs. *The Plant Genome*, *14*(3), e20158. https://doi.org/10.1002/tpg2.20158.

Merrick, L.F., Burke, A.B., Zhang, Z. and Carter, A.H. (2021). Comparison of single-trait and multi-trait genome-wide association models and inclusion of correlated traits in dissecting genetic architecture of a complex trait in a breeding program. *Frontiers in Plant Science*, *12*, 772907. https://doi.org/10.3389/fpls.2021.772907.

Mesterhazy, A., Bartok, T., Mirocha, C.G. and Komoroczy, R. (1999). Nature of wheat resistance to Fusarium head blight and the role of deoxynivalenol for breeding. *Plant Breeding*, *118*(2), 97–110.

Meuwissen, T.H., Hayes, B.J. and Goddard, M.E. (2001). Prediction of total genetic value using genome-wide dense marker maps. *Genetics*, *157*(4), 1819–1829. https://doi.org/10.1093/genetics/157.4.1819.

Meuwissen, T., Hayes, B. and Goddard, M. (2016). Genomic selection: A paradigm shift in animal breeding. *Animal Frontiers*, *6*(1), 6–14. https://doi.org/10.2527/af.2016-0002.

Miao, J., Wu, Y. and Lu, Q. (2024). Statistical methods for gene-environment interaction analysis. *Wiley Interdisciplinary Reviews: Computational Statistics*, *16*(1). https://doi.org/10.1002/wics.1635.

Miller, M.J., Song, Q., Fallen, B. and Li, Z. (2023). Genomic prediction of optimal cross combinations to accelerate genetic improvement of soybean (Glycine max). *Frontiers in Plant Science*, *14*, 1171135. https://doi.org/10.3389/fpls.2023.1171135.

Money, D., Gardner, K., Migicovsky, Z., Schwaninger, H., Zhong, G.-Y. and Myles, S. (2015). LinkImpute: Fast and accurate genotype imputation for nonmodel organisms. *G3: Genes|Genomes|Genetics*, *5*(11), 2383–2390. https://doi.org/10.1534/g3.115.021667.

Montesinos-López, A., Rivera, C., Pinto, F., Piñera, F., Gonzalez, D., Reynolds, M., Pérez-Rodríguez, P., Li, H., Montesinos-López, O.A. and Crossa, J. (2023). Multimodal deep learning methods enhance genomic prediction of wheat breeding. *G3: Genes|Genomes|Genetics*, *13*(5), jkad045. https://doi.org/10.1093/g3journal/jkad045.

Montesinos-López, A., Runcie, D.E., Ibba, M.I., Pérez-Rodríguez, P., Montesinos-López, O.A., Crespo, L.A., ... and Crossa, J. (2021). Multi-trait genomic-enabled prediction enhances accuracy in multi-year wheat breeding trials. *G3: Genes|Genomes|Genetics*, *11*(10). https://doi.org/10.1093/g3journal/jkab270.

Montesinos-López, O.A., Montesinos-López, A., Crossa, J., Gianola, D., Hernández-Suárez, C.M. and Martín-Vallejo, J. (2019). Benchmarking deep learning, support vector machine, and Bayesian threshold best linear unbiased prediction for predicting ordinal traits in plant breeding. *G3: Genes|Genomes|Genetics*, *9*(2), 601–618. https://doi.org/10.1534/g3.118.200998.

Montesinos-López, O.A., Montesinos-López, A., Crossa, J., Gianola, D., Hernández-Suárez, C.M. and Martín-Vallejo, J. (2018). Multi-trait, multi-environment deep learning modeling for genomic-enabled prediction of plant traits. *G3: Genes|Genomes|Genetics*, *8*(12), 3829–3840. https://doi.org/10.1534/g3.118.200728.

Mural, R.V., Sun, G., Grzybowski, M., Tross, M.C., Jin, H., Smith, C., ... and Schnable, J.C. (2022). Association mapping across a multitude of traits collected in diverse environments in maize. *GigaScience*, *11*. https://doi.org/10.1093/gigascience/giac080.

Muvunyi, B.P., Zou, W., Zhan, J., He, S. and Ye, G. (2022). Multi-trait genomic prediction models enhance predictive ability of grain trace elements in rice. *Frontiers in Genetics*, *13*, 883853. https://doi.org/10.3389/fgene.2022.883853.

Nicolaou, N. and Constandinou, T.G. (2016). A nonlinear causality estimator based on non-parametric multiplicative regression. *Frontiers in Neuroinformatics*, *10*, 19. https://doi.org/10.3389/fninf.2016.00019.

Ogutu, J.O., Schulz-Streeck, T. and Piepho, H.-P. (2012). Genomic selection using regularized linear regression models: Ridge regression, lasso, elastic net, and extensions. *BMC Proceedings*, *6*(Suppl 2), S10. https://doi.org/10.1186/1753-6561-6-S2-S10.

Ornella, L., Pérez, P., Tapia, E., González-Camacho, J.M., Burgueño, J., Zhang, X. ... and Crossa, J. (2014). Genomic-enabled prediction with classification algorithms. *Heredity*, *112*(6), 616–626. https://doi.org/10.1038/hdy.2013.144.

Ortiz, R. (2023). Challenges for crop improvement. *Emerging Topics in Life Sciences*, *7*(2), 197–205. https://doi.org/10.1042/ETLS20230106.

Pantazi, X.E., Moshou, D., Alexandridis, T., Whetton, R.L. and Mouazen, A.M. (2015). Wheat yield prediction using machine learning and advanced sensing techniques. *Computers and Electronics in Agriculture*, *121*, 57–65. https://doi.org/10.1016/j.compag.2015.11.018.

Paulsen, V.I. and Raghupathi, M. (2016). *Cambridge Studies in Advanced Mathematics: An Introduction to the Theory of Reproducing Kernel Hilbert Spaces* (Series No. 152). https://doi.org/10.1017/cbo9781316219232.

Pérez, P. and de los Campos, G. (2014). Genome-wide regression and prediction with the BGLR statistical package. *Genetics*, *198*(2), 483–495. https://doi.org/10.1534/genetics.114.164442

Pérez-Rodríguez, P., Crossa, J., Rutkoski, J., et al. (2015). Single-step genomic and pedigree genotype × environment interaction models for predicting wheat lines in international environments. *The Plant Genome*, *8*(2), e00054.

Persa, R., Bernardeli, A. and Jarquin, D. (2020). Prediction strategies for leveraging information of associated traits under single- and multi-trait approaches in soybeans. *Agriculture*, *10*(8), 308. https://doi.org/10.3390/agriculture10080308

Persa, R., Canella Vieira, C., Rios, E., Messina, C.D., Runcie, D. and Jarquin, D. (2023). Improving predictive ability in sparse testing designs in soybean populations. *Frontiers in Genetics*, *14*, 1269255. https://doi.org/10.3389/fgene.2023.1269255.

Rahman, C.R., Arko, P.S., Ali, M.E., Iqbal Khan, M.A., Apon, S.H., Nowrin, F. and Wasif, A. (2020). Identification and recognition of rice diseases and pests using convolutional neural networks. *Biosystems Engineering*, *194*, 112–120. https://doi.org/10.1016/j.biosystemseng.2020.03.020.

Ramadoss, B.R., Premnath, A., Venkatesan, T. and Thirunavukkarasu, N. (2024). Genomic selection and its application in pearl millet improvement. In: *Pearl Millet in the 21st Century*. Singapore: Springer Verlag, pp. 141–162. https://doi.org/10.1007/978-981-99-5890-0_6.

Ramcharan, A., Baranowski, K., McCloskey, P., Ahmed, B., Legg, J. and Hughes, D.P. (2017). Deep learning for image-based cassava disease detection. *Frontiers in Plant Science*, *8*, 1852. https://doi.org/10.3389/fpls.2017.01852.

Rashid, Z., Singh, P.K., Vemuri, H., Zaidi, P.H., Prasanna, B.M. and Nair, S.K. (2018). Genome-wide association study in Asia-adapted tropical maize reveals genomic regions for sorghum downy mildew resistance. *Scientific Reports*, *8*(1), 366. https://doi.org/10.1038/s41598-017-18690-3.

Rio, S., Mary-Huard, T., Moreau, L. and Charcosset, A. (2019). Genomic selection efficiency and a priori estimation of accuracy in a structured dent maize panel. *Theoretical and Applied Genetics*, *132*(1), 81–96. https://doi.org/10.1007/s00122-018-3196-1.

Rosado, R.D.S., Cruz, C.D., Barili, L.D., de Souza Carneiro, J.E., Carneiro, P.C.S., Carneiro, V.Q., … and Nascimento, M. (2020). Artificial neural networks in the prediction of genetic merit for flowering traits in bean cultivars. *Agriculture*, *10*(12), 638. https://doi.org/10.3390/agriculture10120638.

Rozek, L.S., Dolinoy, D.C., Sartor, M.A. and Omenn, G.S. (2014). Epigenetics: Relevance and implications for public health. *Annual Review of Public Health*, *35*, 105–122. https://doi.org/10.1146/annurev-publhealth-032013-182513.

Runcie, D.E., Qu, J., Cheng, H. and Crawford, L. (2021). MegaLMM: Mega-scale linear mixed models for genomic predictions with thousands of traits. *Genome Biology*, 22, Article 213. https://doi.org/10.1186/s13059-021-02461-4

Rutkoski, J., Poland, J., Mondal, S., Autrique, E., Pérez, L.G., Crossa, J., Reynolds, M. and Singh, R. (2016). Canopy temperature and vegetation indices from high-throughput phenotyping improve accuracy of pedigree and genomic selection for grain yield in wheat. *G3: Genes|Genomes|Genetics*, *6*(9), 2799–2808. https://doi.org/10.1534/g3.116.032888.

Sampaio Filho, J.S., Olivoto, T., Campos, M. de S. and de Oliveira, E.J. (2023). Multi-trait selection in multi-environments for performance and stability in cassava genotypes. *Frontiers in Plant Science*, *14*, 1282221. https://doi.org/10.3389/fpls.2023.1282221.

Shamshad, M. and Sharma, A. (2018). The usage of genomic selection strategy in plant breeding. In: *Next Generation Plant Breeding*. Springer Nature Singapore Pte Pvt. Ltd. https://doi.org/10.5772/intechopen.76247.

Simko, I. and Piepho, H.P. (2012). The area under the disease progress stairs: Calculation, advantage, and application. *Phytopathology*, *102*(4), 381–389.

Sinha, D., Maurya, A.K., Abdi, G., Majeed, M., Agarwal, R., Mukherjee, R., … and Chen, J.-T. (2023). Integrated genomic selection for accelerating breeding programs of climate-smart cereals. *Genes*, *14*(7). https://doi.org/10.3390/genes14071484.

Sinha, P., Singh, V.K., Bohra, A., Kumar, A., Reif, J.C. and Varshney, R.K. (2021). Genomics and breeding innovations for enhancing genetic gain for climate resilience and nutrition traits. *Theoretical and Applied Genetics*, *134*(6), 1829–1843. https://doi.org/10.1007/s00122-021-03847-6.

Sørensen, E.S., Jansen, C., Windju, S., Crossa, J., Sonesson, A.K., Lillemo, M. and Alsheikh, M. (2023). Evaluation of strategies to optimize training populations for genomic prediction in oat (Avena sativa). *Plant Breeding*, *142*(1), 41–53. https://doi.org/10.1111/pbr.13061.

Spindel, J.E., Begum, H., Akdemir, D., Collard, B., Redoña, E., Jannink, J. and McCouch, S. (2016). Genome-wide prediction models that incorporate *de novo* GWAS are a powerful new tool for tropical rice improvement. *Heredity*, *116*(4), 395–408. https://doi.org/10.1038/hdy.2015.113.

Spindel, J. and McCouch, S. (2016). Genomic selection in rice breeding: Progress and prospects. *Rice*, *9*(1), 1–11.

Spindel, J., Begum, H., Akdemir, D., et al. (2015). Genomic selection and association mapping in rice (Oryza sativa): Implications for breeding and gene discovery. *Theoretical and Applied Genetics*, *128*(1), 93–113.

Technow, F., Messina, C.D., Totir, L.R. and Cooper, M. (2015). Integrating crop growth models with whole genome prediction through approximate Bayesian computation. PLOS ONE. https://doi.org/10.1371/journal.pone.0130855

Tibshirani, R. (1996). Regression shrinkage and selection via the lasso. *Journal of the Royal Statistical Society: Series B (Methodological)*, *58*(1), 267–288.

Toomey, L., Lecocq, T., Bokor, Z., Espinat, L., Ferincz, Á., Goulon, C., … and Fontaine, P. (2020). Comparison of single- and multi-trait approaches to identify best wild candidates for aquaculture shows that the simple way fails. *Scientific Reports*, *10*(1), 11564. https://doi.org/10.1038/s41598-020-68315-5.

Tu, H.-N. and Liao, C.-T. (2024). A modified Bayesian optimization approach for determining a training set to identify the best genotypes from a candidate population in genomic selection. *Journal of Agricultural, Biological, and Environmentaltal Statistics*. https://doi.org/10.1007/s13253-024-00632-y.

Ullah, N., Khan, J.A., Almakdi, S., Alshehri, M.S., Al Qathrady, M., El-Rashidy, N., ... and Ali, F. (2023). An effective approach for plant leaf diseases classification based on a novel DeepPlantNet deep learning model. *Frontiers in Plant Science*, *14*, 1212747. https://doi.org/10.3389/fpls.2023.1212747.

Urbut, S.M., Wang, G., Carbonetto, P. and Stephens, M. (2019). Flexible statistical methods for estimating and testing effects in genomic studies with multiple conditions. *Nature Genetics*, *51*(1), 187–195. https://doi.org/10.1038/s41588-018-0268-8.

VanRaden, P.M. (2008). Efficient methods to compute genomic predictions. *Journal of Dairy Science*, *91*(11), 4414–4423.

Varshney, R.K. (2021). The Plant Genome special issue: Advances in genomic selection and application of machine learning in genomic prediction for crop improvement. *The Plant Genome*, *14*(3), e20178. https://doi.org/10.1002/tpg2.20178.

Veerendrakumar, H.V., Barmukh, R., Shah, P., Bomireddy, D., Jamedar, H.V.R., Roorkiwal, M., ... and Pandey, M.K. (2024). Genomic selection in crop improvement. In: *Sustainability Sciences in Asia and Africa*. Springer Nature, pp. 187–222. https://doi.org/10.1007/978-981-99-4673-0_9.

Verma, S., Gupta, A.R.S.S.H., Yalla, S., Shreya, Patel, P., Sharma, R., ... and Donga, A. (2024). Integrating marker-assisted and genomic selection for plant functional trait improvement. In: *Plant Functional Traits for Improving Productivity*. Springer Nature, pp. 203–215. https://doi.org/10.1007/978-981-97-1510-7_11.

Wang, H., Tang, J., Wu, M., Wang, X. and Zhang, T. (2022). Application of machine learning missing data imputation techniques in clinical decision making: Taking the discharge assessment of patients with spontaneous supratentorial intracerebral hemorrhage as an example. *BMC Medical Informatics and Decision Making*, *22*(1), 13. https://doi.org/10.1186/s12911-022-01752-6.

Wang, X., Zeng, H., Lin, L., Huang, Y., Lin, H. and Que, Y. (2023). Deep learning-empowered crop breeding: Intelligent, efficient and promising. *Frontiers in Plant Science*, *14*, 1260089. https://doi.org/10.3389/fpls.2023.1260089.

Wolfe, M.D., Del Carpio, D.P., Alabi, O., Ezenwaka, L.C., Ikeogu, U.N., Kayondo, I.S., ... and Jannink, J.-L. (2017). Prospects for genomic selection in cassava breeding. *The Plant Genome*, *10*(3). lantgenome2017.03.0015. https://doi.org/10.3835/plantgenome2017.03.0015.

Xu, Y., Ma, K., Zhao, Y., Wang, X., Zhou, K., Yu, G., ... and Xu, S. (2021). Genomic selection: A breakthrough technology in rice breeding. *The Crop Journal*, *9*(3), 669–677. https://doi.org/10.1016/j.cj.2021.03.008.

Xu, Y., Wang, X., Ding, X., Zheng, X., Yang, Z., Xu, C. and Hu, Z. (2018). Genomic selection of agronomic traits in hybrid rice using an NCII population. *Rice*, *11*(1). https://doi.org/10.1186/s12284-018-0223-4.

Yang, A.Q., Chen, B., Ran, M.L., Yang, G.M. and Zeng, C. (2020). The application of genomic selection in pig cross breeding. *Yi Chuan = Hereditas*, *42*(2), 145–152. https://doi.org/10.16288/j.yczz.19-253.

Yoosefzadeh-Najafabadi, M., Eskandari, M., Torabi, S., Torkamaneh, D., Tulpan, D. and Rajcan, I. (2022). Machine-learning-based genome-wide association studies for uncovering QTL underlying soybean yield and its components. *International Journal of Molecular Sciences*, *23*(10), 5538. https://doi.org/10.3390/ijms23105538.

Yu, J., Pressoir, G., Briggs, W.H., Vroh Bi, I., Yamasaki, M., Doebley, J.F., ... Buckler, E.S. (2006). A unified mixed-model method for association mapping that accounts for multiple levels of relatedness. *Nature Genetics*, *38*(2), 203–208. https://doi.org/10.1038/ng1702.

Zhang, C.H. (2010). Nearly unbiased variable selection under minimax concave penalty. *The Annals of Statistics*, *38*, 894–942. https://doi.org/10.1214/09-AOS729.

Zhang, F., Kang, J., Long, R., Li, M., Sun, Y., He, F., ... and Yang, Q. (2023). Application of machine learning to explore the genomic prediction accuracy of fall dormancy in autotetraploid alfalfa. *Horticulture Research*, *10*(1), uhac225. https://doi.org/10.1093/hr/uhac225.

Zhou, J., Luan, X., Liu, Y., Wang, L., Wang, J., Yang, S., ... and Yao, D. (2023). Strategies and methods for improving the efficiency of CRISPR/Cas9 gene editing in plant molecular breeding. *Plants*, *12*(7). https://doi.org/10.3390/plants12071478.

Zhu, C., Gore, M., Buckler, E.S. and Yu, J. (2008). Status and prospects of association mapping in plants. *The Plant Genome*, *1*(1). https://doi.org/10.3835/plantgenome2008.02.0089.

Chapter 4

RNAi Technology
A Method for Insect Pest and Disease Management for Sustainable Agriculture

Mahesh Math[1*], Belaghihalli N Gnanesh[2], Mahantesha B.N. Naika[3], Sahana KP[4] and Radha Sivarajan Sajeevan[5*]

1. Introduction

Agriculture remains a cornerstone of global food security and economic stability, but its sustainability is constantly challenged by insect pests and diseases that cause substantial yield losses and quality deterioration (Savary et al., 2019; Li & Wu, 2025). The Food and Agriculture Organization (FAO) estimates that up to 20–40% of global crop production is lost annually to insect pests and diseases (Tripathy et al., 2023), affecting the livelihoods of millions of smallholder farmers and threatening food availability globally. Therefore, effective pest management is essential for improving crop productivity and achieving sustainable agriculture. Conventional pest control strategies, including cultural, mechanical, biological, and chemical methods, have historically been employed to reduce crop damage and protect agricultural output (Kogan, 1998; Peshin & Dhawan, 2009). Cultural and mechanical approaches, such as crop rotation, trap cropping, and hand-picking, are environmentally friendly and locally adaptable but tend to be labour-intensive and limited in scale (Pedigo & Rice, 2014). Biological control measures utilizing naturally occurring predators and parasitoids have also gained attention due to their eco-friendly nature and long-term control benefits (Van Lenteren, 2012). However, the effectiveness of these methods is frequently hindered by environmental fluctuations and pest population dynamics.

Chemical pesticides, despite being highly effective and fast-acting, are associated with numerous challenges. These include non-target effects, environmental contamination, pesticide resistance evolution, and risks to human health (Pimentel, 2005; Sharma et al., 2019). Excessive and repeated use of certain chemical groups, especially neonicotinoids and pyrethroids, has resulted in a marked rise in resistance among insect species globally, thereby diminishing the effectiveness of these compounds in pest management (Sparks & Nauen, 2015). Additionally, overreliance on chemical pesticides has led to residue accumulation in soil and water systems, threatening biodiversity and

[1,4] Sampoorna International Institute of Agri Science & Horticultural Technology, Maddur 571 433, Karnataka, India.
[2] Genetics and Plant Breeding, ICAR JSS - Krishi Vigyan Kendra, Mysuru 571 129, Karnataka, India.
[3] Centre for Biotechnology Research, Department of BCI, College of Horticulture, Bagalkot 587 104, University of Horticultural Sciences, Bagalkot, Karnataka, India.
[5] Department of Plant Protection Biology, Swedish University of Agricultural Sciences, SE 23422, Lomma, Sweden.
* Corresponding author: maheshento@gmail.com; sajeevan.radha.sivarajan@slu.se

DOI: 10.1201/9781003614869-4

ecological balance (Goulson, 2013). To minimize such dependence, breeding crops for resistance to insect pests and diseases has become a cornerstone of Integrated Pest Management (IPM) strategies (St. Clair, 2010). However, traditional breeding is time-consuming and is often limited by the availability of resistance genes in the germplasm pool (Zhang et al., 2017). Furthermore, efforts to transfer resistance traits from wild relatives often bring along undesirable agronomic characteristics due to linkage drag, complicating commercial cultivar development (Sandhu & Kang, 2017).

RNA interference (RNAi) has emerged as a powerful and innovative molecular tool, offering a targeted, environmentally friendly, and effective approach to the management of agricultural pests and pathogens. RNAi is a naturally occurring cellular process that regulates gene expression through post-transcriptional gene silencing, conserved across eukaryotes, including plants, insects, fungi, and nematodes. This process enables gene knockdown by guiding small interfering RNAs (siRNAs) to degrade complementary messenger RNA (mRNA), thereby preventing translation of specific genes into proteins. In pest management, RNAi can be exploited to disrupt key physiological and developmental processes by targeting specific genes critical for pest survival, reproduction, or host interaction. Its application spans a broad range of pests from coleopteran insects such as the Colorado potato beetle to hemipterans like whiteflies, as well as plant-parasitic nematodes and fungal pathogens. The main advantage of RNAi is its species-specific mode of action, which minimizes off-target effects and preserves beneficial organisms. Successful demonstrations of RNAi-based pest control include transgenic maize expressing dsRNA against the *Snf7* gene for Western corn rootworm (*Diabrotica virgifera virgifera*), which showed enhanced resistance in field trials (Baum et al., 2007; Khajuria et al., 2018). Similarly, sprayable RNAi formulations targeting the *β-actin* gene in Colorado potato beetle have been approved for commercial use (Palli, 2014; Rodrigues et al., 2021). These developments highlight the technology's potential to serve as a sustainable alternative to traditional chemical pesticides (Zhong et al., 2025; Mendoza-Alatorre et al., 2025).

This chapter provides an in-depth exploration of RNAi technology for insect pest and disease management. It begins with an overview of RNAi components and mechanisms, followed by

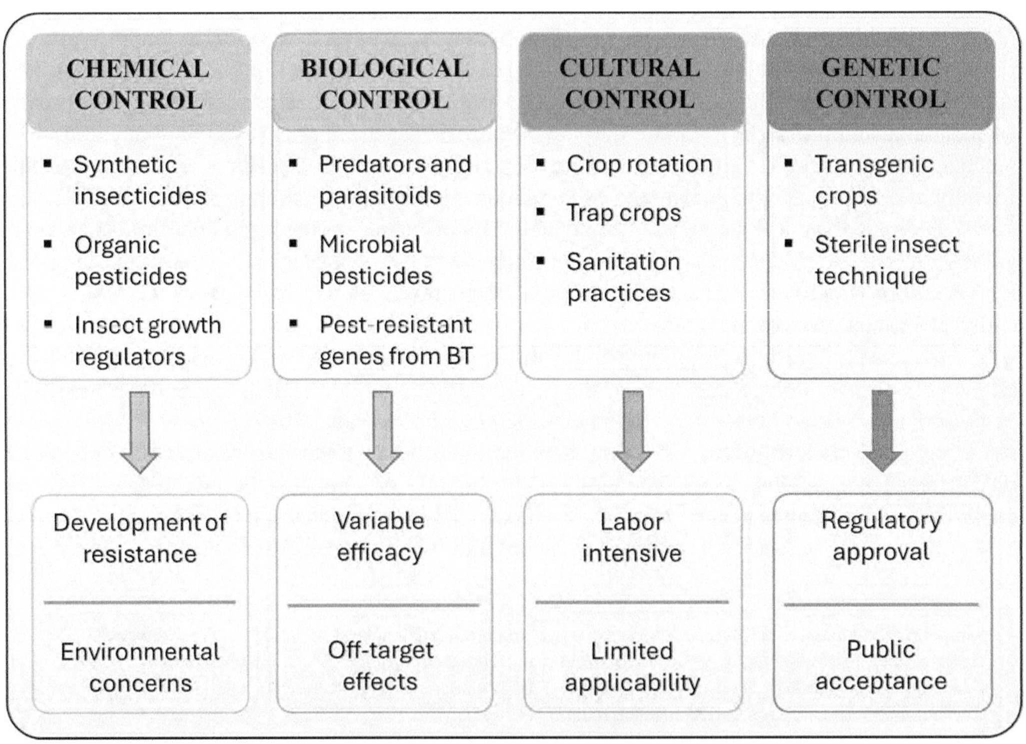

Fig. 1 Overview of current pest control strategies and limitations.

its application in insect, nematode, and pathogen control. Subsequent sections discuss delivery systems, case studies, regulatory challenges, and prospects for integrating RNAi into crop protection strategies. Real-world examples from crops such as potato, maize, and tomato are used to illustrate the practical impact of this transformative technology. An overview of the current pest control strategies and limitations is illustrated in Figure 1.

2. Components of RNAi

RNAi relies on a suite of molecular components that initiate, propagate, and enforce gene silencing. These components are evolutionarily conserved and function cooperatively to detect and degrade target mRNA in a sequence-specific manner. Some of the components identified act as initiators, while others serve as effectors, amplifiers, and transmitters of the gene silencing process (Agrawal et al., 2003). Bernstein et al. (2001) identified a bidentate RNase III-like enzyme in Drosophila extract, later named Dicer, which was shown to have the ability to cleave long double-stranded RNA into 22-nucleotide fragments characteristic of RNAi. This enzyme is involved in the initiation of RNAi by digesting dsRNA into uniformly sized siRNA. These nucleases are evolutionarily conserved across a wide range of organisms, including worms, flies, fungi, plants, and mammals. Table 1 provides key RNAi pathway components and their functions.

Table 1 Key RNAi pathway components and their functions in the *Arabidopsis thaliana* model plant.

Component	*Function*	*Key References*
Dicer-like proteins (DCL1, DCL2, DCL3, DCL4)	Cleave dsRNA into siRNAs of specific lengths; DCL1 for miRNA, DCL2/3/4 for siRNA pathways	Margis et al. (2006); Borges & Martienssen (2015)
Argonaute proteins (AGO1, AGO2, AGO4, AGO10)	Bind siRNA and guide RISC to target mRNA for cleavage or translational repression	Mallory & Vaucheret, (2010); Fang & Qi (2016); Wu et al. (2020)
RNA-dependent RNA polymerases (*RDR1, RDR2, RDR6*)	Synthesize secondary siRNAs from target mRNA templates to amplify silencing	Brosnan & Voinnet (2009); Willmann et al. (2011)
Double-stranded RNA-binding proteins (*DRB1, DRB4*)	Assist DCLs in dsRNA recognition and cleavage specificity	Curtin et al. (2008); Fukudome & Fukuhara (2017)
Small interfering RNAs (siRNAs)	Guide strand is incorporated into RISC; antisense to target mRNA	Baulcombe (2004); Bologna & Voinnet (2014)
RNA-induced silencing complex (RISC)	Core complex responsible for mRNA cleavage guided by siRNAs	Hammond et al. (2000); Bologna & Voinnet (2014)
dsRNases	Degrade exogenous dsRNA; modulate RNAi efficiency	Dubelman et al. (2014); Joga et al. (2016)
SID-like proteins (e.g., *AtSID1, AtSID2*)	Facilitate uptake or intercellular transport of dsRNA in plants	Winston et al. (2002)

2.1 *Double-stranded RNA (dsRNA)*

RNAi begins with the presence of exogenous or endogenous double-stranded RNA (dsRNA) molecules present in a cell. These may occur naturally from viral infections, transposons, or endogenous hairpin RNAs, and also be synthetically designed to target specific genes. In all cases, dsRNA acts as the initiating trigger for RNAi. The recognition of long dsRNA within a cell is often interpreted as a potential viral infection or transposon activity, prompting the RNAi machinery to process and neutralize the threat via Dicer-mediated cleavage into siRNAs (Obbard et al., 2009). In pest management, dsRNAs are tailored to silence genes vital for survival or reproduction, offering high specificity to the target pest without harming beneficial insects. This eco-friendly approach is biodegradable, leaves no chemical residues, and can be customized for any pest gene of interest.

2.2 Dicer Enzyme

Dicer, a member of the ribonuclease III (RNase III) family, is a key endonuclease that recognizes and cleaves dsRNA into siRNAs of approximately 21–25 nucleotides in length. These siRNAs typically have 2-nucleotide 3′ overhangs, a 5′ phosphate, and a 3′ hydroxyl terminus (Elbashir et al., 2001). Dicer is ATP-dependent and possesses multiple domains, including a PAZ domain (binding the 3′ end of siRNA), two RNase III domains (cleaving both strands of dsRNA), and an N-terminal helicase domain involved in unwinding RNA structures (MacRae et al., 2006; Vergani-Junior et al., 2021).

2.3 Small Interfering RNAs

These are short dsRNA molecules, usually 21–25 nucleotides in length, generated by the enzymatic cleavage of longer dsRNA by Dicer-like (DCL) enzymes (Elbashir et al., 2001). These siRNA molecules possess a characteristic 2-nucleotide 3′ overhang on each strand, which is essential for their recognition and processing by downstream components of the RNAi machinery (Elbashir et al., 2001). Once formed, the siRNA duplex is unwound, and one strand, termed the guide strand, is selectively loaded into the RNA-induced silencing complex (RISC), where it directs sequence-specific mRNA recognition and cleavage (Hammond et al., 2000). The choice between guide and passenger strand is determined largely by the relative thermodynamic stability of the duplex ends, with the strand having a less stable 5′ end preferentially loaded into RISC (Schwarz et al., 2003). The passenger strand is actively cleaved and degraded during RISC assembly, a process facilitated by the catalytic activity of Argonaute proteins, which ensures that only the guide strand remains to direct silencing (Matranga et al., 2005). This selective strand retention is essential to maintain high specificity and minimize off-target effects.

2.4 RNA-Induced Silencing Complex

The RISC is a multiprotein complex that executes the RNAi process by identifying and degrading target mRNA transcripts. Its core component is the Argonaute (AGO) protein, which binds the guide siRNA and facilitates target mRNA cleavage when perfect base pairing occurs. When mismatches exist, AGO instead promotes translational repression, thereby reducing protein output without full RNA degradation. Thus, the RISC serves as the central effector of RNAi-mediated gene silencing (Pratt & MacRae, 2009; Filipowicz, 2005).

2.5 Additional Modulators

Beyond the core RNAi machinery, several accessory proteins and cofactors modulate the RNAi efficiency. For instance, SID-1 and SID-2 transmembrane proteins facilitate dsRNA uptake and systemic spread in nematodes and insects. RNA-binding proteins and helicases are also involved in stabilizing siRNAs and facilitating RISC loading. The presence or absence of these cofactors contributes to the pronounced species-specific variability observed in RNAi responsiveness (Zhu & Palli, 2020).

3. Mechanism of RNAi

The RNAi functions through pathways such as Post Transcriptional Gene Silencing (PTGS) and Transcriptional Gene Silencing (TGS) involved in suppressing gene expression (Bernstein et al., 2001; Pal-Bhadra et al., 2002). The process begins when dsRNA is introduced into a cell, either endogenously (transformative) through genetic engineering techniques, or be delivered exogenously (non-transformative) by methods such as infiltration, spraying, soaking, injection, foliar, dropping, spreading, and mechanical inoculation which are widely used onto plants for target gene silencing (Das & Sherif, 2020; Rêgo-Machado et al., 2023). This section elaborates on the molecular events involved in gene silencing through RNAi and highlights its mode of action across various insect and pathogen systems.

3.1 Initiation: Recognition and Cleavage of dsRNA

During the initiation phase of RNAi, long double-stranded RNAs are recognized and cleaved by the RNase III enzyme Dicer into 21–25 nucleotide siRNAs with hallmark 2-nucleotide 3′ overhangs, 5′ phosphate groups, and 3′ hydroxyl termini (Bernstein et al., 2001; Elbashir et al., 2001). In nematodes and some insects, dsRNA uptake occurs either through endocytosis or via SID-1-like transmembrane transporters, which facilitate systemic RNAi by enabling the movement of dsRNA between cells (Winston et al., 2002; Feinberg & Hunter, 2003).

3.2 RISC Assembly and Target Recognition

During RISC assembly, the siRNA duplex is unwound, and the guide strand is selectively incorporated into the Argonaute-centered RISC, while the passenger strand is discarded. The RISC complex is now primed to identify complementary mRNA sequences in the cytoplasm (Hammond et al., 2000; Schwarz et al., 2003; Hammond et al., 2005). Perfect or near-perfect complementary base pairing between the guide strand and the target mRNA leads to endonucleolytic cleavage by the Argonaute protein within RISC, while near-perfect complementarity results in translational repression without cleavage (Hutvágner & Zamore, 2002; Bartel, 2004).

3.3 Gene Silencing and Phenotypic Outcomes

The targeted degradation of mRNA prevents translation into protein, leading to a loss-of-function phenotype (Fire et al., 1998). Silencing may occur in a cell-autonomous manner, restricted to the initiating cell, or systemically, where signals spread to other cells and tissues via proteins like SD-1 (Winston et al., 2002). In agricultural pests, silencing genes involved in neural signaling, cuticle biosynthesis, reproduction, or digestion has been shown to cause developmental arrest, sterility, behavioural disruption, or death (Gu & Knipple, 2013; Joga et al., 2016). Similarly, in pathogens like fungi or bacteria, genes required for pathogenicity or host colonization are ideal targets for RNAi (Mamta & Rajam, 2017).

3.4 Systemic and Environmental RNAi

Systemic RNAi, observed in coleopterans and nematodes, allows silencing signals to move beyond the initial site of induction, possibly via mobile RNA molecules or RNA-binding protein complexes (Whangbo & Hunter, 2008). Environmental RNAi occurs when dsRNA is acquired directly from the surroundings, such as from plant tissue, feeding solutions, or foliar sprays, and then triggers RNAi within the organism (Baum & Roberts, 2014; Joga et al., 2016).

3.5 Examples in Major Pest Groups

Khajuria et al. (2018) and Bachman et al. (2013) demonstrated the effectiveness of the RNAi-based technology in controlling Western corn rootworm populations, highlighting its potential as a sustainable and eco-friendly pest management strategy. However, not all insects exhibit the same level of responsiveness to ingested dsRNA. While Coleoptera (beetles) generally exhibit consistent and systemic RNAi responses, species belonging to the orders Lepidoptera (moths and butterflies),

Diptera (flies and mosquitoes), and Hemiptera (aphids, hoppers, stinkbugs) tend to show more variable silencing efficiency due to physiological and molecular barriers such as dsRNA degradation and differences in cellular uptake (Whyard et al., 2009; Terenius et al., 2011; Zhu & Palli, 2020).

3.5.1 Coleoptera

This is the first insect group successfully targeted using RNAi, leading to the development of commercial RNAi-based pest control products (Baum et al., 2007). Among these, the Western corn rootworm (*Diabrotica virgifera virgifera*) has emerged as a prominent example due to its high sensitivity to RNAi, availability of well-characterized gene targets, and a gut physiology conducive to efficient dsRNA uptake with minimal off-target effects (Vélez et al., 2016). Effective control was achieved through the use of hairpin dsRNA constructs targeting the *v-ATPase A* gene, which significantly impaired larvae mortality (Baum et al., 2007). Similarly, RNAi has shown promise in managing the Colorado potato beetle (*Leptinotarsa decemlineata*), a major pest of solanaceous crops (Zhu et al., 2011). Transgenic potato plants expressing dsRNA targeting the *β-actin* and *Shrub* genes in the chloroplast or cytoplasm have achieved potent insecticidal effects (Zhang et al., 2015; Ren et al., 2021). In contrast, nuclear or cytoplasmic expression systems tend to exhibit reduced efficacy, likely due to premature processing of dsRNA by plant Dicer enzymes (Dalakouras et al., 2020a). The absence of RNAi machinery in plastids enables long dsRNA to remain stable and unprocessed, facilitating more effective delivery to the insect and enhancing gene silencing outcomes.

3.5.2 Lepidoptera

Lepidoptera include major agricultural pests such as *Spodoptera litura* and *Helicoverpa armigera*, which are generally considered less responsive to RNAi compared to coleopterans (Terenius et al., 2011). This reduced sensitivity is primarily attributed to several biological factors, including the presence of high levels of extracellular dsRNases in the gut, inefficient uptake of dsRNA molecules, and limited systemic RNAi spread across tissues (Yu et al., 2013; Wynant et al., 2014; Singh et al., 2017). Despite these challenges, moderate success has been achieved using oral (feeding-based) delivery methods. For example, dsRNA targeting *acetylcholinesterase* (*AChE*) or *aminopeptidase N* has been administered through artificial diets or plant-expressed constructs, resulting in observable effects such as growth retardation, malformed pupae, and reduced fecundity (Kumar et al., 2009). These results suggest that, while Lepidoptera may require more refined delivery methods, such as protective formulations or higher dsRNA doses, they remain viable targets for RNAi-based pest management when approached with tailored methodologies.

3.5.3 Hemiptera

Aphids, fruit flies, and whiteflies have been evaluated for their responsiveness to RNAi with mixed results. Early studies demonstrated that ingestion of species-specific dsRNA could selectively kill the pea aphid (*Acyrthosiphon pisum*), confirming that hemipterans are amenable to RNAi when delivery is effective (Whyard et al., 2009). Similarly, whiteflies (*Bemisia tabaci*) have shown susceptibility to RNAi, with oral delivery of dsRNA leading to significant gene silencing and reduced survival (Upadhyay et al., 2011). The variability in RNAi efficacy among hemipteran species is often linked to inefficient cellular uptake of dsRNA, the presence of higher amounts of dsRNases in the gut and hemolymph, and limited systemic RNAi response (Christiaens & Smagghe, 2014; Ghodke et al., 2019). Despite these constraints, recent advances in formulation, such as nanoparticle encapsulation and protective carriers, have demonstrated that topical and feeding-based RNAi applications have improved the efficiency (Santos-Ortega & Flynt, 2022). Promising target genes in Hemiptera include those involved in salivary gland function, nutrient transport, and detoxification pathways. Cathepsins, aquaporins, and cytochrome P450s have been successfully silenced, leading to disrupted feeding, reduced fecundity, and lower survival rates (Christiaens et al., 2022; Lu & Shen, 2024).

These findings suggest that while Hemiptera may require enhanced delivery systems or protective carriers (e.g., nanoparticles or chemical stabilizers), they remain amenable to RNAi-based control under optimized conditions.

3.5.4 Diptera

The mosquitoes (*Aedes aegypti, Anopheles gambiae*) have been key targets for RNAi research due to their role as vectors of devastating human diseases such as malaria, dengue, and Zika. RNAi approaches in mosquitoes have primarily focused on silencing genes involved in olfaction, reproduction, immunity, and hormonal regulation. For instance, dsRNA-mediated knockdown of odorant receptors has disrupted host-seeking behaviour and reduced pathogen transmission potential (Whyard et al., 2015; Dong et al., 2021). Similarly, silencing genes in the juvenile hormone or vitellogenin pathways has led to impaired ovarian development, oviposition, and reduced vector competence (Erdelyan et al., 2012; Rivera-Pérez et al., 2020). Delivery methods include microinjection, oral ingestion, and transgenic. Although the RNAi response in Diptera is often transient and varies between species and developmental stages, such interventions offer promising avenues for population suppression and vector control.

3.5.5 Nematodes

Plant-parasitic nematodes, such as root-knot nematodes (*Meloidogyne* spp.) and cyst nematodes (*Heterodera* spp.), pose a major threat to global agriculture, affecting a wide range of crops. RNAi has emerged as a powerful strategy to control these pests by targeting genes essential for parasitism, feeding site establishment, or reproduction. One of the most effective approaches has been Host-Induced Gene Silencing (HIGS), wherein transgenic plants express dsRNA corresponding to nematode genes. During feeding, nematodes ingest the plant-derived dsRNA, triggering silencing of parasitism genes, resulting in reduced invasion, impaired syncytium formation, and decreased reproduction. For example, silencing parasitism genes such as *Mc16D10L* or *Mi-msp2* has been shown to significantly reduce nematode infestation in crops like tomato and soybean (Dinh et al., 2014; Joshi et al., 2019; Joshi et al., 2020).

3.5.6 Fungal Pathogens

RNAi-based approaches have shown increasing promise in the control of fungal pathogens, particularly through Spray-Induced Gene Silencing (SIGS) emerging as one of the most effective strategies (Koch et al., 2016). This non-transgenic method involves the exogenous application of dsRNA or siRNA molecules targeting essential fungal genes, which are then absorbed by the pathogen during host infection. SIGS has been successfully demonstrated against economically important pathogens, such as powdery mildew (*Blumeria graminis)*, fusarium head blight (*Fusarium graminearum*), and grey mould (*Botrytis cinerea)*. Silencing of genes like *CYP51,* which encodes a key enzyme in ergosterol biosynthesis, and *DCL* genes, central to the fungal RNAi machinery, has led to impaired fungal development, reduced virulence, and suppression of disease symptoms (Yoshida, 1988; Koch et al., 2016). Because of its sequence specificity, RNAi minimizes off-target effects on beneficial microbes, making SIGS an attractive and environmentally friendly alternative to conventional fungicides. Recent advances focus on improving dsRNA stability, uptake efficiency, and formulation technologies to enhance field-level efficacy (Mann et al., 2023; Ghag, 2023).

3.5.7 Viral Diseases

RNAi has also been extensively utilized in combating plant viral diseases, primarily through the expression of virus-derived siRNAs in transgenic plants. The siRNAs are designed to target conserved regions of viral genomes, such as coat protein (CP), replicase, or movement protein genes, thereby inhibiting viral replication and movement within the host. For example, plants engineered to express siRNAs against the Tomato yellow leaf curl virus (TYLCV) CP gene have exhibited enhanced resistance, delayed symptom onset, and reduced viral accumulation (Tenllado et al., 2004; Zrachya et al., 2007). Further approach has been extended to Cucumber mosaic virus (CMV), Potato virus Y, and Papaya ringspot virus (Xie et al., 2014; Gaffar & Koch, 2019; Singh et al., 2019; Jahromi et al., 2022; Voloudakis et al., 2022). In addition to transgenic approaches, non-transgenic methods such as SIGS have shown potential. Exogenous application of clay nanosheet-based RNAi formulations targeting the viral CP through SIGS has shown protection against Tobacco mosaic virus, CMV, and Pepper mild mottle virus, offering a biodegradable and non-transgenic alternative (Mitter et al., 2017). Likewise, direct spraying of dsRNA targeting the bean common mosaic virus CP successfully protected *N. benthamiana* and cowpea (*Vigna unguiculata*) plants (Worrall et al., 2019).

4. Applications of RNAi in Insect Pest and Disease Management

RNAi technology is an effective and innovative approach for developing sustainable crop protection solutions. It is an alternative to traditional chemical pesticides, and it offers an eco-friendly nature, specificity, and the ability to overcome insecticide resistance. RNAi pesticides offer a specific approach targeting various pests, nematodes, and pathogens. A summary of RNAi applications across pests, with target genes and outcomes, is provided in Table 2.

Table 2 RNAi applications across pests with target genes and their outcomes.

Pest	*Host plant to Plant*	*Targeted Genes*	*Effect*	*Reference*
Phytophthora infestans	Potato	*PiGPB1, PiCESA2, PiPEC*	Transcript level of all targeted genes was reduced; however, significant disease control were observed only in PigPB1	Jahan et al. (2015)
Aspergillus flavus	Maize	*afR, afC*	A 14-fold reduction of aflatoxin production by *A. flavis* was observed; however, grain size was also affected may be due to off-target effects	Masanga et al. (2015)
Fusarium species	Barley	*CYP51A, CYP51B, CYP51C*	Approximately 80% reduction in transcript level of all three genes and enhanced resistance to *F. Germanium*	Koch et al. (2013)
Puccinia striiformis f. sp. Tritici	Wheat	*PsCPK1*	Stable and heritable resistance to disease throughout the 3rd and 4th generations, in two transgenic lines around 56 and 46% disease reduction was observed.	Qi et al. (2018)
Magnaporthe oryzae	Rice	*MoAP1 MoSSADH, MoACT, MoSOM1*	siRNA targeting *MoAP1* resulted in fungal inhibition and abnormal spore production; however, the rest of the three genes didn't show effective resistance against *M. oryzae* strains.	Guo et al. (2019)
Fusarium graminearum	Wheat	*Chs3b*	Stable resistance to Fusarium head and seed blight in five consecutive generations, reduced mycelial growth	Cheng et al. (2015)
Blumeria graminis	Barley	*Avra10*	Reduced fungal growth and virulence; 21% reduction in fungal conidia germination	Nowara et al. (2010)

Contd.

Table 2 *Contd.*

Pest	*Host plant to Plant*	*Targeted Genes*	*Effect*	*Reference*
Fusarium oxysporum	Banana	*ERG11A, ERG11B, ERG11C; ERG6A, ERG6B*	Around 87% of inoculated wild-type plants died; however, transgenic lines showed no or slight disease symptoms.	Dou et al. (2020)
Verticillium dahliae	Tomato	*Ave1, Sge1, NLP1*	Resistance to Verticillium wilt in tomato	Song & Thomma (2018)
Aspergillus Flavus	Groundnut	*nsdC, veA, aflR, and aflM*	Activity of the aflatoxin biosynthetic pathway enzymes was reduced, resulting in resistance to *A. flavus* infection.	Prasad et al. (2023)
Sclerotinia sclerotiorum	Rapeseed	*ABHYDROLASE-3*	Shift in plant defense mechanism, enhanced resistance	Wytinck et al. (2022)
Magnaporthe oryzae	Rice	*CRZ1, PMC1, MAGB, LHS1, CYP51A CYP51B*	All the rice transgenic lines except PMC1 and LHS1 showed fungal inhibition	Wang & Dean (2022)
Chilo partellus	Maize	*CpCHI*	Developmental abnormalities and mortality	Adeyinka et al. (2023)
Helicoverpa armigera	Tomato, Tobacco	*HaCHI*	The growth and survival of *H. armigera* were affected	Mamta et al. (2016)
Mythimna separate	Maize	*MsChi2, MsActin, MsREase*	Developmental abnormalities were observed in insects	Bao et al. (2021)
Nilaparvata lugens	Rice	*GST*	Rice resistance to the insect	Yang et al. (2021)
Sitobion avenae	Barley	*Shp*	Lower levels of shp protein; decline in growth, development, and reproduction rate	Abdellatef et al. (2015)
Myzus persicae	Tobacco	*V-ATPaseE, TBCD, AChE2*	Reduced reproduction rates, 30% fewer nymphs produced per adult	Mao & Zeng (2014)
Helicoverpa armigera	Tobacco	*CYP6AE14*	Reduced transcript levels of *CYP6AE14* in the midgut retarded larval growth	Mao et al. (2011)
Spodoptera exigua	Tobacco	*EcR*	Defects in molting, reduced larvae, resistance against cotton bollworm, and beet armyworm	Zhu et al. (2012)
Bemisia tabaci	Tobacco	*v-ATPaseA*	60% reduction in transcript level of *v-ATPaseA*; no molting effect upon whitefly infestation	Thakur et al. (2014)
Diabrotica virgifera virgifera	Maize	*v-ATPaseC*	Effective knockdown of Western cotton rootworm larvae and adults with a 35-fold reduction in v-ATPaseC transcript	Li et al. (2015a)
Bemisia tabaci	Tomato	Ecdysone synthesis genes	An effective strategy against sap-sucking insects delays development and growth.	Luan et al. (2013)

Contd.

Table 2 *Contd.*

Pest	*Host plant to Plant*	*Targeted Genes*	*Effect*	*Reference*
Leptinotarsa decemlineata	Potato	*Actin,* *Sec23,* *COPβ* *v-ATPaseB,* *v-ATPaseE*	All five genes targeted for silencing showed significant larval mortality and adult body weight gain; *Actin Sec23,* and *COPβ* were more effective in larval control.	Zhu et al. (2011)
Acyrthosiphon pisum	Tobacco	*MIF*	Reduced ability to ingest phloem sap, decreased survival rate	Naessens et al. (2015)
Aphis glycines	Soybean	*TREH,* *ATPD,* *ATPE,* *CHS1*	Silencing of the target gene and high mortality	Yan et al. (2020)
Brassicogethes aeneus	Rapeseed	*αCOP*	Significant reduction in pest survival	Willow et al. (2021b)
Nilaparvata lugens	Artificial diet	*NlCHSA*	Reduced chitin content, lower reproduction rates, 90% loss of egg-hatching	Li et al. (2017)
Lymantria dispar	Green beans	Gut-specific genes	The body mass of *L. dispar* larvae was reduced by 60%, also a two-fold reduction in egg mass was observed	Ghosh & Gundersen-Rindal (2017)
Phytophthora infestans	Potato	*PiGPB1, PiHmp1,* *PiCut3,* *PiEndo3*	Effective against late blight, strong inhibition of sporulation in *PiGPB1*; however, the number of sporangia decreased in *PiCut3* and *PiEndo3* treatments	Kalyandurg et al. (2021)
Golovinomyces orontii	Arabidopsis	*CYP51*	Limited growth and lower asexual reproduction rates of *G. Orontii*	McRae et al. (2023)
Rhizoctonia cerealis	Wheat	*RcOSP1*	A total of 10 *RcOSP* genes were identified; however, suppression of *RcOSP1* inhibited *R. cerealis* growth.	Li et al. (2022)
Botrytis cinerea	Grapevine	*ChS1,* *EF2,* *CYP51*	Enhanced resistance to *B. cinerea* in lab and field conditions	Nerva et al. (2020)
Botrytis cinerea	Barley	*DCL1,* *DCL2*	Hinders the growth	McLoughlin et al. (2018)
Fusarium graminearum	Barley	*CYP51A,* *CYP51B,* *CYP51C*	Inhibition of necrotrophic fungal growth	Koch et al. (2016)
Fusarium graminearum	Barley	*AGO,* *DCL*	Protection of barley leaves from *F. graminearum* infection, on average 50% reduction in infected area compared to the control.	Werner et al. (2020)
Sclerotinia sclerotiorum	Barley	*Ss-ThioR,* *Ss-TIM44,* *Ss-CHC,* *SS-AP2,* *Ss-Arf72A,* *Ss-FCHO1*	The decreased growth of the pathogen	Wytinck et al. (2020)

Contd.

Table 2 *Contd.*

Pest	*Host plant to Plant*	*Targeted Genes*	*Effect*	*Reference*
Magnaporthe oryzae	Rice	*MoDES1*	Resistance against fungal blast disease; 48% silencing of *MoDES1* was observed in intact leaves	Sarkar et al. (2021)
Phakopsora pachyrhizi	Soybean	*ACAT, GCSH,* 40S Protein	A 75% reduction in biomass accumulation and a 73% reduction in infected pustules were observed in leaves compared to the control	Hu et al. (2020)

Adopted from Choudry et al., 2024

4.1 Insect Pest Management

Numerous studies have demonstrated the success of RNAi in controlling insect pests through the silencing of vital genes. In coleopterans like the Colorado potato beetle (*Leptinotarsa decemlineata*), RNAi has been effectively used to target *β-actin* and vacuolar ATPase subunits, leading to high mortality and feeding reduction (Baum et al., 2007; He et al., 2020; He et al., 2022a). In Western corn rootworm (*Diabrotica virgifera virgifera*), plant mediated RNAi using transgenic maize expressing dsRNA targeting the *Snf7* gene has proven effective for population suppression and is the basis for commercial traits like SmartStax PRO (Fishilevich et al., 2016; Khajuria et al., 2018; Darlington et al., 2022; Reinders et al., 2023). In *lepidopteran* pests like *Spodoptera litura* and *Helicoverpa armigera*, RNAi has been employed to silence genes associated with cuticle development, digestive enzymes, chitin synthase, and hormonal pathways (Kumar et al., 2009; Lin et al., 2025). Although RNAi efficiency is lower in such insects due to high levels of dsRNases in the gut, encapsulated or nanoparticle-based delivery systems are improving silencing outcomes (Cooper et al., 2019; Peng et al., 2020; Arjunan et al., 2024; Lu et al., 2025). Hemipteran pests like aphids (*Myzus persicae*) and whiteflies (*Bemisia tabaci*) are also amenable to RNAi interventions (Mao & Zeng, 2014; Malik et al., 2016; Suhag et al., 2021; Feng et al., 2023). Although systemic RNAi is less efficient due to RNA degradation and limited uptake, topical and feeding-based delivery targeting genes such as vitellogenin, salivary gland proteins, aquaporins, and detoxification enzymes have shown success in reducing fecundity and virus transmission (Mamta & Rajam, 2017; Jain et al., 2021).

4.2 Nematode Management

Plant-parasitic nematodes such as root-knot nematodes (*Meloidogyne* spp.) and cyst nematodes (*Heterodera* spp.) are major threats to global agriculture, causing substantial yield losses across crops. Several studies have demonstrated that genes involved in nematode parasitism, such as *16D10* (a parasitism gene in *Meloidogyne* spp.), *eng-2–4* genes (*β-1,4-endoglucanase*), and *cathepsin B* (a cysteine protease), can be effectively targeted using RNAi to impair nematode digestion and development, invasion efficiency, and infectivity (Sindhu et al., 2009; Hahn et al., 2012; Dinh et al., 2014; Dinh et al., 2015; Li et al., 2015b; Mirghasemi et al., 2021). The HIGS, where plants are genetically engineered to express nematode-targeting dsRNA, is an especially effective and sustainable approach across several nematode species (Dinh et al., 2014; Banerjee et al., 2017).

4.3 Management of Fungal Pathogens

RNAi based approaches such as SIGS and HIGS are gaining significant attention as sustainable alternatives to fungicides for plant protection. Several studies have shown the efficacy of SIGS and HIGS against a broad spectrum of fungal and oomycete pathogens. Efficient dsRNA uptake and disease suppression were observed in the fungal plant pathogens *Botrytis cinerea* (Gray mold), *Sclerotinia sclerotiorum* (Cottont rot), *Rhizoctonia solani* (Root rot), *Aspergillus niger* (Black mold), *Verticillium dahliae* (Verticillium wilt), whereas *Colletotrichum gloeosporioides* showed no uptake, and the beneficial fungus *Trichoderma virens* exhibited only weak uptake (Qiao et al., 2021). In the oomycete pathogen *Phytophthora infestans*, dsRNA uptake was limited and highly variable, depending on the cell type and developmental stage (Song et al., 2018; Qiao et al., 2021). For example, *Botrytis cinerea*, foliar application of dsRNA targeting Dicer-like genes (*DCL1* and *DCL2*) impaired fungal development and pathogenicity (McLoughlin et al., 2018; Capriotti et al., 2024). Similarly, dsRNA sprays targeting *AGO* and *DCL* genes silenced *Fusarium graminearum,* suppressing Fusarium head blight disease in cereals (Werner et al., 2020). SIGS has also been extended to oomycetes such as *Hyaloperonospora arabidopsidis* and *Phytophthora infestans*, where dsRNA targeting effector genes reduced downy mildew and late blight disease severity (Bilir et al., 2019; Kalyandurg et al., 2021; Sundaresha et al., 2022).

Similarly, HIGS has been demonstrated to be effective in protecting several agriculturally important crop plants against fungal pathogens (Mann et al., 2023). For example, suppression of *Fusarium graminearum* genes conferred enhanced resistance in cereals (Wang et al., 2020), while transgenic plants expressing dsRNA against *Botrytis cinerea* and *Puccinia triticina* showed significant reduction in target fungal transcript abundance and reduced fungal biomass accumulation, providing effective disease resistance (Panwar et al., 2018; Xiong et al., 2019). Advances in the understanding of fungal gene uptake and cross-kingdom RNA trafficking have further refined the use of HIGS, positioning it as a durable resistance mechanism that complements conventional resistance breeding.

4.4 RNAi against Plant Viruses

Plants naturally use small RNAs as part of their innate antiviral defense systems, silencing viral RNAs through the RNAi pathway. This mechanism has been exploited by engineering transgenic plants that express siRNAs targeting essential viral genes such as the CP, movement protein (MP), and replicase. Numerous studies have demonstrated the effectiveness of this strategy in conferring virus resistance (Akbar et al., 2022). Some of the most successful applications include the development of resistance to Tomato yellow leaf curl virus, Potato virus Y, and Papaya ringspot virus, where RNAi strategies significantly reduced both infection rates and disease severity (Zrachya et al., 2007; Singh et al., 2019; Jahromi et al., 2022; Banji et al., 2025).

4.5 Integrated Pest Management and RNAi Synergies

RNAi-based strategies can complement existing IPM practices by offering precision targeting of pest species while preserving natural enemies. For example, RNAi formulations that silence insect immunity genes may increase susceptibility to biocontrol agents such as entomopathogenic fungi and nematodes, thereby enhancing their effectiveness (Chen & De Schutter, 2024). Furthermore, combining Bt toxins with dsRNA can delay resistance development in pest populations, thus improving the longevity of the control technologies (Christiaens et al., 2020; Jiang et al., 2021; Menezes et al., 2022).

RNAi-based pest control products are anticipated to enter the market in the coming years. However, expanding their applicability across a broader range of insect taxa requires continued research into RNAi mechanisms, delivery systems, and various biological and physiological barriers affecting silencing efficiency in certain pests. Additionally, key challenges must still be addressed, including evaluating and mitigating the risks of off-target gene silencing and assessing the potential development of resistance to RNAi-based interventions. It is important to emphasize that RNAi strategies are not intended to function as standalone solutions but rather integrated into a comprehensive IPM framework that combine cultural, biological, and chemical tactics. Incorporating RNAi into IPM programs will be essential for achieving long-term, sustainable pest management outcomes (Christiaens et al., 2020; Willow et al., 2021a; Irsad et al., 2023).

5. Delivery Systems for RNAi in Agriculture

The success of RNAi-based pest and disease management strategies depends heavily on the effective delivery of dsRNA molecules into the target organisms in a stable and biologically active form. Delivery systems must overcome various biological and environmental barriers, such as rapid dsRNA degradation by nucleases, limited uptake, and variable sensitivity across pest species. To address these challenges, several delivery methods have been developed, including transgenic expression, foliar sprays, root drenching, trunk injection, and nanotechnology-enhanced carriers. A comprehensive overview of RNAi strategies, different delivery approaches, and target genes in agricultural insect pests is summarized in Tables 3 and 4.

Table 3 Overview of RNAi delivery approaches, target genes in agricultural insect pests.

Target Insect	*Mode of Delivery*	*Concentration of RNA*	*Target Gene*	*Reference*
Anopheles gambiae	Feeding nanoparticles	–	*AgCHS1 AgCHS2*	Zhang et al. (2010)
Apis mellifera	Abdominal injection	1 μl	Vitellogenin	Amdam et al. (2003)
Apis mellifera	Mixed with a natural diet	0.5 μg	Vitellogenin	Nunes & Simoes (2009)
Apis mellifera	Soaking	1.26 μg	Toll-related receptor	Aronstein et al. (2005)
Diabrotica virgifera Virgifera	Artificial diet	5.4 ng/cm^2	*vATPaseA*	Baum et al. (2007)
Diabrotica virgifera Virgifera	Feeding	1,000 ng	*DvSnf7*	Bolognesi et al. (2012)
Epiphyas postvittana	Droplet feeding	1.0 μg	Gut carboxylase	Turner et al. (2006)
Glossina morsitans Morsitans	Blood meal	10 μg	Midgut protein TsetseEP	Walshe et al. (2009)
Helicoverpa armigera	Feeding	25–50 nM siRNA	Acetylcholine esterase	Kumar et al. (2009)
Nilaparvata lugens	Feeding	0.02, 0.1, and 0.5 mg/ml	Trehalose phosphate synthase (TPS)	Chen et al. (2010)
Phyllotreta striolata	Injection	1 mg/ml	*PsOr1*	Zhao et al. (2011)
Phyllotreta striolata	Feeding	0.05–3.2 ng/ml	Arginine kinase	Zhao et al. (2008)
Reticulitermes flavipes	Artificial diet	13 μg	Cellulase	Zhou et al. (2008)
Schistocerca gregaria	Injection	5 μg dsRNA	Vitellogenin	Badisco et al. (2011)
Schistocerca gregaria	Injection	5 μg	Gapdh	Wynant et al. (2012)
Spodoptera frugiperda	Droplet feeding	0.4 mg	*Cytochrome P450 (CYP6BF1v4)*	Griebler et al. (2008)
Spodoptera litura	Soaking and in artificial diet	–	*Aminopeptidase N*	Rajagopal et al. (2002)
Spodoptera litura	Injection	1.5 or 2.5 mg/ml	Vitellogenin receptor (VgR)	Shu et al. (2011)
Tribolium castaneum	Injection	1 μg	*eGFP*	Miller et al. (2012)

Table 4 RNAi-mediated gene silencing in insect pests using various delivery methods.

Insects	*Target Gene*	*Effects*	*Reference*
Microinjection/ Injection of dsRNA/siRNA			
Spodoptera exiqua	Trehalase genes (*SeTre*-1 and *SeTre*-2)	RNAi of *SeTre*-1 & *SeTre*-2 resulted in higher mortality rates during the larva-pupa and pupa-adult stages.	Chen et al. (2010b)
Spodoptera exiqua	Trehalose-6-phosphate synthase gene	Survival rates of the treated insects were significantly lower compared to controls.	Tang et al. (2010)
Blatella germanica (German cockroach)	Broad-complex gene (*BgBR-C*), vital role in metamorphic processes	Reduced expression of all *BgBR-C* isoforms resulted in diverse phenotypes.	Piulachs et al. (2010)
Nilaparvata lugens (brown plant hopper)	Cathepsin.B nicotine acetylcholine receptors(*nAchRS*), Calreticulin gene, *Nlβ*1 & *Nlβ*2.	40% and 53% decrease in gene expression of Calreticulin, Cathepsin.B and *Nlβ*2, respectively.	Liu, et al. (2010)
Spodoptera exiqua	Chitin synthase gene A (*CHSA)*	Disordered cuticle in treated insects, reduced *CHSA* transcript level	Chen et al. (2008)
Tribolium castenum	*CHSA*	All insects died after injection of *CHSA*-dsRNA	Arakane et al. (2008)
Feeding dsRNA/siRNA via artificial diet			
Ostrinia nubilalis (European corn borer)	Chitinase gene (*OnCht*)	The *OnCht* transcript level was reduced by 63-64% in the larval midgut.	Khajuria et al. (2010)
Epiphyas postvittana (Brown apple moth)	*EposCXE1, EposPBP1*	Repression in gene expression is observed	Turner et al. (2006)
Helicoverpa armigera	*AchE*	Reduction in larval growth, pupal weight and fecundity. Mortality and malformation in larvae.	Kumar et al. (2009)
Feeding Bacteria/*E.coli* expressing dsRNA via artificial diet			
Leptinotarsa decemlineata	*Actin, vATPase, Sec 23*	Reduced body weight and significant mortality in the treated beetles	Zhu et al. (2011)
Spodoptera exiqua	Chitin synthase gene (*CHSA*)	The survival rates in the 5th instar larvae, prepupae and pupae stages were significantly low.	Tian et al. (2009)

Contd.

Table 4 *Contd.*

Insects	*Target Gene*	*Effects*	*Reference*
Feeding dsRNA via sucrose solution			
Nilaparvata lugens	Vacuolar ATP Synthase	Maximum reduction found in the target gene mRNA	Li et al. (2011b)
Feeding and injection both			
Rhodnius prolixus	Nitroporin 2 (*NP2*)	75% and 42% reduction in gene expression by microinjection and feeding, respectively.	Araujo et al. (2006)
Lipid based transfection			
Homalodisca vitripennis	*Actin* gene	A 10-fold decrease in the level of actin mRNA was found in targeted cells of *H. vitripennis*.	Rosa et al. (2010)

5.1 Host-Induced Gene Silencing (HIGS)

It is a transgenic RNAi-based strategy in which plants are genetically engineered to produce dsRNA molecules that specifically target and silence essential genes in insect pests, pathogens, or even parasitic nematodes (Koch & Kogel, 2014). The dsRNA expressed in the plant is processed into siRNAs, which are subsequently taken up by the target organism during feeding. Once inside the pest or pathogen, the siRNAs trigger the RNAi pathway, leading to transcript degradation or translational repression. This approach offers several advantages: it is highly specific to the pest species without harming beneficial organisms or the plant itself, reduces the need for chemical pesticides, and provides continuous protection as long as the transgenic plant is growing (Anderson et al., 2019).

Recent advances in plant HIGS have established it as a reliable and transgenerationally stable technique, effective for regulating host genes and silencing genes in insect pests and pathogens (Beernink et al., 2024; Basso et al., 2025). This method has shown particular effectiveness against root-knot nematodes, chewing insects, and fungal pathogens (Nowara et al., 2010; Ibrahim et al., 2011; Lu et al., 2023). One of the most notable examples is SmartStax PRO maize, which is a commercial RNAi-based product expressing dsRNA targeting the *snf7* gene in *Diabrotica virgifera virgifera*, which disrupts intracellular trafficking and results in high mortality in corn rootworms (Head et al., 2017; Khajuria et al., 2018; Darlington et al., 2022; Reinders et al., 2023).

5.2 Spray-Induced Gene Silencing (SIGS)

Is a non-transgenic RNAi-based approach in which dsRNA is applied externally to plants or directly to target pests. Unlike HIGS, which relies on transgenic plants to produce dsRNA internally, SIGS delivers RNA molecules from outside the plant, making it more compatible with regulatory frameworks that limit the use of genetically modified organisms (GMOs) (Beernink et al., 2024). Applied dsRNA can be absorbed by plant tissues through leaf surfaces, stems, seeds, or roots, and subsequently ingested by herbivorous insects or pathogens during feeding. Several delivery methods have been achieved, including foliar sprays, seed coatings, trunk injections, or root drenches, offering flexible integration with conventional pest management practices (Dalakouras et al., 2020b; Ahmad et al., 2025).

5.3 Nanocarrier-mediated Delivery

Nanoparticles have revolutionized RNAi delivery by improving the stability of dsRNA from environmental degradation and facilitating its uptake by target cells (Yang et al., 2022). Several classes of nanoparticle-based systems have been developed, including self-assembled branched copolymer and self-assembled block copolymer nanoparticles, inorganic nanoparticles, dendrimers, and nanoparticles based on natural products. Most delivery systems of dsRNA exploit electrostatic interactions between the negatively charged dsRNA and positively charged carriers. Commonly used nanomaterials such as chitosan, silica, liposomes, and layered double hydroxides (LDHs) are widely employed as carriers to improve the stability, uptake, and controlled release of dsRNA. These nanocarriers enhance dsRNA stability and protect from degradation by UV radiation, nucleases, and harsh pH, enhancing their persistence and efficacy against many pests like mites, mosquitoes, aphids, Asian corn borer, and the German cockroaches (He et al., 2013; Lin et al., 2017; Yan et al., 2021). A notable example is the LDH-based "BioClay", which binds dsRNA onto clay nanosheets and enables sustained release on the leaf surface. Reports on BioClay formulations resulted in protection against CMV, tobacco mosaic virus (TMV), and *Fusarium* spp., with efficacy lasting up to 20 days under field-mimicking conditions (Mitter et al., 2017; Worrall et al., 2019). Similarly, chitosan nanoparticles have been used for foliar and root delivery of dsRNA targeting insect pests such as *Spodoptera litura* and *Helicoverpa armigera*, resulting in improved knockdown and reduced larval survival (Kolge et al., 2023; Mendoza-Alatorre et al., 2025). Silica-based porous nanoparticles have also shown efficient internalization in phloem-feeding insects like aphids and mites, resulting in significant gene silencing effects. A comparative summary of nanoparticle-based delivery systems is provided in Table 5.

5.4 Additional Delivery Platforms

Beyond conventional foliar sprays and transgenic methods, several alternative RNAi delivery platforms have been developed to enhance efficacy across diverse cropping systems and pest species. Trunk injection is particularly suitable for perennial crops like citrus and grapevines, allowing the direct application of dsRNA into the plant's vascular tissue to ensure systemic distribution with minimal environmental exposure. For instance, trunk-injection delivery of dsRNA for pest control in apple trees revealed that the quantities of dsRNA were significantly greater in the leaves of the treated trees than untreated (Wise et al., 2022). Another innovative approach is symbiont-mediated RNAi (SMR), which exploits insect-associated microbial symbionts to deliver dsRNA from within the pest. This strategy has shown potential in sap-sucking insects such as whiteflies and aphids, where engineered symbionts can continuously produce insecticidal dsRNA, leading to sustained gene silencing and pest suppression (Whitten et al., 2016; Dyson et al., 2022). However, concerns related to host-symbiont specificity, horizontal gene transfer, and regulatory approval limit its current field applicability. Additionally, plant viral vectors such as TMV or Potato virus X (PVX) have been engineered to transiently express insect- or pathogen-targeting dsRNAs in host plants. This system offers a rapid and scalable alternative to transgenic expression, especially for research and short-term protection, and has shown efficacy against aphid-transmitted viruses and fungal pathogens (Wuriyanghan & Falk, 2013; Necira et al., 2024). While these systems offer advantages in specificity and delivery efficiency, their practical implementation must be evaluated against crop type, pest biology, scalability, and regulatory constraints. Collectively, they expand the RNAi delivery toolkit and support the development of tailored, sustainable crop protection strategies. Each system presents unique advantages and limitations based on crop type, pest biology, scalability, environmental persistence, and regulatory frameworks.

6. RNAi in Precision Pest Management and Biological Control Synergies

The specificity and modularity of RNAi make it an attractive tool in precision pest management. Unlike conventional chemical pesticides that have a broad-spectrum effect, RNAi-based solutions can be tailored to target species-specific genes, minimizing ecological disturbance and impacts on beneficial organisms. This allows RNAi technologies to be integrated into pest management systems that prioritize sustainability and biological diversity.

A promising strategy that has been investigated is the use of "stacked treatments", where two pest control methods are applied concurrently to amplify their effectiveness. For example, in the Asian citrus psyllid (*Diaphorina citri*), a major pest of citrus crops, researchers combined dsRNA targeting the glutathione S-transferase gene with the insecticides thiamethoxam and fenpropathrin. This combined approach resulted in a higher mortality rate compared to either treatment alone (Yu & Killiny, 2018). Similar synergies have been explored through co-delivery of dsRNA with biopesticides, entomopathogens, or conventional insecticides, improving pest suppression while reducing chemical use (Wang et al., 2025).

The development of dsRNAi insecticides, supported by advances in biopesticide genomics and nanocarrier systems, has further opened new possibilities. Recent work has highlighted not only the design of dsRNA molecules but also the importance of assessing their effects on non-target organisms, which remains central to regulatory and biosafety considerations (Hoang et al., 2022). By targeting critical pest genes, RNAi exploits genetic vulnerabilities, offering a highly specific and

Table 5 Different types of nanoparticles used for RNAi in crop protection.

Nanoparticle Type	*Target Genes*	*Insect/Plant Species*	*RNAi Efficiency*	*Additional Findings*	*References*
Guanidinium-functionalized polymers	*SfV-ATPase, AgCHSB*	*S. frugiperda, Spodoptera exigua*	>80% knockdown, 53% mortality	Protects dsRNA at alkaline pH, improves cellular uptake	Christiaens et al. (2018); Parsons et al. (2018)
PAMAM dendrimer	*TCTP, Hemocytin*	*Bombyx mori, Drosophila S2 cells*	95.4% transcript reduction	Systemic delivery, visualization of uptake, multi-species potential	Lu et al. (2019); Zheng et al. (2019)
Layered double hydroxide clay nanosheets	*CMV2b, PMMoVIR54*	*Vigna unguiculata, N. tabacum*	Reduction in lesions after viral challenge	Sustained release on plant leaves, long-term protection	Mitter et al. (2017)
Polymer-coated calcium phosphate	*Luciferase*	*Luciferase-expressing S. frugiperda cells*	35% transcript knockdown	Facilitates endosomal escape, enhances stability	Elhaj Baddar et al. (2020)
Polymer-coated gold	*Luciferase*	*Luciferase-expressing S. frugiperda cells*	31% transcript knockdown	Improves dsRNA binding, reduces aggregation	Laisney et al. (2020)
Branched amphiphilic nano-capsules	*BiP, Armet*	*A. pisum, T. castaneum*	75% mortality	Eco-friendly, biodegradable, stable	Avila et al. (2018)
Cell-penetrating peptides	*YFP, CHS; Renilla luciferase, GFP; AgCHS2*	*A. thaliana, Populus tremula, N. benthamiana, A. thaliana, Anthonomus grandis*	Up to 80% transcript reduction	Rapid, targeted delivery; peptide stability dependent	Gillet et al. (2017); Lakshmanan et al. (2013); Numata et al. (2014)
Chitosan nanoparticles	*IAP, AgCHS1, AgCHS2*	*S. frugiperda larvae; Anopheles gambiae*	47% mortality; 62.8% transcript reduction	Improved solubility, suitable for aquatic pests	Gurusamy et al. (2020); Zhang et al. (2010)
Star Polycation (SPc)	*SfATP-d*	*S. frugiperda*	Up to 50% knockdown	Effective in lepidopteran via multiple delivery methods	Chao et al. (2022)

environmentally friendly approach to pest management (He et al., 2022b). The success of RNAi insecticides depends on the efficient delivery and uptake of dsRNA by the target pest, ensuring activation of the gene-silencing pathway. Ongoing research is focused on making these dsRNAi insecticides better, cheaper, and more compliant with regulatory standards (Zotti et al., 2018). With progress in formulation and delivery technologies, the future of RNAi in agriculture looks promising. It could play a pivotal role in addressing major pest challenges while supporting the shift toward more sustainable and biodiversity-friendly farming systems.

6.1 Enhancing Biological Control Agents

RNAi can be used to weaken pest immune defenses, thereby increasing their susceptibility to entomopathogens, nematodes, and microbial biocontrol agents. For instance, dsRNA targeting immune-related genes such as *Relish* (a key *NF-κB* transcription factor) or antimicrobial peptide genes in insects has been shown to enhance the efficacy of *Beauveria bassiana* and *Metarhizium anisopliae* (Moreira-Pinto et al., 2021; Ye et al., 2022; Shao et al., 2024). Such synergistic applications reduce reliance on chemical pesticide use and support ecosystem-based pest regulation. Targeting multiple essential genes with different dsRNAs has also been investigated to improve gene silencing efficiency. In the emerald ash borer (*Agrilus planipennis*), simultaneous delivery of dsRNA against *Hsp70 and shibire* genes resulted in enhanced mortality rates of up to 90% (Rodrigues et al., 2018). A similar multiplex approach was demonstrated in *Tribolium castaneum*, where dsRNAs targeting *Bip* and *Armet* were delivered in combination with BAPC (Branched Amphiphilic Peptide Capsule) nanoparticles. This treatment significantly increased mortality to approximately 60%, compared to only 25% when dsRNA was used alone (Avila et al., 2018).

6.2 Compatibility with Bt Toxins

RNAi can be combined with Bt toxins to achieve synergistic pest mortality and delay resistance development. For example, maize varieties expressing both *Cry* toxins and dsRNA targeting *Snf7* have shown enhanced protection against *Diabrotica virgifera virgifera* (Head et al., 2017). The use of stacked traits, Bt proteins for acute toxicity and RNAi for gene-specific suppression, creates multiple barriers to pest adaptation, extending the durability of pest control technologies (Moar et al., 2017; Ni et al., 2017).

Similarly, the combination of dsRNA with transgenic plants has been explored as a strategy to enhance pest control efficacy. In this approach, transgenic plants expressing both Bt toxins and dsRNA were evaluated for improvements in RNAi effectiveness. A notable example is the U.S. Environmental Protection Agency's recent approval of the SmartStax® PRO maize, which expresses multiple Bt toxins along with dsRNA targeting the *Snf7* gene in the Western corn rootworm. Field evaluation demonstrated 80–95% mortality and reduced root damage, while also improving resistance management compared to Bt-only hybrids (Head et al., 2017; Moar et al., 2017).

Beyond corn rootworm, similar combinatorial approaches have been tested in other species. For example, cotton plants expressing Bt toxins combined with dsRNA targeting juvenile hormone pathway genes in *Helicoverpa armigera* delayed resistance development in the pest compared to plants expressing Bt alone. This highlights the potential of RNAi to extend the lifespan of Bt crops and draws attention to the ongoing challenge of adaptive resistance (Ni et al., 2017).

6.3 Predator-safe Pest Control

RNAi operates via sequence-specific gene targeting; non-target organisms such as pollinators, predators, and parasitoids remain largely unaffected. Numerous laboratory and field studies have confirmed that dsRNAs designed for pests like Colorado potato beetle and Western corn rootworm do not negatively impact beneficial species such as lady beetles, lacewings, or honeybees. This non-toxic profile enables RNAi to integrate seamlessly with conservation biological control practices (Petrick et al., 2013; Christiaens et al., 2022; Willow et al., 2022).

6.4 Integrative Role in IPM

RNAi can be adopted as a core component of IPM strategies due to its specificity, safety, and adaptability. Unlike conventional insecticides, RNAi can complement cultural, mechanical, and biological methods of control practices without disrupting beneficial organisms. Through rotation with other control tactics or co-application with microbial biocontrol agents, RNAi reduces reliance on chemical insecticides and delays the emergence of resistance (Zotti et al., 2018; Hanamasagar et al., 2024). Furthermore, its compatibility with precision agriculture platforms (e.g., drone spraying, targeted delivery systems) underscores its utility in next-generation IPM frameworks (Willow et al., 2022).

7. Challenges in RNAi-based Crop Protection and Prospects

Despite its promise, the implementation of RNAi-based pest and disease control faces several biological, technical, and logistical challenges. One of the major limitations is the variability of RNAi efficiency among different insect taxa. Coleopterans are highly susceptible to orally delivered dsRNA, whereas lepidopterans and hemipterans often show poor responses due to factors such as rapid degradation by nucleases, inefficient uptake, and absence of systemic RNAi mechanisms (Terenius et al., 2011; Zhu & Palli, 2020). This inconsistency necessitates tailored approaches for each pest group. Delivery also remains a critical bottleneck, especially for field applications. SIGS is limited by dsRNA instability under environmental conditions such as UV radiation, heat, and rain. Encapsulation using nanocarriers, although promising, raises concerns regarding cost, scalability, and biosafety (Lu et al., 2023; Mendoza-Alatorre et al., 2025). HIGS, while effective, requires transgenic crops, which are subject to strict regulatory oversight and limited public acceptance in some regions (Koch & Kogel, 2014). Another major challenge is the potential development of resistance in pest populations. Continuous exposure to dsRNA targeting a single gene may exert selection pressure leading to gene mutations, RNAi pathway suppression, or enhanced dsRNase expression (Khajuria et al., 2018). Strategies to mitigate this include rotating target genes, combining RNAi with other modes of action (e.g., Bt toxins), and deploying gene stacking approaches.

Regulatory frameworks for RNAi products are still evolving and show notable differences globally. For example, the United States Environmental Protection Agency (EPA) has established emerging guidelines for RNAi-based pesticides, while the European Union remains in the early stages of risk assessments for non-GMO RNAi applications. Achieving harmonization of international regulations and expediting approval processes will be essential for product innovation in this field. Additionally, public acceptance plays a crucial role in commercialization. Unlike GMOs, RNAi, especially SIGS-based solutions, tends to be received more positively by regulators and the public due to its non-transgenic nature (Taning et al., 2020). Moreover, as reviewed by Rodrigues and Petrick (2020), biological safeguards such as digestive degradation, limited uptake, and rapid metabolic clearance greatly reduce the risk of systemic exposure to externally applied dsRNA in humans and other vertebrates, supporting the strong safety profile of these products.

However, widespread adoption will depend on transparent communication, education on biosafety, and demonstrable efficacy in real-world settings. From an industry perspective, RNAi has attracted significant interest from both major agricultural biotechnology corporations and emerging startups. For instance, GreenLight Biosciences has secured substantial investment, including €35 million, to advance RNA-based alternatives to conventional chemical pesticides (AgTechNavigator, 2025). Large corporations are also actively investing in RNAi technologies; for example, Syngenta's acquisition of DevGen in 2013 expanded its capacity to develop sprayable RNAi-based crop protection solutions. Commercial formulations are currently in development or undergoing field trials by companies such as GreenLight Biosciences, Bayer CropScience, and Syngenta. Prospects include advancements in formulation technologies, cost-effective dsRNA production platforms (e.g., bacterial or plant-based expression), and integration with precision agriculture tools. Extending RNAi applications beyond insect pest management to target plant pathogens, nematodes, weeds, and even abiotic stress resilience will further strengthen its role in sustainable agriculture.

8. Conclusions and Future Prospects

RNAi represents a precise, adaptable, and environmentally sustainable strategy for the management of agricultural insect pests and diseases. Its proven efficacy against diverse targets, including insects, nematodes, fungi, and viruses, RNAi is positioned as a viable alternative to conventional pesticides, especially when integrated into IPM frameworks. However, for widespread adoption, several challenges must be addressed, such as variability in RNAi efficiency across taxa, rapid degradation of dsRNA under environmental stressors, inefficient uptake in certain species, and potential resistance development remain major constraints. Advances in formulation technologies, nanocarrier-mediated delivery, symbiont-based systems, and multiplex gene targeting are essential to enhance field-level efficacy. Furthermore, integrating RNAi with complementary tools, such as Bt toxins, microbial biocontrol agents, and precision agriculture platforms, can improve control outcomes and delay resistance evolution.

Regulatory harmonization and public acceptance will be critical to the successful commercialization of RNAi-based products. Non-transgenic approaches, such as SIGS, may progress more rapidly in regions with strict GMO regulations, while cost-effective dsRNA production systems, identification of robust and conserved gene targets, and effective incorporation into IPM programs remain key research priorities. Expanding the application of RNAi beyond pest control to include weed management, plant–microbe interaction modulation, and enhancement of abiotic stress resilience further underscores its transformative potential. Coupled with precision agriculture tools such as drone-based spraying and sensor-guided pest monitoring, RNAi could enable highly targeted, responsive, and sustainable crop protection strategies. With continued innovation, interdisciplinary collaboration, and science-based regulatory frameworks, RNAi can significantly contribute to reducing reliance on synthetic pesticides, preserving biodiversity, and ensuring long-term global food security in the coming decades.

References

Abdellatef, E., Will, T., Koch, A., Imani, J., Vilcinskas, A. and Kogel, K.H. (2015). Silencing the expression of the salivary sheath protein causes transgenerational feeding suppression in the aphid, *Sitobion avenae*. *Plant Biotechnology Journal*, *13*(6), 849–857.

Adeyinka, O.S., Nasir, I.A. and Tabassum, B. (2023). Host-induced silencing of the *CpCHI* gene resulted in developmental abnormalities and mortality in maize stem borer (*Chilo partellus*). *PLoS One*, *18*(1), e0280963.

Agrawal, N., Dasaradhi, P.V., Mohmmed, A., Malhotra, P., Bhatnagar, R.K. and Mukherjee, S.K. (2003). RNA interference: Biology, mechanism, and applications. *Microbiology and Molecular Biology Reviews*, *67*(4), 657–685.

AgTechNavigator. (2025). *GreenLight Biosciences gets €35m to develop RNA-based biopesticides*. Retrieved August 13, 2025, from https://www.agtechnavigator.com/Article/2025/03/03/greenlight-biosciences-gets-35m-to-develop-rna-based-biopesticides.

Ahmad, S., Jamil, M., Lodhi, A. F., Barati, Z., Kakar, M.U., Gao, Y. and Zhang, W. (2025). RNAi revolution in agriculture: unlocking mechanisms, overcoming delivery challenges, and advancing sustainable Pest control. *Pest Management Science*. doi: 10.1002/ps.70040. Advance online publication.

Akbar, S., Wei, Y. and Zhang, M.Q. (2022). RNA Interference: Promising Approach to Combat Plant Viruses. *International Journal of Molecular Sciences*, *23*(10), 5312.

Amdam, G.V., Simões, Z.L., Guidugli, K.R., Norberg, K. and Omholt, S.W. (2003). Disruption of vitellogenin gene function in adult honeybees by intra-abdominal injection of double-stranded RNA. *BMC Biotechnology*, *3*, 1.

Anderson, J.A., Ellsworth, P.C., Faria, J.C., Head, G.P., Owen, M.D.K., Pilcher, C.D., Shelton, A.M. and Meissle, M. (2019). Genetically engineered crops: Importance of diversified integrated pest management for agricultural sustainability. *Frontiers in Bioengineering and Biotechnology*, *7*, 24.

Arakane, Y., Specht, C.A., Kramer, K.J., Muthukrishnan, S. and Beeman, R.W. (2008). Chitin synthases are required for survival, fecundity, and egg hatch in the red flour beetle, *Tribolium castaneum*. *Insect Biochemistry and Molecular Biology*, *38*(10), 959–962.

Araujo, R.N., Santos, A., Pinto, F.S., Gontijo, N.F., Lehane, M.J. and Pereira, M.H. (2006). RNA interference of the salivary nitrophorin 2 in the blood-sucking bug *Rhodnius prolixus* (Hemiptera: Reduviidae). *Insect Biochemistry and Molecular Biology*, *36*(9), 683–693.

Arjunan, N., Singh, I.K. and Sharma, H.C. (2024). Advances in nanoparticle-mediated RNAi delivery for pest control: Overcoming dsRNA degradation barriers. *Pest Management Science*, *80*(4), 1550–1564.

Aronstein, K. and Saldivar, E. (2005). Characterization of a honey bee Toll-related receptor gene *Am18w* and its potential involvement in antimicrobial immune defense. *Apidologie*, *36*(1), 3–14.

Avila, L.A., Chandrasekar, R., Wilkinson, K.E., Balthazor, J., Heerman, M., Bechard, J., Brown, S., Park, Y., Dhar, S., Reeck, G.R. and Tomich, J.M. (2018). Delivery of lethal dsRNAs in insect diets by branched amphiphilic peptide capsules. *Journal of Controlled Release*, *273*, 139–146.

Bachman, P.M., Bolognesi, R., Moar, W.J., Mueller, G.M., Paradise, M.S., Ramaseshadri, P. and Levine, S.L. (2013). Characterization of the spectrum of insecticidal activity of a double-stranded RNA with targeted activity against Western corn rootworm (*Diabrotica virgifera virgifera* LeConte). *Transgenic Research*, *22*(6), 1207–1222.

Badisco, L., Marchal, E., Wielendaele, P.V., Verlinden, H., Vleugels, R. and Broeck, J.V. (2011). RNA interference of insulin-related peptide and neuroparsins affects vitellogenesis in the desert locust *Schistocerca gregaria*. *Peptides*, *32*(3), 573–580.

Bao, W., Li, A., Zhang, Y., Diao, P., Zhao, Q., Yan, T., Zhou, Z., Duan, H., Li, X. and Wuriyanghan, H. (2021). Improvement of host-induced gene silencing efficiency via polycistronic-tRNA-amiR expression for multiple target genes and characterization of RNAi mechanism in *Mythimna separata*. *Plant Biotechnology Journal*, *19*(7), 1370–1385.

Bartel, D.P. (2004). MicroRNAs: Genomics, biogenesis, mechanism, and function. *Cell*, *116*(2), 281–297.

Banerjee, S., Banerjee, A., Gill, S.S., Gupta, O.P., Dahuja, A., Jain, P.K. and Sirohi, A. (2017). RNA interference: A novel source of resistance to combat plant parasitic nematodes. *Frontiers in Plant Science*, *8*, 834.

Banji, A., Ajose, T., Matthew, J.O., Azeez, S.O., Arogundade, O. (2025). RNA interference (RNAi)-Based Techniques for Management of Plant Viruses. In: Goswami, K., Gelaw, T.A., Sanan-Mishra, N. (Eds.), *Non-coding RNAs for Crop Improvement*. Springer, Singapore.

Basso, M.F., Vásquez, D.D.N., Campos-Pinto, E.R., Pinheiro, D.H., Cruz, B., Maktura, G.C., Guidelli, G.V., Marques-Souza, H. and Grossi-de-Sa, M.F. (2025). Progress and opportunities of *in planta* and topical RNAi for the biotechnological control of agricultural pests. *Agronomy*, *15*(4), 859.

Baulcombe, D. (2004). RNA silencing in plants. *Nature*, *431*(7006), 356–363.

Baum, J.A. and Roberts, J.K. (2014). Progress towards RNAi-mediated insect pest management. *Advances in Insect Physiology*, *47*, 249–295.

Baum, J.A., Bogaert, T., Clinton, W., Heck, G.R., Feldmann, P., Ilagan, O., Johnson, S., Plaetinck, G., Munyikwa, T., Pleau, M., Vaughn, T. and Roberts, J. (2007). Control of coleopteran insect pests through RNA interference. *Nature Biotechnology*, *25*(11), 1322–1326.

Beernink, B.M., Amanat, N., Li, V.H., Manchur, C.L., Whyard, S. and Belmonte, M.F. (2024). SIGS vs. HIGS: Opportunities and challenges of RNAi pest and pathogen control strategies. *Canadian Journal of Plant Pathology*, *46*(6), 675–689.

Bernstein, E., Caudy, A.A., Hammond, S.M. and Hannon, G.J. (2001). Role for a bidentate ribonuclease in the initiation step of RNA interference. *Nature*, *409*(6818), 363–366.

Bilir, Ö., Telli, O., Norman, C., Budak, H., Hong, Y. and Tör, M. (2019). Small RNA inhibits infection by downy mildew pathogen *Hyaloperonospora arabidopsidis*. *Molecular Plant Pathology*, *20*(11), 1523–1534.

Bologna, N.G. and Voinnet, O. (2014). The diversity, biogenesis, and activities of endogenous silencing small RNAs in *Arabidopsis*. *Annual Review of Plant Biology*, *65*, 473–503.

Bolognesi, R., Ramaseshadri, P., Anderson, J., Bachman, P., Clinton, W., Flannagan, R., Ilagan, O., Lawrence, C., Levine, S., Moar, W., Mueller, G., Tan, J., Uffman, J., Wiggins, E., Heck, G. and Segers, G. (2012). Characterizing the mechanism of action of double-stranded RNA activity against *Western Corn Rootworm* (*Diabrotica virgifera virgifera* LeConte). *PLoS One*, *7*(10), e47534.

Borges, F. and Martienssen, R.A. (2015). The expanding world of small RNAs in plants. *Nature Reviews Molecular Cell Biology*, *16*(12), 727–741.

Brosnan, C.A. and Voinnet, O. (2009). The long and the short of noncoding RNAs. *Current Opinion in Cell Biology*, *21*(3), 416–425.

Capriotti, L., Molesini, B., Pandolfini, T., Jin, H., Baraldi, E., Cecchin, M., Mezzetti, B. and Sabbadini, S. (2024). RNA interference-based strategies to control Botrytis cinerea infection in cultivated strawberry. *Plant Cell Reports*, *43*(8), 201.

Chao, Z., Ma, Z., Zhang, Y., Yan, S. and Shen, J. (2022). RNA interference cannot be operated in lepidopteran insect? A nanocarrier breaks bottlenecks at all developmental stages of *Spodoptera frugiperda*. https://doi.org/10.21203/rs.3.rs-2127285/v1 .

Chen, J., Zhang, D., Yao, Q., Zhang, J., Dong, X., Tian, H., Chen, J. and Zhang, W. (2010a). Feeding-based RNA interference of a trehalose phosphate synthase gene in the brown planthopper, *Nilaparvata lugens*. *Insect Molecular Biology*, *19*(6), 777–786.

Chen, J., Tang, B., Chen, H.X., Yao, Q., Huang, X.F., Chen, J., Zhang, D.W. and Zhang, W.Q. (2010b). Different functions of the insect soluble and membrane-bound trehalase genes in chitin biosynthesis revealed by RNA interference. *PLoS One,* 5(5), e10133.

Chen, X., Tian, H., Zou, L., Tang, B., Hu, J. and Zhang, W. (2008). Disruption of *Spodoptera exigua* larval development by silencing chitin synthase gene A with RNA interference. *Bulletin of Entomological Research*, *98*(6), 613–619.

Chen, Y. and De Schutter, K. (2024). Biosafety aspects of RNAi-based pest control. *Pest Management Science, 80*(8), 3697–3706.

Cheng, W., Song, X., Li, H., Cao, L., Sun, K., Qiu, X., Xu, Y. and Yang, P. (2015). Host-induced gene silencing of an essential chitin synthase gene confers durable resistance to *Fusarium* head blight and seedling blight in wheat. *Plant Biotechnology Journal*, *13*(9), 1335–1345.

Choudry, M.W., Nawaz, P., Jahan, N., Riaz, R., Ahmad, B., Raza, M.H., Fayaz, Z., Malik, K. and Afzal, S. (2024). RNA based gene silencing modalities to control insect and fungal plant pests: Challenges and future prospects. *Physiological and Molecular Plant Pathology*, *130*, 102241.

Christiaens, O., Niu, J. and Taning, C.N.T. (2020). RNAi in insects: A revolution in fundamental research and pest control applications. *Insects*, *11*(7), 415.

Christiaens, O., Sweet, J., Dzhambazova, T., Urru, I., Smagghe, G. and Taning, C.N.T. (2022). Implementation of RNAi-based arthropod pest control: Environmental risks, potential for resistance and regulatory considerations. *Journal of Pest Science, 95*(1), 1–16.

Christiaens, O., Tardajos, M.G., Martinez Reyna, Z.L., Dash, M., Dubruel, P. and Smagghe, G. (2018). Increased RNAi efficacy in *Spodoptera exigua* via the formulation of dsRNA with guanylated polymers. *Frontiers in Physiology*, *9*, 316.

Cooper, A.M., Silver, K., Zhang, J., Park, Y. and Zhu, K.Y. (2019). Molecular mechanisms influencing efficiency of RNA interference in insects. *Pest Management Science, 75*(1), 18–28.

Curtin, S.J., Watson, J.M., Smith, N.A., Eamens, A., Blanchard, C.L. and Waterhouse, P.M. (2008). The roles of plant dsRNA-binding proteins in RNAi-like pathways. *FEBS Letters*, *582*(18), 2753–2760.

Dalakouras, A., Wassenegger, M., Dadami, E., Ganopoulos, I., Pappas, M. L. and Papadopoulou, K. (2020a). Genetically Modified Organism-Free RNA Interference: Exogenous Application of RNA Molecules in Plants. *Plant Physiology*, *182*(1), 38–50.

Dalakouras, A., Wassenegger, M., McMillan, J.N., Cardoza, V., Maegele, I., Dadami, E., Runne, M., Krczal, G. and Wassenegger, M. (2020b). Induction of silencing in plants by high-pressure spraying of *in vitro*–synthesized small RNAs. *Frontiers in Plant Science*, *7*, 1327.

Darlington, M., Reinders, J.D., Sethi, A., Lu, A.L., Ramaseshadri, P., Fischer, J.R., Boeckman, C.J., Petrick, J.S., Roper, J.M., Narva, K.E. and Vélez, A.M. (2022). RNAi for Western Corn Rootworm Management: Lessons Learned, Challenges, and Future Directions. *Insects*, *13*(1), 57.

Das, P.R. and Sherif, S.M. (2020). Application of Exogenous dsRNAs-induced RNAi in Agriculture: Challenges and Triumphs. *Frontiers in Plant Science*, *25*, 11:946.

Dinh, P.T., Brown, C.R. and Elling, A.A. (2014). RNA Interference of Effector Gene Mc16D10L Confers Resistance Against *Meloidogyne chitwoodi* in Arabidopsis and Potato. *Phytopathology*, *104*(10), 1098–1106.

Dinh, P.T., Zhang, L., Mojtahedi, H., Brown, C.R. and Elling, A.A. (2015). Broad Meloidogyne Resistance in Potato Based on RNA Interference of Effector Gene 16D10. *Journal of nematology*, *47*(1), 71–78.

Dong, S., Ye, Z., Tikhe, C.V., Tu, Z.J. and Zwiebel, L.J. (2021). Pleiotropic odorant-binding proteins promote *Aedes aegypti* reproduction and flavivirus transmission. *mBio*, *12*(6), e0253121.

Dou, T., Shao, X., Hu, C., Liu, S., Sheng, O., Bi, F., Deng, G., Ding, L., Li, C., Dong, T., Gao, H., He, W., Peng, X., Zhang, S., Huo, H., Yang, Q. and Yi, G. (2020). Host-induced gene silencing of *Foc TR4 ERG6/11* genes exhibits superior resistance to Fusarium wilt of banana. *Plant Biotechnology Journal*, *18*(1), 11–13.

Dubelman, S., Fischer, J., Zapata, F., Huizinga, K., Jiang, C., Uffman, J. and Levine, S.L. (2014). Environmental fate of double-stranded RNA in agricultural soils. *PLoS One, 9*(3), e93155.

Dyson, P., Figueiredo, M., Andongma, A.A. and Whitten, M.M.A. (2022). Symbiont-mediated RNA interference (SMR): Using symbiotic bacteria as vectors for delivering RNAi to insects. *Methods in Molecular Biology*, *2360*, 295–306.

Elbashir, S.M., Lendeckel, W. and Tuschl, T. (2001). RNA interference is mediated by 21- and 22-nucleotide RNAs. *Genes and Development*, *15*(2), 188–200.

Elhaj Baddar, Z., Gurusamy, D., Laisney, J., Tripathi, P., Palli, S.R. and Unrine, J.M. (2020). Polymer-coated hydroxyapatite nanocarrier for double-stranded RNA delivery. *Journal of Agricultural and Food Chemistry*, *68*(24), 6811–6818.

Erdelyan, C.N., Mahood, T.H., Bader, T.S. and Whyard, S. (2012). Functional validation of the carbon dioxide receptor genes in *Aedes aegypti* mosquitoes using RNA interference. *Insect Molecular Biology, 21*(1), 119–127.

Fang, X. and Qi, Y. (2016). RNAi in plants: An Argonaute-centered view. *Plant Cell*, *28*(2), 272–285.

Feng, H., Chen, W., Hussain, S., Shakir, S., Tzin, V., Adegbayi, F., Ugine, T., Fei, Z. and Jander, G. (2023). Horizontally transferred genes as RNA interference targets for aphid and whitefly control. *Plant Biotechnology Journal*, *21*(4), 754–768.

Feinberg, E.H. and Hunter, C.P. (2003). Transport of dsRNA into cells by the transmembrane protein SID-1. *Science* (*New York, N.Y.*), *301*(5639), 1545–1547.

Filipowicz, W. (2005). RNAi: the nuts and bolts of the RISC machine. *Cell*, *122*(1), 17–20.

Fire, A., Xu, S., Montgomery, M.K., Kostas, S.A., Driver, S.E. and Mello, C.C. (1998). Potent and specific genetic interference by double-stranded RNA in *Caenorhabditis elegans*. *Nature*, *391*(6669), 806–811.

Fishilevich, E., Vélez, A.M., Storer, N.P., Li, H., Bowling, A.J., Rangasamy, M., Worden, S.E., Narva, K.E. and Siegfried, B.D. (2016). RNAi as a management tool for the western corn rootworm, *Diabrotica virgifera virgifera*. *Pest Management Science*, *72*(9), 1652–1663.

Fukudome, A. and Fukuhara, T. (2017). Plant dicer-like proteins: Double-stranded RNA-cleaving enzymes for small RNA biogenesis. *Journal of Plant Research*, *130*(1), 33–44.

Gaffar, F.Y. and Koch, A. (2019). Catch Me If You Can! RNA Silencing-Based Improvement of Antiviral Plant Immunity. *Viruses*, *11*(7), 673.

Ghag, S.B. (2023). RNAi-based biofungicides. In: *Biofungicides: Eco-Safety and Future Trends* (pp. 21–41). CRC Press.

Ghodke, A.B., Good, R.T., Golz, J.F., Russell, D.A., Edwards, O.R. and Robin, C. (2019). Extracellular endonucleases in the midgut of *Myzus persicae* may limit the efficacy of orally delivered RNAi. *Scientific Reports, 9*, 11898.

Ghosh, S.K.B. and Gundersen-Rindal, D.E. (2017). Double-strand RNA-mediated RNA interference through feeding in larval gypsy moth, *Lymantria dispar* (Lepidoptera: Erebidae). *European Journal of Entomology*, *114*, 170–178.

Gillet, F.X., Garcia, R.A., Macedo, L.L., Albuquerque, E.V., Silva, M.C. and Grossi-de-Sa, M.F. (2017). Investigating engineered ribonucleoprotein particles to improve oral RNAi delivery in crop insect pests. *Frontiers in Physiology*, *8,* 256.

Goulson, D. (2013). An overview of the environmental risks posed by neonicotinoid insecticides. *Journal of Applied Ecology*, *50*(4), 977–987.

Griebler, M., Westerlund, S.A., Hoffmann, K.H. and Meyering-Vos, M. (2008). RNA interference with cytochrome P450 monooxygenase gene expression in *Spodoptera frugiperda*. *Archives of Insect Biochemistry and Physiology*, *69*(3), 146–160.

Gu, L. and Knipple, D.C. (2013). Recent advances in RNA interference research in insects: Implications for future insect pest management strategies. *Crop Protection, 45,* 36–40.

Guo, H., Song, X., Wang, G., Yang, K., Wang, Y., Niu, L., Chen, X. and Fang, R. (2014). Plant-generated artificial small RNAs mediated aphid resistance. *PLoS One, 9*(4), e97410.

Guo, X., Li, Y., Fan, J., Xiong, H., Xu, F., Shi, J., Shi, Y. and Wang, W. (2019). Host-induced gene silencing of *MoAP1* confers broad-spectrum resistance to *Magnaporthe oryzae*. *Frontiers in Plant Science*, *10*, 433.

Gurusamy, D., Mogilicherla, K. and Palli, S.R. (2020). Chitosan nanoparticles help double-stranded RNA escape from endosomes and improve RNA interference in the fall armyworm, *Spodoptera frugiperda*. *Archives of Insect Biochemistry and Physiology*, *104*(4), e21677.

Hahn, B.-S., Ajjappala, H. and Sim, J.-S. (2012). RNA interference silencing in root-knot nematodes. *Journal of the Korean Society of International Agriculture, 24(*4), 485–493.

Hammond, S.M. (2005). Dicing and slicing: The core machinery of the RNA interference pathway. *FEBS Letters*, *579*(26), 5822–5829.

Hammond, S.M., Bernstein, E., Beach, D. and Hannon, G.J. (2000). An RNA-directed nuclease mediates post-transcriptional gene silencing in *Drosophila* cells. *Nature*, *404*(6775), 293–296.

Hanamasagar, Y., Ramesha, N.M., Mahapatra, S., Panigrahi, C.K., Vidhya, C.S., Agnihotri, N. and Satapathy, S.N. (2024). Advancing RNAi for sustainable insect management: Targeted solutions for eco-friendly pest control. *Journal of Experimental Agriculture International, 46*(6), 740–775.

He, G. (2022). Engineering chloroplasts for insect pest control. *PNAS*, *119*(25), e2205125119.

He, B., Chu, Y., Yin, M., Müllen, K., An, C. and Shen, J. (2013). Fluorescent nanoparticle delivered dsRNA toward genetic control of insect pests. *Advanced materials (Deerfield Beach, Fla.)*, *25*(33), 4580–4584.

He, W., Xu, W., Xu, L., Fu, K., Guo, W., Bock, R. and Zhang, J. (2020). Length-dependent accumulation of double-stranded RNAs in plastids affects RNA interference efficiency in the Colorado potato beetle. *Journal of Experimental Botany, 71*(9), 2670–2677.

He, W., Xu, W., Fu, K., Guo, W., Kim, D.S. and Zhang, J. (2022a). Positional effects of double-stranded RNAs targeting *β-actin* gene affect RNA interference efficiency in Colorado potato beetle. *Pesticide Biochemistry and Physiology, 184,* 105121.

He, L., Huang, Y. and Tang, X. (2022b). RNAi-based pest control: Production, application, and the fate of dsRNA. *Frontiers in Bioengineering and Biotechnology, 10*, 1080576.

Head, G.P., Carroll, M.W., Evans, S.P., Rule, D.M., Willse, A.R., Clark, T.L., Storer, N.P., Flannagan, R.D., Samuel, L.W. and Meinke, L.J. (2017). Evaluation of SmartStax and SmartStax PRO maize against western corn rootworm and northern corn rootworm: Efficacy and resistance management. *Pest Management Science, 73*(9), 1883–1899.

Hoang, B.T.L., Fletcher, S.J., Brosnan, C.A., Ghodke, A.B., Manzie, N. and Mitter, N. (2022). RNAi as a foliar spray: Efficiency and challenges to field applications. *International Journal of Molecular Sciences*, *23*(12), 6639.

Hu, D., Chen, Z.Y., Zhang, C. and Ganiger, M. (2020). Reduction of *Phakopsora pachyrhizi* infection on soybean through host- and spray-induced gene silencing. *Molecular Plant Pathology*, *21*(6), 794–807.

Hutvágner, G. and Zamore, P.D. (2002). A microRNA in a multiple-turnover RNAi enzyme complex. *Science* (*New York, N.Y.*), *297*(5589), 2056–2060.

Ibrahim, H.M., Alkharouf, N.W., Meyer, S.L., Aly, M.A., Gamal El-Din, A.K., Hussein, E.H. and Matthews, B.F. (2011). Post-transcriptional gene silencing of root-knot nematode in transformed soybean roots. *Experimental Parasitology*, *127*(1), 90–99.

Irsad, I., Shahid, M., Haq, E., Mohamed, A., Rizvi, P.Q. and Kolanthasamy, E. (2023). Entomopathogen-based biopesticides: Insights into unravelling their potential in insect pest management. *Frontiers in Microbiology, 14*, 1208237.

Jahan, S.N., Åsman, A.K.M., Corcoran, P., Fogelqvist, J., Vetukuri, R.R. and Dixelius, C. (2015). Plant-mediated gene silencing restricts the growth of the potato late blight pathogen *Phytophthora infestans*. *Journal of Experimental Botany*, *66*(9), 2785–2794.

Jahromi, M.G., Rahnama, H., Mousavi, A. and Safarnejad, M.R. (2022). Comparative evaluation of resistance to potato virus Y (PVY) in three different RNAi-based transgenic potato plants. *Transgenic Research*, *31*(3), 313–323.

Jain, R.G., Robinson, K.E., Asgari, S. and Mitter, N. (2021). Current scenario of RNAi-based hemipteran control. *Pest Management Science*, *77*(5), 2188–2196.

Jiang, Y.X., Chen, J.Z., Li, M.W., Zha, B.H., Huang, P.R., Chu, X.M., Chen, J. and Yang, G. (2021). The Combination of *Bacillus thuringeiensis* and Its Engineered Strain Expressing dsRNA Increases the Toxicity against *Plutella xylostella*. *International Journal of Molecular Sciences*, *23*(1), 444.

Joga, M.R., Zotti, M.J., Smagghe, G. and Christiaens, O. (2016). RNAi efficiency, systemic properties, and novel delivery methods for pest insect control: What we know so far. *Frontiers in Physiology*, *7*, 553.

Joshi, I., Kumar, A., Singh, A.K., Kohli, D., Raman, K.V., Sirohi, A., Chaudhury, A. and Jain, P.K. (2019). Development of nematode resistance in Arabidopsis by HD-RNAi-mediated silencing of the effector gene Mi-msp2. *Scientific Reports*, *9*(1), 17404.

Joshi, I., Kumar, A., Kohli, D., Singh, A.K., Sirohi, A., Subramaniam, K., Chaudhury, A. and Jain, P.K. (2020). Conferring root-knot nematode resistance via host-delivered RNAi-mediated silencing of four Mi-msp genes in Arabidopsis. *Plant Science: An International Journal of Experimental Plant Biology*, *298*, 110592.

Kalyandurg, P.B., Sundararajan, P., Dubey, M., Ghadamgahi, F., Zahid, M.A., Whisson, S.C. and Vetukuri, R.R. (2021). Spray-induced gene silencing as a potential tool to control potato late blight disease. *Phytopathology*, *111*(12), 2168–2175.

Khajuria, C., Buschman, L.L., Chen, M.S., Muthukrishnan, S. and Zhu, K.Y. (2010). A gut-specific chitinase gene essential for regulation of chitin content of peritrophic matrix and growth of *Ostrinia nubilalis* larvae. *Insect Biochemistry and Molecular Biology*, *40*(8), 621–629.

Khajuria, C., Ivashuta, S., Wiggins, E., Flagel, L., Moar, W., Pleau, M., Miller, K., Zhang, Y., Ramaseshadri, P., Jiang, C., Hodge, T., Jensen, P., Chen, M., Gowda, A., McNulty, B., Vazquez, C., Bolognesi, R., Haas, J., Head, G. and Clark, T. (2018). Development and characterization of the first dsRNA-resistant insect population from western corn rootworm, *Diabrotica virgifera virgifera* LeConte. *PLoS One, 13*(5), e0197059.

Koch, A., Biedenkopf, D., Furch, A., Weber, L., Rossbach, O., Abdellatef, E., Linicus, L., Johannsmeier, J., Jelonek, L., Goesmann, A., Cardoza, V., McMillan, J., Mentzel, T. and Kogel, K. H. (2016). An RNAi-based control of *Fusarium graminearum* infections through spraying of long dsRNAs involves a plant passage and is controlled by the fungal silencing machinery. *PLoS Pathogens*, *12*(10), e1005901.

Koch, A. and Kogel, K.H. (2014). New wind in the sails: Improving the agronomic value of crop plants through RNAi-mediated gene silencing. *Plant Biotechnology Journal*, 12(7), 821–831.

Koch, A., Kumar, N., Weber, L., Keller, H., Imani, J. and Kogel, K. (2013). Host-induced gene silencing of cytochrome P450 lanosterol C14α-demethylase–encoding genes confers strong resistance to *Fusarium* species. In: *Proceedings of the National Academy of Sciences of the United States of America*, *110*(48), 19324–19329.

Kogan, M. (1998). Integrated pest management: Historical perspectives and contemporary developments. *Annual Review of Entomology*, *43*, 243–270.

Kolge, H., Kadam, K. and Ghormade, V. (2023). Chitosan nanocarriers mediated dsRNA delivery in gene silencing for *Helicoverpa armigera* biocontrol. *Pesticide Biochemistry and Physiology*, *189*, 105292.

Kumar, M., Gupta, G.P. and Rajam, M.V. (2009). Silencing of acetylcholinesterase gene of *Helicoverpa armigera* by siRNA affects larval growth and its life cycle. *Journal of Insect Physiology*, *55*(3), 273–278.

Laisney, J., Gurusamy, D., Elhaj Baddar, Z., Palli, S.R. and Unrine, J.M. (2020). RNAi in *Spodoptera frugiperda* Sf9 cells via nanomaterial-mediated delivery of dsRNA: A comparison of poly-L-arginine polyplexes and poly-L-arginine-functionalized Au nanoparticles. *ACS Applied Materials and Interfaces*, *12*(23), 25645–25657.

Lakshmanan, M., Kodama, Y., Yoshizumi, T., Sudesh, K. and Numata, K. (2013). Rapid and efficient gene delivery into plant cells using designed peptide carriers. *Biomacromolecules*, *14*(1), 10–16.

Li, C. and Wu, Y. (2025). Recent advancements in biotic and abiotic stress responses and regulation mechanism in horticultural plants. *Horticulturae*, *11*(4), 408.

Li, H., Khajuria, C., Rangasamy, M., Gandra, P., Fitter, M., Geng, C., Woosely, A., Hasler, J., Schulenberg, G., Worden, S., McEwan, R., Evans, C., Siegfried, B. and Narva, K.E. (2015a). Long dsRNA but not siRNA initiates RNAi in western corn rootworm larvae and adults. *Journal of Applied Entomology*, *139*(6), 432–445.

Li, Y., Wang, K., Xie, H., Wang, D.W., Xu, C.L., Huang, X., Wu, W.J. and Li, D.L. (2015b). Cathepsin B Cysteine Proteinase is Essential for the Development and Pathogenesis of the Plant Parasitic Nematode *Radopholus similis*. *International Journal of Biological Sciences*, *11*(9), 1073–1087.

Li, T., Chen, J., Fan, X., Chen, W. and Zhang, W. (2017). MicroRNA and dsRNA targeting chitin synthase A reveal great potential for pest management of the hemipteran insect *Nilaparvata lugens*. *Pest Management Science*, *73*(7), 1529–1537.

Li, X., Zhang, M. and Zhang, H. (2011). RNA interference of four genes in adult *Bactrocera dorsalis* by feeding their dsRNAs. *PLoS One, 6*(3), e17788.

Li, J., Chen, Q., Lin, Y., Jiang, T., Wu, G. and Hua, H. (2011b). RNA interference in *Nilaparvata lugens* (Homoptera: Delphacidae) based on dsRNA ingestion. *Pest Management Science*, *67*(7), 852–859.

Li, X., Mu, K., Yang, S., Wei, J., Wang, C., Yan, W., Yuan, F., Wang, H., Han, D., Kang, Z. and Zeng, Q. (2022). Reduction of *Rhizoctonia cerealis* infection on wheat through host- and spray-induced gene silencing of an orphan secreted gene. *Molecular Plant-Microbe Interactions*, *35*(9), 803–813.

Lin, Y., Huang, J., Liu, Y., Belles, X. and Lee, H. (2017). Oral delivery of dsRNA lipoplexes to German cockroach protects dsRNA from degradation and induces RNAi response. *Pest Management Science*, *73*(5), 960–966.

Lin, Y.-C., Lu, Y.-H., Liu, Y., Su, Y.-J., Lin, Y.-H. and Wu, Y.-L. (2025). Differential RNAi efficacy of siRNA and dsRNA targeting key genes for pest control in *Spodoptera litura*. *Frontiers in Insect Science, 5*, 1574585.

Liu, S., Ding, Z., Zhang, C., Yang, B. and Liu, Z. (2010). Gene knockdown by intra-thoracic injection of double-stranded RNA in the brown planthopper, *Nilaparvata lugens*. *Insect Biochemistry and Molecular Biology*, *40*(9), 666–671.

Lu, C., Li, Z., Chang, L., Dong, Z., Guo, P., Shen, G., Xia, Q. and Zhao, P. (2019). Efficient Delivery of dsRNA and DNA in Cultured Silkworm Cells for Gene Function Analysis Using PAMAM Dendrimers System. *Insects*, *11*(1), 12.

Lu, J. and Shen, J. (2024). Target genes for RNAi in pest control: A comprehensive overview. *Entomologia Generalis*, *44*(1), 95–114.

Lu, Y., Deng, X., Zhu, Q., Wu, D., Zhong, J., Wen, L. and Yu, X. (2023). The dsRNA Delivery, Targeting and Application in Pest Control. *Agronomy*, *13*(3), 714.

Lu, Y.-H., Liu, Y., Lin, Y.-C., Su, Y.-J., Henry, D., Lacarriere, V., Haller, S., Lin, Y. H., Chiang, P.-C., Hsu, C.-I., Hsu, W.-L., Chan, Y.-C., Chen, L. and Wu, Y.-L. (2025). Lipoplex-based RNA delivery system enhances RNAi efficiency for targeted pest control in *Spodoptera litura*. *International Journal of Biological Macromolecules, 319*(3), 145091.

Luan, J.B., Ghanim, M., Liu, S.S. and Czosnek, H. (2013). Silencing the ecdysone synthesis and signaling pathway genes disrupts nymphal development in the whitefly. *Insect Biochemistry and Molecular Biology*, *43*(8), 740–746.

Macrae, I.J., Li, F., Zhou, K., Cande, W.Z. and Doudna, J.A. (2006). Structure of Dicer and mechanistic implications for RNAi. *Cold Spring Harbor Symposia on Quantitative Biology, 71,* 73–80.

Malik, H. J., Raza, A., Amin, I., Scheffler, J. A., Scheffler, B. E., Brown, J. K. and Mansoor, S. (2016). RNAi-mediated mortality of the whitefly through transgenic expression of double-stranded RNA homologous to acetylcholinesterase and ecdysone receptor in tobacco plants. *Scientific Reports, 6*, 38469.

Mallory, A.C. and Vaucheret, H. (2010). Form, function, and regulation of ARGONAUTE proteins. *Plant Cell*, *22*(12), 3879–3889.

Mamta and Rajam, M.V. (2017). RNAi technology: A new platform for crop pest control. *Physiology and Molecular Biology of Plants*, *23*(3), 487–501.

Mamta, Reddy, K.R.K. and Rajam, M.V. (2016). Targeting chitinase gene of *Helicoverpa armigera* by host-induced RNA interference confers insect resistance in tobacco and tomato. *Plant Molecular Biology*, *90*(3), 281–292.

Mann, C.W.G., Sawyer, A., Gardiner, D.M., Mitter, N., Carroll, B.J. and Eamens, A.L. (2023). RNA-Based Control of Fungal Pathogens in Plants. *International Journal of Molecular Sciences*, *24*(15), 12391.

Mao, J. and Zeng, F. (2014). Plant-mediated RNAi of a gap gene enhances tobacco tolerance against *Myzus persicae*. *Transgenic Research, 23*(1), 145–152.

Mao, Y.B., Tao, X.Y., Xue, X.Y., Wang, L.J. and Chen, X.Y. (2011). Cotton plants expressing CYP6AE14 double-stranded RNA show enhanced resistance to bollworms. *Transgenic Research*, *20*(3), 665–673.

Margis, R., Fusaro, A.F., Smith, N.A., Curtin, S.J., Watson, J.M., Finnegan, E.J. and Waterhouse, P.M. (2006). The evolution and diversification of Dicers in plants. *FEBS Letters*, *580*(10), 2442–2450.

Masanga, J.O., Matheka, J.M., Omer, R.A., Ommeh, S.C., Monda, E.O. and Alakonya, A.E. (2015). Downregulation of transcription factor *aflR* in *Aspergillus flavus* confers reduction to aflatoxin accumulation in transgenic maize with alteration of host plant architecture. *Plant Cell Reports*, *34*(8), 1379–1387.

Matranga, C., Tomari, Y., Shin, C., Bartel, D.P. and Zamore, P.D. (2005). Passenger-strand cleavage facilitates assembly of siRNA into Ago2-containing RNAi enzyme complexes. *Cell*, *123*(4), 607–620.

McLoughlin, A.G., Wytinck, N., Walker, P.L., Girard, I.J., Rashid, K.Y., De Kievit, T., Fernando, W.G.D., Whyard, S. and Belmonte, M.F. (2018). Identification and application of exogenous dsRNA confers plant protection against *Sclerotinia sclerotiorum* and *Botrytis cinerea*. *Scientific Reports*, *8*(1), 7320.

McRae, A.G., Taneja, J., Yee, K., Shi, X., Haridas, S., LaButti, K., Singan, V., Grigoriev, I.V. and Wildermuth, M.C. (2023). Spray-induced gene silencing to identify powdery mildew gene targets and processes for powdery mildew control. *Molecular Plant Pathology*, *24*(9), 1168–1183.

Mendoza-Alatorre, M., Julian-Chávez, B., Solano-Ornelas, S., Siqueiros-Cendón, T.S., Torres-Castillo, J.A., Sinagawa-García, S.R., Abraham-Juárez, M.J., González-Barriga, C.D., Rascón-Cruz, Q., Siañez-Estrada, L.I. and Espinoza-Sánchez, E.A. (2025). RNAi in Pest Control: Critical Factors Affecting dsRNA Efficacy. *Insects*, *16*(7), 737.

Menezes, P.S., Yan, Y., Yang, Y., Mitter, N., Mahony, T.J. and Mody, K.T. (2022). RNAi-Based Biocontrol of Pests to Improve the Productivity and Welfare of Livestock Production. *Applied Biosciences*, *1*(3), 229–243.

Miller, S.C., Miyata, K., Brown, S.J. and Tomoyasu, Y. (2012). Dissecting systemic RNA interference in the red flour beetle *Tribolium castaneum*: Parameters affecting the efficiency of RNAi. *PloS One*, *7*(10), e47431.

Mirghasemi, N., Fanelli, E., Jamali, S., Sohani, M.M. and De Luca, F. (2021). Molecular Characterization of Three B-1,4-Endoglucanase Genes in *Pratylenchus loosi* and Functional Analysis of *Pl-eng*-2 Gene. *Plants*, *10*(3), 568.

Mitter, N., Worrall, E.A., Robinson, K.E., Li, P., Jain, R.G., Taochy, C., Fletcher, S.J., Carroll, B.J., Lu, G.Q. and Xu, Z.P. (2017). Clay nanosheets for topical delivery of RNAi for sustained protection against plant viruses. *Nature Plants*, *3*, 16207.

Moar, W., Khajuria, C., Pleau, M., Ilagan, O., Chen, M. and Jiang, C. (2017). Cry3Bb1-resistant western corn rootworm, *Diabrotica virgifera virgifera* (LeConte) does not exhibit cross-resistance to DvSnf7 dsRNA. *PLoS One, 12*(1), e0169175.

Moreira-Pinto, C.E., Coelho, R.R., Leite, A.G.B., Silveira, D.A., de Souza, D.A., Lopes, R.B., Macedo, L.L.P., Silva, M.C.M., Ribeiro, T.P. and Lucena, W.A. (2021). Increasing *Anthonomus grandis* susceptibility to *Metarhizium anisopliae* through RNAi-induced *AgraRelish* knockdown: A perspective to combine biocontrol and biotechnology. *Pest Management Science, 77*(9), 4054–4063.

Naessens, E., Dubreuil, G., Giordanengo, P., Baron, O.L., Minet-Kebdani, N., Keller, H. and Coustau, C. (2015). A secreted MIF cytokine enables aphid feeding and represses plant immune responses. *Current Biology*, *25*(14), 1898–1903.

Necira, K., Contreras, L., Kamargiakis, E., Kamoun, M.S., Canto, T. and Tenllado, F. (2024). Comparative analysis of RNA interference and pattern-triggered immunity induced by dsRNA reveals different efficiencies in the antiviral response to potato virus X. *Molecular Plant Pathology*, *25*(9), e70008.

Nerva, L., Sandrini, M., Gambino, G. and Chitarra, W. (2020). Double-stranded RNAs (dsRNAs) as a sustainable tool against gray mold (*Botrytis cinerea*) in grapevine: Effectiveness of different application methods in an open-air environment. *Biomolecules*, *10*(2), 200.

Ni, M., Ma, W., Wang, X., Gao, M., Dai, Y., Wei, X., Zhang, L., Peng, Y., Chen, S., Ding, L., Tian, Y., Li, J., Wang, H., Wang, X., Xu, G., Guo, W., Yang, Y., Wu, Y., Heuberger, S., Tabashnik, B.E., Zhang, T., Zhu, Z. (2017). Next-generation transgenic cotton: Pyramiding RNAi and Bt counters insect resistance. *Plant Biotechnology Journal*, *15*(9), 1204–1213.

Nowara, D., Gay, A., Lacomme, C., Shaw, J., Ridout, C., Douchkov, D., Hensel, G., Kumlehn, J. and Schweizer, P. (2010). HIGS: host-induced gene silencing in the obligate biotrophic fungal pathogen *Blumeria graminis*. *The Plant Cell*, *22*(9), 3130–3141.

Numata, K., Ohtani, M., Yoshizumi, T., Demura, T. and Kodama, Y. (2014). Local gene silencing in plants via synthetic dsRNA and carrier peptide. *Plant Biotechnology Journal*, *12*(8), 1027–1034.

Nunes, F.M.F. and Simões, Z.L.P. (2009). A non-invasive method for silencing gene transcription in honeybees maintained under natural conditions. *Insect Biochemistry and Molecular Biology*, *39*(2), 157–160.

Obbard, D.J., Gordon, K.H.J., Buck, A.H. and Jiggins, F.M. (2009). The evolution of RNAi as a defence against viruses and transposable elements. *Philosophical Transactions of the Royal Society B: Biological Sciences*, *364*(1513), 99–115.

Pal-Bhadra, M., Bhadra, U. and Birchler, J.A. (2002). RNAi related mechanisms affect both transcriptional and posttranscriptional transgene silencing in Drosophila. *Molecular Cell*, *9*(2), 315–327.

Palli, S.R. (2014). RNA interference in Colorado potato beetle: Steps toward development of dsRNA as a commercial insecticide. *Current Opinion in Insect Science*, 1–8.

Panwar, V., Jordan, M., McCallum, B. and Bakkeren, G. (2018). Host-induced silencing of essential genes in Puccinia triticina through transgenic expression of RNAi sequences reduces severity of leaf rust infection in wheat. *Plant Biotechnology Journal*, *16*(5), 1013–1023.

Parsons, K.H., Mondal, M.H., McCormick, C.L. and Flynt, A.S. (2018). Guanidinium-functionalized interpolyelectrolyte complexes enabling RNAi in resistant insect pests. *Biomacromolecules*, *19*(4), 1111–1117.

Pedigo, L.P. and Rice, M.E. (2014). *Entomology and Pest Management* (6th Edn.). Waveland Press.

Peng, Y., Wang, K., Zhu, G., Han, Q., Chen, J., Elzaki, M.E.A., Sheng, C., Zhao, C., Palli, S.R. and Han, Z. (2020). Identification and characterization of multiple dsRNases from a lepidopteran insect, the tobacco cutworm, *Spodoptera litura* (Lepidoptera: Noctuidae). *Pesticide Biochemistry and Physiology, 162*, 86–95.

Peshin, R. and Dhawan, A.K. (2009). *Integrated Pest Management: Innovation-Development Process*. Springer.

Petrick, J.S., Brower-Toland, B., Jackson, A.L. and Kier, L.D. (2013). Safety assessment of food and feed from biotechnology-derived crops employing RNA-mediated gene regulation to achieve desired traits: A scientific review. *Regulatory Toxicology and Pharmacology*, *66*(2), 167–176.

Pimentel, D. (2005). Environmental and economic costs of the application of pesticides primarily in the United States. *Environment, Development and Sustainability*, *7*(2), 229–252.

Piulachs, M.D., Pagone, V. and Bellés, X. (2010). Key roles of the Broad-Complex gene in insect embryogenesis. *Insect Biochemistry and Molecular Biology*, *40*(6), 468–475.

Prasad, K., Yogendra, K., Sanivarapu, H., Rajasekaran, K., Cary, J.W., Sharma, K.K. and Bhatnagar-Mathur, P. (2023). Multiplexed host-induced gene silencing of *Aspergillus flavus* genes confers aflatoxin resistance in groundnut. *Toxins*, *15*(5), 319.

Pratt, A.J. and MacRae, I.J. (2009). The RNA-induced silencing complex: A versatile gene-silencing machine. *Journal of Biological Chemistry*, *284*(27), 17897–17901.

Qi, T., Zhu, X., Tan, C., Liu, P., Guo, J., Kang, Z. and Guo, J. (2018). Host-induced gene silencing of an important pathogenicity factor *PsCPK1* in *Puccinia striiformis* f. sp. *tritici* enhances resistance of wheat to stripe rust. *Plant Biotechnology Journal*, *16*(3), 797–807.

Qiao, L., Lan, C., Capriotti, L., Ah-Fong, A., Nino Sanchez, J., Hamby, R., Heller, J., Zhao, H., Glass, N.L., Judelson, H.S., Mezzetti, B., Niu, D. and Jin, H. (2021). Spray-induced gene silencing for disease control is dependent on the efficiency of pathogen RNA uptake. *Plant Biotechnology Journal*, *19*(9), 1756–1768.

Rajagopal, R., Sivakumar, S., Agrawal, N., Malhotra, P. and Bhatnagar, R.K. (2002). Silencing of midgut aminopeptidase N of *Spodoptera litura* by double-stranded RNA establishes its role as *Bacillus thuringiensis* toxin receptor. *Journal of Biological Chemistry*, *277*(49), 46849–46851.

Rêgo-Machado, C.M., Inoue-Nagata, A.K. and Nakasu, E.Y.T. (2023). Topical application of dsRNA for plant virus control: A review. *Tropical Plant Pathology, 48,* 11–22.

Reinders, J.D., Moar, W.J., Head, G.P., Hassan, S. and Meinke, L.J. (2023). Effects of SmartStax® and SmartStax® PRO maize on western corn rootworm (*Diabrotica virgifera virgifera* LeConte) larval feeding injury and adult life history parameters. *PloS One*, *18*(7), e0288372.

Ren, B., Cao, J., He, Y., Yang, S. and Zhang, J. (2021). Assessment on effects of transplastomic potato plants expressing Colorado potato beetle *β-Actin* double-stranded RNAs for three non-target pests. *Pesticide Biochemistry and Physiology*, *178*, 104909.

Rivera-Pérez, C., Clifton, M.E. and Noriega, F.G. (2020). Juvenile hormone regulation and action. In: *Juvenile Hormones in Invertebrates*. CRC Press, pp. 1–76.

Rodrigues, T.B., Duan, J.J., Palli, S.R. and Rieske, L.K. (2018). Identification of highly effective target genes for RNAi-mediated control of emerald ash borer, *Agrilus planipennis*. *Scientific Reports*, *8*, 5020.

Rodrigues, T.B. and Petrick, J.S. (2020). Safety considerations for humans and other vertebrates to agricultural uses of externally applied RNA molecules. *Frontiers in Plant Science*, *11*, 407.

Rodrigues, T.B., Mishra, S.K., Sridharan, K., Barnes, E.R., Alyokhin, A., Tuttle, R., Kokulapalan, W., Garby, D., Skizim, N.J., Tang, Y.W., Manley, B., Aulisa, L., Flannagan, R.D., Cobb, C. and Narva, K.E. (2021). First sprayable double-stranded RNA-based bio-pesticide product targets proteasome subunit beta type-5 in Colorado potato beetle (*Leptinotarsa decemlineata*). *Frontiers in Plant Science*, *12*, 728652.

Rosa, C., Kuo, Y.W., Wuriyanghan, H. and Falk, B.W. (2018). RNA Interference Mechanisms and Applications in Plant Pathology. *Annual Review of Phytopathology*, *56*, 581–610.

Sandhu, S. and Kang, M.S. (2017). Advances in Breeding for Resistance to Insects. In: Arora, R., Sandhu, S. (Eds.) *Breeding Insect Resistant Crops for Sustainable Agriculture*. Springer, Singapore.

Santos-Ortega, Y. and Flynt, A. (2022). Double-Strand RNA (dsRNA) Delivery Methods in Insects: *Diaphorina citri*. *Methods in Molecular Biology (Clifton, N.J.)*, *2360*, 253–277.

Sarkar, A. and Roy-Barman, S. (2021). Spray-induced silencing of pathogenicity gene *MoDES1* via exogenous double-stranded RNA can confer partial resistance against fungal blast in rice. *Frontiers in Plant Science*, *12*, 733129.

Savary, S., Willocquet, L., Pethybridge, S.J., Esker, P., McRoberts, N. and Nelson, A. (2019). The global burden of pathogens and pests on major food crops. *Nature Ecology and Evolution*, *3*(3), 430–439.

Schwarz, D.S., Hutvágner, G., Du, T., Xu, Z., Aronin, N. and Zamore, P.D. (2003). Asymmetry in the assembly of the RNAi enzyme complex. *Cell*, *115*(2), 199–208.

Sharma, A., Kumar, V., Shahzad, B., Tanveer, M., Sidhu, G.P.S., Handa, N., Kohli, S.K., Yadav, P., Bali, A.S., Parihar, R.D., Dar, O.I., Singh, K., Jasrotia, S., Bakshi, P., Ramakrishnan, M., Kumar, S., Bhardwaj, R. and Thukral, A.K. (2019). Worldwide pesticide usage and its impacts on ecosystem. *SN Applied Sciences*, *1*, 1446.

Shao, Z.-L., Lan, C.-P., Yu, X.-P. and Wang, Z.-L. (2024). RNAi-mediated silencing of *NlGRP3* augments the insecticidal virulence of *Metarhizium anisopliae* to the brown planthopper *Nilaparvata lugens*. *Pesticide Biochemistry and Physiology, 205*, 106141.

Shu, Y., Feng, Y., Zhong, G. and Gao, H. (2011). RNA interference of the *Spodoptera litura* vitellogenin receptor gene causes impaired egg development. *BMC Molecular Biology*, *12*, 31.

Sindhu, A.S., Maier, T.R., Mitchum, M.G., Hussey, R.S., Davis, E.L. and Baum, T.J. (2009). Effective and specific in planta RNAi in cyst nematodes: Expression interference of four parasitism genes reduces parasitic success. *Journal of Experimental Botany*, *60*(1), 315–324.

Singh, K., Dardick, C. and Kumar Kundu, J. (2019). RNAi-Mediated Resistance against Viruses in Perennial Fruit Plants. *Plants*, *8*(10), 359.

Singh, I.K., Singh, S., Mogilicherla, K., Shukla, J.N. and Palli, S.R. (2017). Comparative analysis of double-stranded RNA degradation and processing in insects. *Scientific Reports*, *7*(1), 17059.

Song, X.S., Gu, K.X., Duan, X.X., Xiao, X.M., Hou, Y.P., Duan, Y.B., Wang, J.X. and Zhou, M.G. (2018). A *myosin5* dsRNA that reduces the fungicide resistance and pathogenicity of *Fusarium asiaticum*. *Pesticide Biochemistry and Physiology*, *150*, 1–9.

Song, Y. and Thomma, B.P.H.J. (2018). Host-induced gene silencing compromises Verticillium wilt in tomato and *Arabidopsis*. *Molecular Plant Pathology*, *19*(1), 77–89.

Sparks, T.C. and Nauen, R. (2015). IRAC: Mode of action classification and insecticide resistance management. *Pesticide Biochemistry and Physiology*, *121*, 122–128.

St. Clair, D.A. (2010). Quantitative disease resistance and quantitative resistance loci in breeding. *Annual Review of Phytopathology, 48*, 247–268.

Suhag, A., Yadav, H., Chaudhary, D., Subramanian, S., Jaiwal, R. and Jaiwal, P.K. (2021). Biotechnological interventions for the sustainable management of a global pest, whitefly (*Bemisia tabaci*). *Insect Science, 28*(6), 1515–1535.

Sundaresha, S., Sharma, S., Bairwa, A., Tomar, M., Kumar, R., Bhardwaj, V., Jeevalatha, A., Bakade, R., Salaria, N., Thakur, K., Singh, B.P. and Chakrabarti, S.K. (2022). Spraying of dsRNA molecules derived from *Phytophthora infestans*, along with nanoclay carriers, as a proof of concept for developing a novel protection strategy for potato late blight. *Pest Management Science*, *78*(7), 3183–3192.

Taning, C.N.T., Arpaia, S., Christiaens, O., Dietz-Pfeilstetter, A., Jones, H., Mezzetti, B., Sabbadini, S., Sorteberg, H.G., Sweet, J., Ventura, V. and Smagghe, G. (2020). RNA-based biocontrol compounds: Current status and perspectives to reach the market. *Pest Management Science*, *76*(3), 841–845.

Tang, B., Chen, X., Liu, Y., Tian, H. and Zhang, W. (2010). RNA interference of the *Spodoptera exigua* trehalose-6-phosphate synthase gene reveals its crucial role in larval development. *BMC Molecular Biology*, *11*, 66.

Tenllado, F., Llave, C. and Díaz-Ruíz, J. R. (2004). RNA interference as a new biotechnological tool for the control of virus diseases in plants. *Virus Research*, *102*(1), 85–96.

Terenius, O., Papanicolaou, A., Garbutt, J.S., Eleftherianos, I., Huvenne, H., Kanginakudru, S., Albrechtsen, M., An, C., Aymeric, J.L., Barthel, A., Bebas, P., Bitra, K., Bravo, A., Chevalier, F., Collinge, D.P., Crava, C.M., de Maagd, R.A., Duvic, B., Erlandson, M., Faye, I., … and Smagghe, G. (2011). RNA interference in Lepidoptera: an overview of successful and unsuccessful studies and implications for experimental design. *Journal of Insect Physiology*, *57*(2), 231–245.

Thakur, N., Upadhyay, S.K., Verma, P.C., Chandrashekar, K., Tuli, R. and Singh, P.K. (2014). Enhanced whitefly resistance in transgenic tobacco plants expressing double-stranded RNA of v-*ATPase A* gene. *PLoS One, 9*(3), e87235.

Tian, H., Peng, H., Yao, Q., Chen, H., Xie, Q., Tang, B. and Zhang, W. (2009). Developmental control of a lepidopteran pest *Spodoptera exigua* by ingestion of bacteria expressing dsRNA of a non-midgut gene. *PLoS One, 4*(7), e6225.

Tripathy, V., Sharma, K.K., Gupta, R., Tandekar, K. and Walia, S. (2023). Off Label Use of Agrochemicals: Risks and Trade Implications. In: *Advances in Plant Disease Management* (Chapter 4). Taylor & Francis.

Turner, C.T., Davy, M.W., MacDiarmid, R.M., Plummer, K.M., Birch, N.P. and Newcomb, R.D. (2006). RNA interference in the light brown apple moth, *Epiphyas postvittana* (Walker) induced by double-stranded RNA feeding. *Insect Molecular Biology*, *15*(3), 383–392.

Upadhyay, S.K., Chandrashekar, K., Thakur, N., Verma, P.C., Borgio, J.F., Singh, P.K. and Tuli, R. (2011). RNA interference for the control of whiteflies (*Bemisia tabaci*) by oral route. *Journal of Biosciences*, *36*(1), 153–161.

Van Lenteren, J.C. (2012). The state of commercial augmentative biological control: Plenty of natural enemies, but a frustrating lack of uptake. *BioControl*, *57*, 1–20.

Vega-Arreguín, J.C., Jalloh, A., Bos, J.I.B. and Moffett, P. (2014). Recognition of an *Avr3a* homologue plays a major role in mediating nonhost resistance to *Phytophthora capsici* in *Nicotiana* species. *Molecular Plant-Microbe Interactions*, *27*(8), 770–780.

Vélez, A.M., Fishilevich, E., Storer, N.P. and Narva, K.E. (2016). RNAi as a management tool for the western corn rootworm, *Diabrotica virgifera virgifera*. *Pest Management Science*, *72*(9), 1652–1663.

Vergani-Junior, C.A., Tonon-da-Silva, G., Inan, M.D. and Ramos, C.H.I. (2021). DICER: Structure, function, and regulation. *Biophysical Reviews, 13,* 1081–1090.

Voloudakis, A.E., Kaldis, A. and Patil, B.L. (2022). RNA-based vaccination of plants for control of viruses. *Annual Review of Virology*, *9*, 521–548.

Walshe, D.P., Lehane, S.M., Lehane, M.J. and Haines, L.R. (2009). Prolonged gene knockdown in the tsetse fly *Glossina* by feeding double-stranded RNA. *Insect Molecular Biology*, *18*(1), 11–19.

Wang, W., Ghafar, M. A., Liuyang, L., Haq, I.U., Cui, L., Yuan, H. and Wang, L. (2025). Nanoscale Metal-Organic Frameworks for the Co-Delivery of Cycloxaprid and Pooled siRNAs to Enhance Control Efficacy in *Diaphorina citri*. *Journal of Agricultural and Food Chemistry*, *73*(6), 3353–3362.

Wang, K., Peng, Y., Pu, J., Fu, W., Wang, J. and Han, Z. (2016). Variation in RNAi efficacy among insect species is attributable to dsRNA degradation in vivo. *Insect Biochemistry and Molecular Biology*, *77*, 1–9.

Wang, M. and Dean, R.A. (2022). Host-induced gene silencing of *Magnaporthe oryzae* by targeting pathogenicity and development genes to control rice blast disease. *Frontiers in Plant Science*, *13*, 959641.

Wang, M., Wu, L., Mei, Y., Zhao, Y., Ma, Z., Zhang, X. and Chen, Y. (2020). Host-induced gene silencing of multiple genes of *Fusarium graminearum* enhances resistance to Fusarium head blight in wheat. *Plant Biotechnology Journal*, *18*(12), 2373–2375.

Werner, B.T., Gaffar, F.Y., Schuemann, J., Biedenkopf, D. and Koch, A.M. (2020). RNA-spray-mediated silencing of *Fusarium graminearum* AGO and DCL genes improves barley disease resistance. *Frontiers in Plant Science*, *11*, 476.

Whangbo, J.S. and Hunter, C.P. (2008). Environmental RNA interference. *Trends in Genetics*, *24*(6), 297–305.

Whitten, M.M.A., Facey, P.D., Del Sol, R., Fernández-Martínez, L.T., Evans, M.C., Mitchell, J.J., Bodger, O.G. and Dyson, P.J. (2016). Symbiont-mediated RNA interference in insects. *Proceedings of the Royal Society B*, *283*(1825), 20160042.

Whyard, S., Singh, A.D. and Wong, S. (2009). Ingested double-stranded RNAs can act as species-specific insecticides. *Insect Biochemistry and Molecular Biology*, *39*(11), 824–832.

Whyard, S., Erdelyan, C.N.G., Partridge, A.L., Singh, A.D., Beebe, N.W. and Capina, R. (2015). Silencing the buzz: A new approach to population suppression of mosquitoes by feeding larvae double-stranded RNAs. *Parasites and Vectors*, *8*, 96.

Willmann, M.R., Endres, M.W., Cook, R.T. and Gregory, B.D. (2011). The functions of RNA-dependent RNA polymerases in *Arabidopsis*. *The Arabidopsis Book*, *9*, e0146.

Willow, J., Cook, S.M., Veromann, E. and Smagghe, G. (2022). Uniting RNAi Technology and Conservation Biocontrol to Promote Global Food Security and Agrobiodiversity. *Frontiers in Bioengineering and Biotechnology, 10*, 871651.

Willow, J., Taning, C.N.T., Cook, S.M., Sulg, S., Silva, A.I., Smagghe, G. and Veromann, E. (2021a). RNAi Targets in Agricultural Pest Insects: Advancements, Knowledge Gaps, and IPM. *Front. Agron.*, *3*, 794312.

Willow, J., Soonvald, L., Sulg, S., Kaasik, R., Silva, A.I., Taning, C.N.T., Christiaens, O., Smagghe, G. and Veromann, E. (2020). First Evidence of Bud Feeding-Induced RNAi in a Crop Pest via Exogenous Application of dsRNA. *Insects*, *11*(11), 769.

Willow, J., Sulg, S., Taning, C.N.T., Silva, A.I., Christiaens, O., Kaasik, R., Prentice, K., Lövei, G.L., Smagghe, G. and Veromann, E. (2021b). Targeting a coatomer protein complex-I gene via RNA interference results in effective lethality in the pollen beetle *Brassicogethes aeneus*. *Journal of Pest Science, 94*, 703–712.

Winston, W.M., Molodowitch, C. and Hunter, C.P. (2002). Systemic RNAi in *C. elegans* requires the putative transmembrane protein SID-1. *Science* (*New York, N.Y.*), *295*(5564), 2456–2459.

Wise, J.C., Wise, A.G., Rakotondravelo, M., Vandervoort, C., Seeve, C. and Fabbri, B. (2022). Trunk injection delivery of dsRNA for RNAi-based pest control in apple trees. *Pest Manag. Science, 78*(8), 3528–3533.

Worrall, E.A., Bravo-Cazar, A., Nilon, A.T., Fletcher, S.J., Robinson, K.E., Carr, J.P. and Mitter, N. (2019). Exogenous application of RNAi-inducing double-stranded RNA inhibits aphid-mediated transmission of a plant virus. *Frontiers in Plant Science*, *10*, 265.

Wu, J., Yang, J. and Cho, W.C. (2020). Argonaute proteins: Structural features, functions, and emerging roles. *Journal of Advanced Research*, *24*, 317–324.

Wuriyanghan, H. and Falk, B.W. (2013). RNA interference towards the potato psyllid, *Bactericera cockerelli*, is induced in plants infected with recombinant tobacco mosaic virus (TMV). *PLoS One, 8*(6), e66050.

Wynant, N., Santos, D., Verdonck, R., Spit, J., Van Wielendaele, P. and Vanden Broeck, J. (2014). Identification, functional characterization and phylogenetic analysis of double stranded RNA degrading enzymes present in the gut of the desert locust, *Schistocerca gregaria*. *Insect Biochemistry and Molecular Biology*, *46*, 1–8.

Wynant, N., Verlinden, H., Breugelmans, B., Simonet, G. and Vanden Broeck, J. (2012). Tissue-dependence and sensitivity of the systemic RNA interference response in the desert locust, *Schistocerca gregaria*. *Insect Biochemistry and Molecular Biology*, *42*(12), 911–917.

Wytinck, N., Sullivan, D.S., Biggar, K.T., Crisostomo, L., Pelka, P., Belmonte, M.F. and Whyard, S. (2020). Clathrin mediated endocytosis is involved in the uptake of exogenous double-stranded RNA in the white mold phytopathogen *Sclerotinia sclerotiorum*. *Scientific Reports*, *10*(1), 12773.

Wytinck, N., Ziegler, D.J., Walker, P.L., Sullivan, D.S., Biggar, K.T., Khan, D., Sakariyahu, S.K., Wilkins, O., Whyard, S. and Belmonte, M.F. (2022). Host-induced gene silencing of the *Sclerotinia sclerotiorum* ABHYDROLASE-3 gene reduces disease severity in *Brassica napus*. *PLoS One, 17*(8), e0261102.

Xie, X., Song, Y., Liu, X., Wang, S., Zhu, C. (2014). Different target genes and chimeric-gene hairpin structures affect virus resistance mediated by RNA silencing in transgenic tobacco. *Biologia Plantarum, 58*, 575–581.

Xiong, F., Liu, M., Zhuo, F., Yin, H., Deng, K., Feng, S., Liu, Y., Luo, X., Feng, L., Zhang, S., Li, Z. and Ren, M. (2019). Host-induced gene silencing of BcTOR in *Botrytis cinerea* enhances plant resistance to grey mould. *Molecular Plant Pathology*, *20*(12), 1722–1739.

Yan, S., Qian, J., Cai, C., Ma, Z., Li, J., Yin, M., Ren, B. and Shen, J. (2020). Spray method application of transdermal dsRNA delivery system for efficient gene silencing and pest control on soybean aphid *Aphis glycines*. *Journal of Pest Science*, *93*, 449–459.

Yan, S., Ren, B. and Shen, J. (2021). Nanoparticle-mediated double-stranded RNA delivery system: A promising approach for sustainable pest management. *Insect Science*, *28*(1), 21–34.

Yan, S., Ren, B., Zeng, B. and Shen, J. (2020). Improving RNAi efficiency for pest control in crop species. *BioTechniques*, *68*(5), 283–290.

Yang, J., Qin, X., Keyan, S., Salzman, Z., Ming, Q., Qiang, H., Xiang, F. and Kong, D. (2021). Host-induced gene silencing of brown planthopper glutathione S-transferase gene enhances rice resistance to sap-sucking insect pests. *Journal of Pest Science*, *94*(3), 769–781.

Yang, W., Wang, B., Lei, G., Chen, G. and Liu, D. (2022). Advances in nanocarriers to improve the stability of dsRNA in the environment. *Frontiers in Bioengineering and Biotechnology*, *10*, 974646.

Ye, C., Wang, Z.-W., Sheng, Y.-L., Wang, Z.-G., Smagghe, G., Christiaens, O., Niu, J. and Wang, J.-J. (2022). GNBP1 as a potential RNAi target to enhance the virulence of *Beauveria bassiana* for aphid control. *Journal of Pest Science*, *95*(1), 87–100.

Yoshida, Y. (1988). Cytochrome P450 of fungi: Primary target for azole antifungal agents. *Current Topics in Medical Mycology, 2*, 388–418.

Yu, N., Christiaens, O., Liu, J., Niu, J., Cappelle, K., Caccia, S., Huvenne, H. and Smagghe, G. (2013). Delivery of dsRNA for RNAi in insects: An overview and future directions. *Insect Science*, *20*(1), 4–14.

Yu, X. and Killiny, N. (2018). RNA interference of two glutathione S-transferase genes, *Diaphorina citri* DcGSTe2 and DcGSTd1, increases the susceptibility of Asian citrus psyllid (Hemiptera: Liviidae) to the pesticides fenpropathrin and thiamethoxam. *Pest Management Science*, *74*(3), 638–647.

Zhang, H., Mittal, N., Leamy, L.J., Barazani, O. and Song, B.H. (2017). Back into the wild: Apply untapped genetic diversity of wild relatives for crop improvement. *Evolutionary Applications*, *10*(1), 5–24.

Zhang, J., Khan, S.A., Hasse, C., Ruf, S., Heckel, D.G. and Bock, R. (2015). Pest control. Full crop protection from an insect pest by expression of long double-stranded RNAs in plastids. *Science*, *347*, 991–994.

Zhang, X., Zhang, J. and Zhu, K.Y. (2010). Chitosan/double-stranded RNA nanoparticle-mediated RNA interference to silence chitin synthase genes through larval feeding in the African malaria mosquito (*Anopheles gambiae*). *Insect Molecular Biology*, *19*(5), 683–693.

Zhao, Y.Y., Liu, F., Yang, G. and You, M.S. (2011). PsOr1, a potential target for RNA interference-based pest management. *Insect Molecular Biology*, *20*(1), 97–104.

Zhao, Y.Y., Yang, G., Wang-Pruski, G. and You, M.S. (2008). *Phyllotreta striolata* (Coleoptera: Chrysomelidae): Arginine kinase cloning and RNAi-based pest control. *European Journal of Entomology*, *105*(5), 815–822.

Zheng, Y., Hu, Y., Yan, S., Zhou, H., Song, D., Yin, M. and Shen, J. (2019). A polymer/detergent formulation improves dsRNA penetration through the body wall and RNAi-induced mortality in the soybean aphid *Aphis glycines*. *Pest Management Science*, *75*(7), 1993–1999.

Zhong, X., Yu, X., Zhang, J., Xu, J., Qin, M., Cao, M. and Francis, F. (2025). RNAi technologies for insect control in crop protection. *The Crop Journal*, *13*(4), 1009–1021.

Zhou, X., Wheeler, M.M., Oi, F.M. and Scharf, M.E. (2008). RNA interference in the termite *Reticulitermes flavipes* through ingestion of double-stranded RNA. *Insect Biochemistry and Molecular Biology*, *38*(8), 805–815.

Zhu, F., Xu, J., Palli, R., Ferguson, J. and Palli, S.R. (2011). Ingested RNA interference for managing the populations of the Colorado potato beetle, *Leptinotarsa decemlineata*. *Pest Management Science*, *67*(2), 175–182.

Zhu, J.Q., Liu, S., Ma, Y., Zhang, J.Q., Qi, H.S., Wei, Z.J., Yao, Q., Zhang, W.Q. and Li, S. (2012). Improvement of pest resistance in transgenic tobacco plants expressing dsRNA of an insect-associated gene EcR. *PloS One*, *7*(6), e38572.

Zhu, K.Y. and Palli, S.R. (2020). Mechanisms, applications, and challenges of insect RNA interference. *Annual Review of Entomology*, *65*, 293–311.

Zrachya, A., Kumar, P.P., Ramakrishnan, U., Levy, Y., Loyter, A., Arazi, T., Lapidot, M. and Gafni, Y. (2007). Production of siRNA targeted against TYLCV coat protein transcripts leads to silencing of its expression and resistance to the virus. *Transgenic rResearch*, *16*(3), 385–398.

Zotti, M., dos Santos, E.A., Cagliari, D., Christiaens, O., Taning, C.N.T. and Smagghe, G. (2018). RNA interference technology in crop protection against arthropod pests, pathogens, and nematodes. *Pest Management Science*, *74*(6), 1239–1250.

Chapter 5

Nanotechnology in Crop Protection
Smart Delivery of Agrochemicals and Biopesticides

Pandareesh M.D.[1,2*], Bhumika H.S.[1,2], Dinesha Ramadas[1] and Belaghihalli N Gnanesh[3]

1. Introduction

Crop protection plays a critical role in sustaining agricultural productivity, ensuring food security, and maintaining global economic stability (Kalogiannidis et al., 2022). Despite advancements in agrochemicals and pest management techniques, a staggering 20–40% of global crop yields are lost annually due to pests and diseases (Savary et al., 2012). This issue persists even in regions with high pesticide usage, revealing the fundamental limitations of conventional pest control approaches (Savary et al., 2012). Compounding this challenge are rising global demands for food, fuel, and fiber, along with the escalating impacts of climate change, which intensify pest pressures and destabilize traditional control strategies. The overuse of chemical pesticides has led to the emergence of pesticide-resistant "super pests" and ecological imbalances that result in pest resurgence. Moreover, non-target effects, such as harm to pollinators, beneficial soil microbes, and aquatic ecosystems, as well as long-term contamination and food safety concerns, present significant threats to environmental and human health (Arnot et al., 2004). Although biopesticides have emerged as eco-friendly alternatives, their widespread adoption is hindered by issues like short shelf-life, sensitivity to environmental conditions, inconsistent efficacy, and complex handling requirements (Prabha et al., 2016). These limitations often result in slower pest control and reduced farmer confidence, particularly in developing regions where knowledge of integrated pest management (IPM) and access to advanced technologies remain limited (Botnaru et al., 2025). Consequently, there is a continued overreliance on conventional agrochemicals, exacerbating environmental degradation and raising production costs. Regulatory pressures and increasing consumer demand for organic, residue-free produce further highlight the urgent need for safer, more sustainable pest management solutions that balance productivity with ecological responsibility (Botnaru et al., 2025).

In this context, nanotechnology offers a promising avenue for addressing these pressing challenges through the development of smart, targeted delivery systems for agrochemicals and biopesticides. Nano-enabled formulations can significantly improve the stability, solubility, and bioavailability

[1] Department of Biochemistry, Adichunchanagiri School of Natural Sciences, Adichunchanagiri University, B.G. Nagara, Mandya-571448, India.
[2] Center for Research and Innovation, Adichunchanagiri University, B.G. Nagara, Mandya-571448, India.
[3] Genetics and Plant Breeding, ICAR JSS - Krishi Vigyan Kendra, Mysuru-571 129, Karnataka, India
* Corresponding author: pandareesh@acu.ac.in

DOI: 10.1201/9781003614869-5

of active compounds, enabling controlled release and site-specific action that reduces chemical usage and environmental impact (Younis et al., 2025). This study is significant as it systematically explores how nanotechnology can revolutionize pest and disease management by overcoming the inherent drawbacks of conventional and biopesticide-based approaches. Furthermore, it underscores the transformative potential of integrating nanotechnology with precision farming, artificial intelligence, and nanosensors to enable climate-resilient and sustainable agriculture. By providing a comprehensive examination of nanoformulation techniques, mechanisms of action, and real-world applications, this study contributes valuable insights for researchers, industry stakeholders, and policymakers seeking scalable, science-driven solutions for smart crop protection.

1.1 Overview of Global Challenges in Crop Protection

Crop protection plays a critical role in safeguarding agricultural productivity, food security, and the global economy. However, despite decades of advancement in agrochemical development and pest management strategies, agriculture continues to face severe threats from pests, pathogens, weeds, and other biological stressors. Globally, it is estimated that 20–40% of crop yields are lost annually due to pest infestations and plant diseases. This loss persists even in regions where chemical pesticides are widely used, highlighting the limitations of conventional methods. The global demand for food, fuel, and fiber continues to rise due to population growth, rapid urbanization, and changing dietary patterns. Meanwhile, the effects of climate change and environmental degradation are amplifying pest pressures and disrupting traditional crop protection practices, making it increasingly difficult to sustain agricultural productivity (Savary et al., 2012). One of the most urgent and complex challenges is the development of pest resistance to synthetic pesticides. Over time, many pests and pathogens have evolved mechanisms to tolerate or neutralize chemical treatments, rendering many pesticides ineffective. This phenomenon, driven by the indiscriminate and repeated application of chemicals, has led to the emergence of "super pests" that are difficult to control using existing tools. In addition to resistance, pest resurgence is a growing problem where pest populations, once reduced by chemical treatment, rapidly recover and often become more severe (Georghiou, 1972). This is often the result of ecological imbalance caused by the elimination of natural predators and beneficial insects, which are also affected by broad-spectrum pesticides (Georghiou, 1972).

Another significant concern is the non-target effects and environmental contamination associated with conventional agrochemicals (Kookana et al., 1998). Many pesticides lack specificity and act on a wide range of organisms, including pollinators like bees, beneficial insects, soil microflora, aquatic organisms, and birds (Kamal Uddin, 2018). This non-specific toxicity leads to collateral damage within ecosystems, disrupting biodiversity and essential ecological services such as pollination, soil fertility, and biological pest control. Furthermore, pesticides applied in agricultural fields frequently leach into nearby water bodies or accumulate in the soil, contributing to long-term environmental pollution. The persistence and bioaccumulation of these toxic chemicals can have cascading effects on food webs, ultimately affecting entire ecosystems. The implications of widespread pesticide use extend beyond the environment and have direct consequences for human health. Farmers and agricultural workers are particularly vulnerable to pesticide exposure, often lacking proper protective equipment or training in safe handling practices. Acute exposure can result in poisoning, respiratory distress, and neurological symptoms, while chronic exposure has been associated with endocrine disruption, reproductive issues, carcinogenicity, and neurotoxicity (Ibadi et al., 2024). In many developing regions, the continued use of outdated, unregulated, or banned agrochemicals exacerbates these health risks. Additionally, pesticide residues are often found in food, water, and air, raising concerns for consumer safety and public health worldwide.

Global agriculture is also navigating through a landscape of tightening regulations and changing market dynamics. Many countries are phasing out or restricting the use of hazardous pesticides due to their environmental and health risks. This regulatory pressure is forcing farmers and agro-industries to seek safer and more sustainable alternatives. At the same time, consumer demand for organic, residue-free, and environmentally friendly produce is rising, especially in high-value markets (Borah et al., 2024). Meeting these expectations without compromising crop yields presents a substantial challenge for growers and food producers. Adding to these complications is the impact of climatic variability, which is reshaping pest behavior and disease epidemiology. Climate change affects the geographical distribution, population dynamics, and life cycles of pests and pathogens (Karuppaiah et al., 2012). Warmer temperatures and altered precipitation patterns create favorable conditions for the emergence of new pests and the expansion of existing ones into previously unaffected areas. In addition, unpredictable weather events and prolonged droughts or floods can weaken plant resilience, making them more susceptible to pest attacks and reducing the effectiveness of conventional control strategies.

Another challenge in crop protection arises from the limitations of conventional pesticide formulations and delivery systems. Many chemical agents suffer from low solubility, rapid photodegradation, volatility, and short persistence, requiring frequent and high-dose applications. These characteristics contribute to inefficient use, increased cost for farmers, and greater environmental load. Moreover, traditional formulations do not offer targeted or controlled release, resulting in widespread dispersion of chemicals and reduced site-specific activity. While biopesticides have been recognized as promising eco-friendly alternatives to synthetic agrochemicals, their widespread adoption is still constrained by several factors (Fenibo et al., 2022). Biopesticides, which include microbial formulations, plant-based extracts, and naturally occurring compounds, often suffer from limited stability, low shelf life, sensitivity to environmental conditions, and inconsistent field performance. Their mode of action may be slower or less potent than synthetic chemicals, and many formulations require special handling or storage. As a result, farmers are often reluctant to adopt biopesticides on a large scale unless their performance can be reliably improved (Constantine et al., 2020). In many regions, especially in developing and underdeveloped countries, crop protection efforts are further hindered by knowledge gaps and poor access to technology. Many farmers lack formal education, training, or awareness of IPM practices (Upadhyay et al., 2014). Extension services and infrastructure to support the dissemination of advanced technologies are often inadequate. Consequently, farmers rely on traditional practices or excessive use of chemical pesticides without understanding the long-term consequences. This lack of knowledge transfer limits the adoption of sustainable solutions and perpetuates the cycle of chemical dependency and environmental degradation (Upadhyay et al., 2014). Hence, the global challenges in crop protection are complex, interconnected, and escalating. From pest resistance and environmental toxicity to climate change and technological barriers, the current system of crop protection faces significant strain. There is a growing consensus that traditional approaches are no longer sufficient to meet the needs of modern agriculture. Addressing these challenges requires a shift toward innovative, sustainable, and science-based interventions that can deliver effective pest control while minimizing ecological and health risks. Nanotechnology-based smart delivery systems emerge as one such promising solution, capable of revolutionizing crop protection by enabling precision-targeted applications, controlled release, and improved efficiency of agrochemicals and biopesticides. As we transition toward next-generation agricultural practices, integrating nanotechnology with pest management offers a powerful strategy to overcome current limitations and build a more resilient, sustainable food system (Oerke et al., 2004).

1.2 Limitations of Conventional Agrochemical and Biopesticide Applications

While agrochemicals have significantly enhanced crop productivity since the Green Revolution, their traditional formulations and delivery methods are increasingly misaligned with the needs of modern, sustainable agriculture (Pimentel, 1996). One of the major shortcomings is poor targeting and low use efficiency. Conventional application methods like spraying and broadcasting result in widespread off-target deposition, with less than 0.1% of applied pesticides actually reaching the target pests. Most of the chemicals are lost through drift, volatilization, runoff, or photodegradation, leading to repeated applications that escalate costs, labor, and environmental burden (Pimentel, 1996). Environmental instability further undermines the efficacy of these products. Agrochemicals and biopesticides degrade rapidly under harsh field conditions, such as UV exposure, temperature extremes, and varying humidity or pH. Biopesticides are particularly sensitive, often succumbing to desiccation, oxidation, or microbial degradation, which results in inconsistent performance and reduced adoption by farmers (Abdollahdokht et al., 2022). Similarly, many agrochemicals exhibit short-lived activity due to leaching or rapid breakdown, requiring frequent reapplication, an impractical approach for long-season crops or regions with high pest pressure.

Incompatibility of conventional formulations with IPM strategies can be a significant hindrance in developing a sustainable crop protection system. Broad-spectrum pesticides disrupt ecological balance by harming beneficial insects and natural predators, undermining the selective interventions central to IPM. Furthermore, traditional products lack precision or tunable release profiles, making them ill-suited for nuanced pest management (Sharma, 2012). Physicochemical challenges also hinder efficiency; low water solubility of active ingredients necessitates toxic solvents, while particle instability causes sedimentation, nozzle clogging, and uneven dispersion. In biopesticides, maintaining the viability of bioactive agents like spores or proteins through processing and storage remains a significant hurdle. Residue accumulation is another critical concern (Pouton, 2006). Even when used according to guidelines, conventional pesticides often leave trace residues in produce, soil, and water, leading to potential violations of Maximum Residue Limits (MRLs) and market rejection especially in export scenarios. While biopesticides are generally considered safer, they too require strict standardization to meet regulatory requirements (Khandelwal et al., 2022). Additionally, the performance of biopesticides is often inconsistent due to environmental variability and their slower, non-instantaneous pest control action. This unreliability drives farmers back to synthetic options, stalling the broader adoption of eco-friendly alternatives (Khandelwal et al., 2022). Storage and shelf-life issues further complicate crop protection. Chemical pesticides degrade over time due to oxidation or hydrolysis, while biopesticides demand cold storage to maintain the viability of living organisms. In tropical regions with limited refrigeration infrastructure, this presents a serious barrier to accessibility (Chamberlain et al., 2012). Moreover, conventional agrochemicals are poorly compatible with precision farming tools like drones, automated irrigation, and AI-driven systems due to issues like particle size, viscosity, and lack of environmental responsiveness (Kapoor et al., 2024). These constraints limit the integration of conventional formulations into modern, data-driven agriculture, highlighting the need for smarter, adaptable delivery platforms to meet evolving agricultural challenges (Kapoor et al., 2024).

The limitations of conventional agrochemical and biopesticide applications reflect the pressing need for innovation in crop protection. Inefficiencies in delivery, instability in field conditions, environmental and health concerns, and lack of compatibility with modern agricultural systems underscore the shortcomings of current practices. To overcome these barriers, it is imperative to shift toward smart, targeted, and responsive delivery platforms that can enhance the performance of both synthetic and biological crop protectants (Peteu et al., 2010). Nanotechnology-based systems, offering controlled release, improved solubility, enhanced stability, and precision targeting, provide a promising avenue to revitalize and modernize crop protection strategies for sustainable agriculture (Peteu et al., 2010).

1.3 Emergence of Nanotechnology in Agriculture

Nanotechnology, the manipulation of matter at the atomic and molecular scale, typically below 100 nanometers has emerged as a transformative force across multiple disciplines, including medicine, materials science, and energy (Whitesides, 2005). In recent years, it has also made significant strides in agriculture, offering innovative solutions to long-standing challenges in crop production, protection, and sustainability. This integration of nanotechnology into agriculture, often termed 'nanoagriculture', is redefining traditional farming practices by enabling more precise, efficient, and environmentally responsible food production systems (Malik et al., 2023). The origin of nanotechnology in agriculture is rooted in the need for smarter, more effective delivery systems for agrochemicals, fertilizers, and biopesticides (Zaman et al., 2025). Conventional formulations often suffer from poor solubility, instability under environmental stressors, and lack of specificity, leading to reduced efficacy and increased environmental harm. Nanotechnology provides a paradigm shift by enabling the development of nanoformulations with controlled release, targeted delivery, and improved bioavailability (Gessner et al., 2025). These nano-based systems ensure that active ingredients are delivered directly to the site of action in a sustained and regulated manner, minimizing off-target effects and reducing the required dosages. One of the earliest and most impactful applications of nanotechnology in agriculture has been the development of nanofertilizers and nanopesticides. These formulations encapsulate active compounds within nanocarriers such as liposomes, polymers, micelles, dendrimers, or inorganic nanoparticles (Garg et al., 2023). These carriers protect the compounds from premature degradation and release them in response to specific environmental stimuli like pH, moisture, temperature, or enzymatic activity. For example, zinc oxide and silica nanoparticles have been used to enhance nutrient uptake efficiency, while silver and copper nanoparticles have demonstrated strong antimicrobial properties against phytopathogens (Garg et al., 2023).

Nanotechnology has also advanced the development of nanoemulsions and nanosuspensions that enhance the solubility and dispersion of poorly water-soluble agrochemicals. These improved formulations not only enhance field performance but also reduce reliance on harmful solvents and surfactants, lowering toxicity risks for applicators and surrounding ecosystems. In biopesticide delivery, nanoencapsulation protects bioactive molecules and microbial agents from environmental degradation, significantly extending their shelf life and efficacy (Zabot et al., 2022). This approach is particularly beneficial for RNAi-based biopesticides, botanical extracts, and microbial spores, making them more viable for commercial use (Fletcher et al., 2020). Beyond chemical delivery, nanotechnology has driven innovation in diagnostics and precision agriculture. Nanosensors are being developed to detect pests, pathogens, soil moisture, nutrient levels, and pesticide residues in real time (Fletcher et al., 2020). When integrated with Internet of Things (IoT) platforms and advanced data analytics, these sensors facilitate real-time monitoring and site-specific management (Mohammad El-Basioni et al., 2022). Such smart farming technologies enable more efficient use of resources and support informed decision-making, ushering in an era of data-driven agriculture that aligns with sustainability goals (Mohammad El-Basioni et al., 2022). The role of nanotechnology in promoting sustainable and climate-resilient farming is particularly significant. Nano-enabled agricultural inputs help reduce chemical usage, minimize environmental pollution, and conserve natural resources, an urgent necessity given the growing impacts of climate change, declining soil fertility, and shrinking arable land (Zhao et al., 2024). As global markets and regulatory bodies impose stricter limits on pesticide residues and ecological impacts, nanotechnology provides a means to comply with these requirements without compromising agricultural productivity or profitability.

Over the past decade, research and development in agricultural nanotechnology have gained momentum, reflected in a surge of scientific publications, patent filings, and publicly funded initiatives. Countries like India, China, the United States, and Brazil have launched national programs to explore nanotechnology applications in crop inputs, post-harvest technologies, food packaging, and biosensors (Sekhon, 2014). Private sector engagement is also increasing, with agrochemical companies and startups actively investing in the development of nano-enabled products and technologies. However, the adoption of nanotechnology in agriculture is not without challenges. Issues such as nanotoxicology, regulatory uncertainty, public acceptance, and long-term ecological impact must be carefully addressed (Sekhon, 2014). The behavior of nanoparticles in soil-plant systems is complex and requires detailed investigation. Comprehensive and transparent risk assessments are essential to ensure the safe and effective integration of nanotechnology in farming. Regulatory frameworks for nano-agrochemicals remain under development in many countries, underscoring the urgent need for internationally harmonized standards and guidelines. The emergence of nanotechnology in agriculture represents a significant leap toward the modernization and sustainability of global farming systems. By addressing critical limitations of conventional agrochemical and biopesticide applications, nanotechnology offers tools for precision-targeted delivery, improved efficacy, environmental safety, and real-time diagnostics. As research deepens and regulatory clarity improves, nanoagriculture is poised to play a central role in shaping the future of crop protection and food security (Prasad et al., 2017). The next decade will likely witness broader adoption of nano-enabled innovations across the agricultural value chain, from input management to post-harvest handling and supply chain monitoring.

1.4 Scope and Objectives

This chapter delves into the rapidly evolving field of nanotechnology in agriculture, with a specific focus on its application in crop protection through the smart delivery of agrochemicals and biopesticides. As agriculture grapples with increasing pest resistance, environmental pollution, declining effectiveness of traditional pesticides, and the pressing demand for sustainable practices, nanotechnology offers a timely and promising solution (Alam et al., 2024). The scope of this chapter encompasses a comprehensive exploration of the types of nanomaterials utilized in agricultural formulations, including metallic nanoparticles, nanoemulsions, nanogels, polymeric carriers, and dendrimers. Emphasis is placed on how these materials are engineered for enhanced delivery, improved solubility, prolonged stability, and environmental responsiveness. The chapter also covers mechanisms of controlled release, site-specific targeting, and stimuli-responsive activation, which collectively represent a significant advancement over conventional pesticide and biopesticide applications. Besides, the chapter presents real-world applications and case studies that demonstrate the efficacy of nano-enabled agrochemicals and biopesticides in various crop systems. Comparative insights are provided to highlight the differences in performance, environmental safety, and application efficiency between traditional and nanoformulated products. Furthermore, the discussion extends to emerging innovations such as the integration of nanosensors and AI-based precision farming tools that support data-driven decision-making in pest management (El-Fakharany, 2021). Environmental and safety aspects are also critically evaluated, including the potential risks associated with nanomaterials, their behavior in soil and plant systems, and the need for robust regulatory frameworks to ensure safe deployment. The chapter thereby addresses the broader ecological and regulatory context within which nanotechnology must be developed and adopted (El-Fakharany, 2021). The key objectives of this chapter are to provide a clear understanding of why conventional agrochemical practices are becoming increasingly unsustainable and how nanotechnology can overcome these limitations. It aims to explain the scientific principles behind nano-delivery systems and their practical benefits in terms of enhanced efficacy, reduced dosage, and minimized non-target effects. The chapter seeks to familiarize readers with commercially available nano-agrochemical products, cutting-edge research, and ongoing global efforts to integrate nanotechnology into sustainable agriculture. Additionally, it highlights current challenges, such as formulation stability, toxicity concerns, and regulatory uncertainties, while identifying future research directions that can bridge existing knowledge gaps. By offering a balanced perspective that combines scientific innovation with practical implications, the chapter aspires to serve as a valuable resource for researchers, agronomists, policymakers, industry stakeholders, and students interested in the future of crop protection. It underscores the transformative potential of nanotechnology in building resilient agricultural systems that align with the goals of food security, environmental sustainability, and technological advancement.

2. Nanotechnology: Definition and Key Concepts

Nanotechnology refers to science, engineering, and application of materials and devices at the nanoscale, typically ranging from 1 to 100 nanometers (nm). At this scale, materials exhibit unique physical, chemical, and biological properties that differ significantly from their bulk counterparts, due to quantum effects and increased surface area-to-volume ratios (Whitesides, 2005). The fundamental principle of nanotechnology lies in manipulating matter at the atomic and molecular level to achieve desirable functions and behaviors that are not possible at the macro or microscale. A nanometer is one-billionth of a meter, and to put it in perspective, a human hair is approximately 80,000–100,000 nm wide. At this minuscule scale, materials often demonstrate enhanced reactivity, mechanical strength, thermal stability, optical behavior, and bioavailability, which can be harnessed for a wide range of applications (Whitesides, 2005). In the context of agriculture, nanotechnology primarily involves the development and utilization of nanomaterials such as nanoparticles, nanocapsules, nanogels, nanoemulsions, and nanosuspensions for improved delivery of agrochemicals, nutrients, and biopesticides (Jiang et al., 2024). These materials can be designed to release their active ingredients in a controlled, targeted, and sustained manner, often triggered by environmental cues such as pH, moisture, temperature, or enzymatic activity. This site-specific delivery enhances the effectiveness of crop protection agents while minimizing losses, environmental contamination, and non-target toxicity (Jiang et al., 2024). Nanotechnology also encompasses the use of nanosensors for real-time monitoring of soil health, pest incidence, and plant stress, enabling precision agriculture through timely, data-driven interventions. Key concepts within nanotechnology include bottom-up and top-down fabrication approaches, which describe methods for constructing nanoscale materials either by assembling atoms and molecules (bottom-up) or by reducing larger materials to nanoscale dimensions (top-down). Another core concept is functionalization, where the surface of nanomaterials is modified to improve compatibility, targeting, or responsiveness in specific environments. Understanding these foundational concepts is essential for designing nano-enabled tools that are efficient, safe, and adaptable to agricultural settings. As the field continues to evolve, nanotechnology offers a versatile platform for innovation across the agricultural value chain from input delivery and pest management to diagnostics and post-harvest preservation.

3. Types of Nanomaterials Used in Agriculture

Nanotechnology offers a diverse set of materials that can be tailored for various agricultural applications, especially in crop protection and nutrient delivery. The type, size, surface characteristics, and composition of nanomaterials determine their behavior, efficacy, and interaction with biological systems (Shang et al., 2019). In agriculture, nanomaterials are used to improve the solubility, stability, bioavailability, and controlled release of agrochemicals and biopesticides. Below are the major categories of nanomaterials commonly applied in agricultural systems:

3.1 Nanoparticles

Nanoparticles are solid colloidal particles ranging in size from 1 to 100 nm and are composed of various materials including metals, metal oxides, polymers, or silica. Their unique physicochemical properties such as high surface area, enhanced reactivity, and tunable surface functionality make them highly suitable for agricultural use (Mody et al., 2010).

1. **Metallic nanoparticles** (e.g., silver, copper, zinc oxide, and iron oxide) are widely studied for their antimicrobial properties. They can act directly against phytopathogens such as bacteria, fungi, and viruses. Silver nanoparticles (AgNPs), for example, are known to disrupt microbial membranes and DNA replication, making them effective biocides in crop protection (Dizaj et al., 2014).
2. **Polymeric nanoparticles** are formed using biodegradable polymers like chitosan, PLGA (poly lactic-co-glycolic acid), and alginate. These carriers are used to encapsulate pesticides, fertilizers, or biopesticides, allowing for sustained and controlled release, reduced degradation, and increased delivery to specific target sites (Lai et al., 2014).

3. **Silica nanoparticles** are mesoporous structures with a high surface area and tunable pore size, useful for loading and protecting agrochemicals. They provide excellent stability, are chemically inert, and can be functionalized for targeted delivery (Wanyika et al., 2012).

These nanoparticles are often surface-functionalized or coated to improve their interaction with plant surfaces, control their degradation, and enhance their compatibility with other agricultural inputs.

3.2 Nanoemulsions

Nanoemulsions are thermodynamically or kinetically stable colloidal systems consisting of two immiscible liquids (typically oil and water) stabilized by surfactants, with droplet sizes ranging from 20 to 200 nm (Preeti et al., 2023). They offer numerous advantages in the delivery of agrochemicals, especially hydrophobic active ingredients that are otherwise poorly soluble in water. In agriculture, nanoemulsions are used to enhance the dispersion, penetration, and bioavailability of pesticides, herbicides, and plant extracts. Their small droplet size provides a large surface area, which increases the contact between the active ingredient and the target pest or plant tissue. This improves absorption, reduces application volume, and minimizes wastage (Preeti et al., 2023). Nanoemulsions also offer improved shelf-life stability, resistance to environmental degradation, and uniform coverage on leaf surfaces, making them ideal for foliar applications. Biopesticides based on essential oils and plant extracts, which are volatile and unstable, benefit significantly from nanoemulsion-based formulations.

3.3 Dendrimers

Dendrimers are highly branched, tree-like polymeric nanostructures characterized by a central core, internal layers (generations), and numerous terminal functional groups. Their structure provides multiple sites for conjugating active molecules, which can be chemically bonded or physically entrapped within the internal cavities. In agricultural applications, dendrimers are used as multifunctional delivery platforms for pesticides, herbicides, and genetic materials (Perez-Ferreiro et al., 2023). Their well-defined size and structure allow for precise control over loading capacity and release kinetics. The outer surface of dendrimers can be functionalized to enhance solubility, target-specific binding, and responsiveness to environmental stimuli. For example, pH-sensitive dendrimers can release their payload only under certain soil or plant conditions, improving the precision of agrochemical application. Due to their nanoscopic size and biocompatibility, dendrimers hold significant promise for use in both chemical and biological crop protection strategies (Mittal et al., 2021).

3.4 Nanogels

Nanogels are hydrophilic, three-dimensional polymeric networks that can swell in water and encapsulate a wide range of active compounds, including small molecules, proteins, and nucleic acids (Neamtu et al., 2017). These soft, flexible nanomaterials typically range from 20 to 200 nm in diameter and offer excellent biocompatibility, high water content, and responsiveness to environmental triggers. In agriculture, nanogels serve as smart carriers for pesticides, growth regulators, and biopesticides, offering controlled and sustained release based on environmental cues such as temperature, pH, or moisture. Their porous structure allows for high loading efficiency, while their swelling behavior enables prolonged interaction with plant tissues. For instance, nanogels can be designed to release fungicides gradually during prolonged wet conditions that favor disease development, thus improving timing and efficiency. Additionally, biodegradable nanogels made from natural polymers like chitosan or starch contribute to environmentally sustainable crop protection strategies (Neamtu et al., 2017). The diversity of nanomaterials available for agricultural applications offers unparalleled opportunities to enhance crop protection, reduce environmental impact, and improve input efficiency. From hard nanoparticles with potent antimicrobial effects to soft, responsive nanogels and precision-engineered dendrimers, each type of nanomaterial presents unique functionalities that can be tailored to meet specific crop needs. The strategic use of these nanomaterials enables a shift from conventional, broad-spectrum chemical applications to smart, sustainable, and targeted delivery systems, an essential step toward modernizing global agriculture.

4. Properties of Nanomaterials Relevant to Crop Protection

The effectiveness of nanomaterials in agricultural applications, particularly in crop protection, is largely governed by their unique physicochemical properties that differentiate them from bulk materials. These properties specifically size, surface area, and reactivity enable enhanced interaction with biological systems, controlled release of active ingredients, and targeted delivery to pests or plant tissues. Understanding these properties is critical for designing smart nanoformulations that are efficient, stable, and environmentally compatible.

4.1 Nanoscale Size

The most defining characteristic of nanomaterials is their extremely small size, typically ranging from 1 to 100 nm. This nanoscale dimension allows particles to penetrate plant surfaces, such as leaves, stems, or roots, more effectively than larger particles. For example, nanoparticles can enter plant tissues through stomatal openings, cuticular pores, or even be taken up via root transport systems, facilitating systemic delivery of agrochemicals or biopesticides (Pokropivny et al., 2007). Additionally, their small size enables nanomaterials to closely interact with microbial pathogens, insects, and other pests at the cellular or subcellular level, leading to more effective pest suppression. Smaller particles also allow for better dispersibility in liquid carriers, resulting in improved uniformity during spraying and coating applications. This property is particularly advantageous for foliar application, seed coating, and soil treatment, where even distribution and absorption are critical.

4.2 High Surface Area-to-Volume Ratio

Another key property is the high surface area-to-volume ratio of nanomaterials, which increases dramatically as particle size decreases. This large surface area allows for more active sites for chemical interaction, leading to improved adsorption, binding, and catalytic efficiency. In crop protection, this translates into enhanced loading capacity for pesticides, fertilizers, or bioactives per unit of nanocarrier, thus reducing the overall amount of chemical needed while maintaining or increasing efficacy (Sharma et al., 2018). The high surface area also enhances the adhesion of nanomaterials to plant surfaces, improving the residence time of pesticides or nutrients on leaves and stems. This helps resist wash-off during rainfall and reduces the frequency of reapplication. In the case of metallic nanoparticles like silver or copper, the increased surface contact also enhances antimicrobial activity, as these particles can generate reactive oxygen species or interact directly with microbial membranes and enzymes (Sharma et al., 2018).

4.3 Enhanced Reactivity and Tunable Surface Chemistry

Nanomaterials exhibit enhanced chemical reactivity and surface energy compared to their bulk counterparts, primarily due to the presence of more reactive atoms on their surface. This property allows for rapid and targeted biochemical interactions, making nanoparticles effective in interacting with biological molecules, pest cell membranes, or plant receptors. In agrochemical formulations, enhanced reactivity facilitates quicker and more efficient action against pathogens and pests, often at lower doses than required by conventional formulations (Zhou et al., 2011). Furthermore, the surface of nanomaterials can be functionalized or modified with specific ligands, polymers, or surfactants to improve solubility, stability, and target specificity. This tunable surface chemistry allows for the development of stimuli-responsive systems, where the release of the active ingredient is triggered by environmental cues such as pH, temperature, moisture, or enzymatic activity (Zhou et al., 2011). For instance, pH-sensitive nanocarriers can release their payload specifically in alkaline or acidic microenvironments associated with pest infestation or infection sites, thereby increasing precision and reducing off-target effects (Karimi et al., 2016). Functionalization also improves the biocompatibility and safety of nanomaterials. For example, coating metallic nanoparticles with biopolymers like chitosan can reduce toxicity while enhancing antimicrobial activity. These surface modifications play a key role in tailoring the nanomaterial's interaction with biological targets and the environment, making them ideal for both synthetic and biopesticide delivery (Karimi et al., 2016).

The nanoscale size, high surface area, and enhanced reactivity of nanomaterials collectively contribute to their superior performance in crop protection applications. These properties allow for better penetration, increased loading and release control, greater interaction with pests and pathogens, and reduced environmental losses. By leveraging these characteristics, nanotechnology enables the development of advanced agrochemical and biopesticide delivery systems that are not only more efficient but also aligned with the principles of sustainable and precision agriculture. Future innovations in nanomaterial engineering will continue to refine these properties for safer, smarter, and more effective crop protection strategies (Khot et al., 2012).

5. Smart Delivery Systems for Agrochemicals

5.1 Controlled and Targeted Release Mechanisms

Modern agriculture stands at a critical juncture, challenged by the need to feed a projected global population of 9.8 billion by 2050 while confronting the environmental consequences of intensive agrochemical use (Bodede, 2020). Conventional pesticide formulations suffer from staggering inefficiencies, with less than 0.1% of applied pesticides reaching their target organisms. The remainder contributes to environmental pollution, pest resistance, and human health concerns through mechanisms such as drift, volatilization, and leaching. Smart delivery systems for agrochemicals represent a paradigm shift in crop protection, offering precise control over the timing, location, and rate of active ingredient release. These systems leverage advances in nanotechnology and materials science to create formulations that respond to specific environmental or biological stimuli, dramatically improving efficiency while reducing environmental impact. The global market for these innovative formulations is growing rapidly, reflecting their potential to address sustainability challenges in agriculture (Bodede, 2020).

5.2 Mechanisms of Controlled and Targeted Release

Smart delivery systems employ various mechanisms to control agrochemical release, overcoming the limitations of conventional formulations that rely on passive diffusion or coating erosion. These mechanisms can be broadly categorized into temporal control, spatial targeting, Polysaccharide-Based Carriers, Synthetic Polymer Systems, and Inorganic nanocarriers (Li et al., 2021) (Figure 1).

5.2.1 Temporal Control Mechanisms

Temporal control plays a critical role in regulating the rate and duration of active ingredient release, ensuring that optimal concentrations are maintained over extended periods. This controlled release can be achieved through three primary mechanisms: diffusion-controlled, degradation-controlled, and swelling-controlled systems (Varma et al., 2004). In diffusion-controlled systems, the active

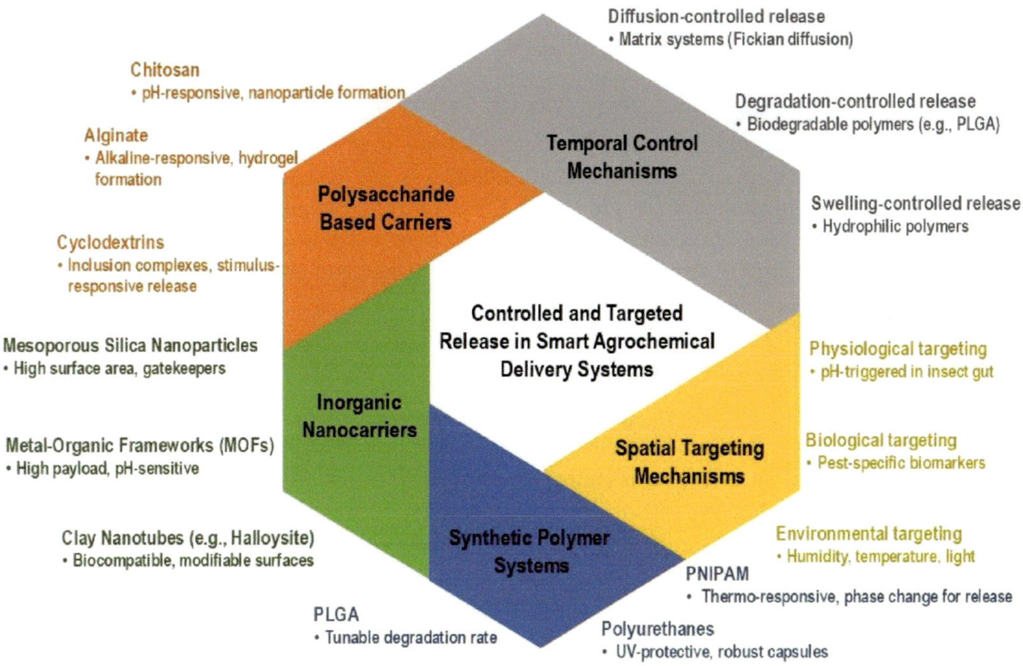

Fig. 1 The diagram depicts smart agrochemical delivery systems integrating polysaccharide carriers (e.g., chitosan, alginate, cyclodextrins), inorganic nanocarriers (e.g., mesoporous silica, MOFs, clay nanotubes), and synthetic polymers (e.g., PLGA, PNIPAM, polyurethanes) for controlled and targeted release. Temporal control is achieved via diffusion, degradation, or swelling mechanisms, while spatial targeting uses physiological, biological, or environmental triggers. These strategies enhance precision, efficiency, and environmental safety in agrochemical application.

ingredient is either dispersed within a polymer matrix or encapsulated in a reservoir design. In matrix systems, the ingredient diffuses through the polymer network, typically following Fickian diffusion laws, while in reservoir systems, it is enclosed within a core surrounded by a rate-controlling membrane, allowing for sustained and predictable release (Varma et al., 2004). Degradation-controlled systems rely on the breakdown of carrier materials through chemical or enzymatic processes. Biodegradable polymers such as PLGA are commonly used, with their degradation rates adjustable by modifying the ratio of lactic to glycolic acid, thus offering customizable release profiles. Swelling-controlled systems utilize hydrophilic polymers that absorb water, causing them to swell and transition from a glassy to a rubbery state, which facilitates the diffusion of the active ingredient. This mechanism is particularly advantageous for applications requiring hydration-triggered release. Together, these systems provide versatile and effective means for achieving precise temporal control in the delivery of agrochemicals or biopesticides (Varma et al., 2004).

5.2.2 Spatial Targeting Mechanisms

Spatial targeting is a strategic approach in agrochemical delivery that directs active ingredients to specific tissues, pests, or infection sites, thereby minimizing off-target effects and enhancing overall efficacy (Hegde et al., 2024). Several key methodologies underpin this precision. Physiological targeting takes advantage of inherent differences in biological environments, such as variations in pH, enzyme activity, or redox conditions. For instance, the alkaline pH of insect midguts (typically pH 8–10) can be used to trigger the release of active ingredients from pH-sensitive carriers, ensuring targeted delivery within the pest's digestive tract. Biological targeting involves the use of systems that interact specifically with pest biomarkers or are selectively transported through plant vascular tissues (Hegde et al., 2024). These formulations may be engineered for preferential uptake by pests or to adhere to particular plant surfaces, enhancing site-specific action. Environmental targeting, on the other hand, relies on external cues such as temperature, humidity, or light intensity that correlate with pest behavior or disease outbreaks, enabling responsive and timely release (Shafiq et al., 2024). The effectiveness of all these smart delivery strategies hinges on the design and functionality of carrier materials, which are responsible for protecting the active ingredients and mediating controlled release. These carriers range from natural polymers to sophisticated engineered nanomaterials, each offering unique advantages in terms of stability, responsiveness, and biocompatibility (Shafiq et al., 2024).

5.2.3 Polysaccharide-Based Carriers

Natural polysaccharides are gaining prominence in smart agrochemical delivery systems due to their inherent biocompatibility, biodegradability, and functional adaptability (Pavithran et al., 2024). Among these, chitosan, a derivative of chitin, is widely utilized for its ability to form nanoparticles via ionotropic gelation and its intrinsic antimicrobial properties. Its amine groups confer pH-responsive behavior, allowing for increased solubility and accelerated release in acidic environments through protonation (Desai, 2016). Chitosan-based composites, such as those with pectin or alginate, have proven effective in reducing soil adsorption of herbicides like paraquat while preserving their biological activity (Desai, 2016). Alginate, extracted from seaweed, forms hydrogels through crosslinking with divalent cations and contains carboxylic acid groups (pKa $\approx$ 4.8), making it suitable for systems that release active ingredients under alkaline conditions (Alshareef, 2019). When combined with chitosan, alginate forms electrostatically stabilized complexes that allow for adjustable release profiles. Cyclodextrins, another class of natural polysaccharides, are cyclic oligosaccharides capable of forming inclusion complexes with hydrophobic molecules (Alshareef, 2019). Their unique toroidal structure encapsulates active ingredients, enhancing their solubility, stability, and protection against degradation. Additionally, cyclodextrin-based systems can provide stimulus-responsive release through mechanisms like competitive displacement. Collectively, these natural polysaccharides offer versatile and sustainable platforms for developing advanced agrochemical delivery systems (Campos et al., 2015).

5.2.4 Synthetic Polymer Systems

Synthetic polymers are highly valued in smart agrochemical delivery systems for their ability to provide precise control over material properties and release kinetics. PLGA, a biodegradable polyester commonly used in biomedical applications, is increasingly being adapted for agricultural use (Singh et al., 2014). Its degradation rate can be finely tuned by altering the lactic-to-glycolic acid ratio, allowing for controlled release profiles that range from several weeks to months, depending on the application. Polyurethanes are another class of synthetic polymers known for forming robust microcapsule walls via interfacial polymerization between isocyanates and polyols (Lubis et al., 2020). These systems have been successfully commercialized for pesticides such as pyraclostrobin, offering added benefits like ultraviolet (UV) protection and reduced toxicity to aquatic organisms. Additionally, temperature-responsive polymers like poly(N-isopropylacrylamide) (PNIPAM) exhibit reversible phase transitions at defined temperatures (Papadakis et al., 2023). This property enables thermally triggered release, making them especially useful for pest control strategies where pest activity is closely linked to ambient temperature fluctuations. Collectively, these synthetic polymers provide customizable, responsive, and environmentally safer delivery systems for next-generation crop protection (Papadakis et al., 2023).

5.2.5 Inorganic Nanocarriers

Inorganic materials offer distinctive structural and functional advantages for agrochemical delivery, making them valuable components in smart crop protection systems. Mesoporous silica nanoparticles are notable for their exceptionally high surface areas (often exceeding 1000 m^2/g) and tunable pore sizes, which enable the encapsulation of large quantities of active ingredients. These particles can be surface-functionalized with 'gatekeeper' molecules that respond to specific environmental stimuli, allowing for controlled release (Shao et al., 2022). For instance, sulfonic acid-modified silica has been effectively used for pH-responsive herbicide delivery, releasing its payload under alkaline conditions commonly found in target environments (Hua et al., 2020). Metal-organic frameworks (MOFs), which are crystalline, porous materials composed of metal ions coordinated to organic ligands, also exhibit remarkable loading capacities and responsiveness to environmental cues (Li et al., 2016). A zinc-based MOF has demonstrated successful delivery of fungicide specifically to citrus plants infected by Botrytis cinerea, with pH-sensitive release triggered by infection-related changes in the plant tissue (Li et al., 2016). Additionally, clay minerals such as halloysite nanotubes offer a low-cost, naturally abundant, and environmentally benign option (Pandey et al., 2020). These aluminosilicate structures possess lumen spaces ideal for loading active ingredients, and their surfaces can be chemically modified to fine-tune release rates or enhance targeting of specific plant tissues. Together, these inorganic materials expand the toolbox for designing efficient, targeted, and sustainable agrochemical delivery platforms (Pandey et al., 2020).

5.3 Nanoformulations of Pesticides, Herbicides, and Fungicides

Nanoformulations represent advanced delivery systems that encapsulate, adsorb, or embed agrochemicals (pesticides, herbicides, fungicides) within nanomaterials (1–100 nm) (Pramod et al., 2024). These systems enhance bioavailability, stability, and target specificity of active ingredients (AIs), addressing limitations of conventional formulations like rapid degradation, runoff, and high organic solvent use. By 2027, the global nanopesticide market is projected to grow at 14.6% CAGR, driven by demands for sustainable agriculture. Nanoformulations reduce AI doses by 70–80% while improving efficacy through controlled release mechanisms (Pramod et al., 2024).

5.3.1 Types of Nano formulations

5.3.1.1 Organic nanoformulations

In nanoemulsions, oil-in-water (O/W) dispersions (20–500 nm) using surfactants (e.g., Tween 80, Span 80) to solubilize hydrophobic AIs (Pramod et al., 2024). Benefits include reduced organic solvents, enhanced foliar adhesion, and UV protection. Example: Neem oil nanoemulsion for mosquito larvicide. Polymeric Nanocapsules/Spheres have core-shell structures (e.g., chitosan, PLGA) encapsulating AIs. Provide slow release and protection from degradation. Example: PCL-loaded tebuconazole for fungal control (Perinelli et al., 2020). Solid Lipid Nanoparticles (SLNs) are biodegradable lipid matrices (e.g., triglycerides) for controlled AI release, Used for herbicides like atrazine (Schöler et al., 2002). Micelles/Dendrimers are amphiphilic polymers forming nanosized cages. Enable stimuli-responsive release (pH/temperature).

5.3.1.2 Inorganic nanoformulations

Metal/Metal oxide NPs: Silver (Ag), copper oxide (CuO), zinc oxide (ZnO), and silicon dioxide (SiO_2) NPs act as dual carriers and bioactive agents. SiO_2 NPs enhance plant resistance and adsorb pesticides for controlled release. Nanoclays & Layered Double Hydroxides (LDHs) are anionic clays (e.g., montmorillonite) for slow release. Example: 2,4-D herbicide-loaded LDHs. Mesoporous silica nanoparticles have high surface area (500–1,000 m^2/g) for large AI loads. Functionalized surfaces improve leaf adhesion.

5.3.1.3 Hybrid and "All-Green" formulations

Combine organic/inorganic materials (e.g., silica-chitosan composites) or use entirely biogenic components (plant extracts + biopolymer carriers). Reduce ecotoxicity and trophic transfer risks (Rosciardi et al., 2022).

Table 1 Comparison of key nano-formulation types.

Type	*Size Range*	*Materials Used*	*Example AIs*	*Key Advantages*
Nanoemulsions	20–500 nm	Oils, nonionic surfactants	Neem oil, permethrin	Solubilizes hydrophobic AIs; low drift
Polymeric NPs	50–300 nm	Chitosan, PLGA, PCL	Tebuconazole, imidacloprid	Slow release; UV protection
Metal Oxide NPs	10–100 nm	SiO_2, ZnO, CuO	N/A (intrinsic activity)	Dual function; low resistance risk
Porous Carriers	100–400 nm	Silica, activated carbon	Avermectin, cyantraniliprole	High loading capacity; sustained release

Table 1 summarizes the major types of nano-enabled delivery systems used in crop protection, categorized by particle size, materials employed, example AIs, and key advantages. Nanoemulsions (20–500 nm) are formulated using oils and nonionic surfactants to enhance the solubility of hydrophobic active ingredients such as neem oil and permethrin, offering low drift and improved dispersion. Polymeric nanoparticles (NPs) (50–300 nm), composed of biodegradable polymers like chitosan, PLGA, and PCL, facilitate slow release and UV protection for active ingredients such as tebuconazole and imidacloprid. Metal oxide nanoparticles (10–100 nm), including SiO_2, ZnO, and CuO, possess intrinsic pesticidal properties, providing dual functionality with a low risk of resistance development. Porous carriers (100–400 nm), such as silica and activated carbon, offer high loading capacity and sustained release of active ingredients like avermectin and cyantraniliprole. This classification underscores the versatility of nanomaterials in enhancing the performance, stability, and precision of agrochemical delivery.

5.4 Environmental and Toxicity Considerations

5.4.1 Advantages

In smart delivery systems for agrochemicals, addressing environmental and toxicity considerations offers several key advantages that contribute to sustainable agriculture. These systems are designed to release active ingredients in a controlled and targeted manner, thereby significantly reducing environmental contamination caused by leaching into soil and runoff into water bodies (Kookana et al., 1998). By ensuring site-specific delivery, smart formulations minimize the exposure of non-target organisms such as beneficial insects, pollinators, soil microbes, and aquatic life, thereby supporting biodiversity and ecosystem health. Additionally, the use of biodegradable carrier materials, such as natural polymers like chitosan or alginate, prevents long-term persistence of harmful residues and eliminates nano-waste accumulation. These systems also enable the use of lower dosages while maintaining efficacy, which reduces the overall toxic load and the risk of forming toxic degradation products. Importantly, this approach minimizes human and animal exposure through reduced volatilization and drift, enhancing occupational and environmental safety. Moreover, by delivering agrochemicals more precisely and over extended periods, smart systems can help mitigate the development of pest and pathogen resistance. Collectively, these benefits align with increasing regulatory demands and public expectations for environmentally responsible and health-conscious agricultural practices (Liu et al., 2024).

5.4.2 *Risks*

The polymer/lipid carriers exhibit combined toxicity with AIs in non-target organisms (e.g., nematode reproduction inhibition). The metal NPs (CuO, Ag) persist in soil, disrupting microbial communities and nutrient cycling. The NPs enter food chains via accumulation in plants/animals (Liu et al., 2024).

Table 2 Environmental behavior of nanoformulations.

Factor	*Impact on Nanoformulations*	*Examples from Studies*
Soil pH/ ionic strength	Alters NP aggregation & AI release kinetics	Cu $(OH)_2$ NPs reduced soil β-glucosidase activity
Microbial activity	Degrades polymeric carriers but may be inhibited by metal NPs	CuO NPs harm bacteria
Ecotoxicity	Nematodes, earthworms, and aquatic organisms at risk	Tebuconazole NPs reduced C. elegans reproduction by 50%

Table 2 highlights key environmental factors affecting the behavior and ecological impact of nano-enabled agrochemical formulations. Soil pH and ionic strength influence nanoparticle (NP) aggregation and the release kinetics of AIs, as demonstrated by Cu $(OH)_2$ NPs reducing soil β-glucosidase activity. Microbial activity plays a dual role by degrading biodegradable polymeric carriers, while being susceptible to toxicity from metal-based NPs such as CuO. Ecotoxicity concerns include potential harm to non-target soil and aquatic organisms; for instance, tebuconazole-loaded NPs have been shown to reduce *C. elegans* reproduction by 50%. Table 2 underscores the importance of evaluating nanoformulations under realistic environmental conditions to ensure their safe and effective deployment.

5.5 *Stimuli-responsive Nanocarriers (pH, Temperature and Enzymes)*

The most significant advancement in smart agrochemical delivery is the development of systems that release active ingredients in response to specific environmental or biological triggers. These stimulus-responsive systems maximize efficiency by synchronizing release with pest presence or optimal application conditions (Camara et al., 2019).

5.5.1 *pH-responsive Systems*

The pH-responsive systems exploit pH variations between different plant tissues, pest digestive systems, and infection sites. Key design strategies include the swell or dissolve at specific pH values (Zhang et al., 2020). Chitosan (pKa $\approx$ 6.3) protonates in acidic environments like insect midguts, increasing solubility and release rates. Conversely, alginate (pKa $\approx$ 4.8) deprotonates in alkaline conditions, disrupting electrostatic crosslinks. They use pH-sensitive molecules to block nanocarrier pores. For example, chitosan-coated mesoporous silica releases herbicides when chitosan dissolves at acidic pH 9. Similarly, polyacrylic acid gates open at alkaline pH, useful for targeting slightly alkaline phloem tissues (Abenaim et al., 2023). The change surface charge in response to pH, altering interaction with targets. A pH-responsive MOF successfully delivered fungicide specifically to citrus disease sites by responding to pH changes from Botrytis cinerea infection (pH 5.5 vs. healthy tissue pH 7.0) (Zhang et al., 2022).

5.5.2 *Temperature-responsive Systems*

The temperature fluctuations in agricultural environments provide natural triggers for release systems. Common approaches include materials like PNIPAM collapse above their lower critical solution temperature (LCST $\approx$ 32°C), expelling encapsulated compounds. This property has been utilized to create formulations that release more rapidly during warm periods when pest activity increases. Thermally labile linkers such as azo bonds or Diels-Alder adducts cleave at specific temperatures. Gold nanoparticle-embedded microcapsules demonstrated near-infrared light-triggered heating and release, offering potential for precision application. The shape-memory polymers change conformation with temperature, useful for foliar adhesives that improve retention during temperature changes (Marvin et al., 2013).

5.5.3 Enzyme-Responsive Systems

Pathogen- or pest-specific enzymes provide highly targeted release triggers the, cellulase-responsive systems degrade in presence of cellulase-producing fungi. Silica-carboxymethyl cellulose microcapsules showed 28.66% release within 1 hour with cellulase versus <10% without. The protease-activated carriers use peptide linkers cleaved by pest-specific proteases. This approach is particularly promising for RNAi delivery, where nucleases in insect guts can trigger release (Kunte et al., 2020). The chitinase-sensitive materials target insect pests that secrete chitinases for molting. Chitosan-based systems can be tuned to degrade specifically by pest chitinases while resisting plant enzymes.

6. Case Studies of Successful Nano-biopesticide Applications

6.1 Case Study 1: Savoy Cabbage Field Trial (Spain)

Researchers compared urea-hydroxyapatite nanoparticles and potassium sulfate nanoparticles with conventional fertilizers in a field study. Remarkably, nanofertilizers applied at 50% of recommended NPK dosage (80 N, 100 P_2O_5, 250 K_2O kg ha^{-1}) achieved comparable biometric parameters and yield to full-dose conventional fertilizers. Soil analysis revealed increased pH and available phosphorus without altering soil structure or nitrogen content. This demonstrates nanofertilizers' potential to reduce nutrient inputs by 50% while maintaining crop productivity, significantly lowering environmental contamination risks (Tei et al., 2005).

6.2 Case Study 2: Zinc Oxide Nanoparticles for Tomato Nutrition

Foliar application of ZnO nanoparticles (500 ppm) increased tomato fruit yield by 51% compared to conventional zinc treatments. The nano-formulation's 2.5 higher solubility enabled superior stomatal penetration and vascular transport, enhancing photosynthetic efficiency and enzymatic activity. This case highlights the efficacy of nano-micronutrients in addressing deficiencies while reducing accumulation in edible parts (Ahmed et al., 2021). The size-dependent solubility of nano-hydroxyapatite (30–50 nm) exhibits 3X higher phosphorus solubility than bulk equivalents via the Ostwald-Freundlich effect. In controlled release, the chitosan-coated potassium nanoparticles extended nutrient availability to 1,176 h, minimizing leaching losses.

Table 3 Comparative performance of nanofertilizers.

Crop	*Nanofertilizer*	*Application Rate*	*Yield Increase*	*Nutrient Reduction*
Savoy Cabbage	Urea-hydroxyapatite NPs	50% conventional	No significant difference	50% N reduction
Tomato	ZnO NPs	500 ppm foliar	51%	3X less Zn applied
Rice	Urea-zeolite nanocomposites	75% conventional	20%	25% N reduction

Table 3 summarizes the effects of different nanofertilizers on crop yield and nutrient input reduction across three major crops. Savoy cabbage treated with urea-hydroxyapatite nanoparticles showed no significant yield improvement but achieved a 50% nitrogen reduction. Tomato plants responded positively to ZnO nanoparticles, with a 51% yield increase using foliar application at 500 ppm and requiring three times less zinc compared to conventional fertilizers. In rice, the application of urea-zeolite nanocomposites at 75% of the conventional rate resulted in a 20% yield increase and 25% nitrogen reduction. These findings highlight the potential of nanofertilizers to reduce nutrient inputs while maintaining or enhancing crop productivity.

6.3 Case Study 3: Abamectin Nanoemulsion for Diamondback Moth Control

A 2% abamectin nanoemulsion with castor oil polyoxyethylene (EL-40) demonstrated exceptional adhesion to cabbage leaves, with a contact angle 50% lower than conventional formulations. Against *Plutella xylostella* larvae, it achieved an LC_{50} of 0.0686 mg/L-40% lower than commercial abamectin. The nanoemulsion's small droplet size (20–50 nm) enhanced cuticular penetration while protecting the active ingredient from UV degradation (Ding et al., 2024).

6.4 Case Study 4: Silica Nanocarriers for Stored Grain Protection

Chlorpyrifos-loaded silica nanoparticles (Ch-SNPs) were tested against warehouse pests *Rhyzopertha dominica* and *Tribolium confusum*. At 2.06 g/kg, Ch-SNPs caused 100% mortality in cowpea weevils by attaching to insect cuticles and disrupting physiological functions. The porous structure of silica nanoparticles provided UV-shielding, reducing active ingredient degradation by < 9% after 24h exposure versus > 90% in non-encapsulated pesticides (Haroun et al., 2020). In innovative delivery systems, the pH-responsive release process, attapulgite-hydrogel spheres released chlorpyrifos under alkaline conditions matching pest gut environments. Whereas, in magnetic recovery, the Fe_3O_4-diatomite nanocarriers enabled post-application retrieval, reducing soil accumulation (Haroun et al., 2020).

6.5 Case Study 5: Biogenic Nanoparticles for Heat-drought Stress Alleviation

Under combined heat-drought stress, chitosan-calcium oxide nanoparticles increased Blackeye bean growth by 8–10.7 cm compared to 6 cm in controls. The nanoparticles stimulated antioxidant enzyme production (SOD, CAT), reducing reactive oxygen species (ROS) damage while improving water retention. Field trials demonstrated 40% higher survival rates in nano-treated plants during heat waves (≥ 35°C), highlighting their role in climate resilience (Prokisch et al., 2024).

6.6 Case Study 6: Silicon Nanoparticles for Salinity-Heat Tolerance in Cucumber

Amorphous silica nanoparticles (200 mg/kg) boosted cucumber yield by 156% under combined salinity and heat stress. The nanoparticles reduced sodium absorption by 60% while enhancing potassium assimilation. Microscopy revealed nanoparticle deposition in trichomes creating a protective 'nano-film' that reduced transpiration and improved thermal regulation (Abdi et al., 2023).

Table 4 Nano-interventions for combined stresses.

Stress Combination	*Nano-Agrochemical*	*Mechanism*	*Efficacy*
Heat + Drought	Chitosan-CaO NPs	ROS scavenging, stomatal regulation	40% higher plant survival
Salinity + Heat	Silica NPs	Na^+ exclusion, cuticular protection	156% yield increase
Pathogen + Heat	Silver-lignin nanocomposites	Thermal-stable antifungal activity	70% disease reduction

Table 4 outlines the effectiveness of various nano-agrochemical formulations in addressing complex stress conditions encountered in agriculture. Under heat and drought, chitosan-CaO nanoparticles enhanced plant survival by 40%, primarily through ROS scavenging and stomatal regulation. For salinity and heat stress, silica nanoparticles improved yield by 156% via sodium exclusion and cuticular protection. Against pathogen and heat stress, silver-lignin nanocomposites demonstrated 70% disease reduction, attributed to their thermal-stable antifungal activity. These findings demonstrate the potential of nano-agrochemicals to improve plant resilience under multiple concurrent stress factors.

6.7 Case Study: Eco-Friendly Nanoparticle Systems for Mosquito Control in Aquatic Environments

Green-synthesized nanoparticles offer a sustainable solution for mosquito control, especially in aquatic habitats where larvae thrive. Using plant extracts to reduce metal salts, these nanoparticles are cost-effective, eco-friendly, and active at ultra-low doses (1–30 ppm). They exhibit strong larvicidal and pupicidal effects, particularly hydrophobic nanosilica against Anopheles stephensi. These nanomaterials can impair larval motility, enhancing predation by biological control agents. Importantly, they show minimal toxicity to non-target aquatic organisms and no genotoxicity at effective doses.

7. Environmental and Industrial Implementation Challenges

Despite the considerable promise of nanotechnology in agriculture, its practical application is constrained by scientifically substantiated concerns related to toxicity, commercialization, and regulatory oversight. Toxicological evaluations have demonstrated that nanoparticles such as silver (Ag) and copper oxide (CuO) can disrupt the soil microbiome by inhibiting nitrogen-fixing bacterial populations at concentrations exceeding 100 ppm, thereby compromising soil fertility and ecosystem function. Furthermore, bioaccumulation studies have revealed that titanium dioxide (TiO_2) nanoparticles can translocate from soil into edible plant tissues, with subsequent mammalian exposure potentially inducing adverse effects such as cardiac arrhythmia, raising food safety concerns (Abdulsada et al., 2021). On the commercialization front, economic and technical barriers persist. Nano-formulations are reported to be 30–100% more expensive than conventional agrochemical products, primarily due to high-energy synthesis methods and complex production protocols. Additionally, scalability remains limited, with green synthesis approaches largely confined to laboratory settings; only approximately 5% of nano-fertilizer-related patents have successfully progressed to commercial-scale deployment.

Regulatory frameworks are also inadequate and fragmented. For instance, inconsistencies in nanoparticle classification criteria exist: the European Union mandates disclosure for materials containing > 50% nanoparticles, while the International Fertilizer Association (IFA) sets a size-based threshold at < 100 nm without specifying content proportion, leading to ambiguity in compliance. Moreover, current risk assessment protocols are insufficient for capturing nano-specific behaviors, such as pH-dependent ion release, aggregation dynamics, and interactions with soil organic matter factors that critically influence environmental fate and toxicity (Ludemann et al., 2022). These scientific and regulatory challenges must be systematically addressed to enable the safe, effective, and sustainable implementation of nanotechnology in agriculture.

8. Role of Nanocarriers in Enhancing Biopesticide Efficacy

Nanotechnology offers innovative approaches to stabilize, protect, and efficiently deliver biopesticides through the use of specialized nanocarriers. These include nanoparticles, nanocapsules, nanoemulsions, dendrimers, and lipid-based systems that can encapsulate or adsorb biopesticidal agents, shielding them from environmental degradation and enabling controlled release. Nanocarriers improve solubility and dispersibility of hydrophobic botanical compounds, extend shelf life, and enable targeted delivery to specific pest or plant tissues. They can also facilitate foliar or soil uptake, enhance penetration into insect cuticles or fungal spores, and reduce volatilization and leaching (Kumar et al., 2025). For microbial biopesticides, nanoencapsulation improves survival of spores or cells during storage and transport, and ensures their release only under favorable conditions such as high humidity or specific pH. In RNAi-based products, nanocarriers like chitosan nanoparticles or layered double hydroxides protect dsRNA molecules from enzymatic degradation and enhance cellular uptake in insect pests, increasing gene silencing efficiency. In addition to passive protection, some nanocarriers can be functionalized with ligands that allow receptor-mediated targeting, enabling pest-specific delivery while reducing off-target effects. This makes nanoformulated biopesticides suitable for precision pest management, minimizing ecological impact and maximizing efficacy (Kumar et al., 2025).

Examples of Nanoformulated Microbial, Botanical, and RNAi-based Biopesticides

1. **Microbial biopesticides,** such as *Bacillus thuringiensis* have been successfully formulated into chitosan-based nanoparticles, which enhance the persistence of Bt toxins on leaf surfaces and improve larvicidal activity. Similarly, nanoencapsulation of fungal spores like *Beauveria bassiana* in silica or polymeric matrices improves their UV stability and adhesion to insect cuticles (Kaur et al., 2024).
2. **Botanical biopesticides,** particularly essential oils like neem, citronella, eucalyptus, and clove oil, have been formulated into nanoemulsions or nanocapsules. These formulations reduce volatility, improve solubility, and enable slow release, resulting in prolonged activity against insect pests and pathogens. For instance, nanoemulsions of neem oil have shown enhanced ovicidal and repellent effects on *Helicoverpa armigera* and *Spodoptera litura* compared to conventional formulations (Acheuk et al., 2022).
3. **RNAi-based biopesticides** have gained attention for their specificity and eco-safety. However, their practical application has been limited by poor environmental stability. Nanoformulations using LDH clay nanoparticles, cationic lipids, or polymeric nanoparticles like PEI (polyethylenimine) and chitosan have significantly improved dsRNA protection and uptake by insect midgut cells. These systems have been used to silence key genes in pests such as *Myzus persicae* (aphids), *Leptinotarsa decemlineata* (Colorado potato beetle), and *Spodoptera frugiperda* (fall armyworm) (Fletcher et al., 2020).

9. Benefits of Nanotechnology in Crop Protection

The application of nanotechnology in crop protection offers numerous advantages over conventional agrochemical methods, addressing many of the pressing challenges in modern agriculture (Figure 2). One of the foremost benefits is the improved stability and bioavailability of active compounds. Many conventional pesticides, biopesticides, and nutrients suffer from rapid degradation due to environmental factors such as UV radiation, heat, microbial activity, or oxidation (Smith et al., 2013). Nanocarriers such as polymeric nanoparticles, nanogels, or lipid-based systems encapsulate these bioactives, providing a protective shield that enhances their shelf life, environmental persistence, and delivery efficiency. Additionally, the nanoformulation of poorly soluble or hydrophobic compounds increases their aqueous dispersibility, improving their absorption by plant tissues or pest targets and enhancing their overall bioavailability. Another major advantage is the reduction in chemical load and environmental impact (Smith et al., 2013). By facilitating controlled release, slow degradation, and site-specific delivery, nano-enabled systems allow for lower quantities of agrochemicals to be used without compromising efficacy. This minimizes runoff, leaching, and volatilization, significantly reducing contamination of soil, water bodies, and non-target organisms, including beneficial insects and microbes. This aspect aligns nanotechnology with global efforts toward sustainable, eco-friendly agriculture and reduces the human and ecological health risks associated with overuse of pesticides.

Nanotechnology also enables enhanced target specificity and systemic activity. Functionalized nanoparticles can be engineered to respond to specific environmental triggers such as pH, moisture, or enzymes, ensuring that the active ingredient is released only under pest-infestation or infection-prone conditions (Smith et al., 2013). Additionally, nanocarriers improve the penetration of agrochemicals into plant tissues, allowing them to be absorbed and translocated throughout the plant systemically. This systemic action ensures protection of both treated and untreated parts of the plant, offering a more comprehensive defense against pests and pathogens. An important yet often overlooked benefit is the reduced development of pest resistance (Whelan et al., 2020). Conventional pesticides, when applied in large doses and repeated over time, exert high selective pressure on pest populations, leading to rapid development of resistance. Nanoformulations, by offering sustained and targeted delivery at lower dosages, reduce this selective pressure and can be engineered to deliver combination therapies (e.g., two AI or a synergist with a pesticide), thereby delaying or mitigating resistance mechanisms. Some nanoformulations can also incorporate RNAi molecules that silence essential pest genes, offering a mode of action distinct from traditional chemicals (Whelan et al., 2020). Nanotechnology contributes to cost-effectiveness and yield enhancement. Although initial formulation and material costs may be higher, the long-term benefits of reduced input requirements, lower application frequency, minimized crop loss, and increased agrochemical efficiency translate into better return on investment for farmers (Channab et al., 2024). Furthermore, the enhanced efficacy of nanoproducts in protecting crops from stressors ensures better plant health, leading to higher yields and improved crop quality. When integrated into precision agriculture practices, nanotechnology can reduce labor and operational costs while optimizing input use efficiency (Channab et al., 2024). Hence, nanotechnology presents a paradigm shift in crop protection by offering precise, sustainable, and intelligent solutions that not only address the limitations of conventional methods but also align with the future goals of food security, environmental conservation, and climate-smart agriculture.

Figure 2 illustrates the multifaceted advantages of applying nanotechnology in crop protection. The central theme emphasizes how nano-enabled innovations contribute to smarter, safer, and more sustainable agricultural practices. The benefits are categorized into four main areas: (1) Improved Stability and Bioavailability nanotechnology enhances the solubility, protects active ingredients from degradation, and enables controlled, sustained release, improving systemic absorption and extending field effectiveness; (2) Reduction in Chemical Load and Environmental Impact—nanoformulations require lower doses, reduce runoff and environmental toxicity, protect non-target organisms, and promote eco-friendly practices; (3) Reduced Development of Pest Resistance sustained release limits resistance buildup, reduces repeated applications, disrupts pest adaptation mechanisms, and improves efficacy even at lower doses; and (4) Cost-Effectiveness and Yield Enhancement by optimizing resource use, lowering input costs, improving nutrient uptake, and reducing crop losses, nanotechnology contributes to increased yields and farm profitability. Together, these benefits demonstrate nanotechnology's transformative role in advancing climate-resilient and sustainable agriculture.

10. Environmental and Health Considerations

While nanotechnology offers numerous advantages in crop protection, its application in agriculture also raises important concerns regarding environmental safety, ecological balance, and human health. The widespread introduction of engineered nanomaterials (ENMs) into agroecosystems necessitates a critical evaluation of their potential risks, particularly in terms of persistence, bioaccumulation, and unintended interactions with non-target biological systems. Understanding and addressing these risks is essential to ensure the sustainable and responsible use of nano-enabled agrochemicals (Marmiroli et al., 2016).

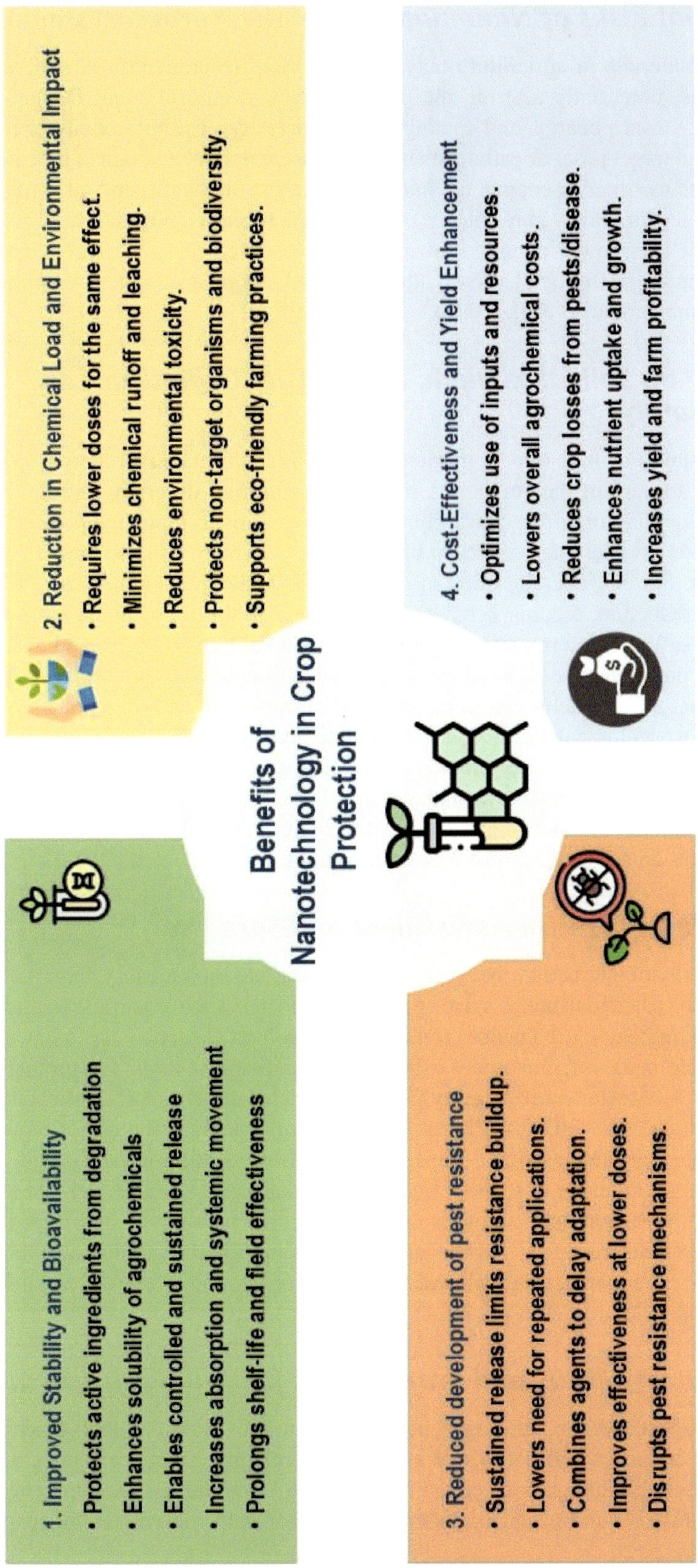

Fig. 2 Benefits of Nanotechnology in Crop Protection.

10.1 Potential Risks of Nanomaterials in the Agroecosystem

The use of nanomaterials in agriculture may result in their accumulation in soil, water bodies, and plant systems, potentially altering the physicochemical environment. Depending on their composition, size, surface charge, and coating, nanoparticles can exhibit toxicity or reactivity that may not only affect target pests or pathogens but also disrupt natural ecological processes. Metallic nanoparticles, such as silver or copper, are known for their strong antimicrobial activity; however, their unintended accumulation may impair beneficial microbial communities, suppress nitrogen fixation, or inhibit the growth of mycorrhizal fungi. Moreover, the persistence and mobility of nanoparticles in the soil or through leaching into groundwater could lead to long-term environmental contamination (Mohammadian et al., 2021).

10.2 Impact on Soil Microbiota, Non-Target Organisms, and Food Safety

Soil microbiota play a central role in nutrient cycling, organic matter decomposition, and plant health. Exposure to certain nanomaterials may alter microbial diversity and enzyme activity, disrupting these essential functions. Similarly, non-target organisms including pollinators like bees, earthworms, aquatic invertebrates, and even birds may be adversely affected through direct contact or trophic transfer. For instance, ingestion of nanoparticle-coated plant materials or residues could interfere with reproduction, feeding behavior, or immunity in these organisms (Ren et al., 2023). In terms of food safety, concerns arise regarding the uptake, translocation, and accumulation of nanoparticles in edible plant tissues, which may ultimately enter the food chain. Although some studies indicate that nanoparticles can be metabolized or immobilized within plant tissues, the lack of comprehensive toxicological data makes it difficult to predict their behavior and long-term effects in humans. Potential risks include cytotoxicity, oxidative stress, and genotoxic effects, particularly from persistent or non-biodegradable nanomaterials. These issues underscore the need for thorough investigation of nanoparticle bioavailability, residue analysis, and dietary exposure to ensure that nano-agrochemicals do not compromise food quality and safety (Ren et al., 2023).

10.3 Guidelines for Risk Assessment and Safe Use

Given the complexity of interactions between nanomaterials and biological systems, risk assessment frameworks must be adapted to address nano-specific parameters such as particle size, shape, surface modification, and aggregation behavior. Traditional toxicological models are often inadequate to predict nanoparticle behavior; thus, new nano-specific assessment tools and protocols are being developed. These include standardized assays for nanoparticle fate, transport, ecotoxicity, and human exposure (Committee et al., 2021). Risk mitigation strategies include the use of biodegradable or environmentally benign nanocarriers, proper encapsulation to minimize off-target release, and controlled application methods to reduce exposure to humans and wildlife. Best practices such as dose optimization, precision application, and use of personal protective equipment (PPE) during handling can further ensure safety. Public and private sector collaboration is essential to develop training programs for farmers and agrochemical handlers, promoting awareness and informed usage of nanoformulations (Committee et al., 2021).

10.4 Regulatory Status and Frameworks for Nano-Agrochemicals

The regulatory landscape for nano-agrochemicals is still evolving globally, with significant disparities across countries and regions. While some regulatory bodies recognize nanomaterials under existing chemical legislation, others are working to establish nano-specific registration pathways and labeling requirements. For example, the European Union has included nanomaterials under REACH (Registration, Evaluation, Authorization and Restriction of Chemicals) and requires detailed characterization and safety data for any product containing nanoparticles. Similarly, the United States Environmental Protection Agency (EPA) evaluates nanoscale pesticide products on a case-by-case basis, emphasizing product-by-product safety reviews (Singh et al., 2021). In India and several other developing countries, nano-agrochemical regulation is still in its nascent stages. The lack of harmonized definitions, testing protocols, and risk evaluation standards presents a major challenge to both manufacturers and regulators. However, international initiatives led by the Organisation for Economic Co-operation and Development (OECD), International Organisation for Standardisation (ISO), and Food and Agriculture Organisation (FAO)/ World Health Organisation (WHO) are working toward establishing guidelines for nanomaterial characterization, safety testing, and environmental risk assessment. To ensure public confidence and environmental protection, there is an urgent need for robust, science-based regulatory frameworks that balance innovation with precaution (Singh et al., 2021).

The environmental and health implications of nanotechnology in agriculture must be addressed proactively to ensure that the benefits of nano-enabled crop protection do not come at the expense of ecological integrity or human well-being. This involves multi-level efforts encompassing scientific research, risk assessment, regulatory oversight, and stakeholder education. By adopting a precautionary yet innovation-friendly approach, it is possible to develop and deploy nanotechnology-based agrochemicals that are safe, sustainable, and socially acceptable within the broader goals of modern agriculture (Singh et al., 2021).

11. Conclusion and Future Prospects

Nanotechnology is poised to revolutionize crop protection by addressing critical challenges such as pest resistance, environmental degradation, and the limitations of conventional agrochemicals and biopesticides (Abdollahdokht et al., 2022). Traditional chemical inputs face increasing scrutiny for their ecological impact and declining efficacy, prompting the need for smarter, safer, and more efficient alternatives. Nanotechnology enables smart delivery systems that enhance the performance and sustainability of agrochemicals and biopesticides through the unique properties of nanomaterials high surface area, small size, and tunable surface chemistry that facilitate controlled, targeted release of active ingredients in response to environmental triggers like pH, moisture, or enzymes. This targeted delivery reduces chemical use, minimizes off-target effects, and improves efficacy, while nanoencapsulation and other nanostructures protect sensitive compounds, extend shelf- life, and ensure sustained activity under field conditions. In biopesticides, nanoformulations of microbial agents, botanical oils, and RNAi molecules offer improved stability, solubility, and plant uptake, with real-world efficacy shown across regions (Fletcher et al., 2020). Environmentally, nanotechnology reduces chemical runoff and residue accumulation, preserving non-target organisms and water quality, although concerns about nanoparticle persistence and food safety require comprehensive risk assessments and regulatory oversight. Despite promising outcomes, adoption is limited by challenges in formulation, scalability, production costs, and farmer awareness. Bridging this gap demands affordable, farmer-friendly products, strong regulations, training, and public trust. Looking forward, integrating nanotechnology with precision agriculture, nanosensors, and AI could enable real-time detection and automated, data-driven interventions, ushering in a new era of sustainable, resilient, and climate-smart crop protection.

Nanotechnology holds immense promise for revolutionizing crop protection, yet its widespread application is hindered by several technical, economic, and regulatory challenges. One major obstacle lies in the complexity of nanoformulation development, which demands precise material engineering to achieve stability, high loading efficiency, environmental responsiveness, and scalability. Stimuli-responsive nanocarriers, while offering controlled release, are sensitive to production and environmental conditions, complicating their large-scale, reproducible manufacture. Moreover, unpredictable interactions between nanoparticles and environmental matrices like soil or water can impact field performance. The need for biodegradable, non-toxic, and cost-effective nanomaterials adds another layer of difficulty. Scaling up production remains a barrier, particularly in resource-limited settings, due to high costs and the lack of infrastructure. Additionally, limited farmer awareness and the absence of adequate training programs slow adoption. Regulatory uncertainty further compounds these issues, as harmonized guidelines for nanoagrochemicals are still lacking in many regions, creating obstacles in product approval and global trade. Public skepticism surrounding nanotechnology's safety, fueled by concerns over toxicity and environmental impact, underscores the need for transparent risk communication, evidence-based policy, and inclusive regulatory frameworks.

Looking ahead, the integration of nanotechnology with precision agriculture and digital technologies presents transformative opportunities. Smart nanosensors can enable real-time, site-specific monitoring of crop health, pests, and soil conditions, feeding hyper-local data into AI-driven platforms that automate agrochemical delivery through drones or smart equipment. This closed-loop system enhances efficiency, minimizes environmental exposure, and supports sustainable practices. Innovations in bioinspired nanomaterials, multifunctional nanocomposites, and environmentally responsive nanoformulations are poised to shape the next generation of crop protection tools. However, achieving this vision requires coordinated efforts across scientific disciplines, regulatory bodies, and agricultural stakeholders. Investments in farmer education, infrastructure, and regulatory clarity are essential to bridge the gap between lab-scale innovation and field-level implementation. The convergence of nanotechnology with data-driven farming offers a scalable, intelligent, and eco-conscious pathway to securing global food production in the face of climate and resource challenges.

References

Abdi, M.J., M. Ghanbari Jahromi, S.N. Mortazavi, S. Kalateh Jari and M.J. Nazarideljou. (2023). Foliar-applied silicon and selenium nanoparticles modulated salinity stress through modifying yield, biochemical attribute, and fatty acid profile of Physalis alkekengi L. *Environ. Sci. Pollut. Res. Int., 30*, 100513–100525.

Abdollahdokht, D., Y. Gao, S. Faramarz, A. Poustforoosh, M. Abbasi, G. Asadikaram and M.H. Nematollahi. (2022). Conventional agrochemicals towards nano-biopesticides: An overview on recent advances. *Chemical and Biological Technologies in Agriculture, 9*, 13.

Abdulsada, Z., R. Kibbee, B. Ormeci, M. DeRosa and J. Princz. (2021). Impact of anaerobically digested silver and copper oxide nanoparticles in biosolids on soil characteristics and bacterial community. *Chemosphere, 263*, 128173.

Abenaim, L. and B. Conti. (2023). Chitosan as a control tool for insect pest management: A review. *Insects, 14*, 949.

Acheuk, F., S. Basiouni, A.A. Shehata, K. Dick, H. Hajri, S. Lasram, M. Yilmaz, M. Emekci, G. Tsiamis, M. Spona-Friedl, H. May-Simera, W. Eisenreich and S. Ntougias. (2022). Status and Prospects of Botanical Biopesticides in Europe and Mediterranean Countries. *Biomolecules, 12*.

Ahmed, R., M. Yusoff Abd Samad, M.K. Uddin, M.A. Quddus and M.M. Hossain. (2021). Recent trends in the foliar spraying of zinc nutrient and zinc oxide nanoparticles in tomato production. *Agronomy, 11*, 2074.

Alam, M.W., P.M. Junaid, Y. Gulzar, B. Abebe, M. Awad and S.A. Quazi. (2024). Advancing agriculture with functional NM: Pathways to sustainable and smart farming technologies. *Discov. Nano, 19*, 197.

Alshareef, S.A. (2019). *Alginate from Brown Seaweeds: Extraction, Characterisation, Modification and the Studies on its Applications for Hydrogels*. Bangor University, United Kingdom.

Arnot, J.A. and F.A. Gobas. (2004). A food web bioaccumulation model for organic chemicals in aquatic ecosystems. *Environmental Toxicology and Chemistry, 23*, 2343–2355.

Bodede, A. (2020). *An Assessment of the Impact of Pesticide Use by Urban Cultivators Inoyo State, South-Western Nigeria*. University of Gloucestershire.

Borah, A., S. Sahu, R. Srivastava, M. Singh and D. Tyagi. (2024). Exploring the Economic Challenges Threatening Global Agriculture and Food Security. *Ecology, Environment, and Conservation, 0971765X*, 30.

Botnaru, A.A., A. Lupu, P.C. Morariu, O.L. Pop, A.H. Nedelcu, B.A. Morariu, O. Cioancă, M.L. Di Gioia, V.V. Lupu and L. Avasilcai (2025) Balancing Health and Sustainability: Assessing the Benefits of Plant-Based Diets and the Risk of Pesticide Residues. *Nutrients, 17*, 727.

Camara, M.C., E.V.R. Campos, R.A. Monteiro, A. do Espirito Santo Pereira, P.L. de Freitas Proença and L.F. Fraceto. (2019). Development of stimuli-responsive nano-based pesticides: Emerging opportunities for agriculture. *Journal of Nanobiotechnology, 17*, 100.

Campos, E.V.R., J.L. de Oliveira, L.F. Fraceto and B. Singh. (2015). Polysaccharides as safer release systems for agrochemicals. *Agronomy for Sustainable Development, 35*, 47–66.

Chamberlain, E., H. Shi, T. Wang, Y. Ma, A. Fulmer and C. Adams. (2012). Comprehensive screening study of pesticide degradation via oxidation and hydrolysis. *Journal of Agricultural and Food Chemistry, 60*, 354–363.

Channab, B.E.A., El Idrissi, A. Ammar, O. Dardari, S.E. Marrane, A. El Gharrak, A. Akil, Y. Essemlali and M. Zahouily. (2024). Recent advances in nano-fertilizers: Synthesis, crop yield impact, and economic analysis. *Nanoscale, 16*, 4484–4513.

Committee, E.S., S. More, V. Bampidis, D. Benford, C. Bragard, T. Halldorsson, A. Hernandez-Jerez, S. Hougaard Bennekou, K. Koutsoumanis, C. Lambre, K. Machera, H. Naegeli, S. Nielsen, J. Schlatter, D. Schrenk, V. Silano Deceased, D. Turck, M. Younes, J. Castenmiller, Q. Chaudhry, F. Cubadda, R. Franz, D. Gott, J. Mast, A. Mortensen, A.G. Oomen, S. Weigel, E. Barthelemy, A. Rincon, J. Tarazona and R. Schoonjans. (2021). Guidance on risk assessment of nanomaterials to be applied in the food and feed chain: Human and animal health. *EFSA J., 19*, e06768.

Constantine, K.L., M.K. Kansiime, I. Mugambi, W. Nunda, D. Chacha, H. Rware, F. Makale, J. Mulema, J. Lamontagne-Godwin and F. Williams. (2020). Why don't smallholder farmers in Kenya use more biopesticides? *Pest Management Science, 76*, 3615–3625.

Desai, K.G. (2016). Chitosan nanoparticles prepared by ionotropic gelation: An overview of recent advances. *Critical Reviews™ in Therapeutic Drug Carrier Systems, 33*.

Ding, X., F. Gao, B. Cui, Q. Du, Z. Zeng, X. Zhao, C. Sun, Y. Wang and H. Cui. (2024). The key factors of solid nanodispersion for promoting the bioactivity of abamectin. *Pesticide Biochemistry and Physiology, 201*, 105897.

Dizaj, S.M., F. Lotfipour, M. Barzegar-Jalali, M.H. Zarrintan and K. Adibkia. (2014). Antimicrobial activity of the metals and metal oxide nanoparticles. *Mater. Sci. Eng. C Mater. Biol. Appl., 44*, 278–284.

El-Fakharany, E.M. (2021). Nanoformulation approach for improved stability and efficiency of lactoperoxidase. *Prep. Biochem. Biotechnol., 51*, 629–641.

Fenibo, E.O., G.N. Ijoma and T. Matambo. (2022). Biopesticides in sustainable agriculture: Current status and future prospects. *New and Future Development in Biopesticide Research: Biotechnological Exploration*, 1–53.

Fletcher, S.J., P.T. Reeves, B.T. Hoang and N. Mitter. (2020). A Perspective on RNAi-Based Biopesticides. *Front. Plant Sci., 11*, 51.

Garg, D., K. Sridhar, B. Stephen Inbaraj, P. Chawla, M. Tripathi and M. Sharma. (2023). Nano-Biofertilizer Formulations for Agriculture: A Systematic Review on Recent Advances and Prospective Applications. *Bioengineering (Basel), 10*.

Georghiou, G.P. (1972). The evolution of resistance to pesticides. *Annual Review of Ecology and Systematics*, 133–168.

Gessner, K.H., J.S. Preisser, E. Pfaff, R. Wang, K. Walters, R. Bradford, M. Clark, M. Ehlers and M. Nielsen. (2025). Predictors of new persistent opioid use after surgery in adults. *Anesthesiol. Perioper. Sci., 3*, 2.

Haroun, S.A., M.E. Elnaggar, D.M. Zein and R.I. Gad. (2020). Insecticidal efficiency and safety of zinc oxide and hydrophilic silica nanoparticles against some stored seed insects. *Journal of Plant Protection Research, 60*.

Hegde, V., M.P. Bhat, D. Shin and K.-H. Lee. (2024). Recent advancement in direct delivery of active ingredients into plants: A shift from traditional approach to targeted precision: A review. *Precision Agriculture Science and Technology, 6*, 278–295.

Hua, Y., L. Chen, C. Hou, S. Liu, Z. Pei and Y. Lu. (2020). Supramolecular vesicles based on amphiphilic pillar [n] arenes for smart nano-drug delivery. *International Journal of Nanomedicine*, 5873–5899.

Ibadi, E.A., H.K. Awad and L.I. Hussain. (2024). Environmental toxins and their organ-specific effects: A comprehensive review of human exposure and accumulation. *Journal of Agricultural, Environmental and Veterinary Sciences, 8*, 10–24.

Jiang, X., F. Yang, W. Jia, Y. Jiang, X. Wu, S. Song, H. Shen and J. Shen. (2024). Nanomaterials and Nanotechnology in Agricultural Pesticide Delivery: A Review. *Langmuir, 40*, 18806–18820.

Kalogiannidis, S., D. Kalfas, F. Chatzitheodoridis and O. Papaevangelou. (2022). Role of crop-protection technologies in sustainable agricultural productivity and management. *Land, 11*, 1680.

Kamal Uddin, M. (2018). Agrochemicals and environmental risks. *Environmental Policy and Law, 48*, 91–96.

Kapoor, D.U., Sharma, J.B., Ganhi, S.M., Prajapati, B.G., Thanawuth, K. Limmatvapirat, S. and Sriamornsak, P. (2024). AI-driven design and optimization of nanoparticle-based drug delivery systems. *Science, Engineering and Health Studies,* 18, 24010003.

Karimi, M., M. Eslami, P. Sahandi-Zangabad, F. Mirab, N. Farajisafiloo, Z. Shafaei, D. Ghosh, M. Bozorgomid, F. Dashkhaneh and M. R. Hamblin. (2016). pH-Sensitive stimulus-responsive nanocarriers for targeted delivery of therapeutic agents. *Wiley Interdisciplinary Reviews: Nanomedicine and Nanobiotechnology, 8*, 696–716.

Karuppaiah, V. and G. Sujayanad. (2012). Impact of climate change on population dynamics of insect pests. *World Journal of Agricultural Sciences, 8*, 240–246.

Kaur, M., M. Nagpal, G.A. Dhingra and A. Rathee. (2024). Exploring chitin: Novel pathways and structures as promising targets for biopesticides. *Z. Naturforsch C. J. Biosci., 79*, 125–136.

Khandelwal, A., R. Joshi, M. Shrivastava and R. Singh. (2022). *Maximum Residue Limit (MRL) of Pesticides and Their Global Significance*. New Delhi Publishers.

Khot, L.R., S. Sankaran, J.M. Maja, R. Ehsani and E.W. Schuster. (2012). Applications of nanomaterials in agricultural production and crop protection: A review. *Crop Protection, 35*, 64–70.

Kookana, R.S., S. Baskaran and R. Naidu. (1998). Pesticide fate and behaviour in Australian soils in relation to contamination and management of soil and water: A review. *Soil Research, 36*, 715–764.

Kumar, R., R. Kaur, A.S. Ranout and G. Nadda. (2025). A Review on Novel Formulations and Delivery Systems of Botanical Insecticides for Enhanced Efficacy. *Neotrop. Entomol., 54*, 81.

Kunte, N., E. McGraw, S. Bell, D. Held and L.A. Avila. (2020). Prospects, challenges, and current status of RNAi through insect feeding. *Pest Management Science, 76*, 26–41.

Lai, P., W. Daear, R. Lobenberg and E.J. Prenner. (2014). Overview of the preparation of organic polymeric nanoparticles for drug delivery based on gelatine, chitosan, poly(d,l-lactide-co-glycolic acid), and polyalkylcyanoacrylate. *Colloids Surf. B Biointerfaces, 118*, 154–163.

Li, B., H.M. Wen, Y. Cui, W. Zhou, G. Qian and B. Chen. (2016). Emerging multifunctional metal–organic framework materials. *Advanced Materials, 28*, 8819–8860.

Li, N., C. Sun, J. Jiang, A. Wang, C. Wang, Y. Shen, B. Huang, C. An, B. Cui and X. Zhao. (2021). Advances in controlled-release pesticide formulations with improved efficacy and targetability. *Journal of Agricultural and Food Chemistry, 69*, 12579–12597.

Liu, Y., D. Kong, G. Tang, Z. Zhou, Y. Huang, X. Zhang, G. Yan, J. Wang, G. Hu and W. Yan. (2024). Fabrication of ionic liquid self-assemblies based on dicamba with improved herbicidal activity and reduced environmental risks. *Colloids and Surfaces B: Biointerfaces, 242*, 114077.

Lubis, M., L. Suryanegara and M. Pramesti. (2020). Synthesis of polymeric isocyanate microcapsules via interfacial polymerization and their characterisation using spectroscopy and microscopy techniques. In: *IOP Conference Series: Materials Science and Engineering*, 012053. IOP Publishing.

Ludemann, C.I., A. Gruere, P. Heffer and A. Dobermann. (2022). Global data on fertilizer use by crop and by country. *Sci. Data, 9*, 501.

Malik, S., K. Muhammad and Y. Waheed. (2023). Nanotechnology: A Revolution in Modern Industry. *Molecules, 28*.

Marmiroli, N. and J.C. White. (2016). Editorial: Nanotoxicology and Environmental Risk Assessment of Engineered Nanomaterials (ENMs) in Plants. *Front. Plant. Sci., 7*, 1370.

Marvin, H.J., G.A. Kleter, M.Y. Noordam, E. Franz, D.J. Willems and A. Boxall. (2013). Proactive systems for early warning of potential impacts of natural disasters on food safety: Climate-change-induced extreme events as case in point. *Food Control, 34*, 444–456.

Mittal, P., A. Saharan, R. Verma, F.M.A. Altalbawy, M.A. Alfaidi, G.E. Batiha, W. Akter, R.K. Gautam, M.S. Uddin and M.S. Rahman. (2021). Dendrimers: A New Race of Pharmaceutical Nanocarriers. *Biomed. Res. Int., 2021*, 8844030.

Mody, V.V., R. Siwale, A. Singh and H.R. Mody. (2010). Introduction to metallic nanoparticles. *J. Pharm. Bioallied. Sci., 2*, 282–289.

Mohammad El-Basioni, B.M., E.S. Mohamed, A.A. Belal, M.E.M. Jalhoum, S.M. Abd El-Kader and M.B. Zahran. (2022). A case study of a real-time internet of things system for site-specific potato crop management in El-Salhia Area-Egypt. *Sci. Rep., 12*, 18600.

Mohammadian, S., B. Krok, A. Fritzsche, C. Bianco, T. Tosco, E. Cagigal, B. Mata, V. Gonzalez, M. Diez-Ortiz, V. Ramos, D. Montalvo, E. Smolders, R. Sethi and R.U. Meckenstock. (2021). Field-scale demonstration of *in situ* immobilization of heavy metals by injecting iron oxide nanoparticle adsorption barriers in groundwater. *J. Contam. Hydrol., 237*, 103741.

Neamtu, I., A.G. Rusu, A. Diaconu, L.E. Nita and A.P. Chiriac. (2017). Basic concepts and recent advances in nanogels as carriers for medical applications. *Drug Deliv., 24*, 539–557.

Oerke, E.-C. and H.-W. Dehne. (2004). Safeguarding production: Losses in major crops and the role of crop protection. *Crop Protection, 23*, 275–285.

Pandey, G., M. Tharmavaram and D. Rawtani. (2020). Halloysite nanotubes: An 'aluminosilicate nanosupport' for energy and environmental applications. In: *Nanotechnology for Energy and Environmental Engineering*. Springer, pp. 125–144.

Papadakis, C.M., B.-J. Niebuur and A. Schulte. (2023). Thermoresponsive Polymers under Pressure with a Focus on Poly (N-isopropylacrylamide) (PNIPAM). *Langmuir, 40*, 1–20.

Pavithran, R.K., S.G. Reddy, B.S. Kumar and S. Kugabalasooriar. (2024). Enhancing sustainability in agriculture: Natural polymer-based controlled release systems for effective pest management and environmental protection. *ES Food and Agroforestry, 18*, 1276.

Perez-Ferreiro, M., M.A.A., A. Criado, I.J. Gomez and J. Mosquera. (2023). Dendrimers: Exploring Their Wide Structural Variety and Applications. *Polymers (Basel), 15*.

Perinelli, D.R., G.F. Palmieri, M. Cespi and G. Bonacucina. (2020). Encapsulation of flavours and fragrances into polymeric capsules and cyclodextrins inclusion complexes: An update. *Molecules, 25*, 5878.

Peteu, S.F., F. Oancea, O.A. Sicuia, F. Constantinescu and S. Dinu. (2010). Responsive polymers for crop protection. *Polymers, 2*, 229–251.

Pimentel, D. (1996). Green revolution agriculture and chemical hazards. *Science of the Total Environment, 188*, S86–S98.

Pokropivny, V., R. Lohmus, I. Hussainova, A. Pokropivny and S. Vlassov. (2007). *Introduction to Nanomaterials and Nanotechnology*. Estonia: University of Tartu.

Pouton, C.W. (2006). Formulation of poorly water-soluble drugs for oral administration: Physicochemical and physiological issues and the lipid formulation classification system. *European Journal of Pharmaceutical Sciences, 29*, 278–287.

Prabha, S., A. Yadav, A. Kumar, A. Yadav, H.K. Yadav, S. Kumar, R. Yadav and R. Kumar. (2016). Biopesticides: An alternative and eco-friendly source for the control of pests in agricultural crops. *Plant Arch., 16*, 902–906.

Pramod, I.U., S.S. Sudhir, U. Nimbalkar, G.K. Aniket, T. Minkina, R.D. Vishnu, P. Golińska, M. Rai and B.K. Jayanta. (2024). Nanomaterials for Controlled and Targeted Delivery of Agrochemicals for Cleaner Environment. In: *Nanotechnology for Environmental Management*. CRC Press, pp. 223–243.

Prasad, R., A. Bhattacharyya and Q.D. Nguyen. (2017). Nanotechnology in Sustainable Agriculture: Recent Developments, Challenges, and Perspectives. *Front. Microbiol., 8*, 1014.

Preeti, S. Sambhakar, R. Malik, S. Bhatia, A. Al Harrasi, C. Rani, R. Saharan, S. Kumar, Geeta and R. Sehrawat. (2023). Nanoemulsion: An Emerging Novel Technology for Improving the Bioavailability of Drugs. *Scientifica (Cairo), 2023*, 6640103.

Prokisch, J., A. Ferroudj, S. Labidi, H. El-Ramady and E.C. Brevik. (2024). Biological Nano-Agrochemicals for Crop Production as an Emerging Way to Address Heat and Associated Stresses. *Nanomaterials (Basel)*, 14.

Ren, Z., T. Cai, Y. Wan, Q. Zeng, C. Li, J. Zhang, K. Ma, S. He, J. Li and H. Wan. (2023). Unintended consequences: Disrupting microbial communities of Nilaparvata lugens with non-target pesticides. *Pestic. Biochem. Physiol., 194*, 105522.

Rosciardi, V., D. Chelazzi and P. Baglioni. (2022). "Green" biocomposite poly (vinyl alcohol)/starch cryogels as new advanced tools for the cleaning of artifacts. *Journal of Colloid and Interface Science, 613*, 697–708.

Savary, S., A. Ficke, J.-N. Aubertot and C. Hollier. (2012). Crop losses due to diseases and their implications for global food production losses and food security. *Food Security, 4*, 519–537.

Schöler, N., H. Hahn, R. Müller and O. Liesenfeld. (2002). Effect of lipid matrix and size of solid lipid nanoparticles (SLN) on the viability and cytokine production of macrophages. *International Journal of Pharmaceutics, 231*, 167–176.

Sekhon, B. S. (2014). Nanotechnology in agri-food production: An overview. *Nanotechnol. Sci. Appl., 7*, 31–53.

Shafiq, S., M.Z.U. Haq, S. Shafique, S. Khalid, M. Hamza, H. Ali and W. Sarfaraz. (2024). Climate-adaptive pest management for sustainable agriculture: Navigating temperature, precipitation, and evolving pest dynamics. In: *Revolutionizing Pest Management for Sustainable Agriculture*. IGI Global, pp. 31–52.

Shang, Y., M.K. Hasan, G.J. Ahammed, M. Li, H. Yin and J. Zhou. (2019). Applications of Nanotechnology in Plant Growth and Crop Protection: A Review. *Molecules, 24*.

Shao, C., H. Zhao and P. Wang. (2022). Recent development in functional nanomaterials for sustainable and smart agricultural chemical technologies. *Nano Convergence, 9*, 11.

Sharma, H. (2012). *Conventional and Biotechnological Approaches for Pest Management: Potential and Limitations*. Studium Press, LLC.

Sharma, J.N., D.K. Pattadar, B.P. Mainali and F.P. Zamborini. (2018). Size determination of metal nanoparticles based on electrochemically measured surface-area-to-volume ratios. *Analytical Chemistry, 90*, 9308–9314.

Singh, G., T. Kaur, R. Kaur and A. Kaur. (2014). Recent biomedical applications and patents on biodegradable polymer-PLGA. *Int. J. Pharmacol. Pharm. Sci., 1*, 30–42.

Singh, H., A. Sharma, S. Bhardwaj, S.K. Arya, N. Bhardwaj and M. Khatri. (2021). Recent advances in the applications of nano-agrochemicals for sustainable agricultural development. *Environ. Sci.: Processes Impacts , 23*, 213.

Smith, D.M., J.K. Simon and J.R. Baker, Jr. (2013). Applications of nanotechnology for immunology. *Nat. Rev. Immunol., 13*, 592–605.

Tei, F., E. Pannacci, A. Cirujeda, A. Dobrzanski, R. Froud-Williams, F. Rocha, P. Ruuttunen, T. Sanseovic, A. Simoncic and M. Torma., (2005. Weeds and weed management in cabbages: A review. In: *13th European Weed Res. Symposium* at Bari, Italy, 19–23 June.

Upadhyay, S., J. Sharma and T. Bhardwaj. (2014). Knowledge level of farmers about integrated pest management practices. *Journal of Community Mobilization and Sustainable Development,* 9, 1–5.

Varma, M.V., A.M. Kaushal, A. Garg and S. Garg. (2004). Factors affecting mechanism and kinetics of drug release from matrix-based oral controlled drug delivery systems. *American Journal of Drug Delivery, 2*, 43–57.

Wanyika, H., E. Gatebe, P. Kioni, Z. Tang and Y. Gao. (2012). Mesoporous silica nanoparticles carrier for urea: Potential applications in agrochemical delivery systems. *J. Nanosci. Nanotechnol., 12*, 2221–2228.

Whelan, C.J. and J.J. Cunningham. (2020). Resistance is not the end: Lessons from pest management. *Cancer Control, 27*, 1073274820922543.

Whitesides, G.M. (2005). Nanoscience, nanotechnology, and chemistry. *Small, 1*, 172–179.

Younis, S., M. Yousaf, A. Nazir and S. Shakoor. (2025). Advances in the Encapsulation and Delivery of Food Functional Factors: Micro-and Nano-Delivery Carrier Platforms in Bio-manufacturing. *Journal of Food and Agricultural Technology Research, 4*, 1–25.

Zabot, G.L., F. Schaefer Rodrigues, L. Polano Ody, M. Vinicius Tres, E. Herrera, H. Palacin, J.S. Cordova-Ramos, I. Best and L. Olivera-Montenegro. (2022). Encapsulation of Bioactive Compounds for Food and Agricultural Applications. *Polymers (Basel), 14*.

Zaman, W., A. Ayaz and S. Park. (2025). Nanomaterials in Agriculture: A Pathway to Enhanced Plant Growth and Abiotic Stress Resistance. *Plants (Basel), 14*.

Zhang, X., X. Tang, C. Zhao, Z. Yuan, D. Zhang, H. Zhao, N. Yang, K. Guo, Y. He and Y. He. (2022). A pH-responsive MOF for site-specific delivery of fungicide to control citrus disease of Botrytis cinerea. *Chemical Engineering Journal, 431*, 133351.

Zhang, Y., J. Yan, A. Avellan, X. Gao, K. Matyjaszewski, R.D. Tilton and G.V. Lowry. (2020). Temperature-and pH-responsive star polymers as nanocarriers with potential for *in vivo* agrochemical delivery. *ACS Nano, 14*, 10954–10965.

Zhao, L., X. Zhou, Z. Kang, J.R. Peralta-Videa and Y.G. Zhu. (2024). Nano-enabled seed treatment: A new and sustainable approach to engineering climate-resilient crops. *Sci. Total Environ., 910*, 168640.

Zhou, Z.-Y., N. Tian, J.-T. Li, I. Broadwell and S.-G. Sun. (2011)., Nanomaterials of high surface energy with exceptional properties in catalysis and energy storage. *Chemical Society Reviews, 40*, 4167–4185.

CHAPTER 6

Precision Pest Management
Integrating Remote Sensing, Artificial Intelligence, and Site-Specific Technologies in Crop Protection

Raju Sapkota[1] and Megha N Parajulee[2*]

1. Introduction

The Food and Agriculture Organization of the United Nations emphasized the urgency of crop protection vis-à-vis production sustainability and food security as arthropod pests, weeds, and diseases are responsible for ~40% of global crop losses each year, a figure that could rise further without effective pest management strategies (Tang et al., 2023). Precision agriculture (PA) utilizes advanced hardware and software tools to enable farmers to make data-driven, tailored decisions for key farming operations, including planting, fertilization, pest management, and harvesting. PA is the agricultural process that involves collecting accurate spatial data on crops and effectively managing the farm operation. Different technologies such as remote sensing, Internet of Things (IoT), geographic information system (GIS), and artificial intelligence (AI) help farmers in pest monitoring. These technologies are highly useful and significantly important for pest forecasting and management of insect pests, nematodes, weeds, and plant diseases. Pest monitoring tools at present include global positioning system (GPS), global navigation satellite system (GNSS), and AI. AI-based monitoring includes deep learning (DL) for pest identification. One study proposes the integration of smart IoT that enables pest monitoring, image-based machine learning (ML), and cloud reporting into PA systems to enhance pest management efficiency. For instance, they integrate smart pheromone traps with embedded cameras for real-time pest monitoring and explore convolutional neural networks (CNNs) for automated pest detection from trap images (Ahmed et al., 2024). Sensor-based monitoring includes the use of radar to track the migration of pests, thermal infrared imaging, GNSS for telemetry of wildlife, video equipment to monitor flying insects, chemiluminescent tags to track insect movement at night, and detection of larval movement through echo-sounding and habitat mapping (Kanwal et al., 2022). These technologies are very crucial for farmers to make data-driven decisions, which can enhance crop yields and conserve resources like water and soil nutrients. Utilizing a heat sensor employs thermal imaging that identifies localized

[1-2] Texas A&M University, Department of Entomology, Texas A&M AgriLife Research and Extension Center, Lubbock, Texas, USA.
* Corresponding author: m-parajulee@tamu.edu

DOI: 10.1201/9781003614869-6

temperature variations in the field that signal potential pest activity. Different models, such as climate models, machine learning models, and physiologically based demographic models (PBDMs), are utilized for forecasting the pest attack. AI and ML help further refine the data that predicts yield and potential pest outbreaks (Balaska et al., 2023). A crucial component of pest population management is that climate influences both insect abundances and natural enemy dynamics. These climate models predict the spectrum of potential changes on a global scale when we pair them with the environmental needs of a specific pest species. For the PBDM model, we can effectively utilize such a model to observe the weather pattern, location, and pest abundance. The insect population dynamics can be examined by considering both the local weather conditions and worldwide climate indicators (Parajulee et al., 1996, 1998, 2001). Combining exogenous and endogenous components, such as regional climate conditions and intrinsic population dynamics, provides a model for pest forecasting (Parajulee 2007, 2014). Some management techniques of insect pests using PA tools include site-specific insect pest management (Parajulee et al., 2005), variable rate technologies (VRT) (Bronson et al., 2006), AI-based application of pesticides (Banerjee et al., 2024), and nano-bio-pesticides (Vinci et al., 2025). Similarly, Paccioretti et al. (2020) demonstrated a 15% increase in corn yield while simultaneously reducing nitrogen runoff by using precision application of nitrogen fertilizer. Also, another PA tool, soil electrical conductivity mapping, helps in the identification of different soil zones, which is essential for site-specific soil treatment, and is very effective for different agricultural crops including alfalfa, wheat, corn, and millet, and rotational fallow crop land (Johnson et al., 2001; Lesch et al., 2005; Luck et al., 2009; Paccioretti et al., 2020).

Integrated Pest Management (IPM) is a comprehensive strategy where different techniques, such as biological, chemical, mechanical, and biorational methods, are utilized to manage pest populations. When IPM is integrated with PA, it offers a promising approach to enhance crop yields while minimizing environmental impact. Using multispectral cameras in drones can provide real-time data on pest infestations, which enables timely interventions. For example, Kebe et al. (2023) reported that the use of drone technology for pest monitoring in an IPM framework resulted in 30% reduction of pesticide. In addition, the use of soil sensors helps with real-time monitoring of soil moisture and nutrient levels. Kebe et al. (2023) documented a 20% increase in corn yields after the synergy of PA and IPM. Kebe et al. (2023) also reported a reduction in 30% water usage in addition to yield improvement in California's vineyards. However, an effective integration of IPM and PA can only be achieved via reliable and cost-effective sampling plans that are specific to host crop, pest, and pest management goals. For example, Parajulee et al. (2006a) developed fixed precision sequential and binomial sampling plans for early season pest sampling in cotton. These sampling plans are faster, and the sampling cost is reduced by 40–60% compared to conventional sampling of the western flower thrips, *Frankliniella occidentalis,* and the cotton fleahopper, *Pseudatomoscelis seriatus*, in cotton. Similarly, precision management of pests must also consider agronomic input variables such as irrigation and nitrogen fertility as these two key inputs are largely responsible for affecting insect pest populations, natural enemies, plant growth characteristics, yield, and lint quality in cotton (Parajulee et al., 2005, Sapkota et al., 2024).

1.1 Technologies in Precision Agriculture

Sensing technologies in PA include a range of tools, such as remote sensing methods like satellite imagery, aerial photography, and drones, along with ground-based instruments like weather stations, soil moisture sensors, and nutrient sensors (Sishodia et al., 2020). Additionally, the integration of GPS and GIS plays a vital role in improving data collection, mapping, and precise location tracking, ultimately supporting informed decision-making and efficient resource management in crop production. Variable rate technologies (VRT) have transformed modern agriculture by allowing the site-specific application of inputs like fertilizers, seeds, pesticides, and irrigation water, based on the spatial variability of soil conditions and crop requirements (Aziz et al., 2025). A study by Mooney et al. (2009) reported that adopting VRT in cotton production led to reduced usage of inputs such as fertilizer, lime, and seed, resulting in cost savings and improved input efficiency. In contrast, inputs like irrigation and spray-applied substances showed increased usage, indicating the need for a more precise and targeted application that may contribute to better crop performance and higher yields. Variable rate technologies include VRS, variable rate nutrient, and variable rate irrigation. The analysis and interpretation of data in precision agriculture support the creation of management zones, which help optimize crop management practices. These zones, developed using data analysis and geostatistical methods, enable the use of variable rate input applications, thereby improving the efficiency and effectiveness of crop production (Sharma, 2023). GIS applications play a vital role in this process, aiding in land suitability evaluation, water resource planning, soil health and fertility management, assessment of biotic and abiotic stresses, crop monitoring, yield forecasting, and biomass estimation. AI has also become a valuable tool in crop management, with AI-powered chatbots providing farmers with real-time guidance, recommendations, and solutions. By analyzing user queries and accessing relevant data, these tools offer tailored advice on issues such as crop diseases, insect pest and nematode control, and irrigation practices, helping farmers make informed decisions quickly and more efficiently. Automation and robotics have become a core component of PA, receiving significant attention in research and practical applications. These technologies are being widely used in operations like planting, monitoring, spraying, and harvesting to increase agricultural productivity while reducing environmental impact. Robots equipped with cameras can monitor crop growth, assess plant health, and detect pest presence in the field. When used for fertilizer and pesticide applications, these robots can help reduce pesticide usage by up to 80%. Additionally, their ability to navigate around obstacles like trees, rocks, and water bodies allows for more efficient and precise field operations (Sachithra & Subhashini, 2023).

1.2 Site-Specific Management Zones

Unlike other forms of precision agriculture that heavily depend on costly variable rate technologies (such as GPS-guided tractors and automated seeders or sprayers), site-specific management zones (SSMZs) allow for the use of more traditional tools, offering a more cost-effective approach particularly beneficial for small- and medium-sized farmers. Delineating SSMZs within crop fields typically involves five key steps, which do not always follow a strict sequence: (1) mapping response variables such as yield, (2) mapping relevant zoning factors like soil pH and terrain slope, (3) identifying the most suitable zoning factors, (4) dividing the field based on these selected factors, and (5) choosing the most appropriate partitioning model for the field (Mendez-Vazquez et al., 2019). Although the five-step protocol for delineating SSMZs appears straightforward in theory, its practical application can be challenging due to both technical and conceptual limitations. This is especially evident in agricultural pest management, where tasks such as mapping zoning factors (e.g., environmental variables) and response variables (e.g., insect pests or diseases) become complex and resource intensive. The main challenge arises from the highly dynamic nature of pest populations, which vary significantly across space and time (Venette et al., 2010). Accurately capturing and interpreting these variations requires high-resolution data and specialized expertise. Moreover, clustering algorithms commonly used in PA to divide crop fields (such as fuzzy c-means and ISODATA) do not always produce zones that are biologically or agronomically relevant (Córdoba et al., 2016). As a result, these methods may not accurately identify sub-field areas that reflect real pest populations or generate zone boundaries with simple, practical shapes that can be effectively implemented during field management activities. Species distribution models (SDMs), combined with high-resolution environmental mapping, enhance traditional PA methods by enabling the identification of SSMZs for pest control. These tools also offer opportunities to refine and strengthen previously established techniques and older strategies used in implementing site-specific pest management (SSPM) practices. Reay-Jones et al. (2019) found that thrips populations, thrips-induced damage, and plant health measured by normalized difference vegetation index (NDVI) varied significantly across the cotton field. When comparing thrips related variables, significant aggregation was found in thrips numbers (adults + immature thrips, 17%) than for thrips injury (13%). The study revealed that field-wide control measures might be more appropriate than site-specific management since most of the datasets showed random patterns of thrips presence and the poor correlation of NDVI with thrips densities and crop injury, suggesting the lack of reliability of the remote sensing tool for thrips management in cotton (Reay-Jones et al., 2019). Unlike NDVI, the positive relationships between thrips variables and soil electrical conductivity suggest soil characteristics as a potential site-specific management tool because of the variability of soil in the southeastern Coastal Plains in the United States. Also, another study found that the damage by stink bugs and stink bug populations is closely associated; the spatial relationship between NDVI and the injury is inconsistent, which means that NDVI is not always a good predictor of boll injury by stink bugs (Reisig et al., 2015), whereas in a study by Reay-Jones et al. (2016), they found a stronger association between NDVI and stink bug injury compared to stink bug densities. Bronson et al. (2006) documented no effect of landscape position and irrigation amount on nitrogen rate. Consistent lint yield was achieved through variable nitrogen rate compared to blanket application in all three years of the study. The research also highlighted no difference in lint yield in bottomslope and sideslopes in response to different irrigation rates suggesting no benefits of variable rate irrigation in the semi-arid Southern High Plains of Texas.

2. Remote Sensing

Remote sensing refers to the process of observing and identifying objects, groups of objects, or landscapes without the sensor physically touching them. In agriculture, remote sensing is commonly used to detect electromagnetic signals, such as heat and light, emitted or reflected by crops and soil. Sensors can detect energy at wavelengths outside the visible spectrum, such as ultraviolet, near infrared, and thermal infrared, allowing them to capture subtle variations in a field or crop that are not visible to the human eye and otherwise be difficult to observe or measure. Digital imaging involves three key types of resolution: spatial, spectral, and temporal. Spatial resolution refers to the spacing between individual data points, for example, a 10-meter resolution provides 100 data points per hectare. Remote sensing offers diverse applications, such as identifying crop stress; monitoring variability in crops, soils, weeds, pests, and diseases; detecting abnormal conditions like damaged drainage systems or crop injury from cultivation; and estimating yield—though yield prediction largely depends on the crop type and variety—as well as supporting various GIS-based analyses. The broader availability of high-resolution satellite imagery across spatial, spectral, and temporal dimensions has greatly expanded the application of remote sensing in PA, supporting practices such as crop monitoring, irrigation scheduling, nutrient management, pest and disease control, and yield forecasting. The vast influx of satellite data has driven researchers to adopt advanced data storage and processing technologies, including cloud computing and ML, to efficiently manage, analyze, and extract valuable insights from these large datasets (Sishodia et al., 2020).

Tang et al. (2023) highlighted several emerging technologies that are transforming the detection and monitoring of agricultural pests and diseases. Advances in M|L, particularly CNNs, have enabled the development of lightweight and efficient models capable of accurately identifying pests in crops such as maize and rice. Additionally, improvements to object detection algorithms, such as YOLO-DPD [You Only Look Once - Diseases and Pests Detection], have enhanced the speed and accuracy of pest injury symptoms in rice (Li et al., 2022).

Midgarden et al. (1997) evaluated how site-specific integrated pest management (IPM) affects insecticide resistance and populations of natural enemies in commercial potato fields in Pennsylvania. Their study, conducted during the 1994 growing season, involved six 1.2-hectare potato fields, three managed with standard IPM and three with site-specific IPM. Fields under standard IPM followed scouting and threshold guidelines from the PotatoES IPM system, while the site-specific IPM fields were divided into 0.04-hectare blocks, each sampled and treated independently. Statistical results showed that all standard IPM fields experienced a significant increase in Colorado potato beetle

(*Leptinotarsa decemlineata*) resistance from pre- to post-season. However, resistance remained stable in the unsprayed control and two site-specific IPM fields. Additionally, the density of natural enemies—including parasitoids and predators—was higher in site-specific IPM fields. These findings suggest that site-specific IPM can help reduce insecticide resistance and better preserve beneficial arthropod populations.

PA is the art and science of applying advanced technologies to improve crop yields while reducing environmental risks (Natikar et al., 2016). In traditional farming systems, pesticides are applied uniformly across entire fields without accounting for spatial variability, often leading to inefficiencies and environmental concerns. In contrast, precision pest management (PPM) addresses this variability by applying pesticides only where needed and in appropriate amounts, enabling more targeted and sustainable pest control at the micro level (Natikar et al., 2016).

Integrating spatially detailed ecological layers, such as SDMs based on high-resolution environmental data, can complement traditional PA approaches to field partitioning by enabling the delineation of SSMZs for more effective implementation of SSPM strategies. The integration of SDMs and high-resolution environmental data enhances traditional PA techniques by enabling the delineation of ecologically meaningful SSMZs for pest control. The SDM-based approach outperforms previous methods in most simulated pest infestation scenarios, especially when incorporating biological filtering to merge zones with low pest suitability, thus avoiding excessive fragmentation (Cordoba et al., 2016). By redefining environmental clusters through SDM, the resulting SSMZs align more closely with actual pest distribution patterns and exhibit practical, simplified geometries. In cases where no single method yields clear zoning outcomes, a combination of multiple approaches may be necessary. The effectiveness of SSMZ delineation is also influenced by pest spatial behaviors; species with broad field distributions support finer subdivisions while localized pests do not. Finally, while validation tools like Schoener's D and environmental equivalence tests need more accessible applications, their use is encouraged to ensure conceptual alignment with SDM-based zoning strategies in SSPM (Mendez-Vazquez et al., 2019). Insect infestations are mapped by combining remote sensing with GPS and GIS technologies. Direct application of remotely sensed data is increasing with the availability of GPS and sensor technology. Integrating aerial videography with GPS and GIS helps create regional maps of insect infestations across the large agricultural landscape (Robertson et al., 2008). Benchtop remote sensing is now increasingly used in insect systematics, behavior, and physiology (Nansen & Elliott, 2016). Entomological radar is a type of remote sensing that can provide monitoring of the abundance of pests and their movement in agricultural, medical, and forestry. It offers the advantages of detecting insects at different altitude ranges, even over 1 km above the ground, and has a large sampling volume (Abd El-Ghany et al., 2020). In addition, it offers a unique ability to study insect migration and their direction during migration. Using different remote sensing tools such as entomological radar and NDVI, monitoring and forecasting of locust species have been carried out (Rani & Jyothi, 2017). Environmental factors such as soil moisture and vegetation can be tracked using satellites to predict locust outbreaks. Various disease-carrying insects, including *Anopheles*, *Culex*, and *Aedes* mosquitoes, as well as the tsetse fly (*Glossina morsitans*), are tracked using satellite systems such as Landsat, SPOT, MODIS, and AVHRR, which help in vector-borne disease monitoring and management (Lacaux et al., 2007).

3. Precision Pest Management (PPM)

PPM is a targeted approach that is based on spatial and temporal variability within the fields, which helps in monitoring, predicting, and managing pest populations effectively. There is often an aggregated or patchy distribution of insects in the field (Parajulee et al., 2006) because of crop conditions and environmental factors, and hence, spatial mapping is very critical. Unlike the conventional method of management, it uses modern technology such as satellite imagery, drones, GPS, GIS, and variable rate application for site-specific management.

3.1 Vegetation Indices and Pest Management

Different vegetation indices, including NDVI, are key indicators to determine plant health and vigor. NDVI is regarded as the proxy of green biomass and is related to canopy photosynthesis and vegetation biodiversity (Waithaka et al., 2023). There is insufficient research about NDVI and pest presence. One study found the spatial association between areas of high cotton growth (NDVI) and tarnished plant bug, *Lygus lineolaris* (Liakos et al., 2025). Jokar et al. (2024) used normalized difference water index (NDWI) data from Sentinel 1 satellite to study the distribution of cotton bollworm, *Helicoverpa armigera*. Similarly, detection of invasive fall armyworm extracting NDWI data was also conducted using Sentinel 2A satellite imagery (Prabhakar et al., 2022). Figure 1 illustrates the cotton crop vigor using NDVI map from PlanetScope satellite within a 2-hectare drip irrigated field in Halfway, Texas. The NDVI map is the most used indices to identify spatial variability in crop vigor.

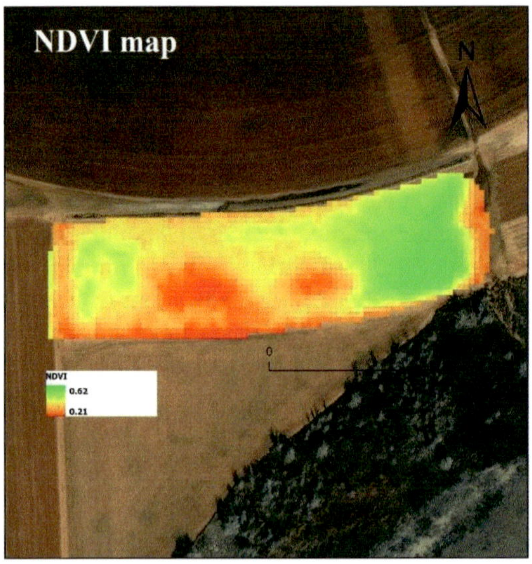

Fig. 1 NDVI map showing spatial variability in vegetation vigor, Texas Southern High Plains, Texas, USA.

3.2 Use of Drones in Precision Pest Management

There are two components in PPM. The first one is identifying hotspots through reflectance-based crop monitoring using remote sensing technologies such as ground-based, aerial, and satellite. The second one is implementing site-specific interventions using tools such as distributors of natural enemies and pesticide spray rigs. Different equipment, such as irrigation equipment, manned or unmanned aerial vehicles, or aerial drones, can be used to mount both technologies (Iost Filho et al., 2019). Drones that are used for identification of pest hotspots are known as sensing drones, whereas the drones that are used for precision distributions of solutions are known as actuation drones. Both drones work together to create an integrated, automated pest management system where there is a close link and coordination between detection and treatment. There are specific stressors, such as arthropod infestations, that trigger physiological responses affecting photosynthetic activity, resulting in altered leaf reflectance within the light spectrum. Aerial remote sensing uses RGB sensors, multispectral sensors (covering 3–12 broad-spectrum bands), and hyperspectral sensors that capture hundreds of narrow spectral bands. In addition to the application of drones in yield predictions and crop phenology, they are used in monitoring abiotic (e.g., drought, nutritional deficiencies) and biotic plant stressors (e.g., pathogens, insects, nematodes, and weeds) (Nansen & Elliott, 2016). There are only a few studies available that concern arthropod-induced stress detection using drone-based technology, whereas most studies have used ground-based remote sensing methods (Iost Filho et al., 2019). Remote sensing doesn't directly detect the pest, but the changes in canopy reflectance are captured, which indicate arthropod-induced plant stress. Vegetation spectral reflectance, also known as spectral signature, is examined to detect plant stress through remote sensing. The availability of higher spectral resolution offers more detailed plant spectral characteristics. This information will help to analyze specific reflectance features directly linked to plant physiology or indirect stress assessment, creating vegetation indices.The Red edge is one of the important spectral features that represents the slope between the red and near infrared region of the spectrum, occurring around 700 nm. This spectral region is associated with chlorophyll concentration in plant leaves and leaf area index (LAI). The steepest change in reflectance between red and near infrared wavelengths, which is known as red edge position (REP), is considered to be an important indicator of stress and senescence that decreases in chlorophyll concentration. For example, higher reflectance around 740 nm is linked to spider mite susceptibility in corn (Nansen et al., 2013). Severtson et al. (2016) used a multispectral sensor on an octocopter to detect susceptibility of canola to green peach aphid, *Myzus persicae*. With a classification accuracy of 69–100%, this drone-based remote sensing technology is promising. Huang et al. (2018) using a quadcopter drone with a multispectral sensor quantified stress in cotton with classification accuracies of 74–95%. There are different novel uses of drones in PPM, including pest outbreak intervention, reducing pest populations through sterile insect techniques and mating disruption, and pest population monitoring. The strong link between abiotic stress, such as drought and nutrient deficiency, and biotic stress, such as arthropod pest outbreaks, requires precision management of these stress factors. This can be achieved through targeted irrigation or fertilizer use, which lowers the risk of pest infestations (Nansen et al., 2013). Releasing sterile insects using drones could be another application in pest management. In New Zealand, Canada, and the United States, drone-assisted release of sterile insects successfully managed codling moth, *Cydia pomonella*, a serious pest in apple orchards (Seymour, 2018). In addition, mating disruptors like SPLAT (Specialized Pheromone & Lure Application Technology) can be used with drones in agricultural fields. The release of pheromones confuses the insect to locate their mates (Miller & Gut, 2015), and there is also the presence of an attracting and killing agent. Monitoring mobile insect populations is another application of this technology. For example, attaching small transponders to locusts helps to track their movements and distribution (Tahir & Brooker, 2009).

3.3 Spatial Distribution of Insects

The way an insect is distributed across a specific area can reveal important insights into its preference for certain hosts, overall fitness, potential for evolutionary adaptation, response to limited resources, and the occurrence and abundance of its natural enemies. Various factors, including landscape variability and the availability of food sources, shape the distribution pattern of prey. Human interventions can significantly alter this pattern, especially in agricultural settings and across different spatial scales of analysis. Pest management decisions are typically based on the average pest density across a field. The study by Parajulee et al. (2006b) documented variation in within-plant distribution of insects at different phenological stages in cotton. They found that thrips numbers were higher in the upper one third stratum (51%), followed by the middle stratum (33%), and the lower stratum (16%). Similarly, in cotton fleahopper, the number was higher in the upper stratum (65%), followed by middle (24%) and lower strata (11%). The type of cotton variety influenced the microclimate, especially the canopy temperature which affected arthropod distribution within the plant. Information on seasonal (temporal), within-field (spatial), and within-plant distribution of pest organisms and their aggregation behavior would be helpful in designing an effective sampling method for PPM. Control measures like chemical pesticides are applied uniformly if the average surpasses a predefined economic threshold. However, pest populations are often unevenly distributed within the field, making uniform application inefficient and potentially wasteful. A deeper understanding of how pests and their natural enemies are distributed over space and time can help enhance the precision of both biological and chemical pest control. This, in turn, can lead to reduced pesticide use, better conservation of beneficial organisms, and more effective overall pest management strategies (Karimzadeh et al., 2011). The development of precision technologies and analytical tools such as GIS, GNSS, geostatistics, and spatial analysis by distance indices (SADIE) helps to create a map of the spatial distribution of pests and their natural enemies. These tools are essential for integrating spatial data into ecological studies and pest management strategies. Geostatistics uses the data in two steps: estimation of spatial autocorrelation, followed by spatial predictions. Spatial association between pests and natural enemies can be determined using spatial sampling with location information. Tools such as GIS and GNSS help to provide high spatial resolution information along with geographic coordinates of sample sites. With the increase in accuracy of GNSS, it is contributing more to the pest management process. The horizontal accuracy ranges from 0.003–10 m and vertical accuracy from 0.0045–15 m (Karimzadeh & Sciarretta, 2022). The first study to use a spatial statistics method was in winter oilseed rape, where pest predator spatial association was conducted (Warner et al., 2000). A significant spatial relationship was obtained between *Dasineura brassicae* larvae and the adults of predators such as *Harpalus rufipes* and *Agonum dorsale*, but not with *Amara similata* and *Nebria brevicollis*. Similarly, the spatial association between aphids and their predators has been studied in many field crops such as winter wheat, chili, and alfalfa, where the studies also included carabid and coccinellid beetles. For example, a study using a grid of 256 sampling points spaced 12 m apart found a strong spatiotemporal relationship between two aphid species, *Metopolophium dirhodum* and *Sitobion avenae*, and the generalist predator *Pterostichus melanarius* in a wheat field with a conventional management system (Karimzadeh & Sciarretta, 2022). The study by Park and Obrycki (2004) revealed a lack of spatial alignment between lady beetles and aphids despite their occurrence at the same time. However, another study found the positive correlation between aphids and lady beetles in chili and alfalfa fields (Shayestehmehr et al., 2017). Uneven crop growth is due to different factors such as soil temperature, differences in ground surface, nutrient availability, and other microclimatic factors in a habitat, and hence this variation leads to the spatial variation in herbivorous pest distribution within the fields. Wind is also one of the factors for the spatial distribution of pests and natural enemies. The findings by Reay-Jones et al. (2019) showed a stronger linkage of thrips density and injury with soil electrical conductivity (ECa) than NDVI, suggesting greater significance of ECa for site-specific thrips management in cotton. The following example (Figure 2) shows the differences in topography and soil type affecting pest pressure and the ecological interactions among insect pests, pathogens, and natural enemies in citrus (Campos Herrera et al., 2013).

One of the studies focused on the spatial stability of insects, where they found the association index for two consecutive samplings was greater than zero in many datasets, which suggest the spatial stability of *Therioaphis maculata* is consistent. The limited mobility of aphids may have caused this stability. Insect mobility affects site-specific management, but these findings highlight the opportunity for site-specific application of insecticides in agricultural systems (Shayestehmehr et al., 2017).

5. Use of AI in Agriculture

AI encompasses different technologies such as ML, DL, and computer vision. The primary component of AI is ML. In agricultural entomology, AI is used in supporting different tasks such as identifying insect species, understanding the behavior and ecology, and pest management.

The development of robots for spraying insecticides and fertilizers offers general crop monitoring capabilities with low operational costs. A wheeled model with cameras helps to detect the presence of pests in the field. There are many advantages to using robotics in agriculture, including the reduction of pesticide usage by 80% when farmers used robots for spraying fertilizers (Sachithra

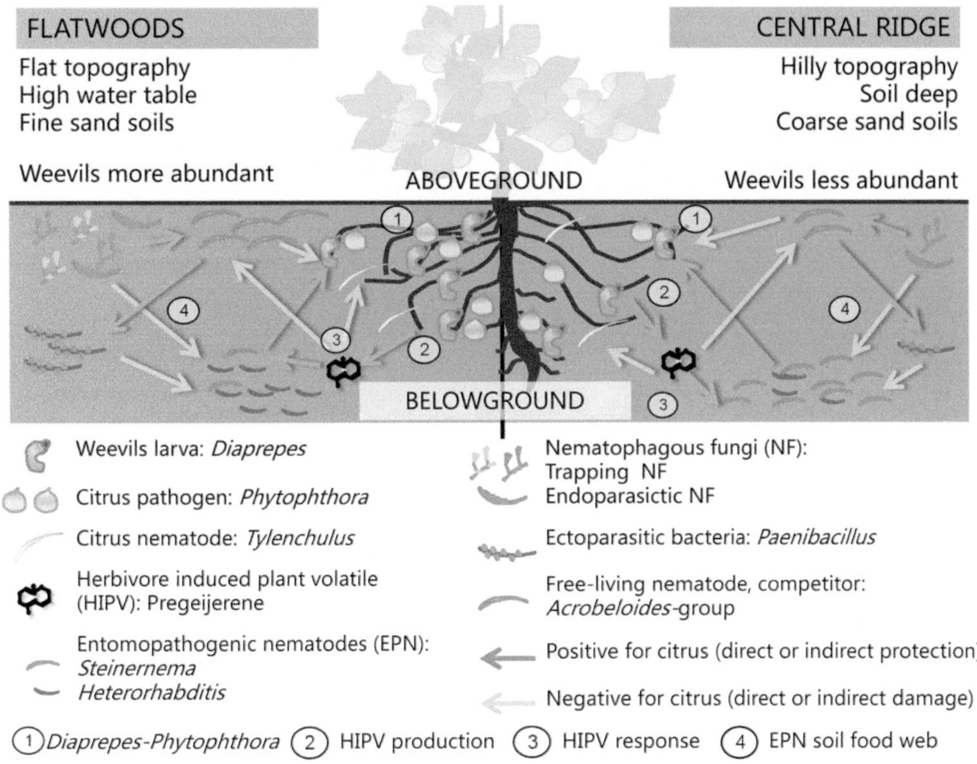

Fig. 2 Soil food web interactions in citrus systems (Campos Herrera et al., 2013).

& Subhashini, 2023). DL has been used in various agricultural applications such as seed and pest identification, identifying nitrogen content both in leaf and soil, detecting water stress, herbicides usage, assessing food defects and damage from hail, and tracking greenhouse environments. AI tools, including DL and robotics, help in different agriculture functions, including plant eco-phenotyping as illustrated in Figure 3.

5.1 Use of AI in Pest Management

For effective management of pests, various tools and mobile applications such as Plantix, Leaf Byte, Bioleaf, Cotton Ace, and Apizoom are used for the identification and diagnosis of insect pests (Rahman & Ravi, 2022). Achieving 90% accuracy, Chen et al. (2020) used AIoT technology to detect *Tessaratoma papillosa*. In this study, they combined different components, including AI, image recognition, environmental sensors, and the IoT for detecting pests. The number of images of the insect included 687, and labeling was done within the images, and a model for identification was created. Using image preprocessing techniques, they further enhanced the images. GPS data was utilized to map the pest locations that provided the model with a recognition accuracy of 90%. Five groups of insects, including aphids, leaf hoppers, flax budworm, flea beetles, and red spider mites, were detected with 99% accuracy using a mobile application (Karar et al., 2021). In a pheromone-based pest management system, accurate monitoring of insect pests is essential. Using images from pheromone traps, a powered automatic moth detection method has been created that offers an approach for timely pest management compared to traditional counting techniques (Ding & Taylor, 2016). This model can be helpful for automatic detection and counting of pests in traps, which sets the economic threshold level (ETL) for targeted pest control.

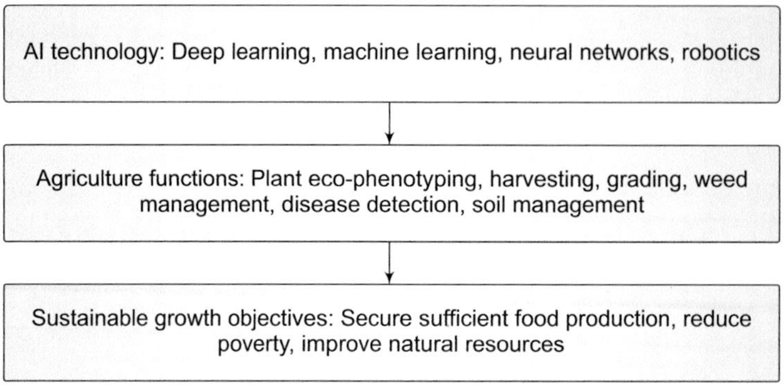

Fig. 3 AI in Agriculture for Sustainability (Sachithra & Subhashini, 2023).

6. Conclusion

The major research challenge in utilizing PPM techniques is to create stress detection algorithms that work well across different locations, seasons, and years for target pests. Remote sensing technology will advance once there is an increase in spatial, spectral, and temporal resolutions of the imagery. The applications can be direct, but most commonly, they are indirect. Therefore, timely responses and sensitivity of these technologies are very crucial in accurately assessing crop conditions to allow management calibration based on these responses. Imagery helps to identify spatial variability of plant health, but before further recommendations, it is necessary to understand well the underlying cause of the variability. Using different technologies such as remote sensing, drones, and AI, PPM helps transform crop production. To implement site-specific interventions, it is very important to understand the variability of the spatial distribution of insect pests, which will help to optimize pesticide use and enhance biological controls. Real-time monitoring and timely decisions will happen due to the use of tools such as vegetation indices and geospatial data. In addition, the use of AI will assist in image recognition, predictive modeling, and effective pest detection and control. Hence, an appropriate combination and effective use of these precision technologies should lead toward reduced pesticide use, optimization of resources, conservation of natural enemies and increased natural pest suppression, sustainable crop production, environmental conservation, and stabilized farm productivity.

References

Abd El-Ghany, N.M., Abd El-Aziz, S.E. and Marei, S.S. (2020). A review: Application of remote sensing as a promising strategy for insect pests and diseases management. *Environmental Science and Pollution Research*, *27*(27), 33503–33515.

Ahmed, S., Marwat, S.N.K., Brahim, G.B., Khan, W.U., Khan, S., Al-Fuqaha, A. and Koziel, S. (2024). IoT-based intelligent pest management system for precision agriculture. *Scientific Reports*, *14*(1), 31917.

Aziz, D., Rafiq, S., Saini, P., Ahad, I., Gonal, B., Rehman, S.A., Rashid, S., Saini, P., Rohela, G.K., Aalum, K., Singh, G., Gnanesh, B.N. and Iliya, M.N. (2025). Remote sensing and artificial intelligence: revolutionizing pest management in agriculture. *Frontiers in Sustainable Food Systems*, *9*, 1551460.

Balaska, V., Adamidou, Z., Vryzas, Z. and Gasteratos, A. (2023). Sustainable crop protection via robotics and artificial intelligence solutions. *Machines*, *11*(8), 774.

Banerjee, D., S. Adhikary, S. Bhattacharya, A. Chakraborty, S. Dutta, S. Chatterjee, A. Ganguly, S. Nanda and P. Rajak. 2024. Breaking boundaries: Artificial intelligence for pesticide detection and eco-friendly degradation. *Environmental Research*, *241*, 117601, https://doi.org/10.1016/j.envres.2023.117601.

Bronson, K.F., Booker, J.D., Bordovsky, J.P., Keeling, J.W., Wheeler, T.A., Boman, R.K., ... and Nichols, R.L. (2006). Site-specific irrigation and nitrogen management for cotton production in the southern High Plains. *Agronomy Journal*, *98*(1), 212–219.

Campos-Herrera, R., Ali, J.G., Diaz, B.M. and Duncan, L.W. (2013). Analyzing spatial patterns linked to the ecology of herbivores and their natural enemies in the soil. *Frontiers in Plant Science*, *4*, 378.

Chen, C.J., Huang, Y.Y., Li, Y.S., Chang, C.Y. and Huang, Y.M. (2020). An AIoT-based smart agricultural system for pest detection. *IEEE Access*, *8*, 180750–180761.

Córdoba, M.A., Bruno, C.I., Costa, J.L., Peralta, N.R. and Balzarini, M.G. (2016). Protocol for multivariate homogeneous zone delineation in precision agriculture. *Biosystems Engineering*, *143*, 95–107.

Ding, W. and Taylor, G. (2016). Automatic moth detection from trap images for pest management. *Computers and Electronics in Agriculture*, *123*, 17–28.

Huang, H., Deng, J., Lan, Y., Yang, A., Deng, X., Zhang, L., ... and Chen, P. (2018). A two-stage classification approach for the detection of spider mite-infested cotton using UAV multispectral imagery. *Remote Sensing Letters*, *9*(10), 933–941.

Iost Filho, F.H., Heldens, W.B., Kong, Z. and de Lange, E.S. (2020). Drones: Innovative technology for use in precision pest management. *Journal of Economic Entomology*, *113*(1), 1–25.

Johnson, C.K., Doran, J.W., Duke, H.R., Wienhold, B.J., Eskridge, K.M. and Shanahan, J.F. (2001). Field-scale electrical conductivity mapping for delineating soil condition. *Soil Science Society of America Journal, 65,* 1829–1837.

Jokar, M., López-Bernal, Á. and Kamkar, B. (2024). The effect of spring flooding on management and distribution of cotton bollworm (Helicoverpa armigera) by flood mapping using SAR sentinel-1 and optical imagery landsat-8: A case study in golestan province, Iran. *International Journal of Pest Management*, *70*(4), 778–788.

Kanwal, S., Khan, M.A., Saleem, S., Tahir, M.N., Muntaha, S.T., Samreen, T., ... and Shahzad, B. (2022). Integration of precision agriculture techniques for pest management. *Environmental Sciences Proceedings*, *23*(1), 19.

Karar, M.E., Alsunaydi, F., Albusaymi, S. and Alotaibi, S. (2021). A new mobile application for agricultural pests recognition using deep learning in a cloud computing system. *Alexandria Engineering Journal*, *60*(5), 4423–4432.

Karimzadeh, R. and Sciarretta, A. (2022). Spatial patchiness and association of pests and natural enemies in agro-ecosystems and their application in precision pest management: A review. *Precision Agriculture*, *23*(5), 1836–1855.

Karimzadeh, R., Hejazi, M.J., Helali, H., Iranipour, S. and Mohammadi, S.A. (2011). Assessing the impact of site-specific spraying on control of Eurygaster integriceps (Hemiptera: Scutelleridae) damage and natural enemies. *Precision Agriculture*, *12*(4), 576–593.

Kebe, A.A., Hameed, S., Farooq, M.S., Sufyan, A., Malook, M.B., Awais, S., ... and Abbas, N. (2023). Enhancing crop protection and yield through precision agriculture and integrated pest management: A comprehensive review. *Asian Journal of Research in Crop Science*, *8*(4), 443–453.

Lacaux, J.P., Tourre, Y.M., Vignolles, C., Ndione, J.A. and Lafaye, M. (2007). Classification of ponds from high-spatial resolution remote sensing: Application to Rift Valley Fever epidemics in Senegal. *Remote sensing of environment*, *106*(1), 66–74.

Lesch, M., Corwin, D.L. and Robinson, D.A. (2005). Apparent soil electrical conductivity mapping as an agricultural management tool in arid zone soils. *Computers and Electronics in Agriculture, 46*, 351–378.

Li, S., Feng, Z., Yang, B., Li, H., Liao, F., Gao, Y., Liu, S., Tang, J. and Yao, Q. (2022). An intelligent monitoring system of diseases and pests on rice canopy. *Frontiers in Plant Science*, *13*. https://doi.org/10.3389/fpls.2022.972286.

Liakos, V., Koutsogeorgiou, E.I., Charouli, S., Navrozidis, I.E., Proias, G. and Andreadis, S.S. (2025). The Delineation of Management Zones of the Halyomorpha halys (Hemiptera: Pentatomidae) Population Based on Its Spatiotemporal Distribution for Precision Agriculture Purposes. *Insects*, *16*(4), 336.

Luck, E., Gebbers, R., Ruehlmann, J. and Spangenberg, U. (2009). Electrical conductivity mapping for precision farming. *Near Surface Geophysics, 7*, 15–26.

Méndez-Vázquez, L.J., Lira-Noriega, A., Lasa-Covarrubias, R. and Cerdeira-Estrada, S. (2019). Delineation of site-specific management zones for pest control purposes: Exploring precision agriculture and species distribution modeling approaches. *Computers and Electronics in Agriculture*, *167*, 105101.

Midgarden, D., Fleischer, S.J., Weisz, R. and Smilowitz, Z. (1997). Impact of site-specific IPM on the development of esfenvalerate resistance in Colorado potato beetle (Coleoptera: Chrysomelidae) and densities of natural enemies. *J. Econ. Entomol.*, *90*, 855–867.

Miller, J.R. and Gut, L.J. (2015). Mating disruption for the 21st century: Matching technology with mechanism. *Environmental Entomology*, *44*(3), 427–453.

Mooney, D.F., Roberts, R.K., English, B.C., Lambert, D.M., Larson, J.A., Velandia, M.M., ... and Reeves, J.M. (2010). Precision farming by cotton producers in twelve southern states: Results from the 2009 southern cotton precision farming survey.

Nansen, C. and Elliott, N. (2016). Remote sensing and reflectance profiling in entomology. *Annual Review of Entomology*, *61*(1), 139–158.

Nansen, C., Sidumo, A.J., Martini, X., Stefanova, K. and Roberts, J.D. (2013). Reflectance-based assessment of spider mite "bio-response" to maize leaves and plant potassium content in different irrigation regimes. *Computers and Electronics in Agriculture*, *97*, 21–26.

Natikar, P.K., Balikai, R.A. and Anusha, C. (2016). Pest management strategies in precision farming. *Journal of Experimental Zoology India*, *19*(1), 1–8.

Paccioretti, P., Córdoba, M. and Balzarini, M. (2020). FastMapping: Software to create field maps and identify management zones in precision agriculture. *Computers and Electronics in Agriculture*, *175*, 105556.

Parajulee, M.N. (2007). Influence of constant temperatures on life history parameters of the cotton aphid, *Aphis gossypii*, infesting cotton. *Environmental Entomology*, *36*, 666–672.

Parajulee, M.N. (2014). Life table and population dynamics of the cotton aphid, *Aphis gossypii*, on upland cotton. *Southwestern Entomologist*, *39*, 401–410.

Parajulee, M.N., Slosser, J.E. and Boring, III, E.P. (1998). Seasonal activity of *Helicoverpa zea* and *Heliothis virescens* (Lepidoptera: Noctuidae) detected by pheromone traps in the Rolling Plains of Texas. *Environmental Entomology*, *27*, 1203–1219.

Parajulee, M.N., Slosser, J.E., Carroll, S.C. and Rummel, D.R. (2001). A model for predicting boll weevil (Coleoptera: Curculionidae) overwintering survivorship. *Environmental Entomology*, *30*, 550–555.

Parajulee, M.N., Wilson, L., Rummel, D.R., Carroll, S.C. and Trichilo, P.J. (1996). Climatic data-based analysis of boll weevil overwintering survival and spring emergence. *Environmental Entomology*, *25*, 882–894.

Parajulee, M.N., R.B. Shrestha, S.C. Carroll, P.L. Bommireddy, A.M. Cranmer and K.F. Bronson. (2005). Variable rate of irrigation water and nitrogen: Potential of site-specific management of cotton aphids. In: *Proceedings, Beltwide Cotton Conferences*, National Cotton Council, Memphis, TN, pp. 1735–1742.

Parajulee, M.N., Shrestha, R.B. and Leser, J.F. (2006a). Sampling methods, dispersion patterns, and fixed precision sequential sampling plans for western flower thrips (Thysanoptera: Thripidae) and cotton fleahoppers (Hemiptera: Miridae) in cotton. *Journal of Economic Entomology*, *99*(2), 568–577.

Parajulee, M.N., Shrestha, R.B. and Leser, J.F. (2006b). Influence of tillage, planting date, and Bt cultivar on seasonal abundance and within-plant distribution patterns of thrips and cotton fleahoppers in cotton. *International Journal of Pest Management*, *52*(3), 249–260.

Park, Y.L. and Obrycki, J.J. (2004). Spatio-temporal distribution of corn leaf aphids (Homoptera: Aphididae) and lady beetles (Coleoptera: Coccinellidae) in Iowa cornfields. *Biological Control*, *31*(2), 210–217.

Prabhakar, M., Gopinath, K.A., Kumar, N.R., Thirupathi, M., Sravan, U.S., Kumar, G.S., ... and Vennila, S. (2022). Detecting the invasive fall armyworm pest incidence in farm fields of southern India using Sentinel-2A satellite data. *Geocarto International*, *37*(13), 3801–3816.

Rahman, S.M. and Ravi, G. (2022). Role of artificial intelligence in pest management. *Current Topics in Agricultural Sciences*, *7*, 64–81.

Rani, A.S. and Jyothi, S. (2017). A study on hyper spectral remote sensing pest management. *Int J. Recent Innov. Trends Comput. Commun.*, *5*(6), 497–503.

Reay-Jones, F.P., Greene, J.K. and Bauer, P.J. (2016). Stability of spatial distributions of stink bugs, boll injury, and NDVI in cotton. *Environmental Entomology*, *45*(5), 1243–1254.

Reay-Jones, F.P., Greene, J.K. and Bauer, P.J. (2019). Spatial distributions of thrips (Thysanoptera: Thripidae) in cotton. *Journal of Insect Science*, *19*(6), 3.

Reisig, D.D., Reay-Jones, F.P.F. and Meijer, A. (2015). Aggregation of stink bug injury and association with remotely sensed data in cotton. *J. Insect Sci.*, *15*, 134.

Robertson, C., Wulder, M.A., Nelson, T.A. and White, J.C. (2008). Risk rating for mountain pine beetle infestation of lodgepole pine forests over large areas with ordinal regression modelling. *Forest Ecology and Management*, *256*(5): 900–912.

Sachithra, V. and Subhashini, L.D.C.S. (2023). How artificial intelligence is used to achieve agricultural sustainability: Systematic review. *Artificial Intelligence in Agriculture*, *8*, 46–59.

Sapkota, R., Parajulee, M.N. and Cradock, K.R. (2024). Interactive Effect of Cover Crop, Irrigation Regime, and Crop Phenology on Thrips Population Dynamics and Plant Growth Parameters in Upland Cotton. *Agriculture*, *14*(7), 1128.

Severtson, D., Callow, N., Flower, K., Neuhaus, A., Olejnik, M. and Nansen, C. (2016). Unmanned aerial vehicle canopy reflectance data detects potassium deficiency and green peach aphid susceptibility in canola. *Precision Agriculture*, *17*(6), 659–677.

Seymour, R. (2018). Drones tested for moth drops in Okanagan orchards. *Kelowna Daily Courier*.

Sharma, S. (2023). Precision agriculture: Reviewing the advancements, technologies, and applications in precision agriculture for improved crop productivity and resource management. *Reviews in Food and Agriculture*, *4*(2), 45–49.

Shayestehmehr, H., Karimzadeh, R. and Hejazi, M.J. (2017). Spatio-temporal association of T herioaphis maculata and Hippodamia variegata in alfalfa fields. *Agricultural and Forest Entomology*, *19*(1), 81–92.

Sishodia, R.P., Ray, R.L. and Singh, S.K. (2020). Applications of remote sensing in precision agriculture: A review. *Remote sensing*, *12*(19), 3136.

Tahir, N. and Brooker, G. (2009, December). Feasibility of UAV based optical tracker for tracking Australian plague locust. In: *Austlalasian Conference on Robotics and Automation* (*ACRA 2009*). Sydney, Australia.

Tang, Y., Chen, C., Leite, A.C. and Xiong, Y. (2023). Precision control technology and application in agricultural pest and disease control. *Frontiers in Plant Science*, *14*, 1163839.

Venette, R.C., Kriticos, D.J., Magarey, R.D., Koch, F.H., Baker, R.H., Worner, S.P., ... and Pedlar, J. (2010). Pest risk maps for invasive alien species: A roadmap for improvement. *BioScience*, *60*(5), 349–362.

Vinci, G., Marco, S., Restuccia, D. and Ruggeri, M. (2025). Nanobiopesticides: Sustainability aspects and safety concerns. *Environments*, *12*, 74. doi:10.3390/environments12030074.

Waithaka, N.A., Kasina, M., Samita, N.E., Guantai, M.M., Omuse, E.R., Toukem, N.K., ... and Dubois, T. (2023). Interactions between integrated pest management, pollinator supplementation, and normalized difference vegetation index in pumpkin, Cucurbita maxima (Cucurbitales: Cucurbitaceae), production. *Environmental Entomology*, *52*(3), 416–425.

Warner, D.J., Allen-Williams, L.J., Ferguson, A.W. and Williams, I.H. (2000). Pest–predator spatial relationships in winter rape: Implications for integrated crop management. *Pest Management Science: Formerly Pesticide Science*, *56*(11), 977–982.

CHAPTER 7

Impact of Climate Variables on Insect Pest Population Amidst Climate Change

Danishta Aziz[1*], Summira Rafiq[2], Aradhana Panda[3], Pawan Saini[4], Belaghihalli N Gnanesh[5], Ishtiyaq Ahad[6], Basanagouda Gonal[7], Mohd Salim Mir[8], Zahoor Ahmad Shah[9], Eajaz Ahmad Dar[10] and Samy Sayed[11]

1. Introduction

1.1 Climate Influence on Insects

Insects contribute about 70% to the total described animal population on earth and are distributed widely in both terrestrial and as well as aquatic habitats, as a result having diverse life history and occupying a wide range of trophic levels (Chapman, 2006). As insects are piokilotherms, the external temperature and other weather variables influence their activity and development to a great extent, so one could expect a large effect of changing climate on insect biodiversity (Strange & Ayres, 2010). Climate exhibits a great effect on distribution and abundance of insect fauna existing on earth in the same manner as on plants and other animals (Sykes, 2009). The pattern of temperature, precipitation, relative humidity, wind, and sunlight globally has set physiological limits, which determine the existence of a particular insect species in a specific habitat. Apart from its direct impact on insect growth and development, abundance and distribution, some indirect effects are also produced by climate change on insects by exerting effects on food resources, competitors, enemies, and symbionts (Strange & Ayres, 2010). Global climate warming tends to impact the overall population dynamics of insect species, by affecting survival rate, fecundity, generation time, and dispersal (Bale et al., 2002). Changing climate is also seen to effect interactions between predators/ parasites and other insects, like one theoretical possibility is shown by Harrison and Roberts (2000), wherein insect enemies are favoured by predator-prey interaction due to increased mobility which exposes them to more attacks, and mobility is due to muscle performance which is highly temperature sensitive.

[1-3,6] Division of Entomology, Faculty of Agriculture, SKUAST Kashmir, J&K, India.
[4,7] Central Sericultural Research and Training Institute, Central Silk Board Galander, Pampore, J&K, India.
[5] Genetics and Plant Breeding, ICAR JSS-Krishi Vigyan Kendra, Mysuru 571129, Karnataka, India.
[8] Division of Agronomy, Faculty of Agriculture, SKUAST Kashmir, J&K, India.
[9] Division of Agri. Extension, Faculty of Agriculture, SKUAST Kashmir, J&K, India.
[10] Krishi Vigyan Kendra Ganderbal, SKUAST Kashmir, J&K, India.
[11] Department of Economic Entomology and Pesticides, Faculty of Agriculture, Cairo University, GIZA 12613, Egypt.
* Corresponding author: daniaziz9974@gmail.com

DOI: 10.1201/9781003614869-7

Climate parameters, primarily temperature and humidity, exert a multifaceted influence on insects. Directly, they regulate the activity level of both larvae and adults, dictate insect dispersal patterns, impact phenology and life cycle duration, and determine survivability under challenging weather conditions. Additionally, climate plays an indirect role by shaping the environments where insects thrive, influencing plant formations, phenological events, food availability and quality, predator-prey dynamics and the prevalence of entomopathogens. Insects act as real-time indicators of man-induced changes in climate, by responding towards this change in all possible ways right from phenological effects to distribution in different regions. Insects are affected more severely by climate change than other organisms like plants and other vertebrates due to their fast-completing generations and enhanced fecundity (Menéndez, 2007). The long-term effect of climate change tends to spend more time in the overwintering phase, which makes them more prone to predator/parasitoid attack if the latter also overwinters for more time. Also, migratory insects like *Helicoverpa* sp. exploit new habitats and distribute widely to new areas (Sharma, 2005).

2. Temperature

Insects are ectothermic as their body temperature is regulated by external conditions. Insects are so sensitive to extreme temperature changes as metabolic rate of insects doubles with every 10-degree temperature rise (Gillooly et al., 2001; Clarke & Fraser., 2004). Temperature affects insect development, reproduction and activity levels. Warmer climate leads to accelerated insect metabolism, in turn causing quicker growth rate and reproduction, while extreme heat can be lethal. Conversely, cold temperatures can slow down insect development and reproduction and prolonged exposure to freezing conditions can cause mortality (Parajulee et al., 1996; Parajulee et al., 1997). Insects, being poikilothermic creatures, exhibit noticeable changes in activity based on their micro-climate (Bale et al., 2002; Menéndez, et al., 2007). When temperatures rise to optimal levels, insect metabolism accelerates, directly leading to an increase in their activity levels. In temperate climate zones, a rise in average temperature is associated with several observable changes. For instance, the imago stage of many phytophagous species in forest environments shows heightened and prolonged activity during both day and night, involving activities such as feeding, mating, and searching for suitable egg-laying sites (Moore & Allard, 2008; Netherer & Schopf, 2010). This in turn tends to enhance insect dispersal within forest environments, more frequent egg laying, and greater potential for colonizing a wider range of hosts. In high temperature regimes, various developmental stages of insects like eggs, larvae and pupae tend to shorten, a characteristic phenomenon observed among many forest insect populations (Szujecki, 1998). This accelerated development of pre-adult stages, translates to reduced exposure time to unfavourable micro-climate such as low temperatures, excessive or insufficient humidity, predation, parasitism, and activity of entomopathogens. Consequently, many insect species experience increased reproductive success. Laboratory studies have demonstrated the temperature's impact on the duration of larval development in two significant native foliophagous species: the nun moth (*Lymantria monacha*) and the gypsy moth (*Lymantria dispar*) (Karolewski et al., 2007). In each case, a slight rise in temperature resulted in a shortened growth time from the egg stage to pupal development. However, differences were noted in larval survivability between the two species. With higher average environmental temperatures, increased mortality was observed among *L. monacha* caterpillars, however survival percentage among larvae of *L. dispar* increased. These variations likely stem from the species-specific optimal thermal range, reflecting their diverse micro-climates. The findings from this experiment underscore the diverse impacts of different climate variables on insect development, even in sister species.

In temperate zones, the periodicity of insect activity is intricately linked to the sequence of seasons with temperature serving as a key factor that regulates insect activity levels. Changes in average temperature are closely tied to shifts in insect phenology, a well-documented indicator of global warming. One of the visible signs of global warming is the early emergence of certain insect populations in spring and their extended active periods (Walther et al., 2002; Logan et al.,

2003; Parmesan & Yohe, 2003; Menéndez, et al., 2007; Moore & Allard, 2008). These changes can significantly impact herbivorous insects that undergo multiple generations within a year. With increasing average temperatures, these species experience accelerated growth, potentially leading to multi-voltinism. Consequently, biological life cycles are shortened, resulting in higher larval densities on host plants and more frequent outbreaks. However, the effects of rising temperatures are not uniformly positive. For instance, the European spruce bark beetle (*Ips typographus*), a voracious pest of European spruce trees, including Poland, is negatively impacted by increasing average temperatures (Jönsson et al., 2007; Netherer & Schopf, 2010). Winter temperatures are particularly crucial for insects prevalent in temperate zones. While higher winter temperatures can enhance survival rates, especially at the Northern and upper range borders, where increased cold typically leads to higher mortality, many species rely on a sufficient number of cold days to complete their developmental cycles or resume feeding in spring (Jönsson et al., 2007; Netherer & Schopf, 2010). Furthermore, the declined snow covers due to rising average temperatures and humidity can negatively affect populations that are quiescent in forest litter and soil (Nupponen et al., 2010). These interconnected factors highlight the complex and multifaceted impacts of changing climate on insect communities and their ecological dynamics inside temperate zones.

Climate change can accelerate evolutionary adaptation in insect populations, as evidenced by studies on various species. Researchers such as Menéndez et al. (2007), Moore and Allard (2008), and Régnière (2009) have documented instances of rapid adaptation in response to shifting environmental conditions. For example, the European butterfly, Brown Argus (*Aricia agestis*) demonstrated swift adaptation to novel thermal regimes by adjusting the temperature threshold for diapause induction. Similarly, the chrysomelid beetle, *Chrysomela aeneicollis* exhibited an enhancement in frequency of alleles, associated with resistance to low-temperature protein synthesis. Hill et al. (1999) observed morphological changes in populations of butterfly, *Pararge aegeria*. Insects from populations that colonized new areas in Great Britain over 20 years prior to the study, displayed larger wing area surfaces and thorax weights compared to individuals from established populations. Furthermore, increased dispersal forms have been noted in bush-cricket species *Conocephalus discolor* and *Metrioptera roeselii* on the British Isles, as they dispersed more from their previous ranges (Thomas et al., 2001). Forest phytophages, such as the Winter Moth (*Operophtera brumata*), have also shown remarkable adaptive capabilities to changing environmental conditions. The hatching time of larvae with respect to bud development and leaf growth is crucial for their survival. Recent research has revealed natural selection favouring genotypes characterized by slower egg development in *O. brumata* populations. Consequently, hatchings occur later, ensuring synchronization with bud burst and reducing the risk of starvation. These examples illustrate the remarkable capacity of insect populations to adapt rapidly to changing environmental conditions, with implications for their survival, distribution, and ecological interactions.

Temperature and humidity changes can indirectly influence insect pests by disturbing the metabolism and physiology of their hosts (Ayres & Lombardero, 2000; Rouault et al., 2006; Moore & Allard, 2008; Netherer & Schopf, 2010). Prolonged and intense droughts, which are a consequence of rising average temperatures, generally have a detrimental effect on host well-being, making them more attractive to phytophagous insects. The widespread die-off of oak stands in Europe caused by water shortages and subsequent attacks from *folivores, cambiophages*, and *xylophages, exemplifies* this interaction (Thomas, 2008). While moderate increases in temperature and CO_2 levels can reduce the food quality for some folivores by lowering nitrogen levels in foliage and increasing the production of secondary compounds like tannins (Buse et al., 1998; Dury et al., 1998; Kuokkanen et al., 2001), this can lead to a deterioration in plant quality as food (Chen et al., 2005) and potentially enhance plant tolerance levels. Huberty and Denno (2004) and Rouault et al. (2006) conducted a study on the effects of plant water stress on various aspects of phytophagous insects like, development, survival, and fertility. The research was motivated by the observed discrepancy between the frequent insect outbreaks following droughts in natural environments and studies showing the negative impact of water shortages on these insects. The analysis revealed that the responses of these insects to reduced moisture levels in plant parts depend on their ecological guilds or groups of species with similar feeding habits. Long-lasting droughts were found to positively affect insects that develop in wood, while decreased moisture and wall pressure levels in cells negatively impacted species that extract sap from tissues (such as aphids) and insects residing inside galls. The effects on other plant-feeding insects, including both feeding externally on leaves or making mines, were found to be ambiguous. The development of various plant-feeding insects is closely associated with the host phenology, primarily regulated by environmental conditions, particularly temperature (Szujecki, 1998; Bale et al., 2002). The same variable, such as an increase in average temperature, can have different effects on hosts and plant-feeding insects.

One notable adverse impact of changing climate is the disruption of synchronization between critical processes at different energy levels of the food chain. For example, higher temperatures with onset of spring can lead to early development of oak leaf, disrupting the synchronization between leaf development and the larval hatching of winter moth (Visser & Holleman, 2001). A same type of relationship has been noted in nun moth, in which the development of first larval instar is typically synchronized with the development of male flowers of Scots pine, which serve as a crucial food source for them, enhancing their survivalability and development rates. Any disruption of this synchronization negatively affects the species (Withers & Keena, 2001). Another example is the tortricid moth, *Zeiraphera griseana* (Hbn.), which shows a similar dependency on plant phenology. However, quicker growth of plants in spring and extended growing seasons can benefit certain phytophagous species that develop within plant tissues. For instance, many European spruce bark beetle communities can take advantage of these conditions to produce an additional generation each year (Netherer & Schopf, 2010). Higher temperatures can act as a stimulating factor, increasing the activity and accelerating the development of natural enemies (Netherer & Schopf, 2010). Changing temperature can directly affect plant-feeding insects, making them more susceptible to enemy attacks. Under higher temperature regimes, aphids have been observed to exhibit a weakened response to alarm pheromones produced during predator or parasitoid attacks (Awmack et al., 1997). Conversely, higher temperatures can also induce faster development, particularly in larvae exposed to parasitoids, potentially increasing the survival percentage of some phytophagous insects (Petzoldt & Seaman, 2006).

3. Relative Humidity, Moisture

Moisture availability impacts insect pest survival and behaviour (Parajulee et al., 1996). Many insects require water for various physiological functions, such as hydration and egg development (Hakeem & Parajulee, 2015). High moisture levels can create favourable conditions for insect proliferation, especially for species that thrive in humid environments. However, excessive moisture also supports fungal growth that can affect insect populations. Relative humidity refers to the amount of moisture contained by air with respect to the maximum level the air can hold at a specific temperature. It is crucial for insect respiration and water balance. Insects have adaptations to regulate water loss through their cuticle and high humidity levels can reduce water loss, promoting their activity and survival. Conversely, low humidity levels can increase desiccation risk, particularly for insects with limited water conservation mechanisms. Natural enemies represent one more aspect of the ecosystem through which changing climate can indirectly impact plant-feeding insects. The activity and effectiveness of these natural enemies, as well as their impact on phytophagous populations can vary widely. Additionally, the interaction between phytophages and their natural enemies is further complicated by the indirect effects of changing climate on hosts. For example, drought-induced reductions in plant food quality can extend the development period of phytophagous insects, increasing the likelihood of attacks by predators and parasitoids (Coviella & Trumble, 1999; Rouault et al., 2006). Moreover, climate-induced changes in plant chemistry can change the quality (in terms of size and chemical content) of plant-feeding insects as parasitoid hosts. This affects various factors such as the effectiveness of parasitoid host-searching, fecundity, and the size and sex ratio of the emerging parasitoids (Coviella & Trumble, 1999).

Mansoor-ul-Hasan et al. (2017) in their study on impact of temperature and relative humidity (RH) over growth of *Sitophilus oryzae* found that among the tested RH percentages, adult emergence was seen maximum in 65% RH as compared to least RH of 55%. In another study, Tamiru et al. (2012) found that most suitable conditions for development and maximum fecundity in *Chilo partellus* are 26–30 degree celsius temperature and RH per cent of 60–80. Holmes et al. (2012) studied the effect of relative humidity on life history of Black soldier fly, *Hermetia illucens*, which is an insect of great importance in waste management. It was found that egg eclosion and adult emergence rate was increased as relative humidity was increasing gradually, however the developmental time was seen decreasing with increasing RH. Another such effect of RH was seen on longevity, ovary maturation, and fecundity in Olive fruit fly, *Bactrocera oleae* by Broufas et al. (2009), where it was found that longevity of adult females, production of mature oocytes in ovaries, egg production, and egg hatch percentage increased with increasing RH, but only up to a limit. The survivalability of pre-adult stages of fleas was checked at various temperatures and RHs to test the hypothesis of microclimate influence on habitat distribution of fleas (Krasnov et al., 2001). It was found that change of humidity in early life stages (egg to larva) increased survival of flea species, *Xenopsylla conformis* while decreasing the same in *X. ramesis*., whereas humidity change on post larval stages decreased survivability of *X. conformis*, with zero effect on *X. ramesis*.

4. Impact of Sunshine Hours on Insect Population

Climate change is one of the most active, inevitable, and quick threats worldwide to all the biotic flora and fauna, which is a combined outcome of natural and human actions (Karthik et al., 2021). Climate change, described by the Intergovernmental Panel on Climate Change (IPCC) as "any change in climate over time, whether due to natural variability or as a result of human activity". Natural variability is caused due to several factors, one of which is because of the earth's rotatory and revolutionary motion around the sun. Any change in this could bring variations in the quantity of sun energy received by the earth's crust, causing a rise or lowering of temperature, rainfall, and geographic distribution of the biodiversity (Melilo et al., 2014). Like other weather variables, the duration of sunshine hours influences the climate directly by heating up the atmosphere and indirectly by contributing to the cloud formation process. An experiment was carried out by Schneider (2020) under laboratory conditions to determine the impact of light intensity on coupling behaviour of black soldier fly, *Hermetia illucens* by irradiating them with a 100W white light LED and it was observed that mating percentage increased from 23% at 0.92W/m^2 to 70% at 431W/m^2 when the irradiation was increased. This showed that an optimum light intensity is required by the insects for carrying out their basic life activities, one of which is mate finding and reproduction.

Changes in sunshine hours also influence host-parasitoid interaction, especially in those where the parasitoid enters into diapause under the strict influence of its host. Parasitoids having a parasitic as well as free-living state are unique but can be affected by multiple diapause-causing factors. The parasitoids enter into a diapause state while parasitizing sexual morphs because of some metabolic and hormonal factors. The ones parasitizing the asexual morphs enter into diapause by the prevailing temperature or photoperiod (Tougeron et al., 2019). A warmer climate due to increased photoperiods could also reduce the generation of sexual forms in case of aphids (Dedryver et al., 2001). Less sexual morphs and more of asexual morphs could increase the physical influence of sunshine for diapause. However, the obligate diapause is genetically conserved in univoltines (Koštál, 2006). Any climate change, e.g., an increase or decrease in sunshine hours, a non-synchronization between the adverse abiotic factors and the diapausing stage could cause mortality and local extinction of both beneficial as well as harmful insects because of their inability to adapt to it. Multi-voltinism is facilitated to overcome this problem and become more adaptable to such climatic situations (Tougeron et al., 2020). Hence, a slight alteration in the sunshine hours could lead to multi-voltinism and early mortality in insects. Insect predators like dragonflies are one of the most efficient aerial predators. Sunshine hours influence their ability for a swift flight. One of the dragonfly species, *Sympetrum striolatum* leaves its overnight roosting site, present under a cooler environment to predate over water source and for this flight to take place, first it has to warm up the flight muscles. In the absence of a clear sky with no sunshine, it has to do wing whirring for the warm energy to get built up, but in case of a bright sun it does not have to go for wing whirring, which is an energy consuming process before taking off any predatory flight. Increased sunshine indirectly raises the predatory capacity of the dragonflies and hence helps in reduction of insect pests (Borkenstein & Jödicke, 2022).

Migration has evolved independently over the years in different taxa among the various important animal lineages (Dingle, 2014). Insects are the most populous and diversified clan among the migrants inhabiting land ecosystems (Satterfield et al., 2020). Studying the migratory behaviour of insects could give us an idea regarding the biomass movement, ecological impacts, migration routes, migration altitude, or flying speed (Wotton et al., 2019; Chapman et al., 2015). Insects often choose their own suitable environment for migration (Gao et al., 2020) and the weather variables regulating the initiation or termination of migratory (Bauer & Klaassen, 2013). Knoblauch et al. (2021) reported that the influence of temperature and cloud occurrence on the abundance and migratory movement of dragonflies is due to sunshine hours. Increase in sunshine hours increases the temperature which has a positive influence on day-flying insects, whereas more cloud cover has a negative effect reducing the overall movement. The diurnal migratory insects range from very minute to large insects. Some utilize their own energy and some, the prevailing wind. During a clear day with long sunshine hours, the earth surface gets heated up causing the convective wind to move in the vertical manner. This upcurrent or thermal wind carries with it the migrating insects, which get transported to long distances without using their own energy. Migration, being the key for food security, public health, and biomass movement, is necessary for maintaining population dynamics.

The small size and flight ability of insects enable them to survive in a wide range of niches. However, these qualities also make them susceptible for being poikilotherms and hence dependent on the environmental cues to detect seasonal climate change. Accordingly, they prepare themselves physiologically and behaviourally. Photoperiodism is one such phenomenon where the organism's biological response is dependent on the day length. One such insect is the flesh fly, *Sarcophaga similis*, which develops normally under longer day length with more sunshine hours and enters into pupal diapause, whenever there is short sunshine length affecting their reproductive ability (Goto & Numata, 2009). A sudden change in the climatic conditions can also lead to reduced reproductive ability, which directly affects the survivalability of the insects. Flesh flies being one of the major decomposing insects in the environment helps in natural cleaning of the dead and decomposed (Mukai et al., 2021). The sunshine period shows the peak active hours in case of several insects. Pollinators' presence shows a healthy environment. Among the pollinators, honey bees are one of the major pollinating insect species. A research carried by Czekońska et al. (2023) stated that bee hives exposed to sunshine duration exceeding 5 hrs in presence of flowering plants have more hive weight. This showed that under more sunshine duration, honey bees utilize the maximum resources available with them and hence their activity is enhanced. More bee visits would ultimately improve pollination and fruit setting leading to increase in yield in return. Bees are influenced by the environmental conditions of their micro-climate. Human activities like spraying of harmful chemical insecticides and climate change like alteration in temperature, rainfall, sunshine hours, humidity, etc., mainly inhibits bee foraging and food collection.

5. Impact of Wind

Insects can modify their behaviour with the change in the environment. These changes are indicators for predicting the forthcoming major or slight changes in the weather conditions (McClure & McNeil, 2009). Weather parameters like temperature, rainfall, sunshine, humidity, and wind affect the nature and behaviour of the insects like mating, migration, and flight. Out of all factors, winds or air currents significantly impact the insects' lives. They often opt for the optimum wind speed necessary for their survival. Migration is one such important phenomenon shown by the insects which could impact food security, public health, and ecosystem vigour (Florio et al., 2020). Moreover, it also plays a great role in maintaining the ecosystem biogeochemistry with trillions of insects being the carriers of micronutrients migrating across Europe (Hu et al., 2016). Migratory insects like dragonflies, moths, and hoverflies select favourable winds, probably as an adaptation to maximize the distance travelled, enhance trajectories, and lower energy costs in their displacements (Gao et al., 2020). In some weak flying insects, it facilitates their migration, but they sometimes avoid it as the strong wind velocity may disable their flight and could cause mortality (Storer et al., 1999). However, wind-borne migration is mainly due to the effect of the wind movement facilitating the downwind ascend of the insects beyond their flight boundary layer (FBL-iso-velocity surface lying at 1–10 m above the surface where the insect's air speeds equal with wind speed) to higher altitudes (Taylor, 1974). The insect migrants depending on wind vary in body volume, from small flies to big insects like grasshoppers, locusts, and moths (Drake & Farrow, 1988), but the adaptive wind-dependent migration has been mostly seen in the order, Hemiptera, Orthoptera, and Lepidoptera (Pedgley, 1990). The ability of insects to move horizontally depends on the stability of the boundary layer of the planet (PBL) which changes according to the day and night temperature. On a clear day, sunlight falling over the PBL makes it convective as the surface becomes warm causing the up currents or thermals to flow vertically disrupting the horizontal wind flow. This results in reduction in the wind speed below the one expected from the current pressure gradient. However, these up currents are responsible for the ascend of some small insects (aphids) into high altitudes and their dispersal into far-flung areas.

During dusk, the earth surface starts to dissipate its heat causing the surface temperature to lower. This vertical thermal wind ceases to flow bringing down stability in the PBL. A temperature inversion occurs due to radiating heat by the earth surface where an increase in temperature is being noticed at higher altitudes up to 100–300 m above ground level. At this level the near geostrophic winds act at a velocity of 10–20 m/s which can become supergeostrophic wind at a speed of 50–100 m/s. Migrating insects flying during or after dusk have to use their own energy for the utilisation of this wind at a height of 100–2000 m which carries the insects rapidly to far away areas. The coastal front of north-west Africa separates the cool air of the ocean from the hot continental air remaining there for weeks trapping the desert locust swarm. Even the storm winds lead to congregations of the adults of African armyworm, causing mass egg laying and caterpillar outbreak (Pedgley et al., 1989). A similar case was recorded from Canada for fall armyworm (Rose et al., 1985) and bollworm eggs in Sudan Gezira (Haggis, 1981) and Australia. Green clover worm moth (Wolf et al., 1987) and celery looper moths (Peterson et al., 1988) in the U.S.A along with some strong fliers like locusts (Casimir, 1987) and grasshoppers (Reynolds & Riley, 1988). Study on the declining insect population caused due to environmental degradation or climate change and understanding their dispersal capacity along with the behavioural algorithms responsible for it, could be used further to determine the ecological impact on population decline (Van Klink et al., 2020). Often, it has been noticed that two types of movement are noticed under windy situations, i.e., either it is downwind (along the direction of wind or away from wind source) or upwind (opposite to the direction of wind or towards the wind source). Many insects follow an odour-mediated anemotaxis in search of the food source through its odour. Cabbage root fly flies in the upwind direction when no odour has been detected in order to limit itself to the odour source expecting it to be somewhere nearer to the wind source (Finch & Skinner, 1982). However, tsetse flies in a downward direction in the absence of any odour in order to cover maximum distance in search of any odour source (Gibson et al., 1991). Therefore, the wind flow regulates the movement in the insects when no odour source is available with them.

Larger insects like moths are highly dependent on olfaction for their orientation, host finding, sexual communication, mate location, oviposition and herbivory (Plath et al., 2012). Pheromones released by the adult females act as a cue for the adult male to search for their mate following the track downwind. After mating the adult female moths again follow the host odour for its oviposition which is again dependent on-air flow (Beyaert & Hilker, 2014). Since, the transport and dispersion of the odour cues is mainly through wind flow and not through molecular diffusion, excessive wind flow, could affect the female fertility (Jactel et al., 2011), distribution of eggs, mass mortality, and the stability and establishment of the insect population (Cardé & Willis, 2008). Less establishment in return could affect the insect herbivory of the host plant. Insect pollinators serve the ecosystem by their pollinating ability. Half of the pollinating service in important crops is done by bees only during foraging (Kleijn et al., 2015). With the progressing climate change, a rise in the speed of wind has been noticed globally affecting this bee service (Hosking et al., 2018). An observation by Hennessy and co-workers in 2020 depicted that with the increase in wind speed by 2.75 m/s, 37% decrease in foraging activity was observed. Wind could challenge the flora visited by insects like bees moving between two individual flowers making several small flights (Goodwin et al., 2011). Furthermore, it was observed that this is due to the reluctance of bees for foraging under windy conditions. After consuming the nectar, the bees take time to take off when there is excessive wind flow. Before taking off, the insects must orient themselves in such a way that their wing angles should align perfectly for a smooth flight (Jakobi et al., 2018). Any change in the environmental condition like wind movement could affect their orientation and in result affect the flight because of varying air currents (Ren et al., 2018). Similar behaviour was noticed in the parasitic wasp, *Diachasmimorpha longicaudata*. However, this is because of rapid reduction in optimal body temperature of the insects in response to the increased cooling by wind flow (Messing et al., 1997). As the flight activity is an energy consuming process, hence sufficient energy and body temperature is necessary for the flight muscle contraction and their functioning (Pasek, 1988). In case of leaf cutter ants, their walk becomes slower by 55% and more diversion from their path is noticed under windy atmosphere (Alma et al., 2016).

Wind could also indirectly alter the presence of prey which could ultimately influence foraging. More insect abundance has been recorded in vegetation areas and forests, as they act as wind buffers (Whitaker et al., 2000). A research on Euglossine bees showed that the amount of energy spent was enhanced by 30% with rise in wind speed, as the bees extend their appendages to achieve stability (Combes & Dudley, 2009). At the same time, landing of insects becomes much more difficult with rising wind speed and less coordinated in case of bumblebees, *Bombus impatiens*, leading in greater-impact landings (Chang, Crall, & Combes, 2016). Less foraging because of increased wind flow can reduce the pollination service and also the pollinators population because of less food reserve collection. Wind is an important source for environmental noises. They act as carriers for various signals, may it be chemical or vibrational. Some nocturnal or diurnal insects also communicate through air currents. Insects communicating with plant-generated oscillations are forced to change their signalling behaviour or its timing under windy conditions. In the neotrophic katydids, *Copiphora brevirostris* peak signalling was observed under the lowest wind level and in between wind-free gaps whereas, least tremulations were recorded by males during excessive wind conditions rather than in silence (Velila et al., 2020).

6. Impact of Carbon Dioxide

The earth's air blanket contains a mixture of gases, out of which carbon dioxide is the one whose level is increasing at an alarming rate (50% more than before industrialization) because of human activities. According to the Mauna Loa observatory in Hawaii, the annual average carbon dioxide was reported to be around 421.08 ppm in 2023. It is one of the greenhouse gases (GHGs) which traps the harmful UV rays coming from the sun, causing a rise in the atmospheric temperature. With the continuous increase in these GHGs, it was predicted that by the year 2100, the air temperature may rise by 2–4 degrees with double the present concentration of carbon dioxide in the air (Pachauri et al., 2014). Suggestions in the Paris Agreement (2015) stated that a rise of 1.5 degree celsius in the average temperature above the preindustrial level would occur by 2040 or maybe earlier (Rogelj et al., 2016). According to Shrivastava (2016), an increase of 1 degree celsius in temperature could again facilitate the release of 30 petagrams of carbon, which is twice the amount released because of anthropogenic activities in a year. In the terrestrial ecosystem, insects occupy a major position among all the present phytophagous species in terms of absolute numbers and species (Schoonhoven et al., 2015). Insects being ectothermic organisms depend upon the external environment for regulating their internal system. Any change in the external factors can affect the insect's biology and activity at present or future carbon dioxide levels (Jactel et al., 2019; Ziska et al., 2018). Interspecific relationships among different phytophagous insect species exploiting the same resources can also be affected by the changing climatic conditions either directly through physiology, behaviour, and phenology, or indirectly by changes in the food material (Wang et al., 2020). Insects possess specific receptors to monitor and measure the CO_2 level in the atmosphere (Jones, 2013). However, any change in the CO_2 level can indirectly affect the insects through alterations in growth of the plant and its biochemical structure (Li et al., 2018; Shi et al., 2019) and directly through changes in physiology (Choi et al., 2018). Some studies have depicted that an elevated carbon dioxide level increases nutrition in the host plants there in altering the physiology of the insects e.g physiological changes in alligator weed because of elevated CO_2 facilitates enhanced nutrition for increasing reproduction in the alligator weed flea beetle feeding on it by speeding up maturity of their reproductive system (Shi et al., 2019). Some important impacts of rising or falling CO_2 levels on insects are presented in Table 1.

Table 1 Impact of carbon dioxide level on physiology of insects.

S.No.	*Insects*	*Impact of CO_2 in Insects*	*Change in CO_2*	*Reference*
1.	*Riptortus pedestris*	Quick completion of immature stages (eggs) and higher intrinsic rates of increase	Elevated CO_2	Ahn et al. (2022)
2.	*Junonia coenia*	Reduced nutritional quality in host plant	Elevated CO_2 (> 700 ppm)	Fajer et al. (1991)
3.	*Helicoverpa armigera*	Increased mortality, reduced fecundity, intrinsic development and fitness	Raised CO_2	Wu et al. (2006); Yin et al., (2010); Liu et al. (2017)
4.	*Ostrinia furnacalis*	Lower digestion of food by larvae and efficiency of its utilization as well as decreased fitness, increased mortality, less pupal weight, and longer development period	Raised CO_2 (550 ppm and 750 ppm)	Xie et al. (2015)
5.	Insect herbivores	Increase in development time, reduced growth rate and fecundity	Elevated CO_2	Stiling & Cornelissen (2007)
6.	*Operophtera brumata*	Increase in fecundity	Elevated CO_2	Buse et al. (1998)
7.	*Agasicles hygrophila*	Increase in adult male longevity, egg laying capacity, oviposition period, eggs laid in single oviposition day, net reproductive rate, intrinsic rate of increase. Reduced mean generation time, adult preoviposition time and total oviposition periods.	Elevated CO_2 (750 ppm)	Fu et al. (2016)
8.	*Sitobion avenae* *Aulacorthum solani*	Increased fecundity and early induction of reproduction	Elevated CO_2	Awmack et al. (1996); Awmack et al. (1997)

Contd.

Table 1 *Contd.*

S.No.	*Insects*	*Impact of CO_2 in Insects*	*Change in CO_2*	*Reference*
9.	*Rhopalosiphum padi, Aphis gossypii, Myzus persicae, Aphis craccivora*	Increased fecundity	Elevated CO_2	Chen et al., (2005); Oehme et al. (2011); Rao et al. (2013)
10.	Aphids	Reduction in alarm signalling resulting lowered escape behaviour	Elevated CO_2	Boullis et al. (2017)
11.	Honeybees	Reduced memory	Elevated CO_2	Stec & Kuszewska (2020)
12.	*Helicoverpa armigera*	Increase in sex pheromone production in females Reduction in behavioural and electrophysiological response against sex pheromone in males	Elevated CO_2	Choi et al. (2018)
13.	*Cactoblastis cactorum, A. janata*	Reduced egg laying	Elevated CO_2	Stange (1997); Rao et al. (2013)
14.	*Helicoverpa armigera*	Increased susceptibility to Bt due to reduced aminopeptidase activity hence more damage	Elevated CO_2	Sharma (2014)

Elevated CO_2 levels also mediate changes in the temperature. Rise in temperature leads to drought situations which affects plant growth and ultimately its nutritional content by altering the nitrogen content as well as secondary metabolites which ultimately influences food consumption and establishment by the insect pest. After food consumption, midgut releases a mixture of enzymes to break down the complex molecules into smaller ones but it can be influenced easily by elevated CO_2 levels and temperature. Increased CO_2 levels cause a rise in secretion of midgut protease and other enzymes which improves the digestibility and metabolism, but reduces the conversion of the digested material to body matter. This in turn also reduces the larval duration because of increased metabolism. CO_2 has a direct relation with their host plants e.g., it is beneficial to C_3 (e.g., rice, cotton, wheat) and less impactful to C_4 (corn, sorghum) plants resulting in differential effect on their ability to feed on these two different plant species (Chen et al., 2007). CO_2 is involved in the photosynthetic process which results in plant growth and development. Any alteration in its level could result in difference in both quality and quantity of vegetation. Plants grown under elevated CO_2 level cause differential deposits of nutrients e.g sugars, starch, or nitrogen in the leaves affecting its palatability due to alteration in the C:N ratio. Some insects feed more plant tissue to compensate for the reduced nutrient levels in hosts (Skendžić et al., 2021).

7. Altitude

Majority of insects are dispersed widely over earth's surface at different elevations, which creates a difference in environmental conditions under which a particular species thrive. Altitude affects temperature and atmospheric conditions, consequently influencing insect distribution and abundance (Hodkinson, 2005). As altitude increases, temperatures generally decrease, which can limit the distribution of some insect pests adapted to warmer climates. However, certain insect species have adapted to high-altitude environments and may thrive in these conditions. Additionally, altitude can affect atmospheric pressure, oxygen availability, and ultraviolet radiation exposure, which can indirectly influence insect physiology and behaviour. The comparative study of population ecology of insects along varying altitudinal regimes provide signals showcasing the response of species to changing climate. Altitude can directly affect the insect species like polymorphism in wing size and shape, colour reflectance pattern, body size, thermal tolerance, diapause, and life cycle duration. Studies reveal that at higher altitudes the majority of insects develop into brachypterous or apterous forms, which leads to reduced activity of flight muscles (Strathdee & Bale, 1998). Zera and Harshman (2001) have shown a straight connection between brachyptery and height in heterozygous long-winged females, which can form both long- and short-winged types based on their role in migration and colony formation. Insect species like *Oreina sulcata* is actually green in colour at lower altitudes, but develop a darker coloured body with reflective metallic morphs at high altitude due to varying UV radiation intake (Mikhailov, 2001). It has been noted that mean body size of an insect changes as per changing altitude; however, many species become larger in size while some show a gradual decline (Chown & Klok, 2003). Some insects like *Isotomurus alticola* and *Allonemobius fasciatus* usually have a smaller growth period at high altitude, which is attributed to lower number of instars in a developmental cycle (Zettel, 2000). For some insects, univoltine life cycle is seen like heather psyllid has only one cycle in heather growing season at low elevation, however is semivoltine at higher altitude (Butterfield et al., 2001).

Elevation gradient has been found to be an important factor which influences the insect community structure and their distribution and diversity. A study carried out in China along the altitudinal gradients of Guandi mountain (Zhao et al., 2023) has revealed that altitude is significantly having an impact over composition and structure of insect community and also the distribution patterns in this region. The percent infestation and damage of insect pests with respect to varying altitude is also showing a significant variation as reported by Aziz et al. (2021) in their survey on cutworm, Agrotis ipsilon infestation in maize crop at different altitude regimes of Kashmir valley. They have reported a higher percentage of damage in maize crop by cutworm at low altitude followed by mid and least in high altitude. Similar observations were noted by Arifie et al. (2019), who reported maximum percentage of damage to maize crop by different insect pests at low altitude and least at high altitude. Poggetti et al. (2019) analysed the impact of altitude on distribution and abundance of two insect species, *Coptodisca lucifluella* and *Rhagoletis completa* on walnut in Italy and found that both the pest species were not found in higher altitudes, with occurrence of *Coptodisca lucifluella* limited only up to 600 m above sea level and that of *Rhagoletis completa* limited to 700 m above sea level only. One more study on coffee plantation done by Asfaw et al. (2019) revealed that the infestation of coffee berry borer, *Hypothenemus hampei* and percentage of damaged berries was highest at low altitude sites. Modlinger et al. (2015) studied insect pests attacking spruce cones in Norway and the altitude preferences of different pests attacking it and they found that the most predominant insect pest in cones was *Cydia strobilella* infesting around 65% cones and cone damage was seen declining with respect to increasing elevation.

Altitude also has an effect over species' richness and abundance of insect pests. In Mexico, a research conducted by Rubin-Aguirre et al. (2015) on bark beetle abundance in pine and species distribution across altitudinal gradient of 2200–2750 m above sea level revealed that bark beetles were showing more species richness and abundance at lower altitudes and higher damage percentage to pine by bark beetle was attributed to temperature rise in lower altitude areas. Altitude also affects the onset and emergence of different insects. An early onset of aphids was recorded in Norway spruce by Mühlenberg and Stadler (2005) at low altitude compared to that of high, due to warm and dry conditions at low altitude. Altitude was seen to effect the seasonal flight activity of *Rhagoletis cerasi* as reported by Kovanci and Kovanci (2006) in Turkey that there was a variation in forecasted and observed times of appearance of adult flies by 1.4–2 days for every 100 m altitude variation and also prolonged flight activity was seen at high altitudes. The effect of temperature and height of occurrence on feeding behaviour of Mexican beetle, *Zygogramma bicolorata* on *Parthenium hysterophorus* was studied by Bhusal et al. (2020) who observed that in areas of low temperature and high altitude, adults were large sized with greater food consumption rate than adults of lower altitudes.

8. Manipulating Climatic Variables for Purposeful Insect Pest Management

Purposeful manipulation of microclimate rather than only reacting to weather can suppress pest populations, slow their development and strengthen natural enemies. Below are some tactics for different climatic variables with which we can achieve better pest management:

- Temperature governs insect metabolism, development rate, voltinism, dispersal, and survival. Thermal buffering with canopy and mulches can be done to depress canopy and soil temperatures, in order to lengthen pest generation time and reduce heat-stimulated activity. Altering plantation dates and shifting cultivar phenology, so that vulnerable stages avoid peak thermal windows for key pests.
- Humidity not only regulates insect water balance, fecundity, and survival, but also shapes entomopathogen pressure and host plant quality. In order to avoid formation of humid leaf boundary layers that favour sap-suckers, time irrigations to daylight hours to shorten wetness periods and go for drip irrigation. Crops should be widely spaced and vertical trellising should be done in order to provide proper ventilation and lower relative humidity which disrupt pest attacks.
- Light influences diapause, dormancy, foraging, and predator-prey interactions. Short day photoperiods through blackout screens can push certain pests towards diapause or reduce reproductive activity and in addition can boost biocontrol agent activity. Deploying species-specific LED traps to intercept nocturnal pests, whose activity is light regulated.
- Wind drives local orientation, dispersal and long distance migration and it also modulates foraging of both pests and pollinators. Planting shelterbelts or barrier crops can reduce advective influx of aphids and whiteflies and reduce insect dispersal. Aligning trap crops downwind in cash crops to catch upwind flying insects.
- Elevated CO_2 alters plant C: N ratio and changes herbivore performance. By adopting protected cultivation, one can avoid unnecessary CO_2 enrichment. Growing of cultivars along with efficient nitrogen management can maintain adequate foliar nitrogen under high CO_2 conditions and prevent compensatory feeding by lepidopterans.
- Altitude works along with temperature, pressure and UV and many pests are less abundant at high elevations. Placing highly susceptible crops or nurseries at higher elevation can lower pest damage. Using hedgerows and contour plantings to create cooler, more humid microsites that favour parasitoids and slow heat-adapted pest development.

9. Conclusion

A thorough review of literature suggested that insect population dynamics is significantly impacted by climatic variables, with temperature, relative humidity, precipitation, and seasonal patterns playing pivotal roles. Warmer temperatures generally accelerate insect growth and reproduction, but extreme heat or cold can reduce survival rates. Humidity impacts behaviour, breeding, and survival, particularly in species relying on moisture rich environments. Changes in precipitation patterns affect the availability of food resources and breeding sites, while shifts in seasonal patterns can change the occurrence time of different developmental stages like emergence and migration. In general, climate change is tending to shift insect distributions, change population dynamics, and potentially increase the frequency of pest outbreaks, as some species adapt to new conditions faster than others. These changes have broad implications for ecosystems, agriculture, and public health, underscoring the need for monitoring and adaptive strategies to mitigate adverse impacts.

References

Ahn, J.J. and Choi, K.S. (2022). Population parameters and growth of *Riptortus pedestris* (Fabricius) (Hemiptera: Alydidae) under fluctuating temperature. *Insects*, *13*(2), 113.

Alma, A.M., Farji-Brener, A.G. and Elizalde, L. (2016). Gone with the wind: Short- and long-term responses of leaf-cutting ants to the negative effect of wind on their foraging activity. *Behavioral Ecology*, *27*(4), 1017–1024.

Arifie, U., Bano, P., Ahad, I., Singh, P., Dar, Z.A., Badri, Z., ... and Kumar, R. (2019). Insect pests of maize at different altitudes of north Kashmir, J&K. *Journal of Entomology and Zoology Studies*, *7*, 1123–1128.

Asfaw, E., Mendesil, E. and Mohammed, A. (2019). Altitude and coffee production systems influence extent of infestation and bean damage by the coffee berry borer. *Archives of Phytopathology and Plant Protection*, *52*(1–2), 170–183.

Awmack, C.S., Harrington, R., Leather, S.R. and Lawton, J.H. (1996). The impacts of elevated CO_2 on aphid-plant interactions. *Implications of "Global Environmental Change" for Crops in Europe. Aspects of Applied Biology*, *45*, 317–322.

Awmack, C., Harrington, R. and Leather, S. (1997). Host plant effects on the performance of the aphid *Aulacorthum solani* (Kalt.) (Homoptera: Aphididae) at ambient and elevated CO_2. *Global Change Biology*, *3*(6), 545–549.

Awmack, C., Woodcock, C. and Harrington, R. (1997). Climate change may increase vulnerability of aphids to natural enemies. *Ecological Entomology*, *22*(3), 366–368.

Ayres, M.P. and Lombardero, M.J. (2000). Assessing the consequences of global change for forest disturbance from herbivores and pathogens. *Science of the Total Environment*, *262*(3), 263–286.

Aziz, D., Ahmed, S.B., Yaqoob, M., Aalum, K., Rehman, S.A. and Anjum, N. (2022). Population dynamics of cut worm *Agrotis ipsilon* Hufnagel on maize in Kashmir. ResearchGate.

Bale, J.S., Masters, G.J., Hodkinson, I.D., Awmack, C., Bezemer, T.M., Brown, V.K., ... and Whittaker, J.B. (2002). Herbivory in global climate change research: Direct effects of rising temperature on insect herbivores. *Global Change Biology*, *8*(1), 1–16.

Bauer, S. and Klaassen, M. (2013). Mechanistic models of animal migration behaviour: Their diversity, structure, and use. *Journal of Animal Ecology*, *82*(3), 498–508.

Beyaert, I. and Hilker, M. (2014). Plant odour plumes as mediators of plant–insect interactions. *Biological Reviews*, *89*(1), 68–81.

Bhusal, D.R., Ghimire, K.C., Patel, P., Bista, M., Upadhyay, R. and Kumar, B. (2020). Temperature and altitude modulate feeding attributes of Mexican beetle, *Zygogramma bicolorata* Pallister on *Parthenium hysterophorus*. *Journal of Thermal Biology*, *89*, 102540.

Borkenstein, A. and Jödicke, R. (2022). Thermoregulatory behaviour of *Sympetrum striolatum* at low temperatures with special reference to the role of direct sunlight (Odonata: Libellulidae). *Odonatologica*, *51*(1–2), 83–109.

Boullis, A., Fassotte, B., Sarles, L., Lognay, G., Heuskin, S., Vanderplanck, M., ... and Verheggen, F.J. (2017). Elevated carbon dioxide concentration reduces alarm signaling in aphids. *Journal of Chemical Ecology*, *43*, 164–171.

Broufas, G.D., Pappas, M.L. and Koveos, D.S. (2009). Effect of relative humidity on longevity, ovarian maturation, and egg production in the olive fruit fly (Diptera: Tephritidae). *Annals of the Entomological Society of America*, *102*(1), 70–75.

Buse, A., Good, J.E.G., Dury, S. and Perrins, C.M. (1998). Effects of elevated temperature and carbon dioxide on the nutritional quality of leaves of oak (*Quercus robur* L.) as food for the winter moth (*Operophtera brumata* L.). *Functional Ecology*, *12*(5), 742–749.

Butterfield, J., Whittaker, J.B. and Fielding, C.A. (2001). Control of the flexible annual/biennial life cycle of the heather psyllid *Strophingia ericae*. *Physiological Entomology*, *26*(3), 266–274.

Cardé, R.T. and Willis, M.A. (2008). Navigational strategies used by insects to find distant, wind-borne sources of odor. *Journal of Chemical Ecology*, *34*, 854–866.

Casimir, M. (1987). *Plague Locusts in New South Wales: Studies of Migration and Displacement of Populations*. Department of Agriculture, New South Wales, pp. 1–21.

Chang, J.J., Crall, J.D. and Combes, S.A. (2016). Wind alters landing dynamics in bumblebees. *Journal of Experimental Biology*, *219*(18), 2819–2822.

Chapman, A.D. (2009). *Numbers of Living WSpecies in Australia and the World*. 1–78. Australian Biological Resources Study.

Chapman, J.W., Reynolds, D.R. and Wilson, K. (2015). Long-range seasonal migration in insects: Mechanisms, evolutionary drivers, and ecological consequences. *Ecology Letters*, *18*(3), 287–302.

Chen, F.G., Wu, F. Ge., Parajulee, M.N. and Shrestha, R.B. (2005). Effects of elevated CO_2 and transgenic Bt cotton on plant chemistry, performance, and feeding of an insect herbivore, the cotton bollworm. *Entomologia Experimentalis et Applicata*, *115*, 341–350.

Chen, F., Ge, F. and Parajulee, M.N. (2005). Impact of elevated CO_2 on tri-trophic interaction of *Gossypium hirsutum*, *Aphis gossypii*, and *Leis axyridis*. *Environmental Entomology*, *34*(1), 37–46.

Chen, F., Wu, G., Parajulee, M.N. and Ge., F. (2007). Long-term impacts of elevated CO_2 and transgenic Bt cotton on performance and feeding of three generations of cotton bollworm. *Entomologia Experimentalis et Applicata, 124*, 27–35.

Choi, K.S., Ahn, S.J., Kim, S.B., Ahn, J.J., Jung, B.N., Go, S.W. and Kim, D.S. (2018). Elevated CO_2 may alter pheromonal communication in *Helicoverpa armigera* (Lepidoptera: Noctuidae). *Physiological Entomology*, *43*(3), 169–179.

Chown, S.L. and Klok, C.J. (2003). Altitudinal body size clines: Latitudinal effects associated with changing seasonality. *Ecography*, *26*(4), 445–455.

Clarke, A. and Fraser, K.P.P. (2004). Why does metabolism scale with temperature? *Functional Ecology*, *18*(2), 243–251.

Combes, S.A. and Dudley, R. (2009). Turbulence-driven instabilities limit insect flight performance. *Proceedings of the National Academy of Sciences*, *106*(22), 9105–9108.

Coviella, C.E. and Trumble, J.T. (1999). Effects of elevated atmospheric carbon dioxide on insect-plant interactions. *Conservation Biology*, *13*(4), 700–712.

Czekońska, K., Łopuch, S. and Miścicki, S. (2023). The effect of meteorological and environmental variables on food collection by honey bees (*Apis mellifera*). *Ecological Indicators*, *156*, 111140.

Dedryver, C.A., Hullé, M., Le Gallic, J.F., Caillaud, M.C. and Simon, J.C. (2001). Coexistence in space and time of sexual and asexual populations of the cereal aphid *Sitobion avenae*. *Oecologia*, *128*, 379–388.

Dingle, H. (2014). *Migration: The Biology of Life on the Move*. Oxford University Press, USA.

Drake, V.A. and Farrow, R.A. (1988). The influence of atmospheric structure and motions on insect migration. *Annual Review of Entomology*, *33*(1), 183–210.

Dury, S.J., Good, J.E.G., Perrins, C.M., Buse, A. and Kaye, T. (1998). The effects of increasing CO_2 and temperature on oak leaf palatability and the implications for herbivorous insects. *Global Change Biology*, *4*(1), 55–61.

Fajer, E.D., Bowers, M.D. and Bazzaz, F.A. (1991). The Effects of Enriched CO_2 Atmospheres on the Buckeye Butterfly, *Junonia Coenia*. *Ecology*, *72*(2), 751–754.

Finch, S. and Skinner, G. (1982). Upwind flight by the cabbage root fly, *Delia radicum*. *Physiological Entomology*, *7*(4), 387–399.

Florio, J., Verú, L.M., Dao, A., Yaro, A.S., Diallo, M., Sanogo, Z.L., ... and Lehmann, T. (2020). Diversity, dynamics, direction, and magnitude of high-altitude migrating insects in the Sahel. *Scientific Reports*, *10*(1), 20523.

Fu, J.W., Shi, M.Z., Wang, T., Li, J.Y., Zheng, L.Z. and Wu, G. (2016). Demography and population projection of flea beetle, *Agasicles hygrophila* (Coleoptera: Chrysomelidae), fed on alligator weed under elevated CO_2. *Journal of Economic Entomology*, *109*(3), 1116–1124.

Gao, B., Wotton, K.R., Hawkes, W.L., Menz, M.H., Reynolds, D.R., Zhai, B.P., ... and Chapman, J.W. (2020). Adaptive strategies of high-flying migratory hoverflies in response to wind currents. *Proceedings of the Royal Society B*, *287*(1928), 20200406.

Gibson, G., Packer, M.J., Steullet, P. and Brady, J. (1991). Orientation of tsetse flies to wind, within and outside host odour plumes in the field. *Physiological Entomology*, *16*(1), 47–56.

Gillooly, J.F., Brown, J.H., West, G.B., Savage, V.M. and Charnov, E.L. (2001). Effects of size and temperature on metabolic rate. *Science*, *293*(5538), 2248–2251.

Goodwin, R.M., Cox, H.M., Taylor, M.A., Evans, L.J. and McBrydie, H.M. (2011). Number of honey bee visits required to fully pollinate white clover (*Trifolium repens*) seed crops in Canterbury, New Zealand. *New Zealand Journal of Crop and Horticultural Science*, *39*(1), 7–19.

Goto, S.G. and Numata, H. (2009). Alteration of the pupal diapause program and regulation of larval duration by photoperiod in the flesh fly *Sarcophaga similis* Meade (Diptera: Sarcophagidae). *Applied Entomology and Zoology*, *44*(4), 603–609.

Haggis, M.J. (1981). Spatial and temporal changes in the distribution of eggs of *Heliothis armigera* (Hübner) (Lepidoptera: Noctuidae) on cotton in the Sudan Gezira. *Bulletin of Entomological Research*, *71*(2), 183–193.

Hakeem, A. and Parajulee., M.N. (2015). Moisture conditions for laboratory rearing of cotton fleahoppers from overwintered eggs laid on woolly croton plants. *Southwestern Entomologist, 40*, 455–462.

Harrison, J.F. and Roberts, S.P. (2000). Flight respiration and energetics. *Annual Review of Physiology*, *62*(1), 179–205.

Hennessy, G., Harris, C., Eaton, C., Wright, P., Jackson, E., Goulson, D. and Ratnieks, F.F. (2020). Gone with the wind: Effects of wind on honey bee visit rate and foraging behaviour. *Animal Behaviour*, *161*, 23–31.

Hill, J.K., Thomas, C.D. and Huntley, B. (1999). Climate and habitat availability determine 20th century changes in a butterfly's range margin. *Proceedings of the Royal Society of London. Series B: Biological Sciences*, *266*(1425), 1197–1206.

Hodkinson, I.D. (2005). Terrestrial insects along elevation gradients: Species and community responses to altitude. *Biological Reviews*, *80*(3), 489–513.

Holmes, L.A., Vanlaerhoven, S.L. and Tomberlin, J.K. (2012). Relative humidity effects on the life history of *Hermetia illucens* (Diptera: Stratiomyidae). *Environmental Entomology*, *41*(4), 971–978.

Hosking, J.S., MacLeod, D., Phillips, T., Holmes, C.R., Watson, P., Shuckburgh, E.F. and Mitchell, D. (2018). Changes in European wind energy generation potential within a 1.5 C warmer world. *Environmental Research Letters*, *13*(5), 054032.

Hu, G., Lim, K.S., Horvitz, N., Clark, S.J., Reynolds, D.R., Sapir, N. and Chapman, J.W. (2016). Mass seasonal bioflows of high-flying insect migrants. *Science*, *354*(6319), 1584–1587.

Huberty, A.F. and Denno, R.F. (2004). Plant water stress and its consequences for herbivorous insects: A new synthesis. *Ecology*, *85*(5), 1383–1398.

Jactel, H., Birgersson, G., Andersson, S. and Schlyter, F. (2011). Non-host volatiles mediate associational resistance to the pine processionary moth. *Oecologia*, *166*, 703–711.

Jactel, H., Koricheva, J. and Castagneyrol, B. (2019). Responses of forest insect pests to climate change: Not so simple. *Current O(pinion in Insect Science*, *35*, 103–108.

Jakobi, T., Kolomenskiy, D., Ikeda, T., Watkins, S., Fisher, A., Liu, H. and Ravi, S. (2018). Bees with attitude: The effect of gusts on flight dynamics. *arXiv preprint arXiv:1802.03580.*

Jones, W. (2013). Olfactory carbon dioxide detection by insects and other animals. *Molecules and Cells*, *35*(2), 87–92.

Jönsson, A.M., Harding, S., Bärring, L. and Ravn, H.P. (2007). Impact of climate change on the population dynamics of *Ips typographus* in southern Sweden. *Agricultural and Forest Meteorology*, *146*(1–2), 70–81.

Karolewski, P., Grzebyta, J., Oleksyn, J. and Giertych, M.J. (2007). Effects of temperature on larval survival rate and duration of development of *Lymantria monacha* (L.) on needles of *Pinus silvestris* (L.) and of *L. dispar* (L.) on leaves of *Quercus robur* (L.). *Polish Journal of Ecology*, *55*(3), 595.

Karthik, S., Reddy, M.S. and Yashaswini, G. (2021). Climate change and its potential impacts on insect-plant interactions. In: *The nature, causes, effects and nitigation of cilmate change on the environment* pp. 01–23. Doi: 10.577/intechopen.98203.

Kleijn, D., Winfree, R., Bartomeus, I., Carvalheiro, L.G., Henry, M., Isaacs, R., ... and Potts, S.G. (2015). Delivery of crop pollination services is an insufficient argument for wild pollinator conservation. *Nature Communications*, *6*(1), 7414.

Knoblauch, A., Thoma, M. and Menz, M.H. (2021). Autumn southward migration of dragonflies along the Baltic coast and the influence of weather on flight behaviour. *Animal Behaviour*, *176*, 99–109.

Koštál, V. (2006). Eco-physiological phases of insect diapause. *Journal of Insect physiology*, *52*(2), 113–127.

Kovanci, O.B. and Kovanci, B. (2006). Effect of altitude on seasonal flight activity of *Rhagoletis cerasi* flies (Diptera: Tephritidae). *Bulletin of Entomological Research*, *96*(4), 345–351.

Krasnov, B.R., Khokhlova, I.S., Fielden, L.J. and Burdelova, N.V. (2001). Effect of air temperature and humidity on the survival of pre-imaginal stages of two flea species (Siphonaptera: Pulicidae). *Journal of Medical Entomology*, *38*(5), 629–637.

Kuokkanen, K., Julkunen-Tiitto, R., Keinänen, M., Niemelä, P. and Tahvanainen, J. (2001). The effect of elevated CO_2 and temperature on the secondary chemistry of *Betula pendula* seedlings. *Trees*, *15*, 378–384.

Li, Z., Parajulee, M.N. and Chen., F. (2018). Influence of elevated CO_2 on development and food utilization of armyworm *Mythimna separata* fed on transgenic Bt maize infected by nitrogen-fixing bacteria. *Peer J. Life and Environment, 6*, e5138.

Liu, J., Huang, W., Chi, H., Wang, C., Hua, H. and Wu, G. (2017). Effects of elevated CO_2 on the fitness and potential population damage of *Helicoverpa armigera* based on two-sex life table. *Scientific Reports*, *7*(1), 1119.

Logan, J.A., Régnière, J. and Powell, J.A. (2003). Assessing the impacts of global warming on forest pest dynamics. *Frontiers in Ecology and the Environment*, *1*(3), 130–137.

Mansoor-ul-Hasan, A.A., Jafir, M., Javed, M.W., Shehzad, M., Chaudhary, M.Z. and Aftab, M. (2017). Effect of temperature and relative humidity on development of *Sitophilus oryzae* L. (Coleoptera: Curculionidae). *J. Entomol. Zool. Stud.*, *5*(6), 85–90.

McClure, M. and McNeil, J.N. (2009). The effect of abiotic factors on the male mate searching behavior and the mating success of *Aphidius ervi* (Hymenoptera: Aphidiidae). *Journal of insect behavior*, *22*, 101–110.

Melillo, J.M., Richmond, T.T. and Yohe, G. (2014). Climate change impacts in the United States. *Third National Climate Assessment*, *52*, 150–174.

Menéndez, R. (2007). How are insects responding to global warming? *Tijdschrift voor entomologie*, *150*(2), 355.

Menéndez, R., González-Megías, A., Collingham, Y., Fox, R., Roy, D.B., Ohlemüller, R. and Thomas, C.D. (2007). Direct and indirect effects of climate and habitat factors on butterfly diversity. *Ecology*, *88*(3), 605–611.

Messing, R.H., Klungness, L.M. and Jang, E.B. (1997). Effects of wind on movement of Diachasmimorpha longicaudata, a parasitoid of tephritid fruit flies, in a laboratory flight tunnel. *Entomologia Experimentalis et Applicata*, *82*(2), 147–152.

Mikhalov, Y. (2001). Significance of colour polymorphism in moun-tain populations of abundant leaf beetles. *Pirineos*, *156*, 57–68.

Modlinger, R., Holuša, J. and Liška, J. (2015). Insect pests of Norway spruce cones: Incidences and altitude preferences with emphasis on *Cydia strobilella* (L.). *Journal of Applied Entomology*, *139*(9), 701–711.

Moore, B. and Allard, G. (2008). *Climate Change Impacts on Forest Health*. FAO.

Mühlenberg, E. and Stadler, B. (2005). Effects of altitude on aphid-mediated processes in the canopy of Norway spruce. *Agricultural and Forest Entomology*, *7*(2), 133–143.

Mukai, A., Yamaguchi, K. and Goto, S.G. (2021). Urban warming and artificial light alter dormancy in the flesh fly. *Royal Society Open Science*, *8*(7), 210866.

Netherer, S. and Schopf, A. (2010). Potential effects of climate change on insect herbivores in European forests: General aspects and the pine processionary moth as specific example. *Forest Ecology and Management*, *259*(4), 831–838.

Nupponen H., Salokannel S., Thomsses P.-M., Vanhatal A. 2010. The impact of climate change on invasive alien insect distribution and diversity. http://www.helsinki.fi/metsatieteet/tiedotteet/pdf/ME408_Group7_report.pdf [20.10.2010].

Oehme, V., Högy, P., Franzaring, J., Zebitz, C.P.W. and Fangmeier, A. (2011). Response of spring crops and associated aphids to elevated atmospheric CO_2 concentrations. *Journal of Applied Botany and Food Quality*, *84*(2).

Pachauri, R.K., Allen, M.R., Barros, V.R., Broome, J., Cramer, W., Christ, R., ... and van Ypserle, J.P. (2014). *Climate Change 2014: Synthesis Report. Contribution of Working Groups I, II, and III to the Fifth Assessment Report of the Intergovernmental Panel on Climate Change.* IPCC, 151 pp.

Parajulee, M.N., Wilson, L.T., Rummel, D.R, Carroll, S.C. and Trichilo, P.J. (1996). Climatic data-based analysis of boll weevil overwintering survival and spring emergence. *Environmental Entomology, 25*, 882–894.

Parajulee, M.N., Wilson, L.T., Rummel, D.R, Carroll, S.C., Trichilo, P.J., Slosser, J.E. and Fuchs., T.W. (1997). Relationship between ambient and leaf litter temperatures in boll weevil overwintering habitats. *Environmental Entomology, 26*, 135–141.

Parmesan, C. and Yohe, G. (2003). A globally coherent fingerprint of climate change impacts across natural systems. N*ature*, *421*(6918), 37–42.

Pasek, J.E. (1988). 30. Influence of wind and windbreaks on local dispersal of insects. *Agriculture, Ecosystems & Environment*, *22*, 539–554.

Pedgley, D.E. (1990). Concentration of flying insects by the wind. *Philosophical Transactions of the Royal Society of London. B, Biological Sciences*, *328*(1251), 631–653.

Pedgley, D.E., Page, W.W., Mushi, A., Odiyo, P., Amisi, J., Dewhurst, C.F., ... and Rose, D.J.W. (1989). Onset and spread of an African army worm upsurge. *Ecological Entomology*, *14*(3), 311–333.

Peterson, R.K., Higley, L.G. and Bailey, W.C. (1988). Phenology of the adult celery looper, *Syngrapha falcifera* (Lepidoptera: Noctuidae), in Iowa: Evidence for migration. *Environmental Entomology*, *17*(4), 679–684.

Petzoldt, C. and Seaman, A. (2006). Climate change effects on insects and pathogens. Climate Change and Agriculture: Promoting Practical and Profitable Responses. 6–16.

Plath, M., Dorn, S., Riedel, J., Barrios, H. and Mody, K. (2012). Associational resistance and associational susceptibility: specialist herbivores show contrasting responses to tree stand diversification. *Oecologia*, *169*, 477–487.

Poggetti, L., Raranciuc, S., Chiabà, C., Vischi, M. and Zandigiacomo, P. (2019). Altitude affects the distribution and abundance of two non-native insect pests of the common walnut. *Journal of applied entomology*, *143*(5), 527–534.

Rao, M.S., Srinivas, K., Vanaja, M., Manimanjari, D., Rao, C.R. and Venkateswarlu, B. (2013). Response of multiple generations of semilooper, *Achaea janata* feeding on castor to elevated CO_2. *Journal of Environmental Biology*, *34*(5), 877.

Régnière, J. (2009). Predicting insect continental distributions from species physiology. *Unasylva*, *60*(231), 37–42.

Ren, G., Liu, J., Wan, J., Li, F., Guo, Y. and Yu, D. (2018). The analysis of turbulence intensity based on wind speed data in onshore wind farms. *Renewable energy*, *123*, 756–766.

Reynolds, D.R. and Riley, J.R. (1988). A migration of grasshoppers, particularly *Diabolocatantops axillaris* (Thunberg) (Orthoptera: Acrididae), in the West African Sahel. *Bulletin of Entomological Research*, *78*(2), 251–271.

Rogelj, J., Den Elzen, M., Höhne, N., Fransen, T., Fekete, H., Winkler, H. ... and Meinshausen, M. (2016). Paris Agreement climate proposals need a boost to keep warming well below 2 C. *Nature*, *534*(7609), 631–639.

Rose, A.H., Silversides, R.H. and Lindquist, O.H. (1975). Migration flight by an aphid, *Rhopalosiphum maidis* (Hemiptera: Aphididae), and a noctuid, *Spodoptera frugiperda* (Lepidoptera: Noctuidae). *The Canadian Entomologist*, *107*(6), 567–576.

Rouault, G., Candau, J.N., Lieutier, F., Nageleisen, L.M., Martin, J.C. and Warzée, N. (2006). Effects of drought and heat on forest insect populations in relation to the 2003 drought in Western Europe. *Annals of Forest Science*, *63*(6), 613–624.

Rubin-Aguirre, A., Saenz-Romero, C., Lindig-Cisneros, R., del-Rio-Mora, A.A., Tena-Morelos, C.A., Campos-Bolaños, R. and Del-Val, E. (2015). Bark beetle pests in an altitudinal gradient of a Mexican managed forest. *Forest Ecology and Management*, *343*, 73–79.

Satterfield, D.A., Sillett, T.S., Chapman, J.W., Altizer, S. and Marra, P.P. (2020). Seasonal insect migrations: Massive, influential, and overlooked. *Frontiers in Ecology and the Environment*, *18*(6), 335–344.

Schneider, J.C. (2020). Effects of light intensity on mating of the black soldier fly (*Hermetia illucens*, Diptera: Stratiomyidae). *Journal of Insects as Food and Feed*, *6*(2), 111–119.

Schoonhoven, L.M., Van Loon, J.J. and Dicke, M. (2005). *Insect-plant Biology*. Oxford University Press.

Sharma, H.C. (2014). Climate change effects on insects: Implications for crop protection and food security. *Journal of Crop Improvement*, *28*(2), 229–259.

Sharma, H.C. (Ed.). (2005). *Heliothis/Helicoverpa Management: The Emerging Trends and Need for Future Research*. Taylor Francis.

Shi, M.Z., Li, J.Y., Ding, B., Fu, J.W., Zheng, L.Z. and Chi, H. (2019). Indirect effect of elevated CO_2 on population parameters and growth of *Agasicles hygrophila* (Coleoptera: Chrysomelidae), a biocontrol agent of Alligator weed (Amaranthaceae). *Journal of Economic Entomology*, *112*(3), 1120–1129.

Shrivastava, A. (2016). Rising temperature causes soil to release more carbon dioxide. *Down to Earth*. Centre for Science and Environment.

Skendžić, S., Zovko, M., Živković, I.P., Lešić, V. and Lemić, D. (2021). The impact of climate change on agricultural insect pests. *Insects*, *12*(5), 440.

Stange, E.E. and Ayres, M.P. (2010). Climate change impacts: Insects. *Encyclopedia of Life Sciences*, *1*.

Stange, G. (1997). Effects of changes in atmospheric carbon dioxide on the location of hosts by the moth, *Cactoblastis cactorum*. *Oecologia*, *110*, 539–545.

Stec, D. and Kuszewska, K. (2020). CO_2 narcosis influences the memory of honey bees. *Journal of Apicultural Research*, *59*(4), 663–668.

Stiling, P. and Cornelissen, T. (2007). How does elevated carbon dioxide (CO_2) affect plant–herbivore interactions? A field experiment and meta-analysis of CO_2-mediated changes on plant chemistry and herbivore performance. *Global Change Biology*, *13*(9), 1823–1842.

Storer, J.R., Young, S. and Hardie, J. (1999). Three-dimensional analysis of aphid landing behaviour in the laboratory and field. *Physiological Entomology*, *24*(3), 271–277.

Strathdee, A.T. and Bale, J.S. (1998). Life on the edge: Insect ecology in arctic environments. *Annual Review of Entomology*, *43*(1), 85–106.

Sykes, M.T. (2009). Climate change impacts: Vegetation. *Els*.

Szujecki A. (1998). *Entomologia leœna*. Warszawa, Wyd. SGGW.

Tamiru, A., Getu, E., Jembere, B. and Bruce, T. (2012). Effect of temperature and relative humidity on the development and fecundity of *Chilo partellus* (Swinhoe) (Lepidoptera: Crambidae). *Bulletin of Entomological Research*, *102*(1), 9–15.

Taylor, L.R. (1974). Insect migration, flight periodicity and the boundary layer. *The Journal of Animal Ecology*, 225–238.

Thomas, C.D., Bodsworth, E.J., Wilson, R.J., Simmons, A.D., Davies, Z.G., Musche, M. and Conradt, L. (2001). Ecological and evolutionary processes at expanding range margins. *Nature*, *411*(6837), 577–581.

Thomas, F.M. (2008). Recent advances in cause-effect research on oak decline in Europe. *CABI Reviews*, (2008), 12 pp.

Tougeron, K., Brodeur, J., Le Lann, C. and van Baaren, J. (2020). How climate change affects the seasonal ecology of insect parasitoids. *Ecological Entomology*, *45*(2), 167–181.

Tougeron, K., Brodeur, J., van Baaren, J., Renault, D. and Le Lann, C. (2018). Sex makes them sleepy: Change in host reproductive status induces diapause in a parasitoid population experiencing harsh winters. *bioRxiv*, 371385.

Van Klink, R., Bowler, D.E., Gongalsky, K.B., Swengel, A.B., Gentile, A. and Chase, J.M. (2020). Meta-analysis reveals declines in terrestrial but increases in freshwater insect abundances. *Science*, *368*(6489), 417–420.

Velilla, E., Muñoz, M., Quiroga, N., Symes, L., ter Hofstede, H.M., Page, R.A., ... and Halfwerk, W. (2020). Gone with the wind: Is signal timing in a neotropical katydid an adaptive response to variation in wind-induced vibratory noise? *Behavioral Ecology and Sociobiology*, *74*, 1–11.

Visser, M.E. and Holleman, L.J. (2001). Warmer springs disrupt the synchrony of oak and winter moth phenology. *Proceedings of the Royal Society of London. Series B: Biological Sciences*, *268*(1464), 289–294.

Walther, G.R., Post, E., Convey, P., Menzel, A., Parmesan, C., Beebee, T.J., ... and Bairlein, F. (2002). Ecological responses to recent climate change. *Nature*, *416*(6879), 389–395.

Wang, C., Fei, M., Meng, L., Harvey, J. A. and Li, B. (2020). Effects of elevated CO_2 and temperature on survival and wing dimorphism of two species of rice planthoppers (Hemiptera: Delphacidae) under interaction. *Pest Management Science*, *76*(6), 2087–2094.

Whitaker, D.M., Carroll, A.L. and Montevecchi, W.A. (2000). Elevated numbers of flying insects and insectivorous birds in riparian buffer strips. *Canadian Journal of Zoology*, *78*(5), 740–747.

Withers, T.M. and Keena, M.A. (2001). *Lymantria monacha* (nun moth) and L. dispar (gypsy moth) survival and development on improved Pinus radiata. *New Zealand Journal of Forestry Science*, *31*(1), 66–77.

Wolf, R.A., Pedigo, L.P., Shaw, R.H. and Newsom, L.D. (1987). Migration/transport of the green cloverworm, *Plathypena scabra* (F.) (Lepidoptera: Noctuidae), into Iowa as determined by synoptic-scale weather patterns. *Environmental Entomology*, *16*(5), 1169–1174.

Wotton, K.R., Gao, B., Menz, M.H., Morris, R.K., Ball, S.G., Lim, K.S. ... and Chapman, J.W. (2019). Mass seasonal migrations of hoverflies provide extensive pollination and crop protection services. *Current Biology*, *29*(13), 2167–2173.

Wu, G., Chen, F.J. and Ge, F. (2006). Response of multiple generations of cotton bollworm *Helicoverpa armigera* Hübner, feeding on spring wheat, to elevated CO_2. *Journal of Applied Entomology*, *130*(1), 2–9.

Xie, H., Zhao, L., Yang, Q., Wang, Z. and He, K. (2015). Direct effects of elevated CO_2 levels on the fitness performance of Asian corn borer (Lepidoptera: Crambidae) for multigenerations. *Environmental Entomology*, *44*(4), 1250–1257.

Yin, J., Sun, Y., Wu, G. and Ge, F. (2010). Effects of elevated CO_2 associated with maize on multiple generations of the cotton bollworm, *Helicoverpa armigera*. *Entomologia Experimentalis et Applicata*, *136*(1), 12–20.

Zera, A.J. and Harshman, L.G. (2001). The physiology of life history trade-offs in animals. *Annual Review of Ecology and Systematics*, *32*(1), 95–126.

Zettel, J. (1999). Alpine Collembola-adaptations and strategies for survival in harsh environments. *Zoology-Analysis of Complex Systems*, *102*, 73–89.

Zhao, L., Gao, R., Liu, J., Liu, L., Li, R., Men, L. and Zhang, Z. (2023). Effects of environmental factors on the spatial distribution pattern and diversity of insect communities along altitude gradients in Guandi Mountain, China. *Insects*, *14*(3), 224.

Ziska, L.H., Bradley, B.A., Wallace, R.D., Bargeron, C.T., LaForest, J.H., Choudhury, R.A., ... and Vega, F.E. (2018). Climate change, carbon dioxide, and pest biology, managing the future: Coffee as a case study. *Agronomy*, *8*(8), 152.

CHAPTER 8

Beneath the Surface
Biology, Ecology, and Sustainable Management of Nematodes

Sirisha Tadigiri[1], Kiran Kumar Kommu[2], Satish N. Chavan[3], Chinnannan Karthik[4] and Priyank Hanuman Mhatre[5*]

1. Introduction

Nematodes are microscopic, filamentous roundworms that are widespread across various ecosystems, ranging from soil to freshwater and marine environments. These organisms inhabit diverse ecological niches and display a wide range of feeding behaviours, consuming fungi, bacteria, protozoa, other nematodes, or plant tissues. Some nematodes also parasitize insects, animals, and humans. Among them, those that feed on plant tissues are classified as plant-parasitic nematodes (PPNs). PPNs infect a wide range of crop plants, including field crops, vegetables, fruits trees, turfgrass, ornamentals, and forest species (Bridge et al., 2005).

To date, more than 4,100 species of parasitic nematodes have been identified, although only a few genera are responsible for major agricultural damage (Poveda et al., 2020). They are generally considered as "unseen enemies" of plants and significantly reduce crop production, quality, and yield. PPNs primarily damage plant roots, disrupting water and nutrient uptake and resulting in symptoms like stunting, yellowing, wilting, and poor yields (Sikder & Vestergard, 2020). In addition to yield reductions, PPNs may also reduce crop quality and marketability due to visible deformities (Palomares-Rius et al., 2017). Other than inflicting direct damage on crops, PPNs frequently interact synergistically with other soil-borne pathogens, resulting in disease complexes that intensify crop losses. Among the PPNs, sedentary endoparasites like cyst nematodes (*Heterodera*, *Globodera* spp.) and root-knot nematodes (*Meloidogyne* spp.) are the most economically important species responsible for huge damage (Jones et al., 2013). The estimated annual yield loss due to PPN's is around $173 billion globally, surpassing the approximate losses of $70 billion caused by invasive insect pests (Elling, 2013; Bradshaw et al., 2016).

To mitigate these substantial losses, effective management of PPNs is of paramount importance. For decades, PPN management has relied mainly on synthetic chemicals and soil fumigants due to

[1] ICAR-Directorate of Floriculture Research, Regional Station, Rajahmundry-533125, Andhra Pradesh, India.
[2] ICAR-Central Citrus Research Institute, Nagpur-440033, Maharashtra, India.
[3] ICAR Indian Institute of Rice Research, Hyderabad-500030, Telangana, India.
[4] Department of Biology, West Virginia State University Institute, WV, 25112-1000, USA.
[5] ICAR-Central Potato Research Institute, Regional Station, Ooty-643004, The Nilgiris, Tamil Nadu, India.
* Corresponding author: priyank.mhatre-icar@nic.in

DOI: 10.1201/9781003614869-8

their fast and reliable action. However, concerns over environmental pollution, human health risks, and non-target toxicity have led to stricter regulations and the phasing out of many conventional nematicides from use. As a result, the focus is now shifting towards sustainable and eco-friendly alternatives. These include the use of bionematicides, microbial biocontrol agents, cultural practices, and resistant crop varieties. Such approaches are safer, environmentally sustainable, and compatible with organic and climate-resilient farming systems.

Considering the increasing regulatory restrictions on chemical nematicides and the growing awareness of sustainable agricultural practices, there is a pressing need to explore and implement eco-friendly alternatives for nematode control. Therefore, this chapter focuses on sustainable management strategies for plant-parasitic nematodes, aiming to provide an in-depth understanding of biological, cultural, and host resistance approaches. The present chapter highlights the significance, potential, and practical applicability of these sustainable methods in modern agriculture, ensuring long-term crop protection, soil health, and environmental safety.

2. Nematode Biology and Ecology

2.1 Life Cycle of PPNs

The life cycle of PPNs generally includes the egg, four juveniles, and the adult males or female stages (Figure 1). Among these, the second-stage juvenile (J2), emerging from the egg, is the principal infective stage in most species. In PPNs, the hatching is mediated by water (*Meloidogyne* spp.) or the chemical cues released by the host roots (*Globodera* spp.). After hatching, J2 moves across a thin water film on soil surface to locate and infect mostly the plant roots or in some cases the above-ground parts (Siddique & Grundler, 2018; El-Sagheer, 2019). After entering the root, J2 forms specialized feeding sites and develops into adult males and females upon three moults (Williamson & Hussey, 1996; Maleita et al., 2012). These specialized cells in root-knot nematodes are called "giant cells", in cyst nematodes "syncytia", and in the reniform nematode "nurse cell", which provides continuous nutrients and water supply from the host plant to the growing nematodes. Nematode life cycle and reproduction is dependent on optimal host conditions and environmental conditions such as temperature, humidity, soil moisture, soil type, pH, etc., (Abawi & Widmer, 2000). Generally, the nematode completes its life cycle in 5–8 weeks, with optimal temperature ranges between 22–27°C (Anderson, 2000) that helps to complete several generations in a crop season.

2.2 Survival Strategies

Nematodes adopt survival strategies to overcome biological threats such as predation, pathogens, and physical stresses including temperature extremes, water stress, pH fluctuation, osmotic and ionic stress, and anoxia in order to avoid death. The PPNs have different adaptation mechanisms at each stage. The eggs of root-knot nematodes (RKNs) are protected by a gelatinous matrix which prevents water loss and shields against microbes (Eisenback, 1985; Orion & Kritzman, 2001). In extreme temperatures, this mass hardens, creating mechanical pressure that inhibits the hatching of J2 larvae, which are vulnerable to drought (Wright & Perry, 2006). Potato cyst nematodes, *viz*., *Globodera pallida* and *G. rostochiensis* and *G. pallida* are well adapted to unfavourable conditions. Unhatched Juveniles survive in the cyst for several years without the host and hatching triggers only by root exudates (Perry, 2002; Turner & Rowe, 2006; Wright & Perry, 2006). Similar strategies are adopted in other PPNs such as *Heterodera glycines*, *H. carotae*, *Meloidogyne chitwoodi* and *M. triticoryzae* (Ishibashi et al., 1973; Greco, 1981; Ibrahim et al., 1993; Thompson & Tylka, 1997; Wesemael et al., 2006).

In the larval stages, PPNs J2s are more susceptible to environmental stresses and employ various strategies to survive, for example, *Anguina* spp (*A. tritici* and Orion *Anguina amsinckia*) inhabit the

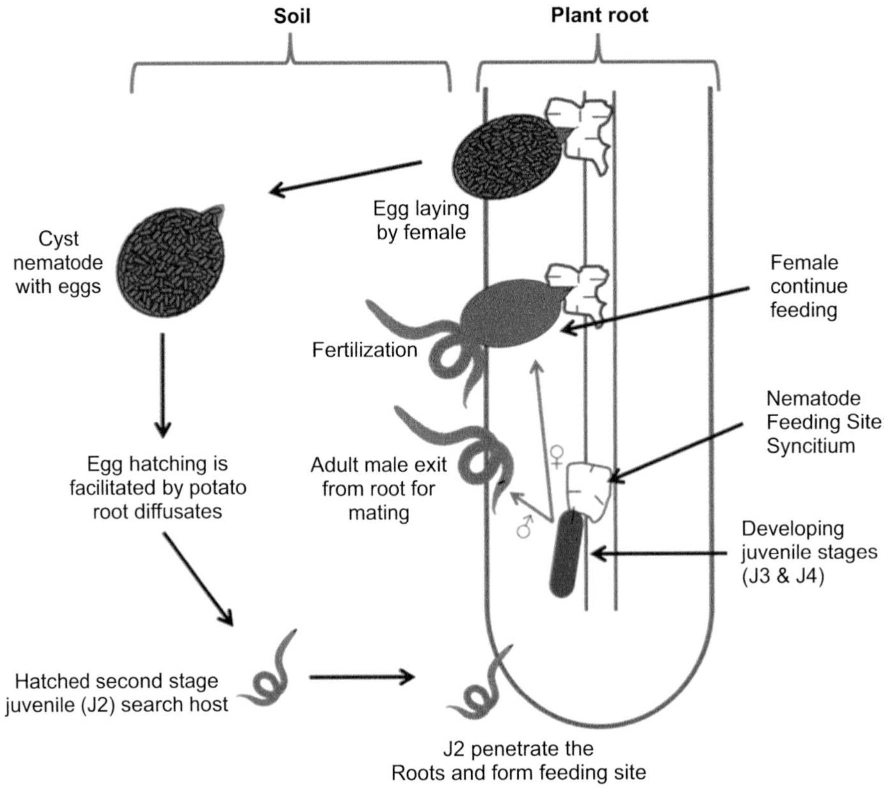

Fig. 1 Life cycle of potato cyst nematode, *Globodera* spp. depicting different life stages from eggs to adult stages.

aerial parts of cereals and forage grasses, and can remain viable for years in desiccated conditions. In contrast, *Ditylenchus dipsaci*, fourth juvenile stage (J4) forms dense aggregations in desiccated form (eelworm wool) (Ellenby, 1969). In *D. angustus*, there is no distinct survival stage; instead, juveniles (J3, J4) and adults are capable of persisting in a desiccated state on rice stems (Ibrahim & Perry, 1993). Certain PPNs' adult stages adopt survival strategies and withstand unfavourable conditions. Foliar nematode *Aphelenchoides besseyi* adults aggregate and cluster under rice grain hulls and can survive in dry, ripened grains for 2–3 years. Understanding the survival strategies will help to choose better management options against a particular PPN.

2.3 Interaction with Host Plants and Soil Microbiota

Soil is a dynamic ecosystem where plants, nematodes, and microbes interact closely, influencing plant health, crop yield, and soil fertility. Usually, PPN parasitize plants and reduce crop yield, but other groups such as free-living nematodes are beneficial as they help in nutrient cycling and microbial regulation. In the farming system, bacterivorous and predatory nematodes contribute 8–19% of nitrogen mineralization by grazing on decomposer microbes (Ingham et al., 1985; Beare et al., 1992; Wardle & Yeates, 1993).

PPNs actively interact within the rhizosphere, where they locate host plant roots by detecting chemical signals released from the roots. This process is facilitated by their chemoreceptors known as amphids (Perry, 1996). Specificity of chemical cues depends on the nematode species' host range (Reynolds et al., 2011). For *P. coffeae*, host compounds like cellulose and xylan influence the regulation of genes by encoding the cell wall and degrading enzymes required for parasitism (Bell et al., 2019).

In contrast, Microbial activity in the rhizosphere may modulate these interactions between plants and nematodes. Specific rhizosphere microbiome assemblages are shaped by the root exudates released by different plant species or genotypes (Badri & Vivanco, 2009; Monchgesang et al., 2016). When under attack, plants recruit beneficial microbes to their roots, enriching the rhizosphere with protective microbial communities (Bakker et al., 2018). A study discovered that nematode challenge in suppressive soils leads to enrichment of specific microbial taxa in both the rhizosphere and root endosphere, indicating a well-coordinated plant-microbiota response (Hussain et al., 2018). Here, the rhizosphere microbiota may act as a frontline defense, while the root endosphere offers a second layer of protection.

3. Nematode Management

3.1 *Challenges in Nematode Management*

Limitations of Conventional Control Methods (Chemical Nematicides)

1. **Non-selective nature:** Affects a wide range of soil organisms, including beneficial non-parasitic nematodes, microbes, and other fauna, disrupting soil health.
2. **Resistance Development:** Selective nematicides are more prone to induce resistance due to repeated use and similar modes of action.
3. **Microbial Degradation:** Few nematicides, particularly older organophosphates and carbamates, degrade more quickly over time due to microbial adaptation in the soil.
4. **Groundwater Contamination:** Toxic residues can leach into groundwater,. and pose serious threats to environment and human health.
5. **High Cost of Newer Nematicides:** New nematicides are expensive, limiting their accessibility.

4. Sustainable Nematode Management Strategies

4.1 *Cultural Practices*

4.1.1 Crop Rotation and Cover Cropping

These strategies play an important role in integrated management of nematodes as they help to decrease the density of plant parasitic nematodes. Crop rotation involving alternate crops such as non-hosts, poor hosts, resistant varieties, and antagonistic plants can effectively reduce soil populations of PPNs thereby minimizing crop damage (LaMondia, 1999).

Rotation with non-host crops is an effective cultural strategy for managing plant-parasitic nematodes. For instance, rotation with non-host crops such as corn, cucurbits, potatoes, and tomatoes effectively suppresses cyst nematodes (Heterodera spp.). Similarly, rotating with asparagus, corn, onions, garlic, small grains, Cahaba white vetch, or 'Nova' vetch helps minimize infestations of root-knot nematodes (Meloidogyne spp.). Other crops, including Crotalaria, velvet bean, soybean, and certain grasses such as rye, also exhibit resistance to root-knot nematodes and are therefore suitable for rotation (Wang et al., 2002). Rotation with resistant potato varieties has been shown to reduce Globodera rostochiensis inoculum levels by 90% or more (Brodie & Mai, 1989).

Cover crops, either grown between main crops or integrated with them, play an additional role in nematode management. They improve soil fertility and structure, reduce soil erosion, and suppress weeds, insects, nematodes, and other pathogens. Furthermore, residues from cover crops can be incorporated as "green manure", providing essential macro- and micronutrients that enhance soil fertility and support the subsequent crop. Important cover crops are cowpea, Sudangrass, sunn hemp, marigolds, jointvetch, grasses, rye, wheat, oats, and lupine. Incorporating "Iron Clay" cowpea (*Vigna unguiculata*) as cover crop and green manure enhanced turnip and basil yield by improving nitrogen-fixing properties and effectively reducing root-knot nematode problem (Wang et al., 2003a). Sunnhemp, as cover crop, reduced nematode populations such as reniform nematode, burrowing nematode, and root-knot nematode (Wang et al., 2002b; Jourand et al., 2004). Amendment of sunnhemp enhanced bacteriovorous and predatory nematodes and improved soil nutrient and organic matter content (Marshall, 2002)

4.1.2 Trap Crops

Trap crops are plants that attract nematodes to their roots. Once the nematodes enter plant roots, they are prevented from reaching maturity, either because the plant itself inhibits their development or the trap crop is destroyed before the nematodes can complete their life cycle, ultimately reducing the overall pest population. Trap crop may stimulate hatching and/or attract hatched juveniles and also may be penetrated by large numbers of nematodes that cannot feed (Trivedi & Barker, 1986). In the case of soybean cyst nematode (SCN), juveniles (J2) penetrated the roots of 31 plant species other than soybean but exhibited minimal to no further development within the roots (Riggs, 1987). In 2008, Chen and colleagues reported that SCN) juveniles hatched in greater numbers in response to chemical leachates from soils where red clover (*Trifolium pratense*) and sunn hemp (*Crotalaria juncea*) were grown, compared with leachates from fallow (nonplanted) sand or from soils planted with the nonhost cover crop corn (*Zea mays*). Similarly, certain cultivars of oilseed radish (*Raphanus sativus* var. *oleiferus*) have been shown to function as trap crops for *Heterodera schachtii* (sugar beet cyst nematode) (Hafez & Sundararaj, 2009). Sticky nightshade or litchi tomato (*Solanum sisymbriifolium*), a non-tuber-bearing solanaceous species, has also been identified as a promising resistant trap crop for potato cyst nematodes (PCN) (Scholte, 2000a, 2000b; Dandurand & Knudsen, 2016). More recently, Mhatre et al. (2021) demonstrated that *S. sisymbriifolium* can reduce populations of both PCN species by more than 97% in the subsequent potato cropping season.

4.1.3 Antagonistic Crops

Antagonistic crops are plant species that release chemicals or alkaloids that possess nematicidal/nematostatic compounds as root exudates which repel/suppress the plant parasitic nematodes. Plants such as Crotalaria, neem (*Azadirachta indica*), marigold (*Tagetes* spp.), and various *Brassica* species have been recognized for their antagonistic effects against PPNs (Chetia et al., 2019).

Marigold effectively reduces nematode populations by releasing a nematicidal compound called alpha-terthienyl. For sustainable nematode management, marigold should be densely planted as a cover crop at least two months before the main crop and cultivated annually for long-term control. Planting marigold has been proposed as an alternative to methyl bromide fumigation, particularly when integrated with other cultural, chemical, and biological management tactics (Noling & Becker, 1994). Species of *Tagetes*, such as *T. patula* and *T. erecta*, are recognized as antagonistic crops capable of suppressing plant-parasitic nematodes. French marigold (*T. patula*) has been shown to suppress lesion and root-knot nematodes, while also limiting the penetration and development of the reniform nematode (Cerruti et al., 2010). It shows a hypersensitive necrotic response to root-knot nematode penetration, preventing giant cell formation and disrupting the nematode's life cycle (Rangaswamy et al., 1993). African marigold suppressed root knot nematode juveniles and also acted as trap crop by hampering the development of J2s that entered inside the root (Ploeg & Maris, 1999).

Brassica species like Indian mustard (*Brassica juncea*) and rapeseed (*B. napus*) are effective in managing root-knot nematodes through a process called biofumigation (Lazzeri et al., 2004). When their plant parts are incorporated into the soil, they release glucosinolates that break down into isothiocyanates, natural compounds toxic to nematodes. Additionally, they act as trap crops by attracting juvenile nematodes and interrupting their life cycle.

4.2 Soil Amendments and Organic Matter Management

Organic amendments such as green manures, slurries, and composts contribute to nematode management in soil by releasing nematicidal compounds during decomposition, enhancing populations of natural enemies of nematodes, and improving plant growth and tolerance (Bonamoni et al., 2010; Thoden et al., 2011). Soil amendment with non-edible deoiled seed cakes of neem (*Azadirachta indica*), karanj (*Pongamia glabra*), mahua (*Madhuca indica*), and castor (*Ricinus communis*) not only improves soil fertility and organic matter status but also suppresses nematode populations through toxic decomposition products and stimulation of antagonistic microbial activity. Additionally, organic inputs derived from sewage sludge, livestock manure, poultry refuse, rice bran, tea waste, sawdust (fresh or dry), farm-waste compost, goat manure, poultry manure, and vermicompost have all been reported as effective in managing RKN. The detailed information on organic amendments along with their mode of actions is presented in Table 1.

Table 1 Effect of organic amendments on plant parasitic nematodes

Plant Name	*Mode of Action against Nematode*	*References*
Neem (neem cake)	Release of toxic metabolites	Ferraz & de Freitas (2004); Oka (2010)
Cruciferous plants	Toxic metabolites	Abdel-Baset et al. (2025)
Castor, Marigod, Velvetbean, Sunnhemp	Tannins and phenolic compounds	Kokalis-Burelle et al. (1994)
Chicken manure	Ammonia (NH_3) volatilization at high pH (>7.5)	Conn & Lazarovits (1999)
Brasicca seed meal	Glucosinolates	Mazzola et al. (2004)
Olive mill waste and its compost	Active antimicrobial and phytotoxic compounds	Cayuela et al. (2008)
Sewage sludge	Nitrogen compounds, organic acids are toxic metabolites	Castagnone-Sereno and Kermarrec (1991)
Chitin, Chitin-urea amendments	Releases Ammonical nitrogen toxic to nematodes Improves chitinolytic fungi (parasitize eggs of PPN's)	Rodriguez-Kabana et al. (1987)

4.3 Sanitation and Field Hygiene

Proper sanitation and field hygiene play a vital role in managing nematode infestations. Crop residues should be removed and destroyed because nematodes are capable of surviving in plant debris. Machinery and equipment from infested fields must be thoroughly cleaned and disinfected to prevent transferring nematodes into uninfested areas. Weed growth should also be controlled, as many weed species act as alternative hosts that sustain nematode populations.

4.4 Biological Control

Biological control of nematodes is defined as the reduction of nematode populations through the activity of living organisms, either naturally occurring, introduced as antagonists, or enhanced by environmental manipulation (Stirling, 1991). Biological suppression of plant pathogens by antagonistic microorganisms can occur either directly, through mechanisms such as antibiosis, competition for nutrients or ecological niches, and siderophore-mediated suppression, or indirectly via the host plant through induced resistance mechanisms such as systemic acquired resistance (SAR) and induced systemic resistance (ISR) (Mhatre et al., 2019).

For PPNs, the egg and second-stage juvenile (J2) are the most vulnerable stages for biological control. These stages remain outside the plant in the water film surrounding soil particles, where they are more readily exposed to antagonistic microorganisms. When these stages are successfully targeted through infection and parasitism by biocontrol agents, the nematode life cycle is disrupted, leading to a reduction in their population density and, ultimately, effective nematode management.

Fungi and bacteria are the most important bioagents for nematode management. Key antagonists include nematode-trapping fungi, endoparasitic fungi, egg-parasitic fungi, bacteria such as *Pasteuria*, plant growth–promoting rhizobacteria, and endophytes. Other organisms—including viruses, mites, collembola, turbellarians, oligochaetes, and protozoans—have also been reported, though their effectiveness is generally limited (Stirling, 2014).

4.4.1 Nematophagous Fungus

Nematophagous fungi are potential biocontrol agents capable of capturing, killing, and digesting nematodes (Hertz et al., 2011). These fungi inhabit both the external and internal environments of the host organism, deriving nutrition from it. Nematophagous fungi (NF) employ diverse strategies to attack nematodes. They use specialized traps to capture prey, penetrate nematode females and eggs with their hyphal tips, or produce conidia that adhere to the host surface and initiate infection. In addition, some species release toxins that immobilize nematodes, facilitating subsequent invasion and death. The detailed information on nematophagous fungi and their application as biocontrol agents is presented in Table 2.

Based on these strategies, NF are traditionally classified into four groups:

1. Fungi that capture nematodes using specialized trapping structures.
2. Egg-parasitic fungi that invade nematode eggs or females via hyphal penetration.
3. Endoparasitic fungi that infect nematodes through their spores.
4. Toxin-producing fungi that paralyse or kill nematodes prior to invasion (Abd et al., 2018).

Table 2 Nematophagous fungi and their application as biocontrol agents.

Nematophagous Fungi	*Application as Biocontrol Agent*	*Reference*
Arthrobotrys spp.	Prey free-living nematode (*Panagrellus* spp.) *in vitro*	Gomes, et al. (2001)
Duddingtonia flagrans	Use for management of parasitic nematodes in domestic animals and plants	Balbino et al. (2022)
Hirsutella rhossiliensis	Effectively control soybean cyst nematode, *Heterodera glycines*	Chen, et al. (2005)
Paecilomyces lilacinus	Infects eggs, juveniles, and females of *Meloidogyne javanica* and also immature cysts of H. avenae and Radopholus similis	Khan et al. (2006)
P. chlamydosporia	Control plant (semi-) endo-parasitic nematodes of the genera *Globodera, Heterodera, Meloidogyne, Nacobbus,* and *Rotylenchulus*	Lopez et al. (2013)
Verticillium chlamydosporium	Biological control agent for root-not nematodes (*Meloidogyne* spp.)	Bourne & Kerry (1998)
Purpureocillium lilacinum	Effective biocontrol agent against the root-knot nematode *M. incognita*	Khan & Tanaka, (2023); Sharma et al. (2020)
Trichoderma harzianum	Potentially control root-knot nematode *M. javanica*	Sahebani & Hadavi (2008); Heidari & Olia (2016)
Dactylellina haptotyla	Traps M. incognita by adhesive knobs and non-constricting loops in environment	Barron (1977); Lei et al. (2023)
Verticillium lecanii	Culture filtrates show toxicity against different stages of *H. glycines*	Shinya (2008)

4.4.2 Predatory Nematodes

Predatory nematodes include members of the orders Mononchida, Rhabditida, Diplogasterida, and Aphelenchida; the superfamilies Nygolaimoidea, Dorylaimoidea, and Actinolaimoidea; as well as the families Monohysteridae, Ironidae, Thalassogeneridae, and Oncholaimidae. The mode of action of the nematophagous fungus is detailed in Table 3. They are generally classified into three groups based on differences in feeding apparatus, prey-capturing methods, and feeding mechanisms:

1. **Slicers and Suckers:** Species that cut open the prey's body and suck out its contents, as they are unable to ingest the prey whole (Diplogasterida).
2. **Cutters, Suckers, and Occasional engulfers:** Species that combine cutting and sucking and sometimes consume the prey as a whole (Mononchida).
3. **Piercers and Suckers:** Species that puncture the prey's cuticle and withdraw body contents without consuming the prey whole (suborders Aphelenchida, Dorylaimina, and Nygolaimina).

Table 3 Mode of action of nematophagous fungus.

Fungus Classification	*Mode of Action and Examples*	*References*
Predatory Fungus	Employ hyphal structures	
a) Adhesive Branches	Adhesive branches act as mechanical traps that secrete a sticky substance to capture, immobilize, and eventually lead to the death of nematodes. *E.g., Dactylella cionopaga, Monacrosporium cionopagum M. gephyrophagum Gamsylella gephyropaga*	Saxena (2018); Zhang & Hyde (2014)
b) Adhesive Hyphal Network	Traps nematodes by forming three-dimensional, adhesive-coated hyphal nets that immobilize prey upon contact, enabling fungal invasion and nutrient absorption, e.g., *A. oligospora*	Martinelli et al., (2010); Niu et al. (2011)
c) Adhesive Knobs	Adhesive knobs are specialized structures which adhere to nematodes and facilitate infection through enzymatic degradation of the nematode's cuticle, e.g., *Dactylellina arcuata, D. asthenopaga, D. leptosphora, D. copepodii*, and *D. ellipsospora* use adhesive knobs to capture nematodes	Xie et al. (2010)
d) Constricting rings	Hyphomycetous fungi like *Drechslerella brochopaga* trap and kill nematodes using constricting ring structures that rapidly swell inward upon contact, triggered by physical or chemical stimuli and driven by water uptake followed by hyphal penetration and assimilation of the nematode. E.g., *D. brochopaga*	Zachariah, K. (1978); Rudek (1975)
f) Non-Constricting rings	Non-constricting rings trap nematodes by loosely encircling their bodies, often detaching from the stalk to remain attached to the host, thereby aiding both capture and fungal dispersal. E.g., *Dactylaria candida and D. lysipaga Dactylellina daliensis.*	Askary, et al. (2015) Su et al. (2008)
2 Egg- and Female-Parasitic Fungi	Fungi form appressoria to attach nematode eggs and mycelial branches penetrates, breaks through the egg and colonize internal tissues of egg and females E.g., *Dactylella ova parasitica, Helicocephalum oligosporum, Lecanicillium psallioatae, Nematophthora gynophila, Purpureocillium spp, Pochonia chlamydosporia, Glomus etunicatum*	Benedetti et al. (2021)

Contd.

Table 3 *Contd.*

Fungus Classification	*Mode of Action and Examples*	*References*
Endoparasitic Fungi	Fungal spores attach and penetrate to the nematode cuticle, absorb the nematode's internal contents Fungi multiply by producing conidia inside the nematode and may also release metabolites that inhibit egg hatching and kill nematodes E.g., *D. coniospora, H. rhossoliensis Drechmeria coniospora*	Braga & Araujo (2014)
Toxin Production	Fungi secrete inhibitory metabolites like fatty acids, lactones, and pyrones that immobilize nematodes, after which their hyphae penetrate the nematode cuticle using enzymes such as proteases and chitinases, disrupt egg development by altering composition, and ultimately consume the nematode body as a nutrient source to complete their lifecycle. E.g., *Coprinus comatus and Stropharia rugosoannulata*	Perez et al. (2021)
Enzymes of NF	Nematophagous fungi (NF) use enzymes like proteases and chitinases, along with mechanical pressure, to penetrate nematode cuticles and eggshells, leading to infection and degradation. Species such as Arthrobotrys oligospora and Monacrosporium thaumasium produce potent serine proteases and chitinases that exhibit strong nematicidal activity against various nematodes and their eggs. *Arthrobotrys oligospora, Haemonchus contortus, Monacrosporium thaumasium*	Yang et al. (2022)

4.4.3 Use of Endophytes and PGPR

Plant Growth-Promoting Rhizobacteria (PGPR) are a diverse group of soil-inhabiting bacteria that colonize plant roots, enhance growth, induce broad-spectrum resistance against pests and pathogens, and improve crop productivity across a wide range of species. Lugtenberg & Kamilova (2009) classified PGPR into four functional groups: biofertilizers (increasing nutrient availability to plants), phytostimulators (promoting plant growth through phytohormone production), rhizoremediators (degrading organic pollutants), and biopesticides (suppressing diseases via antibiotics and antifungal metabolites).

Several research efforts have screened rhizosphere bacteria for antagonistic activity against nematodes (Becker et al., 1988). Isolates of *Agrobacterium radiobacter*, *Bacillus subtilis*, and *Pseudomonas* spp.—already well known as antagonists of soil-borne bacterial and fungal pathogens—have also demonstrated potential as biological control agents against plant-parasitic nematodes. PGPR enhances plant growth through both direct and indirect mechanisms. Directly, they stimulate growth by producing phytohormones such as auxins, cytokinins, and gibberellins, and by improving nutrient availability through processes like nitrogen fixation and phosphorus solubilization. Indirectly, PGPR contributes to plant health by suppressing the harmful effects of plant pathogens. This is achieved through the production of inhibitory compounds (such as antibiotics, antifungal metabolites, siderophores, and cell wall-degrading enzymes), competition for root colonization sites, and by triggering the plants own defense systems through induced systemic resistance (Mhatre et al., 2019).

Several species of the genus *Bacillus* are known to produce cytotoxins and other bioactive compounds that enhance their effectiveness against nematodes (Mejia & Gundy, 1982). The beneficial effects of PGPR are often attributed to their ability to colonize plant roots and exert biocontrol activity against nematodes and other pathogens. Several studies have thoroughly described the mode of action of *P. fluorescens*. *Pseudomonas fluorescens* enhances the production of antibiotics and modifies root exudates, including polysaccharides and amino acids, which subsequently influence nematode behaviour and inhibit early root penetration (Siddiqui et al., 2006). In tomato, a significant reduction in root-knot nematode (*Meloidogyne incognita*) infection has been documented following its application (Cetintas, 2018).

Various *Bacillus* species reported to directly antagonize PPNs (Liu & Sinclair, 1992; Yu et al., 2015). This antagonism may also activate induced systemic resistance (ISR), thereby reducing disease severity caused by a broad spectrum of pathogens and promoting plant growth under field conditions (Kloepper et al., 2004). In banana, *Pseudomonas fluorescens* and *Bacillus subtilis* have both been shown to induce systemic resistance against lesion nematodes (Shanthi & Rajendran, 2003).

A comprehensive review by Saharan and Nehra (2011) highlighted the diverse mechanisms and multifaceted effects of PGPR on a variety of plant types, including ornamentals, forest crops, vegetables, and fruit crops. Beneduzi et al. (2012) demonstrated that ISR triggered by rhizobacteria closely resembles pathogen-induced systemic acquired resistance (SAR), which enhances the resistance of uninfected plants to pathogens across many species. They also proposed that PGPR-induced resistance may be mediated through either the salicylic acid-dependent SAR pathway or by requiring the plant's perception of jasmonic acid and ethylene.

Pseudomonas and *Bacillus* species are especially noted for their antagonistic properties and their capacity to activate ISR. These effects may stem from their production of hydrolytic enzymes, competition for nutrients, and effective colonization of root surfaces. Furthermore, PGPR can regulate plant ethylene levels through the enzyme ACC-deaminase, helping plants cope with stress induced by infections. Their antagonistic activities are also linked to the production of siderophores, antibiotics, phytohormones, and compounds involved in quorum sensing degradation, as well as signaling molecules like indole-3-acetic acid and cytokinins (Wani, 2015). The information on PGPR such as targeted PPN, mode of action and the organism used in suppression of PPN is detailed in Table 4.

Table 4 Plant Growth Promoting Rhizobacteria used in suppression of plant parasitic nematodes.

PGPR	*Target PPN*	*Mode of Action*	*References*
Pasteuria pentrans	*Aphelenchoides besseyi, G. rostochiensis, M. incognita, Pratylenchus penetrans, Radopholus similis*	Parasitism	Ciancio (2018); Abd-Elgawad & Askary (2018); Tian et al. (2007)
Agrobacterium radiobacter (G12)	*Meloidogyne* spp.	Induced systemic resistance	Nikoo et al. (2014)
Bacillus cereus	*M. javanica M. incognita,*	Induced systemic resistance	Jiang et al. (2020)
B. amyloliquefaciens	*M. incognita*	Induced systemic resistance and systemic acquired resistance	Xie et al. (2018)
B. amyloliquefaciens FR203A	*Xiphinema index*	Lipases	Köhl et al. (2019)
B. thuringiensis	*M. incognita*	Bt crystal protein (toxin protein), Thuringiensin (exotoxin)	Mohammed et al. (2008)
B. thuringiensis	*M. hapla*	Cry6Aa2 protoxin	Radhakrishnan et al. (2017)
B. thuringiensis	*Meloidogyne* spp.	Cry5B	Ghahremani et al. (2020)

Contd.

Table 4 *Contd.*

PGPR	*Target PPN*	*Mode of Action*	*References*
B. thuringiensis	*B. xylophilus*	Cry1Ea11	Nascimento et al. (2013)
B. firmus	*M. incognita*	Sep 1 protease Secondary metabolites	Geng et al. (2016)
B. coagulans	*M. incognita*	Hydrolytic enzymes	Xiang et al. (2018)
B. licheniformis	*B. xylophilus*	Protease, Chitinase	El-Nagdi et al. (2019)
B. subtilis	*M. javanica,*	Lipopeptide antibiotics, Hydrolytic enzymes, Secondary metabolites	Basyony & Abo-Zaid (2018)
B. pumilus	*M. incognita*	Induced systemic resistance and systemic acquired resistance	Lastochkina et al. (2019)
B. mojavensis	*M. incognita*	Induced systemic resistance	Xiang et al. (2017)
Bacillus sp.	*M. javanica*	Antibiotics, siderophores, changes in root exudates, and induced resistance	Turatto et al. (2018)
Pasteuria penetrans	*Meloidogyne spp.*	Predation	Bhuiyan et al. (2018)
Pseudomonas fluorescens	*M. incognita*	Sphingosine, Antibiotic production, Chitinase, Protease	Abd El-Aal et al. (2021)
P. fluorescens	*G. rostochiensis*	2,4 diacetylphloroglucinol (DAPG)	Siddiqui & Shaukat (2004)
P. aeruginosa	*M. javanica*	Induced systemic resistance and systemic acquired resistance	Fatima & Anjum (2017)
P. fluorescens	*M. incognita*	Sphingosine, Antibiotic production, Chitinase, Protease	Ali et al. (2021)
P. putida	*M. incognita*	Induced systemic resistance	Almaghrabi et al. (2013)
P. fluorescens	*M. incognita*	Antibiosis, nematicidal toxins, parasite of nematode eggs and adult females, nematode juvenile inside the egg is destroyed by the rapidly growing hyphae	Bharali et al. (2019)
P. stutzeri	*M. incognita*	Hydrogen cyanide (HCN)	Khan et al. (2016)
P. Fluorescent	*M. javanica*	antibiotics, siderophores, changes in root exudates, and induced resistance	Turatto et al. (2018)
P. aeruginosa	*M. incognita, M. javanica*	HCN	Singh & Siddiqui (2010)
P. fluorescens Wood1R	*M. incognita*	2,4-diacetylphloroglucinol (DAPG)	Timper et al. (2009)

Contd.

Table 4 *Contd.*

PGPR	*Target PPN*	*Mode of Action*	*References*
Serratia marcescens	*M. incognita*	Induced systemic resistance	Almaghrabi et al. (2013)
S. marcescens	*M. incognita*	Sphingosine, Antibiotic production, Chitinase, Protease	Ali et al. (2021)
S. marcescens	*M. javanica*	Volatile metabolites, Prodigiosin	Rahul et al. (2014
S. marcescens	*M. incognita*	ISR	Almaghrabi et al. (2013)
S. marcescens	*M. incognita*	Sphingosine, Antibiotic production, Chitinase, Protease	Abd El-Aal et al. (2021)
Rhizobium etli	*Meloidogyne spp.*	Induced systemic resistance	Reitz et al. (2000)
Brevibacillus laterosporus	*B. xylophilus*	Proteases	Abd-Elgawad & Askary (2018)
Photorhabdus luminescens	*A. besseyi*	Toxin A and toxin B, antibiotics	Tülek et al. (2018)

4.5 Resistant and Tolerant Crop Varieties

Utilizing resistant and tolerant crop varieties is a key strategy in sustainable nematode management (Boerma & Hussey, 1992; Fuller et al., 2008; Spychalla & De Jong, 2024). Using resistant crop varieties offers a highly cost-effective and eco-friendly strategy for managing nematodes (Boerma & Hussey, 1992; Fuller et al., 2008). This approach holds immense promise, especially for small and marginal farmers, as it requires no additional inputs once resistant seeds are sown. By ensuring the availability of high-yielding, nematode-resistant varieties through quality seed systems, we can empower resource-poor farmers to sustainably protect their crops, improve productivity, and reduce dependence on chemical nematicides.

4.5.1 Breeding for Nematode Resistance

Traditional breeding programmes continue to exploit natural resistance conferred by R genes (Fuller et al., 2008).. Monogenic resistance has often proven more successful and easier to manage in breeding programmes compared to oligogenic resistance, due to its simpler inheritance pattern and stronger, more specific defence response against target nematodes (Wang et al., 2023). Plant resistance has been mainly identified and exploited against highly specialized parasitic nematodes, including *Meloidogyne*, *Globodera*, *Heterodera*, *Rotylenchulus*, *Ditylenchus*, *Tylenchulus*, etc., which display sedentary endoparasitic behaviour during at least part of their life cycle.

Conventional breeding for nematode resistance involves identifying resistant germplasm and introgressing resistance genes into elite cultivars. Over the decades, host resistance has been successfully deployed in crops such as potato (Gro 1–4, Gpa2 against *Globodera spp.*) tomato (*Mi-1 gene* against *Meloidogyne spp.*) and soybean (*Rhg1* and *Rhg4* against *Heterodera glycines*), etc.

Screening of germplasm collections, landraces, and wild relatives has been instrumental in identifying novel resistance sources. Identification of a strong resistance gene is a key to success in breeding for nematode resistance (Wang et al., 2023). Phenotypic screening, although labour-intensive and environment-dependent, remains essential. Resistance is often inherited as a dominant or polygenic trait, with variability depending on the nematode species and host genotype. Currently, marker-assisted selection (MAS) and QTL mapping are expediting the identification and transfer of nematode resistance loci.

Breeding programmes today leverage molecular markers to track resistance genes during backcrossing and pyramiding (Nadeem et al., 2018). In some crops like rice and potato, where resistance is quantitative, genomic selection and association mapping using diverse panels (e.g., rice 3K genome panel) offer promising tools for predicting resistance traits (Dimpka et al., 2016; Obata et al., 2022). Despite these advances, breeding for resistance is often constrained by linkage drag, nematode pathotype variability, and breakdown of resistance under high pressure or stress conditions (Bimpong et al., 2010; De Carvalho et al., 2015). Although several crop varieties resistant to nematodes have been reported from various countries, cultivars specifically bred for nematode resistance remain limited. As an example, the ICAR-Central Potato Research Institute, Shimla, India, has released potato cultivars such as Kufri Swarna, Kufri Sahyadri, and Kufri Neelima, which are resistant to potato cyst nematodes (Joseph et al., 2019; Mhatre et al., 2025).

4.5.2 Genetic Modification and Gene-Editing Approaches

Genetic engineering and gene editing offer precise tools to develop nematode-resistant crops beyond the limitations of conventional breeding (Abdallah et al., 2015; Dutta et al., 2023; Sirohi & Dutta, 2024). Transgenic approaches have included the expression of anti-nematode proteins like proteinase inhibitors, lectins, chitinases, and cystatins, which interfere with nematode digestion or development (Atkinson et al., 2012). RNAi based gene silencing has emerged as a targeted method where host plants express double-stranded RNAs to knock down nematode effector genes, disrupting nematode parasitism (Banerjee et al., 2017).

The CRISPR/Cas9 genome editing now enables precise modifications in plant genomes to knock out susceptibility (S) genes or enhance expression of resistance genes. Editing regulatory elements to induce defense gene expression or engineer broad-spectrum resistance is an emerging frontier. A recent successful example includes the CRISPR-mediated loss-of-function of a *GmSNAP02* gene confer resistance to *Heterodera glycines* soybean cyst nematode (Usovsky et al., 2023). CRISPR/Cas9-based knockout of susceptibility (S) genes offers a promising approach for the management of PPNs. However, no single S gene has been identified that confers complete resistance to nematodes in any crop. Therefore, achieving absolute resistance through this method remains a distant goal.

4.5.3 Challenges and Future Prospects in Resistant Variety Development

While host resistance remains a pivotal strategy in nematode management, several challenges hinder its widespread deployment. Key limitations include:

- **Nematode Diversity and Pathotype Evolution:** Nematodes exhibit significant genetic diversity, with the ability to overcome single-gene resistance over time, as seen with *Mi-1*-resistant tomatoes. Pyramiding multiple resistance genes and breeding for durable, quantitative resistance is needed.
- **Limited Resistance Sources:** For many crops and nematode species, resistance sources are rare or poorly characterized. This limits the base material for breeding or genetic engineering.
- **Complex Host–Nematode Interactions:** In numerous crops, the molecular basis of resistance has yet to be fully elucidated. Resistance can be environment-dependent, making it less reliable under variable field conditions.
- **Phenotyping Bottlenecks:** Accurate and high-throughput phenotyping for nematode resistance remains a major constraint.

Speed breeding, doubled haploids, and gene editing can accelerate resistant variety development. A multi-pronged approach involving conventional breeding, biotechnology, and system-level understanding of plant–nematode interactions will be key to success in the coming decades.

5. Integrated Nematode Management

Integrated Nematode Management (INM) is a holistic approach that combines several cultural, biological, physical, and chemical strategies to manage PPN in a sustainable and environmentally friendly manner. Rather than relying on a single control method, INM focuses on long-term suppression of nematode populations below economic thresholds by integrating various compatible practices.

Several case studies demonstrate the efficacy of INM strategies in controlling PPN across various crops. Ashraf and Khan (2010) reported that in brinjal, maximum plant growth and the most effective suppression of *M. javanica* were achieved by integrating *P. lilacinus* with groundnut cake, followed by neem cake, linseed cake, castor cake, and mahua cake. Similarly, Muthulakshmi et al. (2012) found that the combined application of *P. chlamydosporia* with *P. fluorescens*, *T. viride*, and carbofuran significantly enhanced plant growth and reduced *Globodera* spp. populations in potato, resulting in a 70.57% increase in tuber weight and a 71.93% reduction in cyst density. In rice, Narasimhamurthy et al. (2017) demonstrated that applying *T. viride* together with carbofuran improved plant growth and grain yield, while reducing *M. graminicola* populations by 69.17%. More recently, Singh et al. (2020) reported that organic amendments such as neem cake, neem seed powder, and vermicompost, when combined with the biocontrol agent *T. harzianum*, significantly suppressed *M. javanica* populations in chickpea compared with untreated controls.

INM offers significant economic and environmental benefits by promoting sustainable crop protection practices. Economically, INM enhances crop productivity and profitability by incorporating cost-effective methods like crop rotation, organic amendments, resistant cultivars, and biocontrol agents. Environmentally, INM minimizes the negative impacts of synthetic nematicides on soil health, non-target organisms, and water resources. By fostering soil biodiversity and promoting the use of renewable, biodegradable inputs such as botanicals and bioagents, INM contributes to ecological balance and long-term agricultural sustainability. Additionally, reduced chemical residues in food and soil enhance food safety and environmental quality, aligning with global goals for environmentally responsible agriculture.

References

Abawi, G.S. and Widmer, T.L. (2000). *Impact of Soil Health Management Practices on Soilborne Pathogens, Nematodes, and Root Diseases of Vegetable Crops*. ResearchGate.

Abdallah, N.A., Prakash, C.S. and McHughen, A.G. (2015). Genome editing for crop improvement: Challenges and opportunities. *GM Crops and Food*, *6*(4), 183–205.

Abd-Elgawad, M.M. (2021). Optimizing safe approaches to manage plant-parasitic nematodes. *Plants*, *10*(9), 1911.

Abd-Elgawad, M.M. and Askary, T.H. (2018), Fungal and bacterial nematicides in integrated nematode management strategies. *Egypt. J., Biol. Pest Control*, *28*, (1), 1–24.

Abdel-Baset, S.H., Eltamany, E.E., Elhady, S.S. and Abdelrazik, E. (2025). Nematicidal Activity of White Mustard Seeds as Promising Biofumigants for Control Root-Knot Nematodes, Meloidogyne spp., Infecting Solanum lycopersicum L. under Field Conditions. *ACS Agricultural Science & Technology*, *5*(8), pp.1753–1766.

Almaghrabi, O.A., Massoud, S.I. and Abdelmoneim, T.S. (2013). Influence of inoculation with plant growth promoting rhizobacteria (PGPR) on tomato plant growth and nematode reproduction under greenhouse conditions. *Saudi J. Biol. Sci.*, *20*(1), 57–61.

Anderson, R.C. (2000). *Nematode Parasites of Vertebrates: Their Development and Transmission*. CABI.

Ashraf, M.S. and Khan, T.A. (2010). Integrated approach for the management of *Meloidogyne javanica* on eggplant using oil cakes and biocontrol agents. *Archives of Phytopathology and Plant Protection*, *43*, 609–614.

Askary, T.H. and Martinelli, P.R.P. (2015). *Biocontrol Agents of Phytonematodes*. CABI: Egham, UK.

Atkinson, H.J., Lilley, C.J. and Urwin, P.E. (2012). Strategies for transgenic nematode control in developed and developing world crops. *Current Opinion in Biotechnology*, *23*(2), 251–256.

Badri, D.V. and Vivanco, J.M. (2009). Regulation and function of root exudates. *Plant, Cell, and Environment*, *32*(6), 666–681.

Bakker, P.A.H.M., Berendsen, R.L., Doornbos, R.F., Wintermans, P.C.A. and Pieterse, C.M.J. (2018). The rhizosphere revisited: Root microbiomics. *Frontiers in Plant Science*, *9*, 841.

Balbino, H.M., de Souza Gouveia, A., Monteiro, T.S.A., Morgan, T. and de Freitas, L.G. (2022). Overview of the nematophagous fungus Duddingtonia flagrans. *Biocontrol Sci. Technol.*, *32*, 911–929.

Banerjee, S., Banerjee, A., Gill, S.S., Gupta, O.P., Dahuja, A., Jain, P.K. and Sirohi, A. (2017). RNA interference: a novel source of resistance to combat plant parasitic nematodes. *Frontiers in plant science, 8*, p.834.

Barron, G.L. and Thorn, R.G. (1987). Destruction of nematodes by species of Pleurotus. *Can. J. Bot.*, *65*, 774–778.

Beare, M. H., Parmelee, R.W., Hendrix, P.F., Cheng, W., Coleman, D.C. and Crossley, D.A. Jr. (1992). Microbial and Faunal Interactions and Effects on Litter Nitrogen and Decomposition in Agroecosystems. *Ecological Monographs*, *62*(4), 569–591.

Becker, J.O., E. Zavaleta-Mejia, S.F. Colbert, M.N. Schroth, A. Weinhold, J.G. Hancock, and S.D. Van Gundy. (1988). Effects of rhizobacteria on root-knot nematodes and gall formation. *Phytopathology 78*: 1466–1469.

Bell, C.A., Lilley, C.J., McCarthy, J., Atkinson, H.J. and Urwin, P.E. (2019). Host-induced expression of cell wall-degrading enzyme genes in the plant-parasitic nematode *Pratylenchus coffeae*. *Molecular Plant Pathology, 20*(4), 495–504. https://doi.org/10.1111/mpp.12775.

Benedetti, T., Antoniolli, Z.I., Sordi, E., Carvalho, I.R. and Bortoluzzi, E.C. (2021). Use of the Glomus etunicatum as biocontrol agent of the soybean cyst nematode. *Res. Soc. Dev.*, *10*, e7310615132.

Bimpong, I.K., Carpena, A.L., Mendioro, M.S., Fernandez, L., Ramos, J., Reversat, G. and Brar, D.S. (2010). Evaluation of *Oryza sativa* x *O. glaberrima* derived progenies for resistance to root-knot nematode and identification of introgressed alien chromosome segments using SSR markers. *African Journal of Biotechnology*, *9*(26), 3988–3997.

Boerma, H.R. and Hussey, R.S. (1992). Breeding plants for resistance to nematodes. *Journal of Nematology*, *24*(2), 242.

Bonamoni, G., Antignani , V., Capodilupo, M. and Scala, F. (2010). Identifying the characteristics of organic soil amendments that suppress soil-borne plant diseases. *Soil Biology and Biochemistry*, *42*, 136–144.

Bourne, J. and Kerry, B. (1998). Effect of the host plant on the efficacy of Verticillium chlamydosporium as a biological control agent of root-knot nematodes at different nematode densities and fungal application rates. *Soil Biol. Biochem.*, *31*, 75–84.

Bradshaw, C.J.A., Leroy, B., Bellard, C., Roiz, D., Albert, C., Fournier, A., Barbet-Massin, M., Salles, J.-M., Simard, F. and Courchamp, F. (2016). Massive yet grossly underestimated global costs of invasive insects. *Nature Communications, 7*, 12986.

Braga, F.R. and de Araújo, J.V. (2014). Nematophagous fungi for biological control of gastrointestinal nematodes in domestic animals. *Appl. Microbiol. Biotechnol.*, *98*, 71–82.

Bridge, J., Luc, M. and Plowright, R.A. (2005). *Plant Parasitic Nematodes in Subtropical and Tropical Agriculture* (2nd Edn.). CABI Publishing.

Brodie, B.B. and Mai, W.F. (1989). Control of the golden nematode in the United States. *Annual Review of Phytopathology*, *27*(1), 443–461.

Castagnone-Sereno, P. and Kermarrec, A. (1991). Invasion of tomato roots and reproduction of Meloidogyne incognita as affected by raw sewage sludge. *Supplement to the Journal of Nematology*, *23*, 724–728.

Cayuela, M.L., Millner, P.D., Meyer, S.L.F. and Roig, A. (2008). Potential of olive mill waste and compost as biobased pesticides against weeds, fungi, and nematodes. *Sci. Total Environ.*, *399*(1–3), 11–18.

Cerruti, R.R., Hooks, Koon-Hui Wang, Antoon Ploeg and Robert McSorley. (2010). Using marigold (Tagetes spp.) as a cover crop to protect crops from plant-parasitic nematodes. *Applied Soil Ecology*, *46*(3), 307–320.

Chen, S., Jhonson, G.A., Warnke, S.A., Wyse, D.L. and Porter, P.M. (2008). Effect of rotation crops on hatch, viability, and development of *Heterodera glycines, Nematology*, *10*, 869–882.

Chen, S. and Liu, X. (2005). Control of the soybean cyst nematode by the fungi Hirsutella rhossiliensis and Hirsutella minnesotensis in greenhouse studies. *Biol. Control*, *32*, 208–219.

Chetia, M., Phukan, M. and Bora, L.C. (2019). Antagonistic plants for the management of plant parasitic nematodes: A review. *Journal of Pharmacognosy and Phytochemistry, 8*(1), 2152–2155.

Ciancio, A. (2018). Biocontrol potential of Pasteuria spp. for the management of plant parasitic nematodes. *CAB Rev.*, *13*, 1–13.

Conn, K.L. and Lazarovits, G. (1999). Impact of animal manures on Verticillium wilt, potato scab, and soil microbial populations. *Can. J. Plant Pathol.*, *21*, 181–192.

Dandurand, L.M. and G.R. Knudsen. (2016). Effect of the trap crop *Solanum sisymbriifolium* and two biocontrol fungi on reproduction of the potato cyst nematode, *Globodera pallida*. *Annals Appl. Biol.*, *169*, 180–189.

de Carvalho, L.M., Benda, N.D., Vaughan, M.M., Cabrera, A.R., Hung, K., Cox, T., ... and Teal, P.E. (2015). Mi-1-mediated nematode resistance in tomatoes is broken by short-term heat stress but recovers over time. *Journal of Nematology*, *47*(2), 133.

Dimkpa, S.O., Lahari, Z., Shrestha, R., Douglas, A., Gheysen, G. and Price, A.H. (2016). A genome-wide association study of a global rice panel reveals resistance in *Oryza sativa* to root-knot nematodes. *Journal of Experimental Botany*, *67*(4), 1191–1200.

Dutta, T. K., Ray, S. and Phani, V. (2023). The status of the CRISPR/cas9 research in plant–nematode interactions. *Planta, 258*(6), 103.

Eisenback, J.D. (1985). Detailed morphology and anatomy of secondstage juveniles, males, and females of the genus Meloidogyne(root-knot nematodes). *An Advanced Treatise on Meloidogyne, 1*, 47–77.

Elling, A.A. (2013). Major emerging problems with minor Meloidogyne species. *Phytopathology, 103*, 1092–1102.

El-Sagheer, A.M. (2019). Plant responses to Phytonematodes infestation. In: *Plant Health Under Biotic Stress*. Springer, Singapore, pp. 161–175.

Ferraz, S. and de Freitas, L.G. 2004. Use of antagonistic plants and natural products. In: Z.X. Chen, S.Y. Chen and D.W. Dickson (Eds.), *Nematology Advances and Perspectives*. Beijing, China: Tsinghua University Press, pp. 931–977.

Fuller, V.L., Lilley, C.J. and Urwin, P.E. (2008). Nematode resistance. *New Phytologist, 180*(1), 27–44.

Geng. C., Nie, X., Tang, Z., Zhang, Y., Lin, J., Sun, M., et al. (2016). A novel serine protease, Sep1, from Bacillus firmus DS-1 has nematicidal activity and degrades multiple intestinal-associated nematode proteins. *Sci. Rep., 6*(1), 1–12.

Ghahremani, Z., Escudero, N., Beltrán-Anadón, D., Saus, E., Cunquero, M., Andilla, J., et al. (2020). Bacillus firmus strain I-1582, a nematode antagonist by itself and through the plant. *Front Plant Sci., 11*, 796.

Gomes, A., Vasconcellos, R., Ramos, M., Guimarães, M., Yatsuda, A., Vieira-Bressan, M. (2001). *In vitro* interaction of Brazilian strains of the nematode-trapping fungi Arthrobotrys spp. on Panagrellus sp. and Cooperia punctata. *Memórias Do Inst. Oswaldo Cruz, 96*, 861–864.

Greco, N. (1981). Hatching of *Heterodera carotae* and *H. avenae*. *Nematologica, 27*, 366–371.

Heidari, F. and Olia, M. (2016). Biological control of root-knot nematode, Meloidogyne javanica, using vermicompost and fungus Trichoderma harzianum on tomato. *Iran. J. Plant Pathol., 52*, 109–124.

Hertz, N.B., Jansson, H.-B. and Tunlid, A. (2011). Nematophagous Fungi. In: ELS. Wiley Online Library: Hoboken, NJ, USA.

Hussain, M., Tian, M.Y., Ahmed, S. and Zheng, J.W. (2018). Suppressive soils and role of rhizosphere microbiome in disease suppression. *Applied Soil Ecology, 123*, 513–526. https://doi.org/10.1016/j.apsoil.2017.09.024.

Ibrahim, S.K. and Perry, R.N. (1993). Desiccation survival of the rice stem nematode *Ditylenchus angustus*. *Fundamental and Applied Nematology, 16*, 31–38.

Ingham, R.E., Trofymow, J.A., Ingham, E.R. and Coleman, D.C. (1985). Interaction of bacteria, fungi, and their nematode grazers: Effects on nutrient cycling and plant growth. *Ecological Monographs, 55*(1), 119–140.

Ishibashi, N., Kondo, R., Muraoka, M. and Yokoo, T. (1973). Ecologoical significance of dormancy in plant parasitic nematodes. Ecological difference between eggs in gelatinous matrix and cysts of Heterodera glycines Ichinohe (Tylenchida:Heteroderidae) *Applied Entomology and Zoology, 8*, 53–63.

Jiang, C., Fan, Z., Li, Z., Niu, D., Li, Y., Zheng, M., et al. (2020). Bacillus cereus AR156 triggers induced systemic resistance against Pseudomonas syringae pv. tomato DC3000 by suppressing miR472 and activating CNLs-mediated basal immunity in Arabidopsis. *Mol. Plant Pathol., 21*(6), 854–870.

Jones, J.T., Haegeman, A., Danchin, E.G., Gaur, H.S., Helder, J., Jones, M.G., et al. (2013). Top 10 plant-parasitic nematodes in molecular plant pathology. *Mol. Plant Pathol. 14*, 946–961.

Joseph, T.A., Sudha, R., Venkatasalam, E.P., Aarti, B., Umamaheswari, R., Divya, K.L. and Priyank, H.M. (2019). Variety Kufri Sahyadri. *Indian. J. Genet., 79*, 517–518.

Jourand, P., Giraud, E., Bena, G., Sy, A., Willems, A., Gillis, M., Dreyfus, B. and de Lajudie, P. (2004). Methylobacterium nodulans sp. nov., for a group of aerobic, facultatively methylotrophic, legume root-nodule-forming and nitrogen-fixing bacteria. *International Journal of Systematic and Evolutionary Microbiology, 54*(6), pp. 2269–2273.

Khan, M. and Tanaka, K. (2023). Purpureocillium lilacinum for plant growth promotion and biocontrol against root-knot nematodes infecting eggplant. *PLoS One, 18*, e0283550.

Kloepper, J.W., Ryu, C.M. and Zhang, S. (2004). Induced systemic resistance and promotion of plant growth by Bacillus spp. *Phytopathology, 94*(11), pp.1259–1266.

Köhl, J., Kolnaar, R. and Ravensberg, W.J. (2019). Mode of action of microbial biological control agents against plant diseases: \relevance beyond efficacy. *Front. Plant Sci., 10*, 845.

Kokalis-Burelle, N., Rodriguez-Kabana, R., Weaver, C.F. and King, P.S. (1994). Evaluation of powdered pine bark for control of Meloidogyne arenaria and Heterodera glycines on soybean. *Plant and Soil, 162*, 163–168.

LaMondia, J.A. (1999). Influence of rotation crops on the strawberry pathogens *Pratylenchus penetrans*, *Meloidogyne hapla*, and *Rhizoctonia fragariae*. *J. Nematol., 31*(4S), 650.

Lastochkina, O., Seifikalhor, M., Aliniaeifard, S., Baymiev, A., Pusenkova, L., Garipova, S., Kulabuhova, D. and Maksimov, I. (2019). Bacillus spp.: efficient biotic strategy to control postharvest diseases of fruits and vegetables. *Plants, 8*(4), p.97.

Lazzeri L., Curto G., Leoni O. and Dallavalle E. (2004). Eff ects of glucosinolates and their enzymatic hydrolysis products via myrosinase on the root-knot nematode Meloidogyne incognita (Kofoid et White) Chitw. *J. Agric. Food Chem., 52*, 6703–6707.

Lei, H.-M., Wang, J.-T., Hu, Q.-Y., Li, C.-Q., Mo, M.-H., Zhang, K.-Q., Li, G.-H. and Zhao, P.-J. (2023). 2-Furoic acid associated with the infection of nematodes by Dactylellina haptotyla and its biocontrol potential on plant root-knot nematodes. *Microbiol. Spectr.*, *11*.

Lopez-Llorca, L., Maciá-Vicente, J. and Jansson, H.-B. (2008). Mode of action and interactions of nematophagous fungi. In: *Integrated Management and Biocontrol of Vegetable and Grain Crops Nematodes*. Springer: Berlin/Heidelberg, Germany, pp. 51–76.

Lugtenberg, B. and Kamilova, F. (2009). Plant-growth-promoting rhizobacteria. *Annual Review of Microbiology*, *63*(1), 541–556.

Maleita, C.M.N., Curtis, R.H.C., Powers, S.J. and Abrantes, I. (2012). Host status of cultivated plants to Meloidogyne hispanica. *European Journal of Plant Pathology*, *133*(2), pp. 449–460.

Marshall, A.J. (2002). *Sunn Hemp (Crotalaria juncea) as An Organic Amendment in Crop Production*. M.Sc. Thesis, Gainesville, FL, USA, University of Florida, 155 p.

Martinelli, P.R. and dos Santos, J.M. (2010). Scanning electron microscopy of nematophagous fungi associated Tylenchulus semipenetrans and Pratylenchus jaehni. *Biosci. J.*, *26*, 809–816.

Mazzola, M., Brown, J., Izzo, A.D. and Cohen, M.F. (2007). Mechanism of action and efficacy of seed meal-induced pathogen suppression differ in a Brassicaceae species and time-dependent manner. *Phytopathology*, *97*(4), 454–460.

Mejia, Z.E. and Gundy, V.S.D. (1982). Effects of rhizobacteria on Meloidogyne infection. *Journal of Nematology*, *14*, 475–476.

Mhatre, P.H., Divya, K.L., Venkatasalam, E.P., Bairwa, A., Sudha, R., Saranya, C., Guru-Pirasanna-Pandi, G. and Sharma, S. (2021). Evaluation of trap crop, *Solanum sisymbriifolium* and antagonistic crops against potato cyst nematodes, *Globodera* spp. *South African Journal of Botany*, *138*, 242–248.

Mhatre, P.H., Lekshmanan, D.K., Patted, P.S., Bairwa, A., Berliner, J. and Watpade, S. (2025). Modern approaches for the management of golden cyst nematodes in potato. *Sustainable Production of Root and Tuber Crops*. Routledge, pp. 296–307.

Mhatre, P.H., Karthik, C., Kadirvelu, K., Divya, K.L., Venkatasalam, E.P., Srinivasan, S., Ramkumar, G., Saranya, C. and Shanmuganathan, R. (2019). Plant growth promoting rhizobacteria (PGPR): A potential alternative tool for nematodes bio-control. *Biocatalysis and Agricultural Biotechnology*, *17*, 119–128.

Mohammed, S., El Saedy, M.A., Enan, M.R., Ibrahim, N.E., Ghareeb, A. and Moustafa, S.A. (2008). Biocontrol efficiency of Bacillus thuringiensis toxins against root-knot nematode, Meloidogyne incognita. *J. Cell Mol. Biol.*, *7*(1), 57–66.

Monchgesang, S., Strehmel, N., Schmidt, S., Westphal, L., Taruttis, F., Müller, E. et al. (2016). Natural variation of root exudates in Arabidopsis thaliana: Linking metabolomic and genomic data. *Scientific Reports*, *6*(1), 1–11.

Muthulakshmi, M., Kumar, S., Subramanian, S. and Anita, B. (2012). Compatibility of *Pochonia chlamydosporia* with other biocontrol agents and carbofuran. *Journal of Biopesticides*, *5*, 243–245.

Nadeem, M.A., Nawaz, M.A., Shahid, M.Q., Doğan, Y., et al., (2018). DNA molecular markers in plant breeding: Current status and recent advancements in genomic selection and genome editing. *Biotechnology and Biotechnological Equipment*, *32*(2), 261–285.

Narasimhamurthy, H.B., Ravindra, H. and Sehgal, M. (2017a). Management of rice rootknot nematode, *Meloidogyne graminicola*. *International Journal of Pure and Applied Biosciences*, *5*, 268–276.

Nascimento, F.X., Vicente, C.S., Barbosa, P., Espada, M., Glick, B.R., Mota, M., et al. (2013). Evidence for the involvement of ACC deaminase from Pseudomonas putida UW4 in the biocontrol of pine wilt disease caused by Bursaphelenchus xylophilus. *Biocontrol.*, *58*(3), 427–433.

Nikoo, F.S., Sahebani, N., Aminian, H., Mokhtarnejad, L and Ghaderi, R. (2014). Induction of systemic resistance and defense-related enzymes in tomato plants using Pseudomonas fluorescens CHAO and salicylic acid against root-knot nematode Meloidogyne javanica. *J. Plant Prot. Res.*, *15*(4).

Niu, X.-M. and Zhang, K.-Q. (2011). Arthrobotrys oligospora: A model organism for understanding the interaction between fungi and nematodes. *Mycology*, *2*, 59–78.

Noling J.W. and Becker J.O. (1994). Th e challenge of research and extension to defi ne and implement alternatives to methyl bromide. *J. Nematol.*, *26*, 573–586.

Obata, N., Tabuchi, H., Kurihara, M., Yamamoto, E., Shirasawa, K. and Monden, Y. (2022). Mapping of nematode resistance in hexaploid sweetpotato using a next-generation sequencing-based association study. *Frontiers in Plant Science*, *13*, 858747.

Oka, Y. (2010). Mechanisms of nematode suppression by organic soil amendments: A review. *Applied Soil Ecology*, *44*, 101–115.

Orion, D. and Kritzman, G. (2001). A role of the gelatinous matrix in the resistance of root-knot nematode (Meloidogyne spp.) eggs to microorganisms. *Journal of Nematology*, *33*, 203.

Palomares-Rius, J.E., Escobar, C., Cabrera, J., Vovlas, A. and Castillo, P. (2017). Anatomical alterations in plant tissues induced by plant-parasitic nematodes. *Front. Plant Sci.*, *8* 1987.

Pérez, C.R.J., Sánchez, J.E., Al-Ani, L.K.T., González-Cortázar, M., Castañeda-Ramírez, G.S., Mendoza-de Gives, P., Sánchez-García, A.D., Millán-Orozco, J. and Aguilar-Marcelino, L. (2021). Biological control of sheep nematode Haemonchus contortus using edible mushrooms. *Biol. Control*, *152*.

Perry, R.N. (1996). Chemoreception in Plant-parasitic Nematodes. *Annual Review of Phytopathology, 34*, 181–199.

Ploeg A. T. and Maris P. C. (1999). Eff ect of temperature on suppression of Meloidogyne incognita by Tagetes cultivars. *J. Nematol.*, *31*, 709–714.

Poveda, J., Abril-Urias, P. and Escobar, C. (2020). Biological Control of Plant-Parasitic Nematodes by Filamentous Fungi Inducers of Resistance: *Trichoderma*, Mycorrhizal and Endophytic Fungi. *Front. Microbiol.*, *11*, 992.

Radhakrishnan, R., Hashem, A. and Abd-Allah, E.F. (2017). Bacillus: A biological tool for crop improvement through bio-molecular changes in adverse environments. *Front. Physiol.*, *8*, 667.

Rahul, S., Chandrashekhar, P., Hemant, B., Chandrakant, N., Laxmikant, S. and Satish, P. (2014). Nematicidal activity of microbial pigment from Serratia marcescens. *Nat. Prod. Res.*, *28*(17), 1399–1404.

Rangaswamy, S.D., Reddy, P.P. and Joshi, S. (1993). Histopathological and histochemical investigation on antagonistic trap crops (marigold and mustard) and susceptible tomato infested with Meloidogyne incognita. *Current Nematology*, *4*, 203–206.

Reitz, M., Rudolph, K., Schroder, I., Hofmann-Hergarten, S., Hallmann, J. and Sikora, R. (2000). Lipopolysaccharides of Rhizobium etli strain G12 act in potato roots as an inducing agent of systemic resistance to infection by the cyst nematode Globodera pallida. *Appl. Environ. Microbiol.*, *66*(8), 3515–3518.

Reynolds, A.M., Dutta, T.K., Curtis, R.H.C., Powers, S.J. and Kerry, B.R. (2011). Chemotaxis can take plant-parasitic nematodes to the source of a volatile phytochemical in the absence of gradients. *Proceedings of the National Academy of Sciences, 108*(42), 17012–17015. https://doi.org/10.1073/pnas.1106139108.

Riggs, J.R. (1987). A Model of Godliness. Biblical and Theological Studies Faculty Publications, 236.

Rodriguez-Kabana, R., Boube, D. and Young, R.W. 1989. Chitinous materials from blue crab for control of root-knot nematode. I. Effect of urea and enzymatic studies. *Nematropica*, *19*, 53–74.

Rudek, W.T. (1975). The constriction of the trapping rings in Dactylaria brochopaga. *Mycopathologia*, *55*, 193–197.

Saharan, B.S. and Nehra, V. (2011). Plant growth promoting rhizobacteria: a critical review. *Life Sci Med Res, 21*(1), p.30.

Sahebani, N. and Hadavi, N. (2008). Biological control of the root-knot nematode Meloidogyne javanica by Trichoderma harzianum. *Soil Biol. Biochem.*, *40*, 2016–2020.

Saxena, G. (2018). Biological control of root-knot and cyst nematodes using nematophagous fungi. In: *Root Biology*. Springer: Berlin/Heidelberg, Germany, pp. 221–237.

Scholte, K. (2000[a]). Effect of potato used as trap crop on potato cyst nematodes and other soil pathogens and on the growth of a subsequent main potato crop. *Ann. Appl. Biol.*, *136*, 229–238.

Scholte, K. (2000[b]). Screening of non-tuber- bearing Solanaceae for resistance and induction of juvenile hatch of potato cyst nematodes and their potential for trap cropping. *Ann. Appl. Biol.*, *136*, 239–246.

Sharma, A., Gupta, A., Dalela, M., Sharma, S., Sayyed, R.Z., Enshasy, H.A.E. and Elsayed, E.A. (2020). Linking Organic Metabolites as Produced by Purpureocillium Lilacinum 6029 Cultured on Karanja Deoiled Cake Medium for the Sustainable Management of Root-Knot Nematodes. *Sustainability*, *12*, 8276.

Shinya, R., Aiuchi, D., Kushida, A., Tani, M., Kuramochi, K. and Koike, M. (2008). Effects of fungal culture filtrates of Verticillium lecanii (Lecanicillium spp.) hybrid strains on Heterodera glycines eggs and juveniles. *J. Invertebr. Pathol.*, *97*, 291–297.

Siddique, S. and Grundler, F.M. (2018). Parasitic nematodes manipulate plant development to establish feeding sites. *Curr. Opi. Microbiol.*, *46*, 102–108.

Siddiqui, I.A., Shaukat, S.S., Sheikh, I.H. and Khan, A. (2006). Role of cyanide production by Pseudomonas fluorescens CHA0 in the suppression of root-knot nematode, Meloidogyne javanica in tomato. *World Journal of Microbiology and Biotechnology, 22*(6), pp. 641–650.

Sikder, M.M. and Vestergard M. (2020). Impacts of root metabolites on soil nematodes. *Front. Plant Sci.*, *10*, 1792.

Singh, B., Hazra, K.K., Singh, U. and Gupta, S. (2020). Eco-friendly management of *Meloidogyne javanica* in chickpea (*Cicer arietinum* L.) using organic amendments and bio-control agent. *Journal of Cleaner Production*, *257*, 120542.

Singh, P. and Siddiqui, Z.A. (2010). Biocontrol of root-knot nematode Meloidogyne incognita by the isolates of Pseudomonas on tomato. *Arch. Phytopathol. Plant Protect.*, *43*(14), 1423–1434.

Sirohi, A. and Dutta, T. (2024). Genome Engineering for Nematode Management. *Indian Journal of Nematology*, *53*(Spl.), 114–121.

Shanthi, A. and Rajendran, G. (2006). Biological control of lesion nematodes in banana. *Nematologia Mediterranea. 34*: 69–75.

Spychalla, P. and De Jong, W.S. (2024). Breeding for potato cyst nematode resistance in Solanum tuberosum. *Crop Science*, *64*(3), 1167–1182.

Stirling, G.R. (1991). *Biological Control of Plant-parasitic Nematodes*. Wallingford, UK: CAB International.

Su, H., Hao, Y.e., Yang, X., Yu, Z., Deng, J. and Mo, M. (2008). A new species of Dactylellina producing adhesive knobs and non-constricting rings to capture nematodes. *Mycotaxon*, *105*, 313.

Thoden, T.C., Korthals, G.W. and Termorshuizen, A.J. (2011). Organic amendments and their influences on plant-parasitic and free-living nematodes: A promising method for nematode management? *Nematology*, *13*, 133–153.

Thompson, J.M. and Tyla, G.L. (1997). Difference in hatching of Heterodera glycines egg-mass and encysted egg *in vitro*. *Journal of Nematology*, *29*, 315–321.

Tian, B., Yang, J. and Zhang, K-Q. (2007). Bacteria used in the biological control of plant-parasitic nematodes: populations, mechanisms of action, and future prospects. *FEMS Microbiol. Ecol.*, *61*(2), 197–213.

Timper, P., Kone, D., Yin, J., Ji, P. and Gardener, B.B.M. (2009). Evaluation of an antibiotic-producing strain of Pseudomonas fluorescens for suppression of plant-parasitic nematodes. *J. Nematol.*, *41*(3), 234.

Trivedi, P.C. and Barker, K.R. 1986. Nematological reviews: Management of nematodes by cultural practices. *Nematropica*, *16*, 213–236.

Tülek, A., Kepenekçi, İI., Oksal, E. and Hazir, S. (2018). Comparative effects of entomopathogenic fungi and nematodes and bacterial supernatants against rice white tip nematode. Egypt. *J. Biol. Pest Control*, *28*(1), 1–6.

Turner, S.J. and Rowe, J.A. (2006) Cyst Nematodes. In: Perry, R.N. and Moens, M., Eds., *Plant Nematology*, CABI, Wallingford, 91–122.

Turatto, M.F., Dourado, F.d.S., Zilli, J.E. and Botelho, G.R. (2018). Control potential of Meloidogyne javanica and Ditylenchus spp. using fluorescent Pseudomonas and Bacillus spp. *Braz. J. Microbiol.*, *49*, 54–59.

Usovsky, M., Gamage, V.A., Meinhardt, C.G., Dietz, N., Triller, M., Basnet, P., ... and Scaboo, A.M. (2023). Loss-of-function of an α-SNAP gene confers resistance to soybean cyst nematode. *Nature Communications*, *14*(1), 7629.

Wang, K.-H., B.S. Sipes and D.P. Schmitt. (2002). Crotalaria as a cover crop for nematode management: A review. *Nematropica*, *32*, 35–57.

Wang, K.-H., Mcsorley, R. and Gallaher, R.N. (2003a). Host status and amendment effects of cowpea on *Meloidogyne incognita* in vegetable cropping systems. *Nematropica*, *33*, 215–224.

Wang, X., Cheng, R., Xu, D., Huang, R., et al., (2023). MG1 interacts with a protease inhibitor and confers resistance to rice root-knot nematode. *Nature Communications*, *14*(1), 3354.

Wani, A.H. (2015). Plant growth-promoting rhizobacteria as biocontrol agents of phytonematodes. In Biocontrol agents of phytonematodes (pp. 339–362). Wallingford UK: CABI.

Wardle, D. A. and Yeates, G.W. (1993). The dual importance of competition and predation as regulatory forces in terrestrial ecosystems: Evidence from decomposer food-webs. *Oecologia*, *93*(2), 303–306.

Wesemael, W.M., Perry, R.N. and Moens, M. (2006). The influence of root diffusate and host age on hatching of the root-knot nematodes, Meloidogyne chitwoodi and M. fallax. *Nematology*, *8*(6).

Williamson, V.M. and Hussey, R.S. (1996). Nematode pathogenesis and resistance in plants. *The Plant Cell, 8*(10), p.1735.

Wright, D.J. and Perry, R.N. (2006). Reproduction, Physiology, and biochemistry *In:* Perry, R.N. & Moens, M. (Eds.), *Plant Nematology*. Wallingford, UK: CAB International, pp. 187–209.

Xiang, N, Lawrence, K.S. and Donald, P.A. (2018). Biological control potential of plant growth-promoting rhizobacteria suppression of Meloidogyne incognita on cotton and Heterodera glycines on soybean: A review. *J. Phytopathol.*, *166*(7–8), 449–458.

Xie, S., Jiang, H., Ding, T., Xu, Q., Chai, W. and Cheng, B. (2018). Bacillus amyloliquefaciens FZB42 represses plant miR846 to induce systemic resistance via a jasmonic acid-dependent signalling pathway. *Mol. Plant Pathol.*, *19*(7), 1612–1623.

Xie, H., Aminuzzaman, F., Xu, L., Lai, Y., Li, F. and Liu, X. (2010). Trap induction and trapping in eight nematode-trapping fungi (Orbiliaceae) as affected by juvenile stage of Caenorhabditis elegans. *Mycopathologia*, *169*, 467–473.

Yang, J., Tian, B., Liang, L. and Zhang, K.Q. (2007). Extracellular enzymes and the pathogenesis of nematophagous fungi. *Appl. Microbiol. Biotechnol.*, *75*, 21–31.

Yu, Z., Xiong, J., Zhou, Q., Luo, H., Hu, S., Xia, L., Sun, M., Li, L. and Yu, Z. (2015). The diverse nematicidal properties and biocontrol efficacy of Bacillus thuringiensis Cry6A against the root-knot nematode Meloidogyne hapla. *Journal of Invertebrate Pathology, 125*, pp.73–80.

Zachariah, K. (1989). Chemical induction of trap closure in Dactylella brochopaga. *Protoplasma*, *148*, 87–93.

Zhang, J., Fu, B., Lin, Q., Riley, I.T., Ding, S., Chen, L., Cui, J., Yang, L. and Li, H. (2020). Colonization of Beauveria bassiana 08F04 in root-zone soil and its biocontrol of cereal cyst nematode (Heterodera filipjevi). *PLoS One*, *15*.

Zhang, K.Q. and Hyde, K.D. eds. (2014). Nematode-trapping fungi (Vol. 23). Springer Science & Business.

CHAPTER 9

Role of Bioactive Metabolites Derived from Algae in Crop Pest and Disease Management

Priyakharini Pragash[1], Arunkumar Nagarathinam[2*], Elango Kolanthasamy[3], Gowrisri Nagaraj[4], Suchitra Rakesh[5] and Sunil Shivananjappa[6]

1. Introduction

Algae consist of a diverse group of photosynthetic organisms living in aquatic or terrestrial systems and are currently receiving attention as a source of biologically active materials for agricultural use. Algal metabolites can be an eco-friendly alternative to agrochemicals, including antimicrobial and antifungal compounds and insecticidal, nematicidal, and antiviral products due to their biodegradability, sustainability, and negligible toxicity to wildlife and environment. This chapter discusses the diversity of algal bioactives, their modes of action, and potential advantages over traditional pesticides, particularly in the context of integrated pest management. It also discusses recent technological development, obstacles in researching algal products, and possible approaches in enhancing their future large-scale use. As global agriculture faces increasing pressure to produce more food while conserving the environment and reducing the risk of pest population adaptation, the reliance on synthetic pesticides has worsened agricultural biodiversity, water pollution, and genetic variability in pest populations leading to resistance (Kashyap, 2024). Consequently, sustainable alternatives are urgently needed for the long-term. Marine macroalgae and freshwater microalgae have been shown to produce a large variety of pesticidal and antimicrobial products (AbdelRaouf et al., 2021; Ahmed et al., 2022) and will likely expand their role in future crop protection strategies.

Algal species have shown potential to suppress bacterial, fungal, viral, and insect pests that commonly affect crop plants. Extracts from the genera such as *Ulva*, *Chlorella*, *Gracilaria*, *Spirulina*, *Scenedesmus*, and *Sargassum* have demonstrated efficacy against a wide range of agricultural pests and diseases in both greenhouse and field conditions (Kumar et al., 2023; Banerjee et al., 2021). Both macroalgae and microalgae, such as, *Ascophyllum nodosum*, *Fucu svesiculosus*, and *Solieria robusta, secrete* polysaccharides (e.g. alginates), polyphenols (phlorotannins), and plant-growth regulators (auxin, cytokinin) with antimicrobial and nematicidal effects (Shukla et al., 2024; Pangestuti & Kim, 2021). The 2024 *Frontiers in Marine Science* review states that *Solieria robusta* was applied as a soil amendment and found to be more effective against *Fusarium solani* than the

[1-3] Kumaraguru Institute of Agriculture, Tamil Nadu Agricultural University, Erode, Tamil Nadu 638315, India.
[4,5] Department of Microbiology, Central University of Tamil Nadu, Thiruvarur, India.
[6] School of Biological Sciences, University of Western Australia, Perth, WA, Australia.
* Corresponding author: drarunmicrobiologylab@gmail.com

DOI: 10.1201/9781003614869-9

fungicide Topsin-M. Cyanobacteria (*Anabaena*, *Nostoc*, *Arthrospira*); Green microalgae (Chlorella, Chlamydomonas) produce a wide range of metabolites such as fatty acids, alkaloids, aldehydes, peptides that have herbicidal, algicidal, nematocidal, and antimicrobial activities (Mhatre et al., 2023; Kumar et al., 2022).

2. Estimation of Bioactive Compounds from Algae

The estimation of bioactive compounds in algal extracts is a critical step for evaluating their potential in agricultural applications. Quantitative analyses often involve spectrophotometric techniques, such as the Folin Ciocalteu method for total phenolic content and the aluminum chloride colorimetric assay for flavonoids (Zhang et al., 2023). Additionally, gas chromatography-mass spectrometry (GC-MS) and high-performance liquid chromatography (HPLC) are employed to identify and quantify specific secondary metabolites with pesticidal or antimicrobial activity. These analytical methods enable the standardization of algal formulations and support their regulatory acceptance in crop protection (Righini et al., 2021). A recent study by Shahin and Fadl (2025) demonstrated the efficacy of *Palmaria palmata* extract through GC-MS analysis, identifying 16 bioactive compounds with notable antibacterial activity against agricultural pathogens such as *Staphylococcus aureus*, *Bacillus cereus*, *Proteus mirabilis*, and *Pseudomonas aeruginosa* (Shahin & Fadl, 2025).

3. Major Algal Biomolecules and Their Applications

Algal organisms through several metabolic pathways such as lipogenesis, photosynthesis, and carotenogenesis produce a remarkable amount of high value primary and secondary metabolites (Gijs et al., 2025). Primary metabolites synthesized by algae include carbohydrates, lipids, and proteins and secondary bioactive compounds include antimicrobial and anti-inflammatory molecules (Hentati et al., 2020). A number of microalgal species, including *Nannochloropsis* sp., *Laminaria japonica*, *Undaria pinnatifida*, and *Dunaliella* sp., contribute significantly to the creation of third-generation biofuels by producing lipids, which account for over half of their total dry biomass (Wang et al., 2020) (Xin et al., 2024). Because of their high protein content and their significance as dietary supplements, *Spirulina* sp., *Arthrospira platensis* (filamentous Cyanobacterium), and *Chlorella vulgaris* have seen a considerable increase in consumption as whole foods (Bleakley et al., 2017).

Algae such as *Anabaena*, *Nanno chloropsis*, *Chlorella* and macroalgae like *Ulva*, *Sargassum*, *Gracilaria* produce a wide spectrum of secondary metabolites, many of which serve defensive roles against herbivores, pathogens, and competitors (Hachicha et al., 2022). These metabolites are mostly in the form of flavonoids, alkaloids, phlorotannins, terpenoids, sterols, lectins, fatty acids, phenolics, and sulphated polysaccharides (Patra et al., 2022). These metabolites work through various mechanisms such as inhibition of microbial growth, disruption of cell membranes, enzyme inhibition, and interference with reproductive or digestive functions of pests. For instance, methanolic extracts of *Gracilaria corticata* were found to inhibit the growth of *Fusarium oxysporum*, a causative agent of wilt disease in tomatoes (Singh et al., 2022). Similarly, aqueous extracts from *Chlorella vulgaris* demonstrated ovicidal and larvicidal effects against Tobacco cutworm (*Spodoptera litura*) and Cotton bollworm (*Helicoverpa armigera*) (Yadav et al., 2023). Also, by generating growth hormones, polysaccharides, minerals, betaines, and sterols, algae extracts are valuable resources that exhibit a wide range of growth-stimulating activities, including seed germination, plant growth, crop performance and yield, post-harvest shelf-life, and protection against biotic and abiotic factors (Lee et al., 2021).

4. Mechanism of Insect Pest and Disease Control

The effectiveness of algal bioactive compounds comes from their ability to interfere with multiple physiological pathways in pests and pathogens (Table 1; Figure 1). Sulphated polysaccharides,

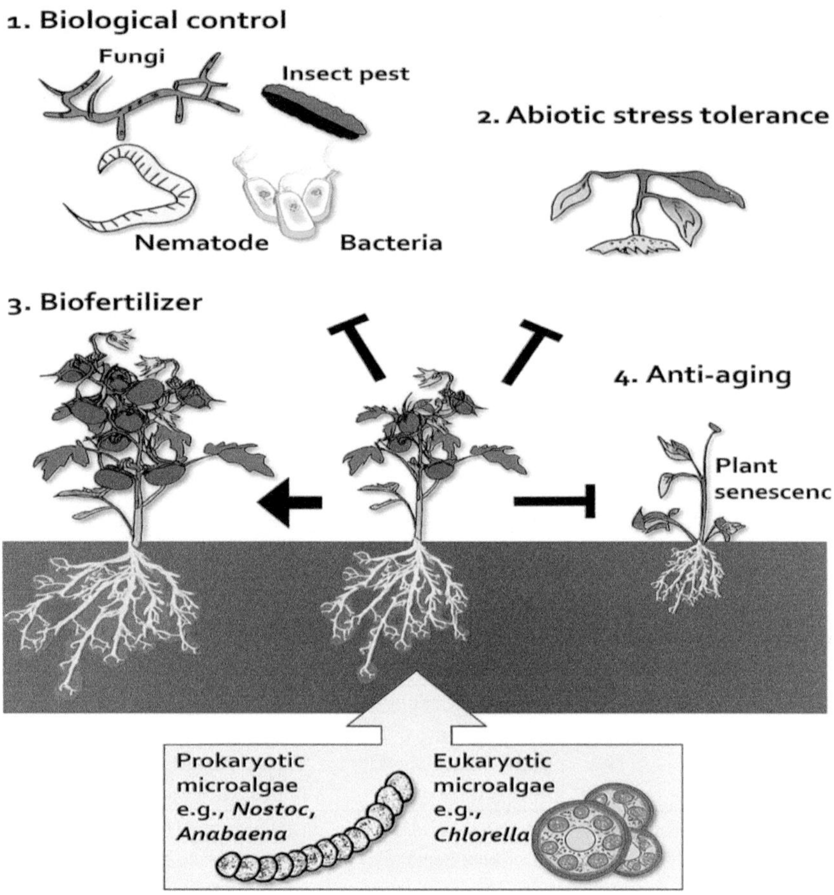

Fig. 1 Mechanism of Algal Biofertilizers in Pest and Disease Control

such as carrageenans and fucoidans derived from red and brown algae, are capable of inducing systemic acquired resistance (SAR) in plants, leading to the activation of pathogenesis related (PR) genes and the production of phytoalexins (De Borba et al., 2022). Terpenoids and phlorotannins act as neurotoxins, affecting ion channels in insect nervous systems, thereby impairing feeding and causing paralysis or death specifically in the cotton leafhopper (*Amrasca devastans*) (Petchidurai et al., 2023). Phenolic compounds disrupt fungal cell walls and membranes, while lectins interfere with root-knot nematode (*Meloidogyne* spp)feeding and development (Khan et al., 2022). These diverse mechanisms reduce the likelihood of resistance development, a common drawback of conventional pesticides (Khan et al., 2022).

5. Advantages over Synthetic Pesticides and Fungicides

Algal bioactive compounds offer several advantages over conventional synthetic pesticides, making them a valuable component of sustainable agriculture. Unlike synthetic chemicals that often target pests indiscriminately, algal metabolites are typically most specific and biodegradable, reducing environmental persistence and the risk to non-target organisms. Their natural origin also means they are less likely to cause pesticide residue buildup in soil and food products. Additionally, algal-based biopesticides have been shown to enhance plant immunity and promote growth by acting as elicitors of defense pathways, a benefit not commonly associated with synthetic alternatives such as mancozeb, chlorothalonil, or carbendazim (Ertani et al., 2022). Importantly, the use of algal

Table 1 Algal Microorganisms in Control of Pest and Disease Control.

Algal Source	*Bioactive Compound*	*Target Pest*	*Mode of Action*	*Reference*
Bio Control of Insect Pests				
Sargassum wightii	Terpenoids	Aphids (*Aphis gossypii*)	Neurotoxic disruption of sodium ion channels, impairing nervous system function	Rajak et al. (2023)
Chlorella vulgaris	Cationic peptides	Root-knot nematode (*Meloidogyne spp.*)	Disruption of nematode cell membranes and inhibition of protein synthesis	Rani et al., (2021)
Spirulina platensis	Fatty acids, alkaloids	Cotton bollworm (*Helicoverpa armigera*)	Feeding deterrence and larval toxicity via oxidative stress	Yadav et al. (2023)
Anabaena flos-aquae	Toxins, peptides	Mosquito larvae (*Culexquinquefasciatus*)	Larvicidal activity through midgut disruption and enzyme inhibition	Kumar et al. (2022)
Scenedesmus obliquus	Saponins	Whiteflies (*Bemisiatabaci*)	Cuticle disruption and molting inhibition	Sharma et al. (2021)
Ulva lactuca	Methanol/ acetone extracts	*Spodoptera littoralis* (cotton pest); *Culexpipiens* (mosquitolarvae)	Toxic or repellent activity affecting larval growth and adult emergence	Sithole et al. (2023)
Spatoglossum variabile	Dry macroalgal powders	*Meloidogyne incognita* (root-knot nematode)	Direct cytotoxic action on nematode eggs and juveniles; induces plant defense responses, reduces gall formation	Sithole et al. (2023)
Stoechospermum polypodioides	Methanol extract	*Meloidogyne javanica* (root-knot nematode)	Almost complete juvenile mortality and inhibition of egg hatching after 72 h exposure	Alessandra et al. (2022)
Corallina spp	Ethanol extracts	*Meloidogyne incognita*	Egg hatch suppression (up to 87%) and juvenile mortality (~85%) via bioactives	Rao et al. (2021)
Biocontrol of Diseases				
Gracilaria edulis	Phenolic compounds	Powdery mildew (*E. nectar*)	Antioxidant and antifungal enzymatic suppression	Singh et al. (2021)
Spirulina platensis	Polysaccharide complexes	*Rhizoctonia solani* (Root rot)	Cell wall disintegration and hyphal inhibition	Khan et al. (2020)
Chlorella sorokiniana	Aqueous/ methanolic extract (phenolics, carotenoids)	*Phytophthora cactorum*(strawberry root and crown rot)	Inhibition of mycelial growth and zoospore germination; antifungal metabolites	Costa et al. (2023)

Contd.

Table 1 *Contd.*

Bio Control of Insect Pests				
Algal Source	*Bioactive Compound*	*Target Pest*	*Mode of Action*	*Reference*
Anabaena sp. (cyanobacterium)	Crude extract (peptides, phenolics)	*Botrytis cinerea* (graymold on fruits)	Inhibition of spore germination and mycelial extension; disrupts fungal cell wall	Righini et al. (2021)
Nannochloropsis sp.	Carotenoids phenolics	*Fusarium graminearum* (cerealhead blight)	Mycelial growth inhibition via oxidative damage and cell membrane disruption	Scaglioni et al. (2021)
Scenedesmus obliquus	Aqueous extract	*Sclerotium rolfsii* (stem rot)	Inhibition of fungal radial growth; antifungal allelochemicals	Schmid et al. (2022)
Phaeodactylum tricornutum	Aqueous extract	*Rhizoctonia solani* (vegetable root rot)	Inhibition of fungal hyphal development	Schmid et al. (2022)
Nostoc commune	Methanol extract	*Fusarium oxysporum f. sp. Lycopersici* (tomato wilt)	Prevents spore germination, reduces infection similar to fungicide; induces plant defense	El-Sheekh et al. (2022)
Chlorella vulgaris	Aqueous extract	*Botrytis cinerea, Alternaria alternata* (postharvest disease)	Reduces spore production, slows colony expansion	Schmid et al. (2022)
Spirulina sp.	Aqueous extract	*Sclerotium rolfsii* (Stem rot)	Limits sclerotia formation and fungal growth	Schmid et al. (2022)

extracts contributes to lower risks of resistance development due to the complex mixture of active compounds that exert multiple modes of action (Radwan et al., 2020). These benefits align well with integrated pest management (IPM) strategies and regulatory trends favoring eco-friendly crop protection tools.

Numerous studies have demonstrated the use of algal compounds in various cropping systems. For instance, sulphated polysaccharides from green microalgae (*Ulva Lactuca*) have shown antiviral activity against Tomato mosaic virus by inhibiting viral attachment and replication (Wijesekara et al., 2011). Phenolic rich extracts of *Gracilaria edulis* have been effective in controlling fungal pathogens such as *Erysiphenecator*, the causative agent of powdery mildew in grapes. *Chlorella vulgaris* extracts have shown promising antifungal activity against Fusarium wilt in tomatoes. Studies reported up to 67% reduction in disease severity, enhanced plant defense enzymes (peroxides, polyphenol oxidase), and disrupted fungal growth.

Haleem et al. (2023) studied the larvicidal effect of eight different methanolic algal extracts from three different algal divisions against the 3rd larval instar of *Culex pipiens* L. (Diptera: Culicidae). In comparative studies, it was revealed that four species of red and green algal extracts revealed good larvicidal activity. Generally, it was often observed that extracts like *Galaxaura elongata* and *Jania rubens* (Rhodophyta), as well as *Codium tomentosum* and *Ulva intestinalis* (Chlorophyta) had a higher larvicidal potency than *Padina boryana, Dictyota dichotoma,* and *Sargassum dentifolium* (Phaeophyta) and *Gelidium latifolium* (Rhodophyta). The best toxicity levels were achieved through exposure to the *G. elongata* extract (LC50 (31.13 ppm) followed by *C. tomentosum* (69.85 ppm), *J.*

rubens (84.82 ppm), and *U. intestinalis* (97.54 ppm). *G. latifolium* (297.38 ppm) was determined to be the least toxic within a 72-hour post-exposure time frame.

Bharadwaj et al. (2025) examined the chemical activity of 15 microalgal strains against seven phytopathogens, with *Chlorella sorokiniana* having the most robust inhibition of the strawberry pathogen, Phytophthora cactorum. Additionally, extracts from *Spirulina sp. and Nanno chloropsis sp.* were explored for their antifungal potential against Fusarium pathogens. It was noted that the aqueous micro algal extract was a more significant fungal growth inhibition than either the phenolic acid extract, or even the synthetic fungicides. Several other investigations into fungi infecting economic crops have also been conducted. Vehapi et al. (2019) examined the antifungal activity of four microalgae: *Chlorella protothecoides, Ulva lactuca, Chlorella minutissima, and Chlorella* vulgaris against three apple-infecting fungi: *Alternaria alternata, Penicillium expansum and Aspergillus niger*. They found *C. vulgaris* was the best fungus mycelial growth inhibition, while *C. minutissima* the worst antifungal. The antifungal activity exhibited by the microalgae was attributable to bioactive compositions. In a related study, *Chlorella fusca* extract applications to a strawberry plant infected with Fusarium, every two weeks, improved plant growth and the pathogen's population.

6. Challenges, Knowledge Gaps, and Future Prospects

Although algae show great promise as eco-friendly alternatives to chemical pesticides, several challenges and knowledge gaps still limit their broader use in agriculture (Ruiz et al., 2016). A key issue is the variation in the types and quantities of bioactive compounds produced by different algal species, which is influenced by growth conditions, extraction methods, and environmental factors (Hentati et al., 2020; Kumar et al., 2022). This variability makes it difficult to produce standardized formulations with consistent results. In addition, large-scale production of algae remains expensive, particularly due to the high cost of cultivation, harvesting, and processing technologies (Xin et al., 2024). Many algal products also have limited stability when applied in open-field conditions, as exposure to sunlight, high temperatures, and microbial degradation can reduce their activity over time. As a result, frequent applications may be needed, which can increase labor costs and limit practicality for widespread use (Radwan et al., 2020).

Regulatory uncertainty is another major hurdle. Algal biopesticides often fall into undefined categories between fertilizers, biostimulants, and pest control agents, making approval processes slow and complex in many regions (Ertani et al., 2022). Moreover, there is limited research on the long-term environmental impacts of these products, especially on beneficial insects, pollinators, soil microbiota, and aquatic life (Righini et al., 2021). Scientifically, although various studies have demonstrated the antimicrobial, nematicidal, and insecticidal effects of algal metabolites, the precise mechanisms through which they affect pests and pathogens remain poorly understood (Patra et al., 2022). For instance, while compounds like phlorotannins and sulphated polysaccharides have been shown to induce plant defenses or disrupt pest physiology, more molecular-level studies are needed to fully explain these interactions (De Borba et al., 2022; Khan et al., 2022). Furthermore, potential synergistic or antagonistic interactions between the different compounds in algal extracts are rarely investigated, which limits the optimization of formulations for maximum effectiveness.

Despite these limitations, the future of algae in pest and disease control remains highly promising. New tools such as genomics, metabolomics, and transcriptomics offer opportunities to identify and select algal strains with enhanced pesticidal activity (Pangestuti & Kim, 2021). Advances in genetic engineering could also allow for improved production of specific bioactive compounds. Additionally, emerging formulation technologies, such as nano-encapsulation and controlled-release systems, may help improve compound stability and reduce the need for repeated applications (Shukla et al., 2024). Algae cultivated from wastewater or carbon capture systems also offer an environmentally sustainable and economically viable approach to biomass production (Ahmed et al., 2022). To support real-world adoption, stronger policy frameworks, investment in research, and farmer education programs will be essential. If these scientific and logistical challenges are addressed, algal bioactives could become a key component of integrated, eco-friendly pest management strategies in future agriculture.

7. Conclusion

Algae provide a rich source of bioactive compounds that play a vital role in sustainable crop protection and disease management. This chapter has highlighted how various algal species produce diverse secondary metabolites such as phenolics, polysaccharides, terpenoids, and peptides that target pests and pathogens through multiple biological mechanisms. These actions not only control pests effectively but also reduce the chance of resistance development, a common issue with synthetic pesticides. Compared to chemical pesticides, algal-based products offer key ecological and agronomic benefits, including biodegradability, low toxicity to non-target organisms, and the ability to boost plant health by activating natural defense responses. They fit well within integrated pest management strategies aimed at minimizing environmental impact. However, challenges remain in standardizing bioactive content, scaling up production, and navigating regulatory frameworks. Translating laboratory successes into consistent field performance requires further research and technological innovation. Rather than simple problems to fix, these challenges present opportunities to advance the field through improved strain selection, metabolic engineering, and formulation methods. As these innovations develop, algae are poised to play an increasingly important role in reducing dependence on synthetic chemicals and supporting more resilient, eco-friendly agricultural systems.

References

Abdel-Raouf, N., Al-Enazi, N.M. and Ibraheem, I.B.M. (2021). Microalgae as a sustainable tool for the management of plant pests and diseases: Current status and future perspectives. *Saudi Journal of Biological Sciences*, *28*(1), 224–232. https://doi.org/10.1016/j.sjbs.2020.09.045.

Ahmed, S., Jahan, R. and Hossain, M.T. (2022). Bioactive compounds from algae as eco-friendly pesticides: A review. *Biocatalysis and Agricultural Biotechnology, 39*, 102249. https://doi.org/10.1016/j.bcab.2022.102249.

Alessandra, R., De Nardi, B., Conti, M., et al. (2022). Nematicidal activity of marine macroalgae methanol extracts against Meloidogyne javanica. *Marine Drugs*, *20*(9), 551. https://doi.org/10.3390/md20090551.

Banerjee, K., Ghosh, S. and Pal, R. (2021). Algal extracts as plant defense elicitors and biopesticides: Advances and prospects. *Journal of Applied Phycology*, *33*(5), 2803–2816. https://doi.org/10.1007/s10811-021-02515-4.

Bhardwaj, R., Yadav, A., Sahoo, A., Kumari, P., Singh, L.A., Swapnil, P. ... and Kumar, S. (2025). Microalgal-based sustainable bio-fungicides: A promising solution to enhance crop yield. *Discover Sustainability*, *6*(1), 39.

Bleakley, S. and Hayes, M. (2017, Apr.). Algal Proteins: Extraction, Application, and Challenges Concerning Production. *Foods*. *6*(5), 33. doi: 10.3390/foods6050033. PMID: 28445408; PMCID: PMC5447909.

Costa, M.J., De Abreu, M.E., Fernández, L. and Hernández, J.A. (2023). Antifungal activity of Chlorella sorokiniana extracts against Phytophthora cactorum affecting strawberry. *Plant Pathology Journal*, *39*(1), 55–65. https://doi.org/10.5423/PPJ.OA.2022.0234.

De Borba, T., Pereira, C.S.G., Jorge, A.O.S. and Barreira, P. (2022). Sulfated oligosaccharides from red algae induce resistance in wheat and apple via elicitation of defense-related enzymes. *Heliyon*, *8*(4), e09405. https://doi.org/10.1016/j.heliyon.2022.e09405.

El-Sheekh, M.M., Eladel, H.M. and El-Naggar, A.H. (2022). Nostoc commune methanolic extract for controlling Fusarium oxysporum wilt in tomato: Efficacy and mechanism. *Egyptian Journal of Biological Pest Control*, *32*(1), 45. https://doi.org/10.1186/s41938-022-00562-2.

Ertani, A., Tamburino, R., Ruperti, B. and Neri, D. (2022). Enhancement of plant defense and growth promotion by a commercial seaweed extract: Insights into transcriptomic and metabolic actions. *Journal of Plant Physiology*, *270*, 153618. https://doi.org/10.1016/j.jplph.2021.153618.

Gijs, E., Verhulst, S. and Clement, L. (2025). Advancing sustainable agriculture: The potential of seaweed-derived biopesticides. *Bioresources and Bioprocessing*, *12*, Article 15. https://doi.org/10.1186/s40643-025-00849-w.

Hachicha, R., Elleuch, F., Ben Hlima, H., Dubessay, P., de Baynast, H., Delattre, C., Pierre, G., Hachicha, R., Abdelkafi, S., Michaud, P. and Fendri, I. (2022). Biomolecules from microalgae and cyanobacteria: Applications and market survey. *Applied Sciences*, *12*(4), 1924.

Haleem, D.R.A., El Tablawy, N.H., Alkeridis, L.A., Sayed, S., Saad, A.M., El-Saadony, M.T. and Farag, S.M. (2022). Screening and evaluation of different algal extracts and prospects for controlling the disease vector mosquito Culex pipiens L. *Saudi Journal of Biological Sciences*, *29*(2), 933– 940.

Hentati, F., Tounsi, L., Djomdi, D., Pierre, G., Delattre, C, ; Ursu, A.V., Fendri, I., Abdelkafi, S. and Michaud, P. (2020). Bioactive Polysaccha rides from Seaweeds. *Molecules*, *25*, 3152.

Hernández, A., González-Moya, M., Márquez, A. and Acevedo, L. (2024). Review microalgae drying: A comprehensive exploration from conventional air drying to microwave drying methods. *Future Foods*, *10*, 100420.

Kashyap, U., Garg, S. and Arora, P. (2024). Pesticide pollution in India: Environmental and health risks, and policy challenges. *Toxicology Reports*, *13*, 101801.

Khan, F., Jeong, G.J., Khan, M.S.A. and Tabassum, N. (2022). Seaweed-derived phlorotannins: A review of multiple biological roles and mechanisms. *Marine Drugs*, *20*(6), 384. https://doi.org/10.3390/md20060384.

Khan, S.A., Gupta, N. and Mohanty, A. (2020). Antifungal effect of Spirulina platensis extract against soil-borne pathogen Rhizoctonia solani. *Plant Archives*, *20*(Supplement 2), 3917–3920.

Kumar, G., Nanda, S., Singh, S.K., Singh, S., Singh, B.N. and Mukherjee, A. (2024). Seaweed extracts: Enhancing plant resilience to biotic and abiotic stresses. *Frontiers in Marine Science, 11*, Article 1457500. https://doi.org/10.3389/fmars.2024.1457500.

Kumar, P., Mishra, R.K. and Singh, A. (2022). Algal metabolites as eco-friendly larvicides against mosquito vectors: A case study of Anabaena flos-aquae. *Ecotoxicology and Environmental Safety*, *238*, 113610. https://doi.org/10.1016/j.ecoenv.2022.113610.

Kumar, R., Singh, S. and Yadav, A.N. (2023). Algal biopesticides: Potential and mechanisms in sustainable pest management. *Frontiers in Microbiology*, *14*, 1198453. https://doi.org/10.3389/fmicb.2023.1198453.

Lee, S.-M. and Ryu, C.-M. (2021). Algae as New Kids in the Beneficial Plant Microbiome. *Front. Plant Sci.*, *12*, 599742.

Martins, A.P. and Silva, J.B. (2021). *Biopesticide Formulations: Current Challenges and Future Perspectives*. In: *Formulations in Integrated Pest Management*. Elsevier, pp. 202–230. https://doi.org/10.1016/B978-0-12-823355-9.00010-9

Mhatre, A., Gore, S. and Mulla, N. (2023). Cyanobacteria and green microalgae: Emerging bioresources for biocontrol applications in agriculture. *Journal of Applied Phycology*, *35*(2), 1345–1360. https://doi.org/10.1007/s10811-023-02895-0.

Pangestuti, R. and Kim, S.-K. (2021). Biological activities and health benefit effects of natural pigments derived from marine algae. *Marine Drugs*, *19*(11), 620. https://doi.org/10.3390/md19110620.

Patra, J.K., Das, G., Baek, K.H. and Shin, H.S. (2022). Role of marine algal-derived secondary metabolites in plant disease control: An emerging sustainable approach. *Marine Drugs*, *20*(4), 215. https://doi.org/10.3390/md20040215.

Petchidurai, G., Sahayaraj, K., Al-Shuraym, L.A., Albogami, B.Z. and Sayed, S.M.S. (2023). Insecticidal activity of tannins from selected brown macroalgae against the cotton leafhopper Amrascadevastans. *Plants*, *12*(18), 3188. https://doi.org/10.3390/plants12183188.

Radwan, O., Saleh, A. and Elkhateeb, W. (2020). Use of Chlorella vulgaris extracts as biopesticide in tomato: effect on growth, yield, and resistance mechanisms. *Biocontrol Science and Technology*, *30*(10), 1002–1018. https://doi.org/10.1080/09583157.2020.1791234.

Rajak, R.C., Singh, A. and Bhowmik, A. (2023). Bioefficacy of seaweed-derived terpenoids against sucking insect pests in cotton. *Journal of Applied Phycology*, *35*(1), 1123–1131. https://doi.org/10.1007/s10811-023-02876-3.

Rani, S., Gunjyal, N., Ojha, C. and Singh, R.P. (2021). Review of challenges for algae-based wastewater treatment: Strain selection, wastewater characteristics, abiotic, and biotic factors. *J. Hazardous Toxic Radioactive Waste*, *25* (2), 03120004. doi: 10.1061/(ASCE)HZ.2153-5515.0000578.

Rani, S., Kumari, R. and Prasad, R. (2021). Nematicidal activity of Chlorella vulgaris extract against *Meloidogyne incognita* in tomato plants. Biocontrol Science and Technology, *31*(5), 514–525. https://doi.org/10.1080/09583157.2020.1871107.

Rao, M. S., Sharma, R. and Patel, D. (2021). Red macroalgal bioactives for management of *Meloidogyne incognita* in solanaceous crops. *Archives of Phytopathology and Plant Protection*, *54*(5–6), 673–688. https://doi.org/10.1080/03235408.2021.1890971.

Righini, H., Puggioni, G., Deligios, M., et al. (2021). Antifungal potential of cyanobacteria extracts against Botrytis cinerea in postharvest fruit decay. *Biological Control*, *161*, 104694. https://doi.org/10.1016/j.biocontrol.2021.104694.

Righini, H., Roberti, R. and Baldoni, G. (2021). Extraction and analytical techniques for bioactive compounds in algae with potential application in agriculture. *Plants*, *10*(9), 1919. https://doi.org/10.3390/plants10091919.

Ruiz, J., Giuseppe, O., Jeroen, deV., Rouke, B., Philippe, W., Reith, J.H., Michel, H.M.E., Dorinde, M.M.K., René, H.W. and Maria, J.B. (2016). Towards industrial products from microalgae. *Energ. Enviro. Sci.* https://doi.org/10.1039/c6ee01493c.

Scaglioni, P.T., Oliveira, L.S., Machado, J.P., et al. (2021). Bioactive compounds from *Nannochloropsis* sp. for fungal control in cereals. *Crop Protection*, *144*, 105624. https://doi.org/10.1016/j.cropro.2021.105624.

Schmid, F., Gomez, J. and Hempel, S. (2022). Microalgal aqueous extracts as sustainable antifungal agents against soil-borne pathogens. *Frontiers in Microbiology*, *13*, 830219. https://doi.org/10.3389/fmicb.2022.830219.

Shahin, A.S. and Fadl, N.M. (2025). GC-MS based profiling and antibacterial potential of *Palmaria palmata* extract against phytopathogenic bacteria. *Algal Research*, *75*, 103195. https://doi.org/10.1016/j.algal.2025.103195.

Sharma, N., Meena, R.L. and Patel, A. (2021). Saponin-rich green algae Scenedesmus obliquus as a biocontrol agent against Bemisia tabaci. *Plant Protection Science*, *57*(4), 250–258. https://doi.org/10.17221/170/2020-PPS.

Shukla, S., Dey, P. and Saha, S. (2024). Exploring the antifungal efficacy of marine red algae Solieria robusta against soilborne pathogens in tomato cultivation. *Frontiers in Marine Science*, *11*, 1192875. https://doi.org/10.3389/fmars.2024.1192875.

Singh, M., Kumar, A. and Tiwari, K.L. (2021). Antifungal potential of marine red alga Gracilaria edulis against Erysiphenecator causing powdery mildew in grapes. *Biocatalysis and Agricultural Biotechnology*, *31*, 101884. https://doi.org/10.1016/j.bcab.2021.101884.

Sithole, T., Dlamini, N., Mamba, B., et al. (2023). Biocontrol potential of selected brown algae against root-knot nematode in tomato crops. *Journal of Applied Phycology*, *35*(1), 101–113. https://doi.org/10.1007/s10811-023-02899-9

Vehapi, M., Koçer, A.T., Yılmaz, A. and Özçimen D. (2019). Investigation of the antifungal effects of algal extracts on apple-infecting fungi. *Arch Microbiol*. *202*(3), 455–471.

Wang, M., Chen, S., Zhou, W., Yuan, W. and Wang, D. (2020). Algal cell lysis by bacteria: A review and comparison to conventional methods. *Algal Res.*, *46*, 101794.

Wijesekara, I., Pangestuti, R. and Kim, S.K. (2011). Biological activities and potential health benefits of sulfated polysaccharides derived from marine algae. *Carbohydrate Polymers*, *84*(1), 14–21. https://doi.org/10.1016/j.carbpol.2010.10.062.

Xin, Y., Wu, S., Miao, C., Xu, T. and Lu, Y. (2024). Towards Lipid from Microalgae: Products, Biosynthesis, and Genetic Engineering. *Life*, *14*(4), 447. https://doi.org/10.3390/life14040447.

Yadav, N., Rani, R. and Thakur, M. (2023). Larvicidal and ovicidal activity of freshwater green microalga Chlorella vulgaris against *Spodoptera litura* and *Helicoverpa armigera*. *Journal of Applied Phycology*, *35*(2), 1467–1476. https://doi.org/10.1007/s10811-023-02987-x.

Zhang, L., Wang, Y., Li, X. and Chen, H. (2023). Quantitative determination of phenolic and flavonoid content in algal extracts using spectrophotometric techniques. *Journal of Applied Phycology*, *35*(2), 789–798. https://doi.org/10.1007/s10811-023-03012-0.

CHAPTER 10

Management of Major Wheat Diseases Under Changing Climatic Scenario for Sustainable Food and Nutritional Security

Ravindra Kumar[1], Barnik Debnath[2], Prashanth Babu H.[3], Meenakshi Rana[4] and Seweta Srivastava[4*]

1. Introduction

Wheat is one of the most important cultivated food crops across the world (Singh et al., 2023b). Diseases and pests that are constantly changing and expanding to new places due to changes in agricultural techniques, climate, and international trade pose a threat to wheat output worldwide (Pequeno et al., 2024). With more than 218 million hectares of agricultural area, wheat is one of the most significant grain crops grown worldwide (Alam et al., 2024). Global warming puts wheat production especially at danger in areas that are thought to be susceptible to rising temperatures and/or falling precipitation. According to Daloz et al. (2021), the yield reduction of wheat crop in India due to climate change is expected to vary from 6 to 23% by 2050.

As a significant staple food crop, wheat is crucial to the security of both food and nutrition. In almost every part of the world, including South Asia, it meets about 20% of the nutritional requirements of humans in terms of calories and protein intake. Therefore, in order to cater for the world's food needs, wheat's production potential needs to be maintained and improved. The world's staple crops continue to have genetic yield advances (Li et al., 2018), but better agronomic methods are needed to fully use these gains at farmer's fields under changing climate circumstances (Fischer & Connor, 2018). According to Fischer (2015), potential yield is the yield of the cultivar that is best suited to the current guidelines in agronomic management, provided that manageable abiotic and biotic challenges are absent. Abiotic factors such as lodging, salt, terminal heat stress, and water scarcity, as well as biotic factors including rusts, foliar blight, aphid infestation, and others, limit wheat's potential yields in various climates (Khan et al., 2023).

[1] ICAR-Indian Institute of Wheat and Barley Research, Karnal-132001, Haryana, India.
[2] ICAR-Central Institute for Cotton Research, Nagpur-440 010, Maharashtra, India.
[3] Division of Genetics, ICAR-Indian Agricultural Research Institute, New Delhi-110012, India.
[4] School of Agriculture, Lovely Professional University, Phagwara-144 411, Punjab, India.
* Corresponding author: seweta.21896@lpu.co.in.

DOI: 10.1201/9781003614869-10

The term "climate change" refers to variations in the average climate resulting from a variety of human activities, such as the burning of fossil fuels, deforestation, industrialization, exploitation of natural resources, and others, with regard to biotic and abiotic (temperature, precipitation, wind, and other elements) factors. Following the 18th century, as the Industrial Revolution grew, so did these kinds of activities. For over 650 thousand years, there has been a notable rise in the ambient concentration of greenhouse gases (GHGs), such as carbon dioxide (Bhatti et al., 2024; Shah et al., 2024). Over the past few decades, it has also been stated that the ambient global temperature on Earth has increased at a pace of 0.2°C each decade (Hansen et al., 2006; Forster et al., 2023). Climate change has also been linked to variations in the water cycle, such as unpredictable rainfall (Ficklin et al., 2022).

Climate change is expected to cause a major loss in global agriculture as long as greenhouse gas emissions are significantly reduced. According to estimates, global warming will cause agricultural productivity to decline by 3–16% by the 2080s (Malhi et al., 2021). Plant diseases are thought to be a significant barrier to attaining global food security since they cause at least 10% of the world's food supply to be lost (Mwangi et al., 2023). For more than a millennium, many biotic and abiotic environmental elements have been recognized as contributing to the development of a particular plant disease. The host (growth and resistance), the pathogen (reproduction, dispersal, survival, and pathogenicity), and their interactions can all be impacted by these parameters. Climate change is expected to necessitate changes to the current phyto-sanitary setup due to the dependence of plant diseases on multiple environmental conditions (Velásquez et al., 2018). The development of diseases in a particular host plant or location may be positively, neutrally, or negatively impacted by climate change (Lahlali et al., 2024). Thus, comprehending these impacts becomes essential for putting better disease control techniques into practice, such as developing plants resistant to disease, and preventing further yield losses (Ghini et al., 2008; Prasad et al., 2021).

Climate change is affecting many facets of our lives and livelihoods, including agriculture production. Climate change is having an ever-greater influence on agricultural land and food supply, and it becomes more important as the world's population continues to increase. Thus, a thorough understanding of the interaction of agricultural production systems and dynamic environmental circumstances is urgently needed to appropriately adapt to these climatic perturbations. This chapter examines the population dynamics of major wheat diseases and their management (Figure 1) vis-à-vis climate change.

Fig. 1 Some major diseases in wheat; (a) Stripe rust, (b) Leaf rust, (c) Stem rust, (d) Spot blotch, (e) Powdery mildew, (f) Fusarium head scab, (g) Loose smut, (h) Karnal bunt

Climate change and wheat crop diseases

Climatic conditions contribute to the disease triangle, which involves the presence of a susceptible host, a pathogen, and suitable environmental conditions for infection to occur, and climate change affects environmental conditions, whether it is in favor of the host or the pathogen. Examples of these favourable environmental conditions include dew, rain, relative humidity, temperature, aeration (wind), soil moisture, and sunlight intensity (Table 1). Predicting the impacts of climate change on plant disease is complex and challenging, as multiple aspects of plants, pathogens and the environment are involved. These factors include the distribution and abundance of taxa (geographical range, niche preference), their fitness and virulence, abiotic interactions, plant–microorganism evolutionary processes, host and vector biology, and environmental conditions. On the other hand, when environmental factors favour pathogen replication and susceptible hosts are available, a variety of soil-borne pathogenic microbes can trigger disease outbreaks (Cheng et al., 2019; Trivedi et al., 2020).

Table 1 Important diseases of wheat, their pathogens, and favourable climatic conditions.

Name of Disease	*Pathogen*	*Favourable Climatic Conditions*	*References*
Air-borne Diseases of Wheat			
Stem rust	*Puccinia graminis* f. sp. *tritici*	Thrives well in warm, moist conditions. Temperatures around 25–30°C (77–86°F) with high humidity are ideal.	Lewis et al. (2018)
Leaf rust	Puccinia triticina	Favors moderate temperatures (15–22°C or 59–72°F) and high humidity. Dew or light rain enhances spore germination.	Rodríguez-Moreno et al. (2020)
Stripe rust	*Puccinia striiformis* f. sp. *tritici*	Prefers cooler temperatures (10–15°C or 50–59°F) and high humidity. Nighttime temperatures are particularly critical, as cool, moist conditions promote infection.	Ali et al. (2023)
Powdery mildew	*Blumeria graminis* f. sp. *tritici*	Develops in cool, dry climates with temperatures between 15–20°C (59–68°F). Although it prefers dry conditions, a high humidity level is needed for spore germination and infection.	Draz et al. (2019); Bapela et al. (2023)
Septoria Tritici Blotch (STB)	*Zymoseptoria tritici* (formerly *Septoria tritici*)	Optimal conditions are cool and wet, with temperatures between 15–20°C (59–68°F). Prolonged periods of leaf wetness from rain or dew are necessary for infection. Under favourable conditions of moisture and temperature, secondary cycles of infection occur every 21–28 days.	Toropova et al. (2020); Navathe et al. (2023)
Fusarium Head Blight (FHB)	*Fusarium graminearum* *Gibberella zeae*	Warm and humid conditions, particularly during flowering (anthesis), increase the risk of infection. Temperatures between 20–30°C (68–86°F) with frequent rainfall or high humidity are ideal for the spread of the pathogen.	Martínez et al. (2022)

Contd.

Table 1 *Contd.*

Name of Disease	*Pathogen*	*Favourable Climatic Conditions*	*References*
Tan spot	*Pyrenophora tritici-repentis Drechslera tritici-repentis*	The disease thrives in cool, wet conditions, with temperatures between 15–28°C (59–82°F). High humidity and prolonged leaf wetness from rain or dew are necessary for infection and disease progression.	Navathe et al. (2023)
Seed-borne Diseases of Wheat			
Loose smut	Ustilago nuda tritici	The disease is more common in regions with cool, wet springs during flowering of the host. Infection occurs during the flowering stage of wheat when temperatures are cool (15–22°C or 59–72°F) and humid conditions prevail.	Ram & Singh (2004); Woldemichael (2019)
Common bunt (Stinking Smut)	*Tilletia caries* and *Tilletia foetida*	Cool, moist soil conditions at the time of seedling emergence are favourable for infection. Optimal temperatures for the development of common bunt are between 5–15°C (41–59°F).	Lunzer et al. (2023)
Karnal bunt	*Tilletia indica*	The disease thrives in cool and humid conditions during the heading and flowering stages. Optimal temperatures for infection are between 15–25°C (59–77°F) with high humidity or frequent rains.	Bishnoi et al. (2020)
Black point or kernel smudge	Various species of *Alternaria, Fusarium*, and *Helminthosporium*, and possibly other fungal genera	Occurrence of low vapour pressure deficit, high humidity, and low temperatures was associated with the formation of black point in susceptible varieties of wheat. High humidity or frequent rainfall during the maturation and harvest period increases the risk of infection.	Khani et al. (2018); Li et al. (2019)
Seedling blight and Root rot	*Fusarium* spp., *Bipolaris sorokiniana*, and other fungi	Common root rot is a disease of dry and warm areas. Disease severity increases when the plant is under stress or grown in warm soil and less moisture. Heavy, poorly drained soils are more prone to seedling blight and root rot.	Acharya et al. (2011); Al-Sadi (2021)
Wheat blast	*Magnaporthe oryzae* pathotype *Triticum* (MoT)	Hot and rainy humid conditions during the heading and flowering stages are ideal for wheat blast development. Temperatures around 25–30°C (77–86°F) with high humidity and frequent rain or dew favour the spread of the disease.	Islam et al. (2020)
Soil-borne Diseases of Wheat			
Take-All	*Gaeumannomyces graminis* var. *tritici*	Cool and moist soil conditions during early plant growth stages. High soil pH (neutral to alkaline). Soils with low organic matter and poor drainage.	Kwak & Weller (2013)

Contd.

Table 1 *Contd.*

Name of Disease	*Pathogen*	*Favourable Climatic Conditions*	*References*
Rhizoctonia root rot	*Rhizoctonia solani* and *R. cerealis*	Cool, wet weather conditions, especially in poorly drained soils. Temperature range between 10–20°C (50–68°F). Excessive soil moisture or poorly drained fields.	Dada (2017)
Fusarium crown rot	Several *Fusarium* species cause crown rot, with *Fusarium culmorum* and *F. pseudograminearum* being the most common	Warm, dry conditions during the growing season (temperature between 20–30°C or 68–86°F) but *F. culmorum* is often found in cooler regions with higher rainfall. Moist conditions at planting, followed by dry weather. Reduced tillage or stubble retention practices can increase the risk.	Poole et al. (2013); Kazan & Gardiner (2018)
Pythium root rot	*Pythium irregulare, P. vanterpoolii,* and *P. spinosum*	Cool, wet soil conditions, particularly in poorly drained fields. Soil temperatures around 10–15°C (50–59°F). Excessive soil moisture or frequent rainfall.	Reeves et al. (2021)
Common root rot	*Cochliobolus sativus*, also known as *Bipolaris sorokiniana*)	Warm temperatures ranging from 16 to 40°C, with ideal soil temperatures of 28–32°C and moderate to low soil moisture. Stressed conditions, such as drought or nutrient deficiencies, can exacerbate the disease.	Acharya et al. (2011); Xiong et al. (2023)
Flag smut	*Urocystis agropyri*	The disease is favoured under sandy light soil conditions and cooler temperature at the time of planting and seedling emergence. Soil temperatures ranging from 10–18°C (50–64°F).	Singh (2017)
Molya disease	Cereal Cyst Nematode (*Heterodera avenae*)	Cool, moist soil conditions during the early growth stages. Well-drained, light-textured soils.	Srivastava & Singh (2019)

Worldwide, more than 100 different diseases that are brought on by different pathogens and pests target wheat crops. These diseases cause around 21.5% of crop yield loss each year (Savary et al., 2019).

The environmental factors influence the microbes that cause these severe diseases as mentioned in Figure 2. Future climate circumstances that favour infections are expected to lead to a rise in multiple important wheat diseases. Through the introduction of novel pathogens, the eradication of antagonistic microbes, and the expansion of already existing pathogen populations, climate change promotes biological invasions (Bajwa et al., 2020; Kumar et al., 2021). Climate changes over the past 25 years also suggest increasingly conducive conditions for infection and, when climatic conditions are conducive, could severely harm wheat production (Lewis et al., 2018).

3. Impact of the Dynamics of Climate Change on Development of Some Important Diseases of Wheat

Disease development is significantly influenced by environmental factors (Iquebal et al., 2021). Worldwide climate variability has been affected by temperature and rainfall patterns, which has hampered wheat crop yields and food security. This has been exacerbated by probable changes in the distribution of diseases in many regions (Zhao et al., 2017; Li et al., 2019; Jägermeyr et al.,

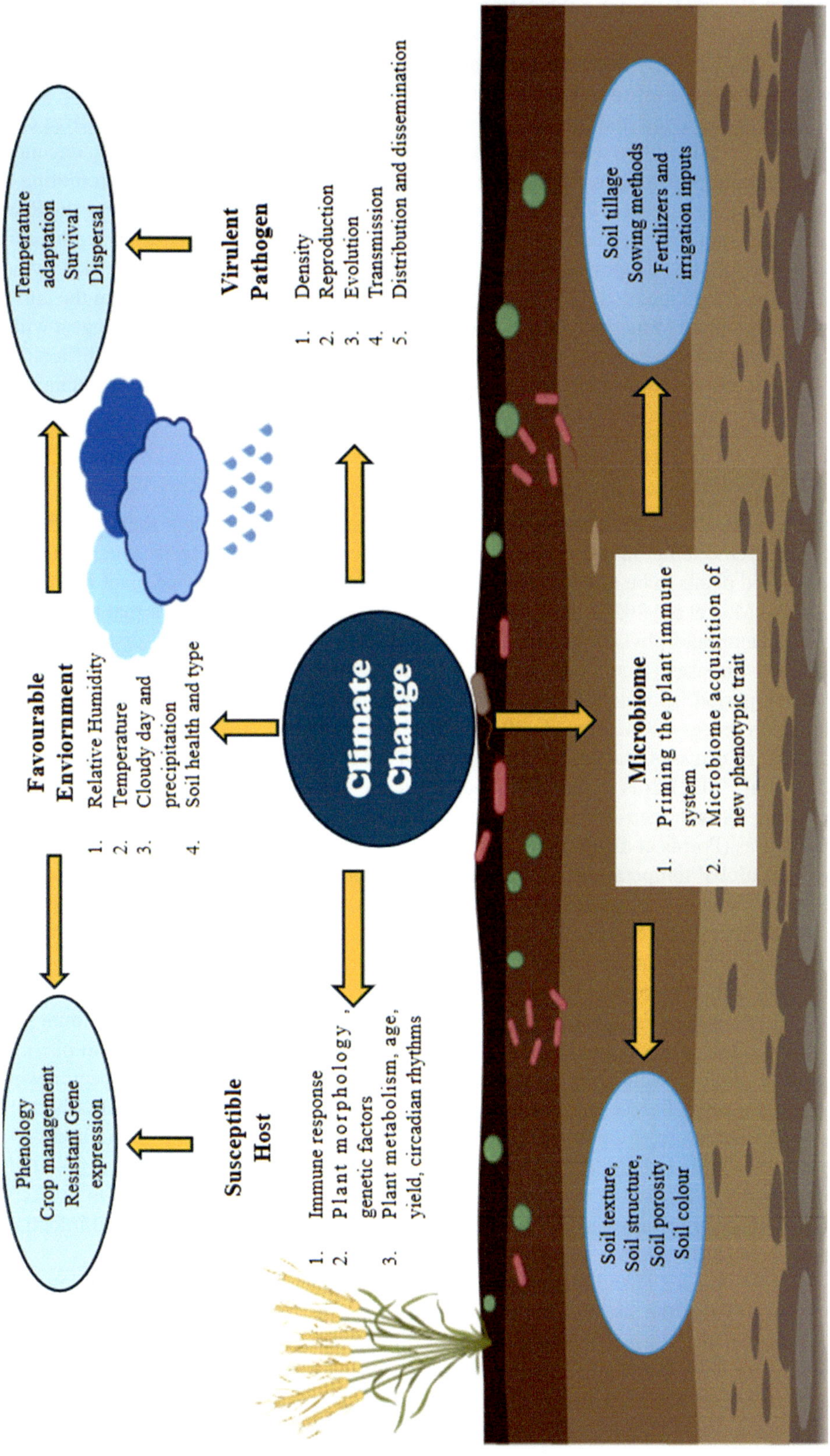

Fig. 2 The "disease triangle" concept simplified with focus on potential climate change effects.
(*Source:* Made by authors)

2021; Eckardt et al., 2023). Among the major staple crops threatened by climate change is wheat (*Triticum aestivum*), which is the source of >20% of the total protein and calories consumed worldwide on a daily basis. By 2050, there is a possibility that nearly 50% of the world's arable land may become salinized as a result of excessive irrigation and global climate change, posing serious obstacles to wheat production and global food security (Ismail & Horie 2017; Bailey-Serres et al., 2019). Furthermore, it should be mentioned that high temperatures have the potential to encourage the growth and spread of pathogens to new areas, greatly affecting wheat crops and promoting the local emergence of new diseases. Fusarium head blight (FHB) and Fusarium crown rot (FCR) are relatively new and rapidly spreading entrants that are currently the focus of resistance breeding efforts, while rust fungus and powdery mildew have been considered the greatest risk to wheat production for decades (Xiao et al., 2022). As a result, boosting wheat production from the current global total of 763 million tonnes annually to the 858 million tonnes that projections suggest will be necessary to feed a global population of 9.7 billion people by 2050 will be extremely difficult due to adverse weather patterns brought on by global climate change (Hickey et al., 2019; Gupta et al., 2020).

3.1 Temperature

Wheat production has been significantly impacted by global warming through the more frequent occurrence of extreme temperature events, which have increased water scarcity, aggravated soil salinization, caused plants to be more vulnerable to diseases, and directly reduced plant fertility and suppressed yield (Mao et al., 2023). Temperature is a critical environmental factor that affects plant growth, development, and physiological processes. It also shapes the life cycle, distribution, and virulence of plant diseases (Hunjan & Lore 2020). Furthermore, temperature influences the rate, duration, latency period, sporulation, inoculum dispersal, and host resistance mechanisms of the pathogen infection process (Kweku et al., 2019). High temperatures typically encourage the growth and spread of plant diseases, particularly bacteria and fungi, while also weakening host defence systems (Devi et al., 2022). It is imperative to acknowledge that there exists a variation in the ideal temperature range for pathogen-host systems, given that certain viruses display a sensitivity to high or low temperatures (Porras et al., 2023).

The average global temperature of an infected field will rise by approximately 40% due to a warmer climate with reduced relative humidity and increased turbulence. Reductions in the linkages between Europe, Africa, and South Asia as well as a rise in the quantity of spores traversing mid-latitude oceans are the two main anticipated changes in atmospheric long-range transport. The fungus's overwintering territories will grow as a result of a decrease in subfreezing circumstances (Prank et al., 2019). Wheat is known to have heat sensitivity during meiosis, which limits plant growth and yield under stress (Chen et al., 2021; Choudhary et al., 2022). In addition to playing a role in the spread of an epidemic, air temperature has been shown to influence how specific wheat cultivars respond to rusts. In recent years, it has become more common and severe, partially due to a warmer climate. Therefore, it is essential to comprehend and foresee how leaf rust will be affected by climate change (Caubel et al., 2017). It has been suggested that rising temperatures could lead to a rise in wheat rust infections (Coakley et al., 1999; Chakraborty et al., 2000). Due to weakened effector-based immune responses, wheat may become more susceptible to other fungal infections, such as Karnal bunt (KB) (Selin et al., 2016).

As a result of climate change, thermophilic fungal species proliferate and infections start to move northward from the south. Climate influences the soil fungi *Pythium, Rhizoctonia*, and *Sclerotinia*. Additionally, warmer temperatures have the potential to enhance the spread of many diseases that are currently constrained by the need to overwinter, such as wheat stem rot caused by *Puccinia graminis* f. sp. *tritici* (Singh et al., 2023a). They form the overwintering structures that protect them from external influences. Raising the temperature has the potential to shorten the latent period and make pathogens more aggressive. The virulence and resistance genes of pathogens in plants are influenced by temperature. Controlling the establishment of new plant diseases, enhancing prevention strategies, and creating cultivars with great adaptability are therefore imperative (Levitin, 2015). Evidence of *Puccinia striiformis* strains exhibiting aggressiveness at higher temperatures shows that wheat rust fungi are able to adapt to and take advantage of warmer temperatures in order to induce severe disease in previously unfavourable environmental conditions (Milus et al., 2009).

3.2 Relative Humidity (RH)

Relative humidity significantly impacts plant diseases by evapotranspiration, transpiration, and stomatal conductance as well as the germination, infection, and survival of plant pathogens (Dixit et al., 2023). The environmental factors that contribute to the severity of foliar and head diseases in wheat are rainfall, high humidity, and moderate temperatures (Agriopoulou, 2021). Furthermore, humidity is important for the production and deposit of dew, serving as a source of free water for many pathogens (Nath, 2021). Elevated humidity generally lowers the effectiveness of host resistance mechanisms and control measures, thereby promoting the growth and spread of plant diseases, especially those brought on by fungi and bacteria (Lim et al., 2023). Effect of relative humidity on plant diseases, however, might differ according to the pathogen type and stage, the host plant, and how it interacts with other weather conditions like temperature and precipitation (Dixit et al., 2023).

Conditions such as high air humidity, high soil wetness, and rain are conducive to the majority of plant diseases. Rain and high humidity, in particular, dramatically enhance the pathogenicity of the microbes that infect aerial tissues. Many devastating plant pathogens, like *Puccinia striiformis* (the fungus that causes stripe rust), require at least five hours of leaf wetness for disease development. For the majority of fungal diseases, this period of time is crucial for the development (Magarey et al., 2005). Some of the effects of high humidity on pathogens result in reduced marketability of products rather than direct reductions in agricultural yields, as demonstrated by various pathogenic fungi that generate mycotoxins, such as *Fusarium* spp. and *Aspergillus* spp. The crop cannot be sold if mycotoxins are present, even in trace amounts. Higher humidity enhanced the concentration of the mycotoxin deoxynivalenol in wheat plants infected with the head blight fungus *Fusarium graminearum* (Cowger et al., 2009), preventing the grain from being sold.

Warm temperatures of around 20–25°C combined with high relative humidity (RH) or leaf wetness create favourable conditions for the occurrence of several major diseases of wheat. For example, Head blight of wheat occurs when the RH reaches 85 to 100% and temperature of around 20–25% during flowering (Reis et al., 2023). Wheat blast disease risk increases when there is warm climate and elevated RH combined with prolonged wetness period (Pequeno et al., 2024). A number of climate change models have projected ongoing, extremely heavy precipitation occurrences, which may be advantageous for teliospore viability and germination. At first, this might lead to a rise in infections and the emergence of diseases (Coakley et al., 1999; Bishnoi et al., 2020). Higher yield losses due to teliospore-mediated yearly recurring infections and wheat growth period extension may result from the increased moisture. However, more wetness might promote suicidal germination, in which teliospores can form in the absence of a host, lowering the density of the inoculum (Carris et al., 2006).

3.3 Effects of Atmospheric CO_2 Level

The threat posed by wheat diseases to the world's food security is both ongoing and dynamic, but nothing is known about how CO_2 levels may affect these diseases and, in turn, the security of the grain supply (Lahlali et al., 2024). Increased CO_2 has been demonstrated to boost wheat growth, flowering time, maturation, and yield, but it also lowers grain nutritional quality and modifies plant primary and secondary metabolism (Högy et al., 2013). Plant susceptibility towards pathogens may be impacted by any of these modifications. This is concerning because it is additionally anticipated that future weather patterns will enhance the virulence, growth rate, and geographic distribution of important pathogens of wheat crop. Particularly, crop-destroying fungal diseases are predicted to pose one of the biggest threats to the world's food supply in the future (Bebber et al., 2014; Bebber & Gurr, 2015; Fisher et al., 2012).

A favourable association between the host-pathogen–environment is the cause of the occurrence of KB disease (Gupta et al., 2020; Iquebal et al., 2021). Therefore, the presence and severity of Karnal bunt disease would also be impacted by changes in the environmental circumstances in the locations where it occurs. Increased temperature, altered precipitation patterns, elevated CO_2 levels, and drought are a few predicted climate changes that could affect how severe the KB infection is (Iquebal et al., 2021). According to several climate change models, there may be a chance for teliospore survival and germination, which could boost infection and pathogen proliferation. Rising CO_2 levels accelerate the growth of leaves and stems, creating canopies that are rather dense and damp, which favours KB disease. Higher CO_2 can reduce the rate of decomposition of plants, resulting in increased crop residue that can harbour inoculum of some phytopathogens leading to an epidemic of diseases. Changes in temperature, humidity, and CO_2 concentration would cause a decrease in the effectiveness of chemical fungicides because of decreased absorption. By creating a more ideal habitat, the temperature and CO_2 changes may hasten the pathogen's main evolutionary phase (Smith & Luna, 2023).

FHB is a highly destructive fungal disease that affects wheat. Numerous closely related fungus species can cause FHB; nevertheless, *Fusarium graminearum* is the main etiological agent in many parts of the world (Thuraga et al., 2024). FHB epidemics can produce dangerous trichothecene mycotoxins in infected grains, making them unfit for human or animal consumption and significantly reducing grain output and quality. Since trichothecenes limit protein synthesis and cause diseases of the nervous system, gastrointestinal tract, and immune system, they pose a major threat to human health (Pestka, 2010; Wu et al., 2014). Nonetheless, the synthesis of trichothecene, particularly deoxynivalenol (DON), is a crucial virulent component that enables the pathogen to break the plant defences and disperse throughout the host (Bai et al., 2002; Hay et al., 2020). Widespread outbreaks of FHB caused by *F. graminearum* over the past few decades have cost the wheat market a great deal of money (Dweba et al., 2017). According to recent research, wheat resistance to FHB and related DON contamination may be jeopardized by rising atmospheric CO_2 concentrations (Váry et al., 2015; Bencze et al., 2017). When winter wheat was cultivated in ambient and raised CO_2 environments, transcriptome studies showed that the expression of key genes involved in plant defense response was significantly decreased by elevated CO_2 (Kane et al., 2013). These results raise worries that, assuming anticipated future climatic circumstances, the prevalence of FHB and severely DON-contaminated grain would increase (Cuperlovic-Culf et al., 2019; Hay et al., 2020).

4. Management of Wheat Diseases in Comparison with Climate Change

The complexity of managing wheat diseases in the context of climate change is growing because of changes in temperature, precipitation patterns, and extreme weather events. Wheat diseases' frequency, distribution, and severity may all change as a result of these modifications. The difficulty of modifying disease management strategies to account for climate change is unlikely to differ significantly from the adjustments needed to make opportunities for technology advancements or modifications to the economic structure that are now necessary for the protection of the wheat crop (Juroszek & von Tiedemann, 2011). Here are some strategies to manage wheat diseases in the context of climate change:

4.1 Breeding for Disease Resistance

1. **Developing Climate-Resilient Varieties:** The most effective, financially advantageous, and environmentally benign method of controlling plant diseases is to grow resistant cultivars (Omara et al., 2021). The goal of breeding projects should be to create wheat cultivars resistant to a wide range of diseases, particularly those that are anticipated to become more common as a result of climate change. Resistance to both current and new diseases fall under this category. Rust spore propagation is impacted by variations in wind direction and speed (Mukherjee et al., 2021). Identification of the more rust races is required, and precautions against new races should be performed, including disease inspections and variety analyses. A range of breeding studies should employ molecular tools to combat the virulence and aggression of rust infections, which can arise in response to temperature increases and other environmental factors. The pathogen creates new races, and these races have the potential to break the resistance that wheat cultivars have. Furthermore, the prevalence of old susceptible cultivars is a major factor in the rise and spread of rust diseases. The decline in the genetic diversity of wheat, in the pursuit of elite high-performing cultivars, may contribute to pathogen emergence to the point at which diseases can seriously affect global wheat yields seriously. Wheat productivity is severely hampered by pathogenic fungus (Figueroa et al., 2013). The capacity to expand production regions and boost productivity is limited by the scarcity of arable land, the pressure on water resources, and climate change. Moreover, a number of biotic and abiotic variables that lower wheat's potential yield are additionally responsible for the crop's low productivity (Mondal et al., 2016; Singh et al., 2016). Thus, in order to maximize yield potential, new-generation wheat cultivars must be created with improved tolerance/resistance to a variety of challenges, such as resistance to diseases, pests, soil salinity and alkalinity, and nitrogen-use efficiency (Bapela et al., 2023).
2. **Pyramiding Resistance Genes:** Combining multiple resistance genes into a single variety can provide more durable resistance against diseases that may evolve or spread due to climate change (Dong & Ronald, 2019). On the whole, it is best to identify numerous new sources of resistance that are feasible and, if applicable, incorporate them into future wheat cultivars that are preferred in resistance gene pyramids. Genetic variability of landraces is the basis for their outstanding adaptation to fluctuating climates (heat, cold, and drought), disease and pest resistance, stable and intermediate to higher yield levels even under low input agricultural systems, and good agronomic wheat traits (Lopes et al., 2015; Mondal et al., 2016; Bapela et al. 2023). Thus, for ensuring food security, valuable breeding qualities that are preferred by farmers and markets can be easily included into wheat varieties that are highly productive and well-suited from landraces (Cseh et al., 2021). In order to address the growing danger posed by the resurgence of warmer temperature-loving pathogens, e.g., *Puccinia graminis tritici*, wheat breeding programmes focusing on stem rust resistance should be strengthened. Yellow rust and stem rust resistances are not correlated; nevertheless certain seedling resistance genes may confer both (Bouvet et al., 2022).

4.2 Integrated Pest Management (IPM)

1. **Biological Control:** Promoting sustainable agriculture with a steady reduction in the application of synthetic agrochemicals and a more significant exploitation of compounds obtained from biological waste as well as the microbial; and genetic potential of crop plants and microbes is an effective strategy to decelerate the rapid environmental deterioration while ensuring high agricultural productivity and better soil health (Basu et al., 2021; Kumar et al., 2022). Encouraging the application of biocontrol agents—such as nematodes, bacteria, and beneficial fungi—can aid in the suppression of pathogens that cause disease in plants. Because of its excessive reliance on chemical fertilizers, use of pesticides, monoculture farming, high water consumption, and susceptibility to climate change, conventional wheat crop management

has a number of negative effects (Gianessi, 2013; Melnikov & Zhelezova, 2019; Zhelezova et al., 2019). Adopting more ecologically safe and sustainable farming practices, such as using biocontrol agents to treat wheat diseases, may be important for addressing these obstacles (Panin et al., 2022). Numerous strains of plant growth-promoting microorganisms (PGPM), such as *Bacillus subtilis* and *Pseudomonas fluorescens*, have been effectively used alone or in consortiums for pathogen's biocontrol. These PGPM are capable of successfully fighting off infections like *Fusarium graminearum*, which is responsible for head blight disease of wheat (Moussa et al., 2013). PFM2 strain of *Pseudomonas fluorescens* used against *Septoria tritici* in wheat causing Septoria tritici blotch (STB) (Levy et al., 1992).

2. **Precision Agriculture (PA):** It is necessary to comprehend the several factors influencing crop quality and quantity in order to manage the wheat crop effectively. Farmers can detect disease risks early and implement targeted treatments by using advanced monitoring techniques like remote sensing and disease prediction models. PA is used more effectively when data from remote sensing is used (Rebouh et al., 2023). Since it offers information and data on crop status during the growing season, remote sensing has emerged as among the most significant agricultural technologies that helps to improve agricultural management techniques (Gavrilovskaya et al., 2023). In order to make farming more sustainable and economically competitive, PA can assist farmers in using chemicals and fertilizers more effectively as per the needs of certain crops. Acquiring the required protein content in cereals like wheat through the use of variable rate application (VRA) of nutrients can improve crop quality (Pal et al., 2023).

4.3 Favourable Cultural Practices

1. **Adjusting Planting Dates:** Some diseases can be avoided by altering the planting schedule in order to prevent instances of high disease pressure, such as planting earlier or later in the season. A change in the planting date might decrease the likelihood that favourable weather conditions for disease will coincide with vulnerable crop growth stages (Jbir et al., 2022). This can be especially helpful in preventing warm, damp spells that encourage the growth of disease. Another efficient prevention method against disease associated with climate change is to modify the planting dates. Selecting the right date for sowing may additionally avoid situations that promote wheat blast (Pequeno et al., 2024). Early planting of spring wheat was found to minimize the incidence of FHB and the severity of leaf spot by 10 and 30%, respectively, in comparison to late planting (Subedi et al., 2007).
2. **Soil Health Management:** The soil microbiome in particular is essential for preserving plant health and biomass production. Long-term improvements in food crop production, nutritional value, and sustainability may be possible by targeted and untargeted management of soil microbial populations (Manda et al., 2023). Reduced tillage, organic amendments, and cover crops are examples of practices that promote soil health and can increase wheat's tolerance to soil-borne phytopathogens and stress conditions brought on by climate change. Encourage the use of cover crops and crop diversification to enhance soil health and give PGPR a consistent habitat. Agroecological techniques can be employed that promote sustainable agriculture and ecological equilibrium. Suppressive soils for numerous significant diseases, such as Pythium damping-off of wheat, were linked to enrichment in *Pseudomonas* spp. (Weller et al., 2002). Intriguing evidence can be seen in the continuous monoculture of barley and wheat, where significant take-all outbreaks have been reported to cause a phenomena known as "take-all decline" (TAD). Various fluorescent *Pseudomonas* species that may produce the antibiotic 2,4-diacetylphloroglucinol (DAPG) are enriched in TAD, which allows for the particular suppression of the pathogen *Gaeumannomyces graminis* var. *tritici*. According to certain research, in combinations like *Fusarium graminearum—Triticum aestivum* (wheat), continual minimum tillage (MT) treatment or long-term no-tillage (NT) can improve soil suppressiveness (Campos et al., 2016). A study on wheat leaf blotch disease caused by *Pyrenophora tritici-*

repentis revealed a decrease in disease severity and an improvement in yield after six years of rotation, particularly in systems utilizing NT and diverse crop rotations (Jalli et al., 2021). Plant diseases can be prevented and plant resilience to climate change can be enhanced by the phytomicrobiome, which is an important part of healthy soil and ecosystems (Chen et al., 2023).

3. **Mechanical Method and Crop Sanitation:** Preventing an outbreak of disease requires the implementation of good sanitation measures.To get rid of any pathogen sources, farmers should remove crop remains as soon as possible after harvest. Furthermore, the unintentional transfer of illnesses from one wheat field to another can be avoided by cleaning and disinfecting the equipment used there (Copes, 2016). An excellent sanitation strategy to control some soilborne plant infections is to deep summer plough beneath contaminated crop detritus. This is because tillage can expose the diseased plant materials to direct sunlight, which can kill some plant pathogens. In an integrated system, this is the initial stage in managing soilborne diseases (Manda et al., 2023). The primary cultural practice to eradicate the resting stages of wheat rust pathogens and limit the severity of the disease is field sanitation and rouging of volunteer plants and alternative hosts (Figueroa et al., 2018). Tan spot of wheat, caused by *Pyrenophora tritici-repentis*, can be managed by removal or destruction of infested residue, or tillage, which buries infested residue (Simón et al., 2013). The quantity of Septoria tritici blotch of wheat spores that can infect the crop the following season will be decreased if stubble is destroyed by grazing or cultivation. Mechanically cleaning of seeds is the most efficient way to control seed gall nematodes (Singhroha et al., 2017).
4. **Crop Rotation:** Crop rotation is one agronomic technique that makes an enormous impact in treating wheat diseases by lowering their occurrence and severity (Vagelas, 2025). The accumulation of pathogen inoculum in soil can be decreased by rotating crops with non-host crops. For some diseases, like take-all, crop rotation, or fallow, with at least one year out of wheat, barley, or other host crops can effectively lower inoculum levels, relying on the climate and the rate at which cereal residues decompose. Other diseases, like eye-spot and Cephalosporium stripe, necessitate longer rotations. For small grain diseases that have many hosts, such as legumes or oilseeds, or for circumstances where several years away from wheat or barley are necessary, long rotations could not be a cost-effective or efficient management strategy (Singh et al., 2023). Crop rotation, for example, will lessen inoculum accumulation and, consequently, wheat stripe rust (Carmona et al., 2020). According to Murray et al. (2000), rotating winter wheat with spring wheat or legumes will slow down the rate of fungal growth and aid in the breakdown of earlier soil-borne inoculum.*Cephalosporium gramineum* is a soil-borne pathogen that causes Cephalosporium stripe, a vascular disease of winter wheat and related cereals. Cephalosporium stripe infection risk can be decreased by rotating crops with non-host crops such as soybeans or maize (Murray, 2006; Singh et al., 2023). Only cropland that continuously grows wheat or barley without taking into account appropriate crop rotation techniques is likely to sustain the take-all illness (Cunfer et al., 2006). Wheat stem rust is not much affected by crop rotation because *Puccinia graminis tritici* resting spores can last for extended periods of time in soil (Salotti et al., 2022).

5. Climate-Smart Disease Monitoring and Forecasting

1. **Disease Surveillance and Thresholds:** Implementing robust disease monitoring systems that take into account changing climatic conditions can help in early disease detection and timely intervention. Mobile apps, satellite imagery, and weather-based disease forecasting models can be integrated to provide real-time alerts to farmers. Proactive measures, including early detection and eradication by frequent disease-scouting, preventive measures, and forecasting of potential outbreaks or new introductions by predictive modelling, could help in the production of wheat crop during climate change (Bajwa et al., 2020). These are all combined

with technologies that use automated spore capture and molecular identification of specific DNA markers of plant pathogens to anticipate the incidence of diseases throughout climate change. Such tools will also be crucial for evaluating the effectiveness of anti-resistance tactics meant to lower disease inocula, like the establishment of cropping-free intervals, which halt the development and dissemination of resistance to both new and existing fungicides (Ceresini et al., 2024).

2. **Climate Modeling and Disease Forecasting:** Utilizing climate models to predict how changes in temperature, humidity, and rainfall patterns will influence disease outbreaks can inform proactive disease management strategies. For example, models can help predict rust epidemics based on climate trends. Adopting a warning system (relying on weather monitoring or aerobiology, for example) to forecast outbreaks of diseases and reduce fungicide applications is advised in order to improve fungicide resistance management. To address these obstacles, more financing, teamwork, mandated reports, and capacity training are needed (Ali et al., 2023).

By making the agricultural system more resilient to climatic variability, the application of crop modeling techniques, such as those identified in the Agricultural Production Systems sIMulator (APSIM) (Keating et al., 2003) and the Decision Support System for Agrotechnology Transfer (DSSAT) (Jones & Thornton, 2003), may enhance the diversity in the agricultural system with regard to plant disease management (Lin, 2011). The wheat rust disease was simulated using the DYMEX-APSIM (Dynamic Model for Crop and Ecosystem Simulation linked with the Agricultural Production Systems SIMulator) crop disease model. It was demonstrated that the model could adapt the progression of the wheat plant to the expanding rust population and forecast the disease proportion in some of the years studied (Whish et al., 2015).

This always results in increased crop output and growth. Two risk models were used to gather decisions on the control of leaf blotch disease in wheat (caused by *Zymoseptoria tritici, Parastagonospora nodorum,* and *Pyrenophora tritici-repentis*) in order to decrease the consumption of chemicals and maximize crop yield during plant disease epidemics (Jørgensen et al., 2020). The two risk models, the humidity model (HM) and Crop Protection Online (CPO), both use relative humidity and precipitation to assess if fungicide application is necessary. In the 2018 trials, 95% of the predictions were accurate since, according to the models, only a small number of treatments decreased the quantity of chemicals used prior to the onset of the disease (Jørgensen et al., 2020). Temperature and moisture are the two main environmental elements that raise the incidence of wheat rust diseases and restrict local cultivation and wheat species yield, thus ideal weather conditions are crucial for their development. Therefore, a thorough understanding of the host plant's needs, the pathogen's disease cycles, and the environmental elements impacting the cycle are necessary for managing these deadly wheat diseases (Rodríguez-Moreno et al., 2020; Ajilogba & Walke, 2023). Rodríguez-Moreno et al. (2020) were able to predict the occurrence of leaf rust (LR) and stripe rust (SR) on wheat in Mexico by using two weather-based models and classification and regression trees (CARTs) for data analysis. As early warning systems, these forecasts are crucial for improving well-informed plans to lower yield losses brought on by rust infections.

6. Chemical Control Adjustments

1. **Rotating Fungicides:** Fungicide resistance is a concerning threat to agricultural systems, having a significant impact on food safety, animal and human health, and the environment. Rotating fungicides with various modes of action is crucial to manage resistance, particularly in a warmer environment where disease pressure may rise (Ceresini et al., 2024). In agriculture, normally management practices for plant disease control frequently depend excessively on a limited number of fungicides that pose a risk. Plant pathogenic fungi with naturally occurring genetic variation for fungicide resistance or newly developing de novo mutations may become

more numerous through selection and adaptation, if these few fungicides are applied frequently and extensively. Restricted alternatives for fungicides also make it more difficult to combine or rotate fungicides, which can slow the emergence of resistance (van den Bosch et al., 2014).

2. **Combined Formulations:** The issue is made worse by the sluggish development of new fungicides or fungicidal combinations as well as the quick evolution of resistance in plant pathogenic fungus populations. The restricted supply of new fungicides limits the alternatives available for managing diseases and makes it more difficult to effectively eradicate resistant strains (Corkley et al., 2022). Co-formulations meet the requirements for reduced environmental impact, public health safety, and agronomical efficacy. Novel fungicide formulations have been approved and released in recent times to address wheat-related diseases. The second-generation carboxamide fluxapyroxad, an SDHI that is thought to be at medium risk of fungicide resistance emergence, is one of these fungicides (FRAC, 2022).
3. **Climate-Responsive Applications:** Timing of fungicidal applications based on weather forecasts and disease development stages can maximize effectiveness and minimize environmental damage. Inadequate doses or scheduling are examples of incorrect application practices that might lead to the development of resistance. Fungal plant pathogen populations may develop fungicide resistance or other adaptive features more quickly as a result of stress brought on by being exposed to sub-lethal fungicide doses, which can also induce genomic instability in fungi (Gambhir et al., 2021). In addition to increased selection in pathogen populations, subtle dosages of fungicides may additionally allow resistant strains to persist and spread. Furthermore, insufficient crop coverage during treatment may leave some regions untreated, which will favor resistant lineages (Hollomon, 2015). For a period of several decades, rusts and other wheat leaf and head diseases have been significantly managed with the use of two fungicidal groups *i.e.*, azoles and QoIs (Poloni et al., 2021; Torres et al., 2022; Vicentini et al., 2022). The continuous use of a fixed fungicide creates high chances for fungicide resistance in fungal phytopathogens.

7. Policy and Extension Support

1. **Research and Development:** Increased investment in research to understand the impacts of climate change on wheat diseases and to develop new management practices is crucial. The policies encouraging research on climate change and its impact on crop and phytopathogens should be created in order to give farmers the help they need to adapt to climate change. In drafting policies that minimize the effects of climate change, farmers' perspectives and preferences for adaptation are critical, underscoring the necessity of looking into the different aspects influencing their decision-making process (Sheikh et al., 2024). Numerous studies have shown that farmers who have access to resources, information, and support were better able to adapt to climate change. Policies ought to concentrate on supplying these resources and subsequently aiding local farmers (Lybbert & Sumner, 2012; Arunrat et al., 2017). In order to produce stress-resilient crops and sustainable systems for their distribution, production, and safe utilization, the research community must collaborate. Even though there are still a lot of unanswered questions, substantial advances are being made in significant fields that point to viable solutions that will only be put into action in the years ahead with a coordinated effort by the scientific community working with policy and regulatory bodies (Benitez-Alfonso et al., 2023).
2. **Farmers' Education and Training:** Adapting to climate change is crucial for minimizing its impact on agriculture, particularly for the farmers of developing countries. Extension services should focus on educating farmers about climate change impacts and providing training on adaptive management practices. This includes how to implement IPM, use climate-smart technologies, and select appropriate wheat varieties. Farmers in different regions should

employ region-specific adaptation measures based on the type of crop taking into account climate change (Sheikh et al., 2024).

The perspective, trend, and drivers of wheat producers' adaptation strategies to climate change, however, are not well covered in the literature. According to several past research, farmers' livelihoods might be preserved and adverse consequences could be lessened by using climate change adaptation techniques (Ahmed & Givens, 2025). The results demonstrate that farmers' choices of climate change adaptation strategies are influenced by a number of factors, including gender, age, household heads' educational attainment, farmer-to-farmer extension, funding availability, and subsidy accessibility. Farmers' ability to adapt to climate change is influenced by a number of factors, including the institutional context, the socioeconomic status of households, and the availability and use of climatic information (Hamer et al., 2025). Wheat producers in the research area don't know enough about climate change and adaptation techniques. By working together, awareness-raising campaigns, and information and communication technology can significantly close the information gap. This suggests that extensive education and training initiatives are required to improve farmers' understanding of climate change adaptation. In order to increase farmers' awareness and understanding, training might be quite important (Raji et al., 2024).

The necessity of incorporating climate change considerations into national agriculture plans is further supported by the reported relationships between disease severity and climatic conditions. In order to help farmers control disease risks and minimize losses, a thorough wheat disease monitoring system that is backed by surveillance, forecasting models, and advisory services would be established. Programmes for training farmers in climate-smart farming techniques, such as crop rotation, soil health management, and precision irrigation, could support these initiatives (Mbanasor et al., 2024). In addition to addressing regional issues, this strategy establishes India as a pioneer in the adoption of climate-resilient farming methods in the area. India can create sustainable, locally appropriate solutions to guarantee wheat production stays stable and resilient in the face of changing climate circumstances by working together with government organizations, academic institutions, non-governmental organizations, and farmers.

8. Agro-Ecological and Integrated Approaches

1. **Agroforestry:** One major threat to agricultural system production is climate change. Extreme events and greater variability in climate conditions will likely have a greater effect on yield production than changes in mean climate alone (Burgess et al., 2022). Global wheat yield is likely to decrease by 6% for each °C increment in mean temperature (Zhao et al., 2017). Agricultural systems are facing additional challenges that interact with climate change and which can exacerbate its negative effects (Aguilera et al., 2020). Agroforestry systems (AFS) are sustainable land uses that deliberately combine trees (or shrubs) with crops or pastures (livestock) to expand farm and forest production. Incorporating trees and shrubs into wheat farming systems can help create microclimates that reduce disease pressure by modifying temperature, humidity, and wind patterns. Microclimate changes can have major effects on crop performance (Rolo et al., 2023). A voluminous literature describes the use of shade trees in temperate and tropical regions to protect heat-sensitive crops including wheat (Arenas-Corraliza et al., 2018; Inurreta-Aguirre et al., 2018).
2. **Diversified Farming Systems:** Diversification can simply imply functionally diversified crops (Lakhran et al., 2017), integration of livestock with crops in mixed farms (Bonaudo et al., 2014) which can enhance ecosystem services like natural pest and disease control and reduce vulnerability to diseases by increasing farm biodiversity. System diversification forms part of the strategy for coping with climate change effects (Altieri et al., 2015). The diversification of practices, such as intercropping and crop rotation, has been shown to be as beneficial as the addition of bio-fertilizers (Mahmud et al., 2021). When compared to monoculture, a diverse

crop rotation enhanced spring wheat yield by up to 30% and helped farmers cope with biotic challenges (Jalli et al., 2021). One effective tactic that farmers can use to adjust to climate change is on-farm diversification. The best strategies to diversify farm systems in light of the socioeconomic and agroecological difficulties farmers confront are not well recommended, nevertheless (van Zonneveld et al., 2020).

Many countries have adopted ambitious climate change mitigation and adaptation targets that will exacerbate the problem, as they require significant changes in current agri-food systems. By adopting the above-mentioned strategies, wheat growers can enhance the resilience of their crops to disease threats posed by a changing climate, ensuring sustainable wheat production in the future.

9. Conclusion

It is imperative to acknowledge the significance of climate perturbations in the field of Phytopathology and the close relationship between the environment and plant diseases. We are better able to predict or forecast the future occurrence of a plant disease when we possess the key details regarding the etiology of the disease as well as the environmental conditions around the pathogen and host. It is well known that disease rates are rising due to climate perturbations. It is challenging to forecast the location and timing of disease outbreaks. The efficiency of biological variables varies with the ecosystem, pathogen distribution is impacted, and the introduction of novel infections is made possible. Variations in plant-pathogen systems can be observed with climate change. Factors affecting both the host and the pathogen contribute to the development of disease. Effective crop protection technologies are available and will provide appropriate tools to adapt to altered climatic conditions, although the complexity of future risks for plant disease management may be considerable, particularly if new crops are introduced in an area. Potential beneficial effects of climate change, such as longer growing seasons, fewer frosts, and shifted precipitation patterns, must not be neglected, as they could counteract the presumed enhancement of particular diseases. Indeed, this information could provide useful insights into the likelihood and severity of disease outbreaks, the likelihood that a plant disease will manifest at all, or the areas of a state or nation where a crop can be grown with a reasonable degree of assurance that the disease won't spread.

References

Acharya, K., A.K. Dutta and P. Pradhan. (2011). *Bipolaris sorokiniana* (Sacc.) Shoem.: The most destructive wheat fungal pathogen in the warmer areas. *Aust. J. Crop Sci.*, *5*, 1064–1071.

Aguilera, E., C. Diaz-Gaona, R. Garcia-Laureano, C. Reyes-Palomo, G.I. Guzmán, L. Ortolani, M. Sanchez-Rodriguez and V. Rodriguez-Estevez. (2020). Agroecology for adaptation to climate change and resource depletion in the Mediterranean region. A review. *Agricultural Systems*, *181*, 102809.

Ahmed, M.N.Q. and J.E. Givens. (2025). Farmers' Climate Change Adaptation Strategies and the Role of Environmental Awareness and Education: A Review on Africa. *In*: Mbah, M.F., Molthan-Hill, P., Molua, E.L. (Eds.), *Practices, Perceptions, and Prospects for Climate Change Education in Africa*. Springer, Cham. doi: https://doi.org/10.1007/978-3-031-84081-4_9.

Ajilogba, C.F. and S. Walker. (2023). Using crop modeling to find solutions for wheat diseases: A review. *Front. Environ. Sci.*, *10*, 987765. doi: 10.3389/fenvs.2022.987765.

Alam, M., P. Stephen Baenziger and K. Frels. (2024). Emerging trends in wheat (*Triticum* spp.) breeding: Implications for the future. *Front. Biosci.* (Elite Edn.), *16*(1), 2.

Ali , Y., Iqbalb, S., Aatifa, H.M., Naveed, K., Khan, A.A., Ijaza, M., Magsi, M.M., Ahmad, S., Syed U.A., Magsi, M.A., Iqbal, R.K., Bukhari, N.A., Hatamleh, A.A. and Raza, A. (2023). Predicting Stripe Rust Severity in Wheat Using Meteorological Data with Environmental Response Modeling. *Journal of King Saud University–Science*, *35*, 102591.

Al-Sadi, A.M. (2021). *Bipolaris sorokiniana*-Induced black point, common root rot, and spot blotch diseases of wheat: A review. *Front. Cell Infect. Microbiol.*, *11*, 584899.

Altieri, M.A., C.I. Nicholls, A. Henao and M.A. Lana. (2015). Agroecology and the design of climate change-resilient farming systems. *Agron. Sustain. Dev.*, *35*(3), 869–890.

Arenas-Corraliza, M.G., M.L. López-Díaz and G. Moreno. (2018). Winter cereal production in a Mediterranean silvoarable walnut system in the face of climate change. *Agric. Ecosyst. Environ.*, *264*, 111–118.

Arunrat, N., C. Wang, N. Pumijumnong, S. Sereenonchai and W. Cai. (2017). Farmers' intention and decision to adapt to climate change: A case study in the Yom and Nan basins, Phichit province of Thailand. *J. Clean. Prod.*, *143*, 672–685.

Bai, G.H., A.E. Desjardins and R.D. Plattner. (2002). Deoxynivalenol-nonproducing *Fusarium graminearum* causes initial infection, but does not cause disease spread in wheat spikes. *Mycopathologia*, *153*, 91–98.

Bailey-Serres, J., J.E. Parker, E.A. Ainsworth, G.E.D. Oldroyd and J.I. Schroeder. (2019). Genetic strategies for improving crop yields. *Nature*, *575*(7781), 109–118.

Bajwa, A.A,, Farooq, M., Al-Sadi, A.M., Nawaz, A., Jabran, K. and Siddique, K.H.M. (2020). Impact of climate change on biology and management of wheat pests. *Crop Protection*, *137*, 105304.

Bapela, T., H. Shimelis, T. Terefe, S. Bourras, J. Sánchez-Martín, D. Douchkov, F. Desiderio and T.J. Tsilo. (2023). Breeding Wheat for Powdery Mildew Resistance: Genetic Resources and Methodologies: A review. *Agronomy*, *13*, 1173.

Basu, A., P. Prasad, S.N. Das, S. Kalam, R.Z. Sayyed, M.S. Reddy and H.E. Enshasy. (2021). Plant growth promoting rhizobacteria (PGPR) as green bioinoculants: Recent developments, constraints, and prospects. *Sustainability*, *13*, 140.

Bebber, D.P. and S.J. Gurr. (2015). Crop-destroying fungal and oomycete pathogens challenge food security. *Fungal Genet. Biol.*, *74*, 62–64.

Bebber, D.P., T. Holmes, D. Smith and S.J. Gurr. (2014). Economic and physical determinants of the global distributions of crop pests and pathogens. *New Phytol.*, *202*, 901–910.

Bencze, S., K. Puskás, G. Vida, I. Karsai, K. Balla, J. Komáromi and O. Veisz. (2017). Rising atmospheric CO_2 concentration may imply higher risk of *Fusarium* mycotoxin contamination of wheat grains. *Mycotoxin Res.*, *33*, 229–236.

Benitez-Alfonso, Y., B.K. Soanes, S. Zimba, B. Sinanaj, L. German, V. Sharma, A. Bohra, A. Kolesnikova, J.A. Dunn, A.C. Martin, M. Khashi u Rahman, Z. Saati-Santamaría, P. García-Fraile, E.A. Ferreira, L.A. Frazão, W.A. Cowling, K.H.M. Siddique, M.K. Pandey, M. Farooq, R.K. Varshney and C.H. Foyer. (2023). Enhancing climate change resilience in agricultural crops. *Curr. Biol.*, *33*(23), R1246–R1261.

Bhatti, U.A., M.A. Bhatti, H. Tang, M.S. Syam, E.M. Awwad, M. Sharaf and Y.Y. Ghadi. (2024). Global production patterns: Understanding the relationship between greenhouse gas emissions, agriculture greening, and climate variability. *Environ. Res.*, *245*, 118049.

Bishnoi, S.K., X. He, R.M. Phuke, P.L. Kashyap, A. Alakonya, V. Chhokar, R.P. Singh and P.K. Singh. (2020). Karnal Bunt: A re-emerging old foe of wheat. *Front Plant Sci.*, *11*, 569057.

Bonaudo, T., A.B. Bendahan, R. Sabatier, J. Ryschawy, S. Bellon, F. Leger, D. Magda and M. Tichit. (2014). Agroecological principles for the redesign of integrated crop–livestock systems. *Eur. J. Agron.*, *57*, 43–51.

Bouvet, L., S. Holdgate, L. James, J. Thomas, I.J. Mackay and J. Cockram. (2022). The evolving battle between yellow rust and wheat: Implications for global food security. *Theor. Appl. Genet.*, *135*(3), 741–753.

Burgess, A.J., M.E.C. Cano and B. Parkes. (2022). The deployment of intercropping and agroforestry as adaptation to climate change. *Crop and Environ.*, *1*, 145–160.

Campos, S.B., B.B. Lisboa, F.A. Camargo, C. Bayer, A. Sczyrba, P. Dirksen, A. Albersmeier, J. Kalinowski, A. Beneduzi, P.B. Costa, L.M.P. Passaglia, L.K. Vargas and V.F. Wendisch. (2016). Soil suppressiveness and its relations with the microbial community in a Brazilian subtropical agroecosystem under different management systems. *Soil Biol. Biochem.*, *96*, 191–197.

Carmona, M., F. Sautua, O. Pérez-Hérnandez and E.M. Reis. (2020). Role of fungicide applications on the integrated management of wheat stripe rust. *Front. Plant Sci.*, *11*, 733. doi: 10.3389/fpls.2020.00733.

Carris, L.M., L.A. Castlebury and B.J. Goates. (2006). Nonsystemic bunt fungi-*Tilletia indica* and *T. horrida*: A review of history, systematics, and biology. *Annu. Rev. Phytopathol.*, *44*, 113–133.

Caubel, J., M. Launay, D. Ripoche, D. Gouache, S. Buis, F. Huard, L. Huber, F. Brun and M.O. Bancal. (2017). Climate change effects on leaf rust of wheat: Implementing a coupled crop-disease model in a French regional application. *Eur. J. Agron.*, *90*, 53–66.

Ceresini, P.C., T.C. Silva, S.N.C. Vicentini, R.P. Leite Júnior, S.I. Moreira, K.C. Ríos, F.R.G. Fiallos, L.D. Krug, S.S. de Moura, A.G. da Silva, A.A. de Paiva Custódio, L.L.M. De Mio, M.C. de Godoy Gasparoto, D. Portalanza and W.C. de Jesus Júnior. 2024. Strategies for managing fungicide resistance in the Brazilian tropical agroecosystem: Safeguarding food safety, health, and the environmental quality. *Trop. Plant Pathol.*, *49*, 36–70.

Chakraborty, S., A.V. Tiedemann and P.S. Teng. (2000). Climate change: Potential impact on plant diseases. *Environ. Pollut.*, *108*, 317–326.

Chen, S., K. Stefanova, K.H.M. Siddique and W.A. Cowling. (2021). Transient daily heat stress during the early reproductive phase disrupts pod and seed development in *Brassica napus* L. *Food Energy Secur.*, *10*, e262.

Chen, W., D. Modi and A. Picot. (2023). Soil and phytomicrobiome for plant disease suppression and management under climate change: A review. *Plants*, *12*(14), 2736.

Cheng, Y.T., L. Zhang and S.Y. He. (2019). Plant–microbe interactions facing environmental challenge. *Cell Host Microbe*, *26*, 183–192.

Choudhary, M., G. Yan, K.H. Siddique and W.A. Cowling. (2022). Heat stress during meiosis has lasting impacts on plant growth and reproduction in wheat (*Triticum aestivum* L.). *Agronomy*, *12*, 987.

Coakley, S.M., H. Scherm and S. Chakraborty. (1999). Climate change and disease management. *Ann. Rev. Phyto.*, *37*, 399–426.

Copes, W.E. (2016). Sanitation for management of Florists' crops diseases. In: McGovern, R. and Elmer, W. (Eds.) *Handbook of Florists' Crop Diseases. Handbook of Plant Disease Management*. Springer, Cham. doi: https://doi.org/10.1007/978-3-319-32374-9_9-1.

Corkley, I., B. Fraaije and N. Hawkins. (2022). Fungicide resistance management: Maximizing the effective life of plant protection products. *Plant Pathol.*, *71*, 150–169.

Cowger, C., J. Patton-Ozkurt, G. Brown-Guedira and L. Perugini. (2009). Post-anthesis moisture increased Fusarium head blight and deoxynivalenol levels in North Carolina winter wheat. *Phytopathol.*, *99*, 320–327.

Cseh, A., P. Poczai, T. Kiss, K. Balla, Z. Berki, A. Horváth, C. Kuti, and I. Karsai. 2021. Exploring the legacy of Central European historical winter wheat landraces. *Sci. Rep.*, *11*, 23915.

Cunfer, B.M., G.D. Buntin and D.V. Phillips. (2006). Effect of crop rotation on take-all of wheat in double-cropping systems. *Plant Dis. 90*, 1161–1166.

Cuperlovic-Culf, M., M.M. Vaughan, K. Vermillion, A. Surendra, J. Teresi and S.P. McCormick. (2019). Effects of atmospheric CO_2 level on the metabolic response of resistant and susceptible wheat to *Fusarium graminearum* Infection. *Mol. Plant Microbe Interactions*, 32(4), 379–391.

Dada, T.E. (2017). Epidemiology of Wheat *Rhizoctonia. International J. Scientific and Res. Pub.*, *7*(5), 463–474.

Daloz, A.S., J.H. Rydsaa, O. Hodnebrog, J. Sillmann, B. van Oort, C.W. Mohr, M. Agrawal, L. Emberson, F. Stordal and T. Zhang. (2021). Direct and Indirect impacts of climate change on wheat yield in the Indo-Gangetic plain in India. *J. Agric. Food Res.*, *4*, 100132.

Devi, R., T. Kaur, D. Kour, A. Yadav, A.N. Yadav, A. Suman, A.S. Ahluwalia and A.K. Saxena. (2022). Minerals solubilizing and mobilizing microbiomes: A sustainable approach for managing minerals' deficiency in agricultural soil. *J. Appl. Microbiol.*, *133*, 1245–1272.

Dixit, S., P.N. Sivalingam, R.K. Murali-Baskaran, M. Senthil-Kumar and P.K. Ghosh. (2023). Plant responses to concurrent abiotic and biotic stress: Unravelling physiological and morphological mechanisms. *Plant Physiol. Rep.*, *29*, 6–17.

Dong, O.X. and P.C. Ronald. (2019). Genetic engineering for disease resistance in plants: Recent Progress and Future Perspectives. *Plant Physiol.*, *180*(1), 26–38.

Draz, I., Esmail, S., Abou-Zeid, M. and Essa, T. (2019). Powdery mildew susceptibility of spring wheat cultivars as a major constraint on grain yield. *Ann. Agric. Sci.*, *64*, 39–45.

Dweba, C., S. Figlan, H. Shimelis, T. Motaung, S. Sydenham, L. Mwadzingeni and T.J. Tsilo. (2017). Fusarium head blight of wheat: Pathogenesis and control strategies. *Crop Prot.*, *91*, 114–122.

Eckardt, N.A., E.A. Ainsworth, R.N. Bahuguna, M.R. Broadley, W. Busch, N.C. Carpita, G. Castrillo, J. Chory, L.R. DeHaan, C.M. Duarte, A. Henry, S.V.K. Jagadish, J.A. Langdale, A.D.B. Leakey, J.C. Liao, K.J. Lu, M.C. McCann, J.K. McKay, D.A. Odeny, E.J. de Oliveira, J.D. Platten, I. Rabbi, E.Y. Rim, P.C. Ronald, D.E. Salt, A.M. Shigenaga, E. Wang, M. Wolfe and X. Zhang. (2023). Climate change challenges, plant science solutions. *The Plant Cell*, *35*(1), 24–66.

Ficklin, D.L., S.E. Null, J.T. Abatzoglou, K.A. Novick and D.T. Myers. (2022). Hydrological intensification will increase the complexity of water resource management. *Earth's Future*, *10*(3), e2021EF002487.

Figueroa, M., K.E. Hammond-Kosack and P.S. Solomon. (2018). A review of wheat diseases: A field perspective. *Mol. Plant Pathol.*, *19*(6), 1523–1536. doi: 10.1111/mpp.12618.

Fischer, R.A. (2015). Definitions and Determination of Crop Yield, Yield Gaps, and of Rates of Change. *Field Crops Res.*, *182*, 9–18.

Fischer, R.A. and D.J. Connor. (2018. Issues for cropping and agricultural science in the next 20 years. *Field Crops Res.*, *222*, 121–142.

Fisher, M.C., D.A. Henk, C.J. Briggs, J.S. Brownstein, L.C. Madoff, S.L. McCraw and S.J. Gurr. (2012). Emerging fungal threats to animal, plant, and ecosystem health. *Nature*, *484*, 186–194.

Forster, P.M., C.J. Smith, T. Walsh, W.F. Lamb, R. Lamboll, M. Hauser, A. Ribes, D. Rosen, N. Gillett, M.D. Palmer, J. Rogelj, K. von Schuckmann, S.I. Seneviratne, B. Trewin, X. Zhang, M. Allen, R. Andrew, A. Birt, A. Borger, T. Boyer, J.A. Broersma, L. Cheng, F. Dentener, P. Friedlingstein, J.M. Gutiérrez, J. Gütschow, B. Hall, M. Ishii, S. Jenkins, X. Lan, J.Y. Lee, C. Morice, C. Kadow, J. Kennedy, R. Killick, J.C. Minx, V. Naik, G.P. Peters, A. Pirani, J. Pongratz, C.F. Schleussner, S. Szopa, P. Thorne, R. Rohde, M. Rojas Corradi, D. Schumacher, R. Vose, K. Zickfeld, V. Masson-Delmotte and P. Zhai. (2023). Indicators of Global Climate Change 2022: Annual update of large-scale indicators of the state of the climate system and human influence. *Earth Syst. Sci. Data*, *15*, 2295–2327.

FRAC–Fungicide Resistance Action Committee. (2022). Recommendations for SDHI fungicides. In: *Crop Life Int. Bruss. Belg.*

Gambhir, N., Z.N. Kamvar, R. Higgins, B.S. Amaradasa and S.E. Everhart (2021) Spontaneous and fungicide-induced genomic variation in *Sclerotinia sclerotiorum. Phytopathol.*, *111*, 160–169.

Gavrilovskaya, N., A. Cheremisin, A. Novikov, A. Lukyanets and E. Moreva. (2023). Advances in space-scale farming: UAV and satellite monitoring of wheat production in Krasnodar, Russian Federation. *Int. J. Environ. Stud.*, *80*(1), 158–171.

Ghini, R., E. Hamadaand and W. Bettiol. (2008). Climate change and plant diseases. *Scientia Agricola*, *65*, 98–107.

Gianessi, L.P. 2013. The increasing importance of herbicides in worldwide crop production. *Pest Manag. Sci.*, *69*(10), 1099–1105.

Gupta, A., A. Kumar, R. Kumar and K. Verma (2020). Evaluation of foliar applications at different growth stages and soil amendment for management of Karnal bunt disease in wheat. *Seed Research*, *48*(1), 25–28.

Gupta, A.K., M. Negi, S. Nandy, M. Kumar, V. Singh, D. Valente, I. Petrosillo and R. Pandey. (2020). Mapping socio-environmental vulnerability to climate change in different altitude zones in the Indian Himalayas. *Ecological Indicators*, *109*, 105787.

Hamer, E.A., T.D.T. Oyedotun, E. Charles and S. Moonsammy. (2025). Livelihood adaptation to climate-related changes in agroecological zones in developing countries: Challenges, prospects, and policy concerns. *Front. Clim.*, *7*, 1594747. doi: 10.3389/fclim.2025.1594747.

Hansen, J., M. Sato, R. Ruedy, K. Lo, D.W. Lea and M. Medina-Elizade. (2006). Global temperature change. *Proc. Natl. Acad. Sci. USA*, *103*(39), 14288–14293.

Hay, W.T., S.P. McCormick, M.P. Hojilla-Evangelista, M.J. Bowman, R.O. Dunn, J.M. Teresi, M.A. Berhow and M.M. Vaughan. (2020). Changes in wheat nutritional content at elevated [CO_2] alter *Fusarium graminearum* growth and mycotoxin production on grain. *J. Agric. Food Chem.*, *68*(23), 6297–6307.

Hickey, L.T., A.N. Hafeez, H. Robinson, S.A. Jackson, S.C.M. Leal-Bertioli, M. Tester, C. Gao, I.D. Godwin, B.J. Hayes and B.B.H. Wulff. (2019). Breeding crops to feed 10 billion. *Nat. Biotechnol.*, *37*, 744–754.

Högy, P., M. Brunnbauer, P. Koehler, K. Schwadorf, J. Breuer, J. Franzaring, D. Zhunusbayeva and A. Fangmeier. (2013). Grain quality characteristics of spring wheat (*Triticum aestivum*) as affected by free-air CO_2 enrichment. *Environ. Exper. Bot.*, *88*, 11–18.

Hollomon, D.W. (2015). Fungicide resistance: Facing the challenge: A review. *Plant Prot. Sc.*, *51*, 170–176.

Hunjan, M.S. and J.S. Lore. (2020). Climate change: Impact on plant pathogens, diseases, and their management. *Crop Protection under Changing Climate*. Springer Nature, pp. 85–100.

Inurreta-Aguirre, H.D., P.É. Lauri, C. Dupraz and M. Gosme. (2018). Yield components and phenology of durum wheat in a Mediterranean alley-cropping system. *Agrofor. Syst.*, *92*(4), 961–974.

Iquebal, M.A., P. Mishra, R. Maurya, S. Jaiswal, A. Rai and D. Kumar. (2021). Centenary of soil and air borne wheat karnal bunt disease research: A review. *Biology* (*Basel*), *10*(11), 1152.

Islam, M.T., D.R. Gupta, A. Hossain, K.K. Roy, X. He, M.R. Kabir, P.K. Singh, Md. A.R. Khan, M. Rahman and G.L. Wang. (2020). Wheat blast: A new threat to food security. *Phytopathol. Res.*, *2*, 28.

Ismail, A.M. and T. Horie. (2017). Genomics, physiology, and molecular breeding approaches for improving salt tolerance. *Annu. Rev. Plant Biol.*, *68*, 405–434.

Jägermeyr, J., C. Müller, A.C. Ruane, J. Elliott, J. Balkovic, O. Castillo, B. Faye, I. Foster, C. Folberth, J.A. Franke, K. Fuchs, J.R. Guarin, J. Heinke, G. Hoogenboom, T. Iizumi, A.K. Jain, D. Kelly, N. Khabarov, S. Lange, T.S. Lin, W. Liu, O. Mialyk, S. Minoli, E.J. Moyer, M. Okada, M. Phillips, C. Porter, S.S. Rabin, C. Scheer, J.M. Schneider, J.F. Schyns, R. Skalsky, A. Smerald, T. Stella, H. Stephens, H. Webber, F. Zabel and C. Rosenzweig. (2021). Climate impacts on global agriculture emerge earlier in new generation of climate and crop models. *Nat. Food*, *2*, 873–885.

Jalli, M., E. Huusela, H. Jalli, K. Kauppi, M. Niemi, S. Himanen and L. Jauhiainen. (2021). Effects of crop rotation on spring wheat yield and pest occurrence in different tillage systems: A multi-year experiment in Finnish growing conditions. *Front. Sustain. Food Syst.*, *5*, 647335.

Jbir, T.G., F.K. Crutcher, J. Rickertsen, D. Fonseka, A.J. Friskop and A.K. Kalil. (2022). Influence of planting date and cultivar on diseases of spring durum wheat. *Plant Dis.*, *106*(12), 3083–3090.

Jones, P.G. and P.K. Thornton. (2003). The potential impacts of climate change on maize production in Africa and Latin America in 2055. *Glob. Environ. Change*, *13*(1), 51–59. doi:10.1016/s0959-3780(02)00090-0.

Jørgensen, L.N., N. Matzen, A. Ficke, G.C. Nielsen, M. Jalli, A. Ronis, B. Andersson and A. Djurle. (2020). Validation of risk models for control of leaf blotch diseases in wheat in the Nordic and Baltic countries. *Eur. J. Plant Pathology*, *157*(3), 599–613. doi:10.1007/s10658-020-02025-6.

Juroszek, P. and A. von Tiedemann. (2011). Potential strategies and future requirements for plant disease management under a changing climate. *Plant Pathol.*, *60*(1), 100–112.

Kane, K., Dahal, K.P., Badawi, M.A., Houde, M., Hüner, N.P. and Sarhan, F. (2013). Long-term growth under elevated CO_2 suppresses biotic stress genes in non-acclimated, but not cold-acclimated winter wheat. *Plant Cell Physiol.*, *54*, 1751–1768.

Kazan, K. and D.M. Gardiner. 2018. Fusarium crown rot caused by *Fusarium pseudograminearum* in cereal crops: recent progress and future prospects. *Mol. Plant Pathol.*, *19*(7), 1547–1562.

Keating, B.A., P.S. Carberry, G.L. Hammer, M.E. Probert, M.J. Robertson, D. Holzworth, N.I. Huth, J.N.G. Hargreaves, H. Meinke, Z. Hochman, G. McLean, K. Verburg, V. Snow, J.P. Dimes, M. Silburn, E. Wang, S. Brown, K.L. Bristow,

S. Asseng, S. Chapman, R.L. McCown, D.M. Freebairn and C.J. Smith. (2003). An overview of APSIM, a model designed for farming systems simulation. *Eur. J. Agron.*, *18*(3–4), 267–288. doi:10.1016/s1161-0301(02)00108-9.

Khan, H., H.M. Mamrutha, C.N. Mishra, G. Krishnappa, R. Sendhil, O. Parkash, A.K. Joshi, R. Chatrath, B.S. Tyagi, G. Singh and G.P. Singh. (2023). Harnessing high yield potential in wheat (*Triticum aestivum* L.) under climate change scenario. *Plants*, *12*, 1271.

Khani, M., J. Cheong, K. Mrva and D. Mares. (2018). Wheat black point: Role of environment and genotype. *J. Cereal Sci.*, *82*, 25–33.

Kumar, L., N. Chhogyel, T. Gopalakrishnan, Md K. Hasan, S.L. Jayasinghe, C.S. Kariyawasam, B.K. Kogo and S. Ratnayake. (2021). Climate change and future of agri-food production. *In*: Rajeev Bhat (Eds.), *Future Foods: Global Trends, Opportunities, and Sustainability Challenges*. Elsevier, New York, USA, pp. 49–79.

Kumar, R., S. Kumar, S. Srivastava, P.L. Kashyap, A. Kumar, R.K. Shekhar and G.P. Singh. (2022). Rhizosphere Microbes and Wheat Health Management. *In*: U.B. Singh et al. (Eds.) *Rhizosphere Microbes, Microorganisms for Sustainability*. Springer Nature Singapore Pte Ltd., pp. 223–242.

Kwak, Y.S. and D.M. Weller. (2013). Take-all of wheat and natural disease suppression: A review. *Plant Pathol J.*, *29*(2), 125–135.

Kweku, D., O. Bismark, A. Maxwell, K. Desmond, K. Danso, E. Oti-Mensah, A. Quachie and B. Adormaa. (2019). Greenhouse effect: Greenhouse gases and their impact on global warming. *J. Sci. Res. Rep.*, *17*, 1–9.

Lahlali, R., M. Taoussi, S.E. Laasli, G. Gachara, R. Ezzouggari, Z. Belabess, K. Aberkani, A. Assouguem, A. Meddich, M. El Jarroudi and E.A. Barka. (2024). Effects of climate change on plant pathogens and host-pathogen interactions. *Crop and Environ.*, *3*(3): 159–170.

Lakhran, H., Kumar, S. and Bajiya, R. (2017). Crop diversification: An option for climate change resilience. *Trends in Bioscience*, *10*(2), 516–518.

Levitin, M.M. (2015). Microorganisms and global climate change. *Selskokhozyaystvennaya Biologiya = Agricultural Biol.*, *50*(5), 641–647.

Levy, E., F.J. Gough, K.D. Berlin, P.W. Guiana and J.T. Smith. (1992). Inhibition of *Septoria tritici* and other phytopathogenic fungi and bacteria by *Pseudomonas fluorescens* and its antibiotics. *Plant Pathol.*, *41*(3), 335–341. doi:10.1111/j.1365-3059.1992.tb02355.x.

Lewis, C.M., A. Persoons, D.P. Bebber, R.N. Kigathi, J. Maintz, K. Findlay, V. Bueno-Sancho, P. Corredor-Moreno, S.A. Harrington, N. Kangara, A. Berlin, R. García, S.E. Germán, A. Hanzalová, D.P. Hodson, M.S. Hovmøller, J. Huerta-Espino, M. Imtiaz, J.I. Mirza, A.F. Justesen, R.E. Niks, A. Omrani, M. Patpour, Z.A. Pretorius, R. Roohparvar, H. Sela, R.P. Singh, B. Steffenson, B. Visser, P.M. Fenwick, J. Thomas, B.B.H. Wulff and D.G.O. Saunders. (2018).. Potential for re-emergence of wheat stem rust in the United Kingdom. *Commun Biol.*, *1*, 13.

Li, H., A. Rasheed, L.T. Hickey and Z. He. (2018). Fast-Forwarding Genetic Gain. *Trends Plant Sci.*, *23*, 184–186.

Li, Q.Y., S.Y. Wang, S.W. Chang, K.G. Xu, M.Y. Li, Q.Q. Xu, Y.M. Jiang and J.S. Niu. (2019). Key periods and effects of meteorological factors affecting incidence of wheat black point in the Yellow and Huai wheat area of China. *Crop Prot.*, *125*, 104882.

Lim, J.A., J.S. Yaacob, S.R.A. Mohd Rasli, J.E. Eyahmalay, H.A. El Enshasy and M.R.S. Zakaria. (2023). Mitigating the repercussions of climate change on diseases affecting important crop commodities in Southeast Asia, for food security and environmental sustainability: A review. *Front. Sustain. Food Syst.*, *6*, 1030540.

Lin, B.B. (2011). Resilience in agriculture through crop diversification: Adaptive management for environmental change. *BioScience*, *61*(3), 183–193. doi:10.1525/bio.2011.61.3.4.

Lopes, M.S., L. El-Basyoni, P.S. Baenziger, S. Singh, C. Royo, K. Ozbek, H. Aktas, E. Ozer, F. Ozdemir, A. Manickavelu, T. Ban and P. Vikram. (2015). Exploiting genetic diversity from landraces in wheat breeding for adaptation to climate change. *J. Exp. Bot.*, *66*, 3477–3486.

Lunzer, M., V. Dumalasová, K. Pfatrisch, H. Buerstmayr and H. Grausgruber. (2023). Common bunt in organic wheat: Unravelling infection characteristics relevant for resistance breeding. *Front. Plant Sci.*, *14*, 1264458.

Lybbert, T.J. and D.A. Sumner. (2012). Agricultural technologies for climate change in developing countries: Policy options for innovation and technology diffusion. *Food Pol.*, *37*, 114–123.

Magarey, R.D., T.B. Sutton and C.L. Thayer. (2005). A simple generic infection model for foliar fungal plant pathogens. *Phytopathol.*, *95*, 92–100.

Mahmud, A.A., S.K. Upadhyay, A.K. Srivastava and A.A. Bhojiya. (2021). Biofertilizers: A nexus between soil fertility and crop productivity under abiotic stress. *Curr. Res. Environ. Sustain.*, *3*, 100063.

Malhi, G.S., M. Kaur and P. Kaushik. (2021). Impact of climate change on agriculture and its mitigation strategies: A review. *Sustainability*, *13*(3), 1318.

Manda, R.R., V.A. Addanki, A. Giabardo, J. Benjamin, M.J. Hossain, S. Khanna, M. Gaddam, R. Kumar and S. Srivastava. (2023). Soil Health Management and Microorganisms: Recent Development. *In*: U. B. Singh et al. (Eds.), *Detection, Diagnosis, and Management of Soil-borne Phytopathogens*. Springer Nature Singapore Pte Ltd., pp. 437–493.

Mao, H., Jiang, C., Tang, C., Nie, X., Du, L., Liu, Y., Cheng, P., Wu, Y., Liu, H., Kang, Z. and Wang, X. (2023). Wheat adaptation to environmental stresses under climate change: Molecular basis and genetic improvement. *Mol. Plant*, *16*(10), 1564–1589.

Martínez, M., Biganzoli, F., Arata, A., Dinolfo, M.I., Rojas, D., Cristos, D. and Stenglein, S.. (2022). Warm nights increase Fusarium Head Blight negative impact on barley and wheat grains. *Agricultural and Forest Meteorology*, *318*, 108909.

Mbanasor, J.A., C.A. Kalu, C.I. Okpokiri, C.N. Onwusiribe, P.O. Nto, N.M. Agwu and M.C. Ndukwu. (2024). Climate smart agriculture practices by crop farmers: Evidence from south east Nigeria. *Smart Agric Technol.*, *8*, 100494.

Melnikov, A. and S. Zhelezova. (2019). Traditional tillage or zero tillage technology: What is more profitable for the production of winter wheat in the non-chernozem zone of Russia? *Theor. Appl. Probl. Agr. Industry*, *39*, 35–40.

Milus, E.A., K. Kristensen and M.S. Hovmøller. (2009). Evidence for increased aggressiveness in a recent widespread strain of *Puccinia striiformis* f. sp. *tritici* causing stripe rust of wheat. *Phytopathol.*, *99*, 89–94.

Mondal, S., J.E. Rutkoski, G. Velu, P.K. Singh, L.A. Crespo-Herrera, C. Guzman, S. Bhavani, C. Lan, X. He and R.P. Singh. (2016). Harnessing diversity in wheat to enhance grain yield, climate resilience, disease and insect pest resistance, and nutrition through conventional and modern breeding approaches. *Front. Plant. Sci.*, *7*, 991.

Moussa, T.A.A., O.A. Almaghrabi and T.S. Abdel-Moneim. (2013). Biological control of the wheat root rot caused by *Fusarium graminearum* using some PGPR strains in Saudi Arabia. *Ann. Appl. Biol.*, *163*, 72–81.

Mukherjee, R., H.A. Gruszewski, L.T. Bilyeu, D.G. Schmale III and J.B. Boreyko. (2021). Synergistic dispersal of plant pathogen spores by jumping-droplet condensation and wind. *Proc. Natl. Acad. Sci. USA*, *118*(34), e2106938118.

Murray, T.D. (2006). Seed transmission of *Cephalosporium gramineum* in winter wheat. *Plant Dis.*, *90*, 803–806. doi: 10.1094/PD-90-0803.

Murray, T., S. Jones and E. Adams. (2000). *Snow mold diseases of winter wheat in Washington*. Washingt. State. Univ. Coop. Ext. USDA, pp. 1–8.

Mwangi, R.W., M. Mustafa, K. Charles, I.W. Wagara and N. Kappel. (2023). Selected emerging and reemerging plant pathogens affecting the food basket: A threat to food security. *Journal of Agriculture and Food Res.*, *14*, 100827.

Nath, S. (2021). Dew as source of emerging contaminants in agricultural system. *Sustainable Agriculture Reviews*, *50*. Springer, Cham, Switzerland, pp. 61–78.

Navathe, S., He, X., Kamble, U., Kumar, M., Patial, M., Singh, G., Singh, G.P., Joshi, A.K. and Singh, P.K. (2023). Assessment of Indian wheat germplasm for Septoria nodorum blotch and tan spot reveals new QTLs conferring resistance along with recessive alleles of Tsn1 and Snn3. *Front. Plant Sci.*, *14*, 1223959.

Omara, R.I., Y. Nehela, O.I. Mabrouk and M.M. Elsharkawy. (2021). The emergence of new aggressive leaf rust races with the potential to supplant the resistance of wheat cultivars. *Biol.* (*Basel*), *10*(9), 925.

Pal, D., S. Saha and S. Banerjee. (2023). Chapter-6, Precision agriculture in view of climate change towards long term sustainability. *In*: *Advances in Agronomy*, *26*. AkiNik Publications® New Delhi, pp. 119–144.

Panin, A.V., D.V. Timokhin, L.A. Golovina and E.P. Lidinfa. (2022). Priorities for the development of domestic crop production in the context of closing the resource and technological cycles of the "Smart Village". *In*: *Smart innovation in Agriculture*. Springer Nature Singapore, Singapore, pp. 323–331.

Pequeno, D.N.L., T.B. Ferreira, J.M.C. Fernandes, P.K. Singh, W. Pavan, K. Sonder, R. Robertson, T.J. Krupnik, O. Erenstein and S. Asseng. (2024). Production vulnerability to wheat blast disease under climate change. *Nat. Clim. Change*, *14*(2), 178–183.

Pequeno, D., T. Ferreira, J. Fernandes, P. Singh, W. Pavan, K. Sonder, R. Robertson, T. Krupnik, O. Erenstein and A. Asseng. (2024). Production vulnerability to wheat blast disease under climate change. *Nature Climate Change*, *14*, 178–183.

Pestka, J. (2010). Toxicological mechanisms and potential health effects of deoxynivalenol and nivalenol. *World Mycotoxin J.*, *3*, 323–347.

Poloni, N.M., G. Carvalho, S. Nunes Campos Vicentini, A.F. Dorigan, J.L. Nunes Maciel, B.A. McDonald, S. Intra Moreira, N. Hawkins, B.A. Fraaije and D.E. Kelly. (2021). Widespread distribution of resistance to triazole fungicides in Brazilian populations of the wheat blast pathogen. *Plant Pathol.*, *70*, 436–448.

Poole, G.J., R.W. Smiley, C. Walker, D. Huggins, R. Rupp, J. Abatzoglou, K. Garland-Campbell and T.C. Paulitz. (2013). Effect of climate on the distribution of Fusarium species causing crown rot of wheat in the Pacific Northwest of the US. *Phytopathology*, *103*, 1130–1140.

Porras, M.F., C.A. Navas, G.A. Agudelo-Cantero, M.G. Santiago-Martínez, V. Loeschcke, J.G. Sørensen, S.G., Crandall, D. Biddinger and E.G. Rajotte. (2023). Extreme heat alters the performance of hosts and pathogen. *Front. Ecol. Evol.*, *11*, 1186452.

Prank, M., Kenaley, S.C., Bergstrom, G.C., Acevedo, M. and Mahowald, N.M. (2019). Climate change impacts the spread potential of wheat stem rust, a significant crop disease. *Environ. Res. Lett.*, *14*, 124053.

Prasad, P., S.C. Bhardwaj, R.K. Thakur, S. Adhikari, O.P. Gangwar, C. Lata and S. Kumar. (2021). Prospects of climate change effects on crop diseases with particular reference to wheat. *J. Cereal Res.*, *13*(2), 118–135.

Raji, E., T.I. Ijomah and O.G. Eyieyien. (2024). Improving agricultural practices and productivity through extension services and innovative training programs. *International J. Appl. Res. Social Sci.*, *6*(7), 1297–1309. doi: https://doi.org/10.51594/ijarss.v6i7.1267.

Ram, B. and K.P. Singh. (2004). Smuts of wheat: A review. *Indian Phytopath.*, *57*(2), 125–134.

Rebouh, N.Y., E.S. Mohamed, P.M. Polityko, P.A. Dokukin, D.E. Kucher, M. Latati, S.E. Okeke and M.A. Ali. (2023). Towards improving the precision agriculture management of the wheat crop using remote sensing: A case study in Central Non-Black Earth region of Russia. *The Egypt. J. Remote Sens. Space Sci.*, *26*(3), 505–517.

Reeves, E.R., J.P. Kerns, C. Cowger and B.B. Shew. 2021. *Pythium* spp. associated with root rot and stunting of winter wheat in North Carolina. *Plant Dis.*, *105*(4), 986–996.

Reis, E.M., S.M. Zoldan and M. Zanata. (2023). Interactions between temperature and wheat head wetting duration on Fusarium head blight intensity. *Summa Phytopathologica*, *49*, e268908.

Rodríguez-Moreno, V.M., A. Jiménez-Lagunes, J. Estrada-Avalos, J.E. Mauricio-Ruvalcaba and J.S. Padilla-Ramírez. (2020). Weather-data-based model: An approach for forecasting leaf and stripe rust on winter wheat. *Meteorol. Appl.*, *27*(2), e1896. doi:10.1002/met.1896.

Rodríguez-Moreno, M., A., Jiménez-Lagunes, JEstrada-Avalos, J.E. Mauricio-Ruvalcaba and J.S. Padilla-Ramírez. (2020). Weather-data-based model: An approach for forecasting leaf and stripe rust on winter wheat. *Meteorol. Appl.*, *27*, e1896.

Rolo, V., D. Rivest, E. Maillard and G. Moreno. (2023). Agroforestry potential for adaptation to climate change: A soil-based perspective. *Soil Use and Manag.*, *39*(3), 1006–1032.

Salotti, I., F. Bove and V. Rossi. (2022). Development and validation of a mechanistic, weather-based model for predicting *Puccinia graminis* f. sp. *tritici* infections and stem rust progress in wheat. *Front Plant Sci.*, *13*, 897680. doi: 10.3389/fpls.2022.897680.

Savary, S., L. Willocquet, S.J. Pethybridge, P. Esker, N. McRoberts and A. Nelson. (2019). The global burden of pathogens and pests on major food crops. *Nat. Ecol. Evol,*. *3*(3), 430–439.

Selin, C., T.R. de Kievit, M.F. Belmonte and W.G.D. Fernando. (2016). Elucidating the role of effectors in plant-fungal interactions: Progress and challenges. *Front. Microbiol.*, *7*, 600.

Shah, I.H., M.A. Manzoor, W. Jinhui, X. Li, M.K. Hameed, A. Rehaman, P. Li, Y. Zhang, Q. Niu and L. Chang. (2024). Comprehensive review: Effects of climate change and greenhouse gases emission relevance to environmental stress on horticultural crops and management. *J. Environ. Manag.*, *351*, 119978.

Sheikh, Z.A., S. Ashraf, S. Weesakul, M. Ali and N.C. Hanh. (2024). Impact of climate change on farmers and adaptation strategies in Rangsit, Thailand. *Environ. Chall.*, *15*, 100902.

Simón, M.R., M.C. Fleitas and S. Schalamuk. (2013). Recent advances on integrated foliar disease management with special emphasis in Argentina wheat production. In: Nita, M. (Ed.), *Fungicides: Showcases of Integrated Plant Disease Management from Around the World*. InTechOpen, pp. 3–34. doi:10.5772/51950.

Singh, B.K., M. Delgado-Baquerizo, E. Egidi, E. Guirado, J.E. Leach, H. Liu and P. Trivedi. (2023a). Climate change impacts on plant pathogens, food security, and paths forward. *Nat. Rev. Microbiol.*, *21*, 640–656.

Singh, D.P. (2017). Flag smut of wheat and its management practices. *In*: *Management of Wheat and Barley Diseases*. Routledge, pp. 231–238.

Singh, J., B. Chhabra, A. Raza, S.H. Yang and K.S. Sandhu. (2023). Important wheat diseases in the US and their management in the 21st century. *Front. Plant Sci.*, *13*, 1010191. doi: 10.3389/fpls.2022.1010191.

Singh, R.P., P.K. Singh, J. Rutkoski, D.P. Hodson, X. He, L.N. Jørgensen, M.S. Hovmøller and J. Huerta-Espino. (2016). Disease impact on wheat yield potential and prospects of genetic control. *Ann. Rev. Phytopath.*, *54*, 303–322.

Singhroha, G., G. Reddy, V. Gupta and S. Kumar. (2017). Wheat diseases and their management. *In*: *Wheat: — A Premier Food Crop*. Kalyani Publishers, pp. 348–372.

Smith, F. and E. Luna. (2023). Elevated atmospheric carbon dioxide and plant immunity to fungal pathogens: Do the risks outweigh the benefits? *Biochem J.*, *480*(22), 1791–1804.

Srivastava, S.K. and A.K. Singh. (2019). Important Barley diseases and their management strategies: A review. *Internat. J. Agri. Ext. Soc. Dev.*, *2*(2), 64–66.

Subedi, K.D., B.L. Ma and A.G. Xue. (2007). Planting date and nitrogen effects on Fusarium head blight and leaf spotting diseases in spring wheat. *Agron. J.*, *99*, 113–121.

Thuraga, V., F. Ghadamgahi, F.A. Dadi, R.R. Vetukuri and A. Chawade. (2024). A new bacterial consortia for management of Fusarium head blight in wheat. *Sci. Rep.*, *14*, 10131.

Toropova, E.Y., Kazakova, O.A. and Piskarev, V.V. (2020). Septoria blotch epidemic process on spring wheat varieties. *Vavilovskii Zhurnal Genet Selektsii*, *24*(2), 139–148. doi: 10.18699/VJ20.609.

Torres, G.A.M., J.R. Ferreira, E. Binneck, J.L.N. Maciel and L. Consoli. (2022). Blast disease and wheat production in Brazil. *Pesquisa Agropecuária Brasileira*, *57*, e02487.

Trivedi, P., J.E. Leach, S.G. Tringe, T. Sa and B.K. Singh. (2020). Plant–microbiome interactions: From community assembly to plant health. *Nat. Rev. Microbiol.*, *18*, 607–621.

Vagelas, I. (2025). Effective strategies for managing wheat diseases: Mapping academic literature utilizing VOSviewer and insights from our 15 years of research. *Agrochemicals*, *4*, 4. doi: https://doi.org/10.3390/agrochemicals4010004.

van den Bosch, F., N. Paveley, F. Van Den Berg, P. Hobbelen and R. Oliver. 2014. Mixtures as a fungicide resistance management tactic. *Phytopathol.*, *104*, 1264–1273.

van Zonneveld, M., M.S. Turmel and J. Hellin. (2020). Decision-making to diversify farm systems for climate change adaptation. *Front. Sustain. Food Syst.*, *4*, 32.

Váry, Z., E. Mullins, J.C. McElwain and F.M. Doohan. (2015). The severity of wheat diseases increases when plants and pathogens are acclimatized to elevated carbon dioxide. *Glob. Change Biol.*, *21*, 2661–2669.

Velásquez, A.C., C.D.M. Castroverde and S.Y. He. (2018). Plant-Pathogen Warfare under Changing Climate Conditions. *Curr. Biol.*, *28*(10), R619–R634.

Vicentini, S.N., P.S. Casado, G. de Carvalho, S.I. Moreira, A.F. Dorigan, T.C. Silva, A.G. Silva, A.A. Custódio, A.C.S. Gomes, J.L. Nunes Maciel, N. Hawkins, B.A. McDonald, B.A. Fraaije and P.C. Ceresini. (2022). Monitoring of Brazilian wheat blast field populations reveals resistance to QoI, DMI, and SDHI fungicides. *Plant Pathol.*, *71*, 304–321.

Weller, D.M., J.M. Raaijmakers, B.B.M. Gardener and L.S. Thomashow. (2002). Microbial populations responsible for specific soil suppressiveness to plant pathogens. *Annu. Rev. Phytopathol.*, *40*, 309–348.

Whish, J.P., N.I. Herrmann, N.A. White, A.D. Moore and D.J. Kriticos. (2015). Integrating pest population models with biophysical crop models to better represent the farming system. *Environ. Model. Softw.*, *72*, 418–425. doi:10.1016/j.envsoft.2014.10.010.

Woldemichael, M.D. (2019). Importance, biology, epidemiology, and management of loose smut (*Ustilago nuda*) of barley (*Hordeum vulgare*): A review. *East African J. Sci.*, *13*(1), 89–108.

Wu, F., Groopman, J.D. and Pestka, J.J. (2014). Public health impacts of foodborne mycotoxins. *Annu. Rev. Food Sci. Technol.*, 5, 351–372.

Xiao, Y., Y. Dong, W. Huang and L. Liu. (2022). Regional prediction of Fusarium head blight occurrence in wheat with remote sensing based Susceptible-Exposed-Infectious-Removed model. *Int. J. Appl. Earth Obs. Geoinf.*, *114*, 103043.

Xiong, Y., C. McCarthy, J. Humpal and C. Percy. (2023). A review on common root rot of wheat and barley in Australia. *Plant Pathol.*, *72*(8), 1347–1364.

Zhao, C., B. Liu, S. Piao, X. Wang, D.B. Lobell, Y. Huang, M. Huang, Y. Yao, S. Bassu, P. Ciais, J.L. Durand and S. Asseng. (2017). Temperature increase reduces global yields of major crops in four independent estimates. *Proc. Natl. Acad. Sci. USA*, *114*, 9326–9331.

Zhelezova, S.V., A.V. Melnikov and A.A. Ananiev. (2019). Pros and cons of no-till technology in a long-term field experiment on sod-podzolic soil. *In*: *IOP Conference Series: Earth and Environmental Science*, *368* (1), 012055. IOP Publishing.

CHAPTER 11

Enhancing Agroforestry Sustainability with Beneficial Arthropods

Chitta Ranjan Satpathi[1*], A.M. Raut[2], Megha N Parajulee[3], Prasakha Dutta[4] and A. Najitha Banu[5]

1. Introduction

Agroforestry is a complex system that integrates the silvopastoral system and farm forestry with crop production for a sustainable farm enterprise. It plays a vital role in linking ecosystem conservation with human development through ecological agriculture and ecological restoration (Tscharntke et al., 2005; Nichols et al., 2008; Xiao & Xiong, 2022). This system facilitates nutrient recycling, energy flow across different trophic levels, ecological signaling within ecosystem services, and provides food resources for many living communities (Godfray & Garnett, 2014; Sparks & Nauen, 2015). Moreover, agroforestry is often regarded as a niche that supports the biodiversity of living organisms. Agroforestry provides a wide range of forest products such as fruits, timber, fodder, resin, and medicinal resources, while also contributing to soil fertility, carbon sequestration, and ecosystem resilience. Protecting and promoting these systems is therefore essential for sustainable farm management, economic security, and long-term ecological balance (Nair et al., 2022). Farmers often cultivate diverse species within agroforestry systems, which directly influence biodiversity in the ecosystem. Proper planning and design of such plantations can create suitable buffer zones that enhance and support natural habitats (Rahman et al., 2023). Therefore, agroforestry knowledge is highly beneficial to local populations for establishing strong and sustainable policies and developing effective frameworks to enhance crop productivity.

Nowadays, crop production sustainability has gained prominence in any farm operation as world population continues to rise and crop acreages shrink. Crop protective measures like insecticides, pesticides, herbicides and fungicides play a very crucial role in crop production. However, the injudicious use of agrochemicals contributes to environmental pollution, development of resistance and resurgence in pests, and a negative impact on functional biodiversity of both natural enemies

[1,4] Department of Agricultural Entomology, Bidhan Chandra Krishi Vishwavidyalaya, Mohonpur, Nadia, West Bengal, India.
[2] Department of Entomology, School of Agriculture, Lovely Professional University, Phagwara, Punjab, India.
[3] Texas A&M University, Department of Entomology, Texas A&M AgriLife Research.
[5] Department of Zoology, School of Bioengineering and Bioscience, Lovely Professional. University, Phagwara, Punjab, India.
* Corresponding author: csatpathi2003@yahoo.co.in

DOI: 10.1201/9781003614869-11

and pollinators (Mishra et al., 2024). In addition, the excessive application of agrochemicals has contributed to habitat loss, particularly reducing nectar- and pollen-rich flowering plants essential for pollinator survival (Azpiazu et al., 2023). A recent meta-analysis by Schulz et al. (2021) examined 381 pesticides encompassing 1,591 substances and their acute toxicity thresholds for eight non-target insect species groups that further illustrates the environmental cost of injudicious use of crop chemicals and pesticides.

To reduce reliance on chemical pesticides in crop production, several sustainable techniques have been adopted that not only manage pests effectively but also enhance ecosystem services. Notable approaches include ecological engineering, push–pull strategies, the use of potential botanical pesticides, deployment of identified semiochemicals, application of pheromones as ecosystem-based pest management strategy to reduce pest pressure below the economic threshold level (Barzman et al., 2015). Agricultural landscape can be significantly improved through habitat diversification via ecological engineering including strip cropping, intercropping, relay cropping or as border crops to promote ecosystem services of predators, parasitoids and pollinators (Parajulee et al., 1997; Westphal et al., 2015). The flowers are the powerhouse of energy source for beneficial insects that significantly enhances predator longevity and fecundity due to availability of pollen and nectar (He et al., 2021). Therefore, establishing flower strips, particularly those with high pollen- and nectar-rich plant species, can attract parasitoids and predators that support natural predation (Alcalá Herrera et al., 2022; Hyder et al., 2025), while simultaneously enhancing pollinator activity for improving crop productivity through pollination. The diversity and abundance of these beneficial insects often depend on floral traits such as color, shape, pollen load, and blooming duration, as well as prevailing environmental conditions (Trunschke et al., 2021). Within an ecological framework, greater plant species diversity further strengthens these interactions by providing suitable habitats, alternative prey, and continuous floral resources, thereby augmenting populations of natural enemies and sustaining ecosystem services (Parajulee & Slosser, 1999; Slosser et al., 2000; Hyder et al., 2025). These natural enemies, including parasitoids and predators, suppress natural pest populations below economic threshold level either by direct predation or releasing plant volatiles. In addition, plant-emitted volatiles not only attract herbivorous insect pests but also lure parasitic and predatory insects that serve as their natural enemies, while simultaneously inducing defense responses in neighboring plants (Paré & Tumlinson, 1999). Collectively, these ecological mechanisms highlight the importance of designing diversified agroecosystems that conserve biodiversity and enhance natural pest regulation (Lewis et al., 1982). Thus, diverse agroforestry ecosystems play a significant role in diversity and conservation of beneficial arthropods. In this chapter, we emphasize the role of beneficial arthropods including predators, parasitoids, spiders, and decomposers in an agroforestry ecosystem.

2. Common Beneficial Arthropods of Agroforestry

The main ecological functions of beneficial insects in forest ecosystems are predation or parasitism, pollination and decomposition. The interaction effect from diverse natural vegetation, soil factors and seasonal effects (Figure 1) and often simultaneous growth promotion of crop and tree species result in the co existence of pests and natural enemies maintaining a balance in the agroecosystem (Altieri & Nicholls, 2008). However, agroforestry practices may result in lower abundance of weeds and higher population of beneficial insects and occasionally lesser pest loads (Pumariño, et al., 2015; Kishore et al., 2024). Beneficial arthropods include predators (wasps, lady beetles, lacewings), pollinators (butterfly, bees), and Acaria (spiders) but, insect parasitoids (tachinid flies, braconid and ichneumonid wasps) and ecological engineers (termites, collembola) and decomposers enhance soil fertility and integrity in functioning of this ecosystem process (Chakraborti et al., 2024).

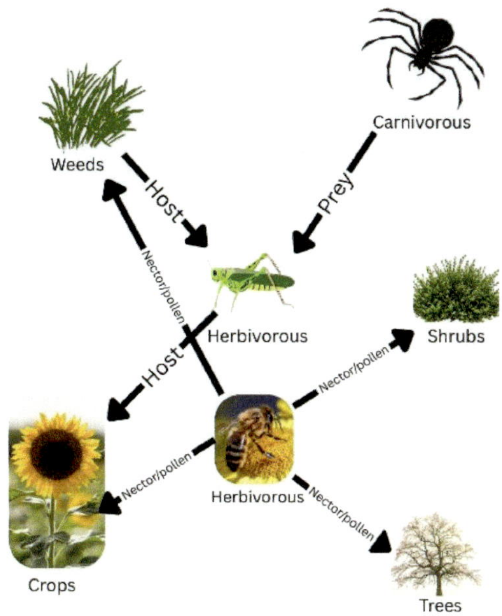

Fig. 1 A complex relationship between the trophic level and host-plant resources

2.1 *Role of Predators and Parasitoids in Agroforestry Ecosystem*

Agroforestry ecosystems provide suitable habitats that enhance the abundance and activity of natural enemies, including predators, parasitoids, and even predatory birds, thereby contributing significantly to conservation biological control (Samal et al., 2024). Predators are free-living organisms that actively search for multiple prey to complete their life cycle, equipped with strong mandibulate mouthparts for capturing and killing hosts. In contrast, parasitoids are generally smaller than their hosts and complete their life cycle within a specific host, using long ovipositors to lay eggs on or inside the host body, often accompanied by stinging or paralyzing behavior to facilitate successful parasitism. Both predators and parasitoids play a significantly positive impact on managing the crop pest (Symondson et al., 2002). However, the common and predominant predators like ground beetle, dragonfly, damselfly, bug, minute pirate bugs, water bug, plant bug, earwig, green lacewings, praying mantis, fireflies, ground beetle, lady bird beetle, syrphid fly, tachinid fly, fire ants, wasp in an agroforest ecosystem (Figure 2). These predominant predators usually prey on soft body insects, lepidopteran pests, and sometimes observed cannibalism, especially in odonata insects (Woodcock et al., 2014; Satpathi, 2021). Predator diversity is generally higher in polycultures than in monocultures because the combination of trees and crops in agroforestry provides more niche diverse and complex habitats (Stamps & Linit, 1997). Although, under organic farming, alley cropping enhanced two-fold of both the activity-density and complementarity of generalist predators especially carabids than the monocropping due to its diverse vegetation available, provided shade maintaining with optimum temperature and less pressure of pesticides and synthetic fertilizers utilization in an ecosystem (Boinot et al., 2020; Martin-Chave et al., 2019). Flowering understory in a silvoarable system increased predator densities in apple trees, decreased aphid colony growth, while increasing ground-based natural enemy diversity in the adjacent arable crop (Staton et al., 2021). As the conclusion, diurnal activity of predominant predator diversity and density will impact on pesticides application, availability of host, crop diversification, and microhabitats.

Parasitoids are smaller in size and parasitize many hosts by ovipositing on/in to the host. These parasitoids had good host search capability due to its multiparous sensilla structure. Adult parasitoids preferred nectar and pollen food sources as high protein contain, enhance the reproductive capability. Majority parasitoids belong to the insect order Hymenoptera which live in colonies and follow an S-shape survival curve. Ferracini et al. (2018) stated that the 4,410 chalcid parasitoids were observed against asian chestnut gall wasp, *Dryocosmus kuriphilus* Yasumatsu in oak-based forestry system where most abundant species were *Torymus flavipes* Walker, *Eupelmus* spp., and *Bootanomyia* (= Megastigmus) *dorsalis* (F.). A most frequent and abundance parasitoid family contributes of Braconidae, Formicidae, Ichneumonidae, Vespidae, Brachonidae, Chalcididae, Pompilidae, Sphecidae, Evaniidae, Halictidae and Anthophoridae are potential to as an effective biological control agents in silvipastoral system (Sivinski et al., 2011; Auad et al., 2012; Marchiori, 2021; Pérez-Benavides et al., 2022). The incorporation of aromatic plants such as basil (*Ocimum basilicum*), French marigold (*Tagetes patula*), or ryegrass (*Lolium perenne*) around or within apple orchards has been shown to increase parasitism rates of *Cydia pomonella* by providing floral resources and volatile organic compounds that attract parasitoids (Laffon et al., 2022). Flower morphology, particularly flower area and density, further supports the abundance of key parasitoid families such as Ichneumonidae, Braconidae, and Chalcididae (Sivinski et al., 2011). In addition to structural traits, nectar availability in flowering fields enhances parasitoid movement towards plants, guided by olfactory cues emitted by flowers (Bianchi et al., 2008). Parasitoids that feed on nectar benefit from increased fecundity, prolonged reproductive lifespan, and accelerated egg maturation (Vattala et al., 2006; Zhu et al., 2023). Therefore, integrating nectar-rich, multi-flowering plants with diverse floral traits, such as extended blooming duration, high flower density, and larger floral area, can significantly enhance parasitoid diversity and represents an important step in conservation biological control.

Fig. 2 Diversity of predators in model agroforestry at BCKV, West Bengal, India.

2.2 Role of Spiders in Agroforestry

Spiders are generalist predators belonging to order Araneae and phylum Arthropoda which act as sensitive bioindicators and are highly valuable arthropods for ecosystem functioning. The spider community abundance is generally enhanced with herb litter cover (Stenchly et al., 2012). There is no record of harmful effects of spiders on agroecosystems, but sometimes they prey on beneficial insects as a higher-level carnivore (Savory, 1928). Indiscriminate predations of spiders generally result in natural suppression of pest populations in the agroecosystem as spiders are long-lived and can withstand starvation and desiccation. Spiders can be broadly grouped into non-web forming and web forming, among the non-web forming spiders *Lycosa pseudoannulata* (Boes and Strand), *Marpissa bengalensis* Tikader, *Oxyopes javanus* Thorell., *O. lineatipes* (C.L. Koch), *O. shweta* Tikader, *Phidippus indicus* Tikader, *Pardosa sumatrana* (Thorell), *Thomisus sikkimensis* Tikader, *T. cherapunjeus* Tikader, *Zygobolus narmadaensis* Tikader are predominant in India (Figure 3). Spiders belonging to the family Araneidae, Liniphidae, Clubionidae, Tetragnathida are unique for building their webs in agroforestry. Among the different web-forming spiders, *Tetragnatha bengalensis* Walck. and T. *maxillosa* Thorell prefer wetlands where they prefer the crop canopy during midday and form ring-shaped webs for maximum trapping of prey in the morning. Similarly, long and circular webs are formed over the crop canopy by *Argiope catenulate* Doleschall and *A. pulchella* Thorell for tapping grasshoppers like *Ceracris nigricornis* Walk., *Chondacris rosea* Degeer, *Oxya hyla hyla* Survill along with adult stages of different lepidopteran insect-pests of field crop. Also, spiders prefer preying on adults of Lepidoptera, Linyphiidae and Dictynidae, Formicidae, and hemipteran insects to fulfill the nutrient requirement, especially amino acid (Nyffeler et al.,1994; Nyffeler, 1999; Jackson & Pollard, 1996).

The diversity and density of spiders is higher in natural farming than the common farmers practices (Satpathi et al., 2005), although an alley tree agroecosystem enhanced the predatory spider abundance especially *Pardosa lugubris* (ground hunter spider) and *Tenuiphantes tenuis* (Web weaver) (Matevski et al., 2024). Canopy thinning supports spider communities by providing sufficient light, whereas shaded and highly humid conditions result in lower spider populations (Hamřík et al., 2023). In contrast, results observed that the coffee grown in shades offered more suitable habitat for arboreal spiders due to a variety of the characteristics of the shade trees (Hajian-Forooshani et al., 2014). Fruit orchards, forestry plantation, and vineyards are the stable agroecosystems where diverse and abundant predator communities are found than annual and greenhouse crops (Isaia et al., 2010; Happe et al., 2019). Spider communities support services like recycling nutrients through the breakdown of organic matter, acting as food sources for predators, or creating habitats for other organisms (Cardoso et al., 2025). However, the biocontrol efficacy of spiders in agroforestry is highly influenced by factors such as pest type, tree species, density of trees, and season (Khum et al., 2025).

Lycosa sp. *Oxyopes* sp. *Zygoballus* sp.

Fig. 3 Common non-web forming spiders in an agroforestry ecosystem.

2.3 Role of Pollinators in Agroforestry

An agroforestry ecosystem is more stable than an ephemeral agroecosystem. It provides shelter for diverse arthropods and pollinators. The complex faunal diversity is largely due to availability of stable food resources (Klein et al., 2007; Gallai et al., 2009). Both bee and non-bee species serve as potential pollinators in an agroforestry ecosystem which contributes to 35% of global food production (Giannini et al., 2015; Pashte & Patil, 2018). Wild bee species, flies, beetles, and butterflies play a significant role as pollinators, including solitary bees, bumblebees, and honeybees (Garibaldi et al., 2013). Eight families of pollinators are encountered in dill (*Anethum graveolens* L.), an aromatic herb, and among these *Apis florea* and *A. mellifera* are the most dominant foragers (Table 1). An agroforestry ecosystem plays a more significant role in providing shelter for many native pollinators as well as migratory pollinators than the crop agriculture ecosystem. Conventional agriculture may cause detrimental impact on pollinator abundance due to indiscriminate use of pesticides, loss of habitat, change in agriculture landscaping, urbanization and climate change (Centeno-Alvarado et al., 2023). Nevertheless, flowering plants with abundant pollen and nectar planted as border crops, intercrops, cover crops, and hedgerows can enhance pollinator populations in non-forestry ecosystem (Parajulee et al., 1997; Lowe et al., 2021; Fijen et al., 2025). Land use patterns and distance from forest edges also influence pollinator communities. For example, bee species were more abundant in agroecosystem at greater distances from forests, while butterflies were more common near the forested areas (Didita et al., 2025). Also, floral functional traits like flower color, type of flower, length of corolla, and density of flowers impact density and diversity of pollinators. Pollinators have been shown to significantly improve crop production and maintain cropping system sustainability (Aviron et al., 2023; Jachowicz & Sigsgaard, 2025).

Table 1 Major arthropod families that serve as important pollinators in a dill (*Anethum graveolens* L.) cropping system, West Bengal, India.

Name of the Species	*Order*	*Family*	*Population% of Total Visitor*
Apis dorsata Fab.	Hymenoptera	Apidae	9.44
Apis florea F		Apidae	30.14
Apis mellifera L		Apidae	15.71
Ceratina sexmaculata Smith		Apidae	2.47
*Xylocopa*sp		Apidae	0.48
Polistes hebraeus(F)		Vespidae	0.43
Camponotus sp.		Formicidae	2.02
Halictus sp.		Halictidae	9.81
Episyrphus balteatus (De Geer)	Diptera	Syrphidae	6.24
Eristalis sp.		Syrphidae	2.82
Musca domestica		Muscidae	2.93
Coccinella septumpunctata L.	Coleoptera	Coccinellidae	5.41
Menochilus maculatus Fab.	Coleoptera	Coccinellidae	2.59
Chrysoperla zastrovisillemi	Neuroptera	Chrysopidae	0.14
Pieries brassicae L	Lepidoptera	Pieridae	0.53

Source: (Meena et al., 2022).

2.4 Role of Decomposers in Agroforestry

An agroforestry ecosystem, initially developed by human effort, has taken the shape of a more natural habitat for complex arthropod fauna. All living fossils dry or die at a certain period of time and potential decomposers help to convert these remnants into useful nutrients. Broadly, decomposers are categorized as mesofauna (100–200 mm) and macrofauna that are larger in size (Galante et al., 2008). The key groups of decomposer arthropods include woodlice (Isopods), millipedes (Diplopoda), springtails (Collembola), oribatid mites (Arachinae), beetles (Tenebrionidae and Scarabaeidae), and ants (Hymenoptera) that play an important role in recycling plant litter into soil nutrients (Figure 4). These organisms enhance the decomposition process by physically breaking down organic material and boosting microbial activity. Along with insect gut microbiota that support in digesting cellulose, yeasts decompose fruit residues through fermentation, facilitating easier nutrient uptake in certain Diptera and Hymenoptera species. Therefore, well-decomposed materials improve nutrient availability for plants and maintain soil structure and fertility in an agroforestry system (Lavelle et al., 1997; Seastedt, 1984). However, termitarium developed by termites is an important component in the agroforestry ecosystem where its soil structure is well-developed which is beneficial for overall sustainability (Dangerfield et al., 1998). In addition, termitarium soil act as biofertilizer especially solubilizing phosphate and potassium as well as producing indole acetic acid which possess plant growth-promoting potentials and as well suppressing plant soil pathogen (Adebajo et al., 2021; Tilahun et al., 2021; Kathbaruah et al., 2024).

Agroforestry has potential for conservation of decomposer by providing diverse habitat and food sources and that supports soil microbiota diversity. The pruning of agroforestry trees used as a mulch will enhance the population of collembola and initiate the decomposition process (Badejo et al., 1998). During the process of decomposing, soil temperature and moisture also impacted on the density of collembola. The soil physical and chemical properties affected collembola diversity; *Eucalyptus* plantations under an agroforestry scheme significantly decreased plant richness, soil pH, and soil temperature (Jorge et al., 2023). Therefore, it is suggested that the less disturbance with tillage operation, crop diversification, and within-crop habitat complexity significantly conserve the functional guild including decomposers diversity in an agroecosystem (Mamabolo et al., 2024).

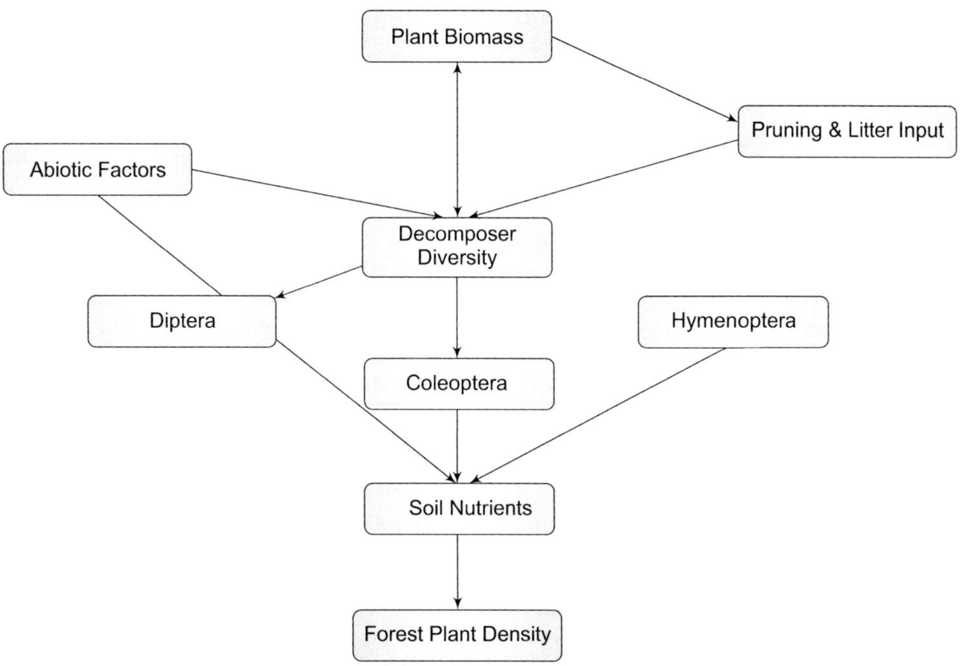

Fig. 4 Ecological cycle of decomposers in enhancing soil nutrients in agroforestry systems.

3. Conclusion

Positive population growth, rapid urbanization, and loss of fertile agricultural land areas have constantly threatened world food security. This critical situation demands efficient utilization of arable lands for improved yield and production sustainability. A robust agroforestry system where farm forestry and silviculture can be integrated for enhanced beneficial fauna, especially the guild of pollinators, that perform valuable ecosystem services for farmscape sustainability. Thus, agroforestry is one of the newer land management practices where forest plants are grown on the same land area with agricultural crops and livestock to meet the needs of human beings. The term *ecosystem guardians* refer to the animal species that naturally protect and restore entire ecosystems without human intervention. This discussion highlights insects as guardians providing valuable services like pest control, pollination, and decomposition that contribute to enhanced farm productivity. Similarly, crop selection and habitat management play a key role in maximizing the utilization of natural resources in agriculture for sustainability. In an ideal farm forestry ecosystem, maximum energy is utilized by consumers while unutilized energy is ultimately used by decomposers and recycled back into the agroforestry ecosystems. Thus, the role of agroforestry ecosystem functions provided by beneficial arthropods is enormous in modulating tri-trophic interactions within the agroforestry ecosystem for increased farm productivity and system sustainability.

References

Adebajo, S.O., Akintokun, P.O., Ezaka, E., Ojo, A.E., Olannye, D.U. and Ayodeji, O.D. (2021). Use of termitarium soil as a viable source for biofertilizer and biocontrol. *Bulletin of the National Research Centre*, *45*(1), 100.

Alcalá Herrera, R., Cotes, B., Agustí, N., Tasin, M. and Porcel, M. (2022). Using flower strips to promote green lacewings to control cabbage insect pests. *Journal of Pest Science*, *95*(2), 669–683.

Altieri, M.A. and Nicholls, C.I. (2004). Effects of agroforestry systems on the ecology and management of insect pest populations. In: *Ecological Engineering for Pest Management: Advances in Habitat Manipulation for Arthropods*. CSIRO, Collingwood, pp. 143–155.

Auad, A.M., Resende, T.T., Da Silva, D.M. and das Graças Fonseca, M. (2012). Hymenoptera (Insecta: Hymenoptera) associated with silvopastoral systems. *Agroforestry Systems*, *85*(1), 113–119.

Aviron, S., Berry, T., Leroy, D., Savary, G. and Alignier, A. (2023). Wild plants in hedgerows and weeds in crop fields are important floral resources for wild flower-visiting insects, independently of the presence of intercrops. *Agriculture, Ecosystems, and Environment*, *348*, 108410. https://doi.org/10.1016/j.agee.2023.108410.

Azpiazu, C., Medina, P., Sgolastra, F., Moreno-Delafuente, A. and Vinuela, E. (2023). Pesticide residues in nectar and pollen of melon crops: Risk to pollinators and effects of a specific pesticide mixture on Bombus terrestris (Hymenoptera: Apidae) micro-colonies. *Environmental Pollution*, *326*, 121451. https://doi.org/10.1016/j.envpol.2023.121451.

Badejo, M.A., Nathaniel, T.I. and Tian, G. (1998). Abundance of springtails (Collembola) under four agroforestry tree species with contrasting litter quality. *Biology and Fertility of Soils*, *27*(1), 15–20.

Barzman, M., Bàrberi, P., Birch, A.N.E., Boonekamp, P., Dachbrodt-Saaydeh, S., Graf, B., Hommel, B., Jensen, J.E., Kiss, J., Kudsk, P. and Lamichhane, J.R. (2015). Eight principles of integrated pest management. *Agronomy for Sustainable Development*, *35*(4), 1199–1215.

Bianchi, F.J. and Wäckers, F.L. (2008). Effects of flower attractiveness and nectar availability in field margins on biological control by parasitoids. *Biological Control*, *46*(3), 400–408. https://doi.org/10.1016/j.biocontrol.2008.04.010.

Boinot, S., Mézière, D., Poulmarc'h, J., Saintilan, A., Lauri, P.E. and Sarthou, J.P. (2020). Promoting generalist predators of crop pests in alley cropping agroforestry fields: Farming system matters. *Ecological Engineering*, *158*, 106041. https://doi.org/10.1016/j.ecoleng.2020.106041.

Cardoso, P., Pekár, S., Birkhofer, K., Chuang, A., Fukushima, C.S., Hebets, E.A., Henaut, Y., Hesselberg, T., Malumbres-Olarte, J., Michálek, O. and Michalko, R. (2025). Ecosystem services provided by spiders. *Biological Reviews*. https://doi.org/10.1111/brv.70044.

Centeno-Alvarado, D., Lopes, A.V. and Arnan, X. (2023). Fostering pollination through agroforestry: A global review. *Agriculture, Ecosystems, and Environment*, *351*, 108478. https://doi.org/10.1016/j.agee.2023.108478.

Chakraborti, J., Mondal, S. and Palit, D. (2024). Agroforestry and arthropod diversity for ecosystem services. In: *Agroforestry for Carbon and Ecosystem Management*. Academic Press, pp. 257–269. https://doi.org/10.1016/B978-0-323-95393-1.00032-4.

Didita, M., Nemomissa, S., Lemessa, D. and Bekele, T. (2025). The diversity of pollinators within different land use types across agricultural landscapes is influenced by the proximity of surrounding forests. *Global Ecology and Conservation*, e03776. https://doi.org/10.1016/j.gecco.2025.e03776.

Ferracini, C., Bertolino, S., Bernardo, U., Bonsignore, C.P., Faccoli, M., Ferrari, E., Lupi, D., Maini, S., Mazzon, L., Nugnes, F. and Rocco, A. (2018). Do Torymus sinensis (Hymenoptera: Torymidae) and agroforestry system affect native parasitoids associated with the Asian chestnut gall wasp? *Biological Control*, *121*, 36–43. https://doi.org/10.1016/j.biocontrol.2018.01.009.

Fijen, T.P., Eeraerts, M., Osterman, J., Beyer, N., Hass, A., Lundin, O. and Westphal, C. (2025). Crop diversification for pollinator conservation. *Landscape Ecology*, *40*(1), 19.

Galante, E. and Marcos-Garcia, M.A. (2008). Decomposer Insects. In: Capinera, J.L. (Eds.), *Encyclopedia of Entomology*. Springer, Dordrecht, pp. 1158–1169. https://doi.org/10.1007/978-1-4020-6359-6_845.

Gallai, N., Salles, J.M., Settele, J. and Vaissiere, B.E. (2009). Economic valuation of the vulnerability of world agriculture confronted with pollinator decline. *Ecological Economics*, *68*, 810–821.

Garibaldi, L.A., Steffan-Dewenter, I., Winfree, R., Aizen, M.A., Bommarco, R., Cunningham, S.A., Kremen, C., Carvalheiro, L.G., Harder, L.D., Afik, O. and Bartomeus, I. (2013). Wild pollinators enhance fruit set of crops regardless of honey bee abundance. *Science*, *339*(6127), 1608–1611. https://doi.org/10.1126/science.1230200.

Giannini, T.C.S., Boff, S., Cordeiro, G.D., Cartolano, E.A., Veiga, A.K., Imperatriz-Fonseca, V.L. and Saraiva, A. (2015). Crop pollinators in Brazil: A review of reported interactions. *Apidologie*, *46*(2), 209–223.

Godfray, H.C.J. and Garnett, T. (2014). Food security and sustainable intensification. *Philosophical Transactions of the Royal Society B: Biological Sciences*, *369*(1639), 20120273. https://doi.org/10.1098/rstb.2012.0273.

Hajian-Forooshani, Z., Gonthier, D.J., Marín, L., Iverson, A.L. and Perfecto, I. (2014). Changes in species diversity of arboreal spiders in Mexican coffee agroecosystems: Untangling the web of local and landscape influences driving diversity. *PeerJ.*, *2*, e623.

Hamřík, T., Košulič, O., Gallé, R., Gallé-Szpisjak, N. and Hédl, R. (2023). Opening the canopy to restore spider biodiversity in protected oakwoods. *Forest Ecology and Management*, *541*, 121064. https://doi.org/10.1016/j.foreco.2023.121064.

Happe, A.K., Alins, G., Blüthgen, N., Boreux, V., Bosch, J., García, D., Hambäck, P.A., Klein, A.M., Martínez-Sastre, R., Miñarro, M. and Müller, A.K. (2019). Predatory arthropods in apple orchards across Europe: Responses to agricultural management, adjacent habitat, landscape composition, and country. *Agriculture, Ecosystems, and Environment*, *273*, 141–150.

He, X., Kiær, L.P., Jensen, P.M. and Sigsgaard, L. (2021). The effect of floral resources on predator longevity and fecundity: A systematic review and meta-analysis. *Biological Control*, *153*, 104476. https://doi.org/10.1016/j.biocontrol.2020.104476.

Hyder, M., Ul Haq, I., Younas, M., Ghafar, M.A., Akhtar, M.R., Ahmed, Z., Bukero, A. and Hou, Y., (2025). Floral resource integration: Enhancing biocontrol of *Tuta absoluta* within sustainable IPM frameworks. *Plants*, *14*(3), 319. https://doi.org/10.3390/plants14030319.

Isaia, M., Beikes, S., Paschetta, M., Sarvajayakesevalu, S. and Badino, G. (2010). Spiders as potential biological controllers in apple orchards infested by *Cydia* spp. (lepidoptera: Tortricidae). In: *Proceedings of the 24th European Congress of Arachnology* (Eds. W. Nentwig, M. Entling and C. Kropf). European Society of Arachnology, Bern, pp. 25–29.

Jachowicz, N. and Sigsgaard, L. (2025). Highly diverse flower strips promote natural enemies more in annual field crops: A review and meta-analysis. *Agriculture, Ecosystems, and Environment*, *381*, 109412. https://doi.org/10.1016/j.agee.2024.109412.

Jackson, R.R. and Polland, S.D. (1996). Predatory behaviour of jumping spiders. *Annual. Review of Entomology*, *41*, 287–308.

Jorge, B.C.S., Winck, B.R., da Silva Menezes, L., Bellini, B.C., Pillar, V.D. and Podgaiski, L.R. (2023). Grassland afforestation with Eucalyptus affect Collembola communities and soil functions in southern Brazil. *Biodiversity and Conservation*, *32*(1), 275–295.

Kathbaruah, S., Bhattacharyya, B., Borkataki, S., Gogoi, B., Hatibarua, P., Gogoi, S., Bhairavi, K.S. and Dutta, P. (2024). Termite mound soil based potting media: A better approach towards sustainable agriculture. *Frontiers in Microbiology*, *15*, 1387434. https://doi.org/10.3389/fmicb.2024.1387434.

Khum, W., Košulič, O. and Michalko, R. (2025). Trait diversity and spider community composition are associated with lower herbivory in young forest plantations. *Biological Control*, *200*, 105666. https://doi.org/10.1016/j.biocontrol.2024.105666.

Kishore, S.M., Khajuria, M., Supraja, K.V. L., Srija, P., Manideep, K.S.N., Harideep, G. and Morabad, P.B. (2024). The role of herbivore-induced plant volatiles in tri-trophic interactions and pest management. *Journal of Advances in Biology and Biotechnology, 27* (11), 763–770. https://doi.org/10.9734/jabb/2024/v27i111659.

Klein, A.M., Vaissière, B.E., Cane, J.H., Steffan-Dewenter, I., Cunningham, S.A., Kremen, C. and Tscharntke, T. (2007). Importance of pollinators in changing landscapes for world crops. In: *Proceedings of the Royal Society B: Biological Sciences*, *274*(1608), 303–313.

Laffon, L., Bischoff, A., Gautier, H., Gilles, F., Gomez, L., Lescourret, F. and Franck, P. (2022). Conservation biological control of codling moth (*Cydia pomonella*): Effects of two aromatic plants, basil (*Ocimum basilicum*) and French marigolds (*Tagetes patula*). *Insects*, *13*(10), 908. https://doi.org/10.3390/insects13100908.

Lavelle, P., Spain, A.V., Blanchart, E., Martin, A. and Martin, S. (1997). Impact of soil fauna on the properties of soils in the humid tropics. In: *Myths and Science of Soils of the Tropics*. Soil Science Society of America (SSSA Special Publication No. 29), pp. 157–185.

Lewis, W.J., Nordlund, D.A., Gueldner, R.C., Teal, P.E.A. and Tumlinson, J.H. (1982). Kairomones and their use for management of entomophagous insects: XIII. Kairomonal activity for *Trichogramma* spp. of abdominal tips, excretion, and a synthetic sex pheromone blend of *Heliothis zea* (Boddie) moths. *Journal of Chemical Ecology*, *8*(10), 1323–1331.

Lowe, E.B., Groves, R. and Gratton, C. (2021). Impacts of field-edge flower plantings on pollinator conservation and ecosystem service delivery: A meta-analysis. *Agriculture, Ecosystems and Environment*, *310*, 107290. https://doi.org/10.1016/j.agee.2020.107290.

Mamabolo, E., Pryke, J.S. and Gaigher, R. (2024). Soil fauna diversity is enhanced by vegetation complexity and no-till planting in regenerative agroecosystems. *Agriculture, Ecosystems and Environment*, *367*, 108973. https://doi.org/10.1016/j.agee.2024.108973.

Marchiori, C.H. (2021). Biodiversity of Parasitoids Present in Agroforestry in Brazil. *European Journal of Biology and Biotechnology*, *2*(3), 89–91.

Martin-Chave, A., Beral, C., Mazzia, C. and Capowiez, Y. (2019). Agroforestry impacts the seasonal and diurnal activity of dominant predatory arthropods in organic vegetable crops. *Agroforestry Systems*, *93*(6), 2067–2083.

Matevski, D., Sagolla, V., Beule, L. and Schuldt, A. (2024). Temperate alley-agroforestry improves pest control potential by promoting spider abundance and functional diversity. *Journal of Applied Ecology*, *61*(12), 3079–3091.

Meena, N.K., Meena, R.S., Singh, R., Verma, A.K., Choudhary, S., Singh, B., Meena, R.D., Ravi, Y. and Meena, M. (2022) Managed pollination is a much better way of increasing productivity and essential oil content of dill seeds crop. *Scientific Reports, 12*, 13134.

Mishra, M.K., Mishra S.K., Panday, L. and Pandey, R. (2024). Agrochemicals and their impact on environment. In: *Recent Trends in Plant Protection,* NIPA Publication, pp. 47–56.

Nair, P. R., Kumar, B.M. and Nair, V. D. (2022). Classification of agroforestry systems. In: *An Introduction to Agroforestry: Four Decades of Scientific Developments*. Cham: Springer International Publishing, pp. 29–44.

Nichols, E., Spector, S., Louzada, J., Larsen, T., Amezquita, S., Favila, M.E. and Network, T.S.R. (2008). Ecological functions and ecosystem services provided by Scarabaeinae dung beetles. *Biological Conservation*, *141*(6), 1461–1474.

Nyffeler, M. (1999). Prey selection of spiders in the field. *Journal of Arachnology*, *27*, 317–324.

Nyffeler, M., Sterling, W.L. and Dean, D.A. (1994). How spiders make a living. *Environmental Entomology, 23*(6), 1357–1367.

Parajulee, M.N., Montandon, R. and J.E. Slosser. (1997). Relay intercropping to enhance predators of the cotton aphids in Texas cotton. *International Journal of Pest Management, 43*, 227–232.

Parajulee, M.N. and Slosser, J.E. (1999). Evaluation of potential relay strip crops for predator enhancement in Texas cotton. *International Journal of Pest Management, 45*, 275–286.

Paré, P.W. and Tumlinson, J.H. (1999). Plant volatiles as a defense against insect herbivores. *Plant Physiology*, *121*(2), 325–332.

Pashte, V.V. and Patil C.S. (2018). Toxicity and poisoning symptoms of selected insecticides to honey bees (Apis mellifera mellifera L.). *Archives of Biological Science ,70*(1), 5–12.

Pérez-Benavides, A.L., Malagon-Aldana, L., Hanson, P. and Serna, F. (2022). New records and a voucher collection of parasitoid wasps (Hymenoptera) inhabiting agroforestry systems in the Colombian Amazon basin. *Papéis Avulsos de Zoologia*, *62*, e202262035. https://doi.org/10.11606/1807-0205/2022.62.035.

Pumariño, L., Sileshi, G.W., Gripenberg, S., Kaartinen, R., Barrios, E., Muchane, M.N., Midega, C. and Jonsson, M., (2015). Effects of agroforestry on pest, disease and weed control: A meta-analysis. *Basic and Applied Ecology*, *16*(7), 573–582. https://doi.org/10.1016/j.baae.2015.08.006.

Rahman, S.A., Samsudin, Y.B., Bhatta, K.P., Aryal, A., Hayati, D., Cahya, M., Trihadmojo, B., Husain, I., Andini, S., Narulita, S. and Rahman, M.A. (2023). The role of agroforestry systems for enhancing biodiversity and provision of ecosystem services in agricultural landscapes in Southeast Asia. In: *Agroforestry for Sustainable Intensification of Agriculture in Asia and Africa*., Singapore: Springer Nature Singapore, pp. 303–319.

Samal, I., Bhoi, T.K., Mahanta, D.K. and Komal, J. (2024). Agroforestry as an effective tool for pest management in the era of climate change. In: *Agroforestry Solutions for Climate Change and Environmental Restoration*. Singapore: Springer Nature Singapore, pp. 221–244.

Satpathi, C.R. (2021). *Ground Beetles and Their Role in Management of Crop Pests*. CRC Press, United Kingdom, 216 pp. https://doi.org/10.4324/9781003261087.

Satpathi, C.R., Mukhopadhyay, A.K., Katti, G., Pasalu, I.C. and Venkateswarlu, B. (2005). Quantification of the role of natural biological control in farmer's rice field in West Bengal. *Indian Journal of Entomology*, *67* (3), 211–214.

Savory, T.H. (1928). *The Biology of Spiders*. Sidwick and Jackson, London, 371 pp.

Schulz, R., Bub, S., Petschick, L.L., Stehle, S. and Wolfram, J. (2021). Applied pesticide toxicity shifts toward plants and invertebrates, even in GM crops. *Science*, *372*(6537), 81–84. https://doi.org/10.1126/science.abe1148.

Seastedt, T.R. (1984). The role of microarthropods in decomposition and mineralization processes. *Annual Review of Ecology and Systematics*, *15*(1), 25–46.

Sivinski, J., Wahl, D., Holler, T., Al Dobai, S. and Sivinski, R. (2011). Conserving natural enemies with flowering plants: Estimating floral attractiveness to parasitic Hymenoptera and attraction's relationship to flower and plant morphology. *Biological Control*, *58*(3), 208–214. https://doi.org/10.1016/j.biocontrol.2011.05.002.

Slosser, J.E., Parajulee, M.N. and Bordovsky, D.G. (2000). Evaluation of food sprays and relay strip crops for enhancing biological control of bollworms and cotton aphids. *International Journal of Pest Management 46*:267–275

Sparks, T.C. and Nauen, R. (2015). IRAC: Mode of action classification and insecticide resistance management. *Pesticide Biochemistry and Physiology*, *121*, 122–128. https://doi.org/10.1016/j.pestbp.2014.11.014.

Stamps, W.T. and Linit, M.J. (1997). Plant diversity and arthropod communities: Implications for temperate agroforestry. *Agroforestry Systems*, *39*(1), 73–89.

Staton, T., Walters, R., Smith, J., Breeze, T. and Girling, R. (2021). Management to promote flowering understoreys benefits natural enemy diversity, aphid suppression, and income in an agroforestry system. *Agronomy*, *11*(4), 651. https://doi.org/10.3390/agronomy11040651.

Stenchly, K., Clough, Y. and Tscharntke, T. (2012). Spider species richness in cocoa agroforestry systems, comparing vertical strata, local management, and distance to forest. *Agriculture, Ecosystems, and Environment*, *149*, 189–194.

Symondson, W.O.C., Sunderland, K.D. and Greenstone, M.H. (2002). Can generalist predators be effective biocontrol agents? *Annual Review of Entomology*, *47*, 561–594. https://doi.org/10.1146/annurev.ento.47.091201.145240.

Tilahun, A., Cornelis, W., Sleutel, S., Nigussie, A., Dume, B. and Van Ranst, E. (2021). The potential of termite mound spreading for soil fertility management under low input subsistence agriculture. *Agriculture*, *11*(10), 1002. https://doi.org/10.3390/agriculture11101002.

Trunschke, J., Lunau, K., Pyke, G.H., Ren, Z.X. and Wang, H. (2021). Flower color evolution and the evidence of pollinator-mediated selection. *Frontiers in Plant Science*, *12*, 617851.

Tscharntke, T., Klein, A.M., Kruess, A., Steffan-Dewenter, I. and Thies, C. (2005). Landscape perspectives on agricultural intensification and biodiversity–ecosystem service management. *Ecology Letters*, *8*(8), 857–874. https://doi.org/10.1111/j.1461-0248.2005.00782.x.

Vattala, H.D., Wratten, S.D., Phillips, C.B. and Wäckers, F.L. (2006). The influence of flower morphology and nectar quality on the longevity of a parasitoid biological control agent. *Biological Control*, *39*(2), 179–185.

Westphal, C., Vidal, S., Horgan, F.G., Gurr, G.M., Escalada, M., Van Chien, H., Tscharntke, T., Heong, K.L. and Settele, J. (2015). Promoting multiple ecosystem services with flower strips and participatory approaches in rice production landscapes. *Basic and Applied Ecology*, *16*(8), 681–689. https://doi.org/10.1016/j.baae.2015.10.004.

Woodcock, B.A., Harrower, C., Redhead, J., Edwards, M., Vanbergen, A.J., Heard, M.S., Roy, D.B. and Pywell, R.F. (2014). National patterns of functional diversity and redundancy in predatory ground beetles and bees associated with key UK arable crops. *Journal of Applied Ecology*, *51*, 142–151. https://doi.org/10.1111/1365-2664.12171.

Xiao, J. and Xiong, K. (2022). A review of agroforestry ecosystem services and its enlightenment on the ecosystem improvement of rocky desertification control. *Science of the Total Environment*, *852*, 158538. https://doi.org/10.1016/j.scitotenv.2022.158538.

Zhu, P., Liang, R., Qin, Y., Xu, H., Zou, Y., Johnson, A.C., Zhang, F., Gurr, G.M. and Lu, Z. (2023). Extrafloras and floral nectar promote biocontrol services provided by parasitoid wasps to rice crops. *Entomologia Generalis*, *43*(5), 971–979.

Chapter 12

Sustainable Management of Foliar Nematodes in Tuberose Cultivation

Challenges and Prospects

Shimpy Sarkar[1*], Arshdeep Singh[2] and Iza Fatima[3]

1. Introduction

Tuberose [*Agave amica* (Medikus) Thiede & Govaerts] is a tropical flowering plant known for its exquisite fragrance and ornamental appeal. Its star-shaped, waxy florets are loosely arranged along the spike, making it a highly sought-after flower in the floriculture industry. The single-flowered varieties are more aromatic compared to the double-flowered types and are often preferred for making garlands and traditional floral ornaments like *Gajara* (hair garland). In contrast, the double and semi-double varieties are widely used for vase arrangements and decorative purposes. Tuberose is primarily propagated through bulbs, which typically measure between 1.5–2.0 cm in diameter and weigh around 30 grams. The plant begins blooming three to four months after planting, with peak flowering occurring during the summer and rainy seasons. Once established, tuberose plants can continue to produce commercially viable flowers for up to three years. Despite its economic significance, tuberose cultivation is frequently challenged by various biotic constraints, the most notable being plant-parasitic nematodes (Iftikhar et al., 2023). Among these pests, the foliar nematode *Aphelenchoides besseyi* Christie, poses a significant threat to tuberose production in regions like West Bengal (Figure 1). This nematode was first reported to cause foliar disease in tuberose by Holtzmann in 1968 in Hawaii. In India, its damaging effects were initially documented in the Ranaghat region of the Nadia district, West Bengal, by Chakraborti and Ghosh in 1993. Beyond India, *A. besseyi* has been identified as a major concern in the Mekong Delta of Vietnam (Cuc & Pilon, 2007) and has also been reported to affect other crops such as rice, onion, and gladiolus (Roy et al., 2010, 2011).

The infestation of *A. besseyi* typically begins with the development of light chlorotic lesions along the midrib of the leaf, which gradually spread upwards but never reach the apex. Over time, these lesions turn yellow, then brown or black, forming elongated necrotic spots. Infected leaves eventually wither and fall off. The scape, or flower stalk, may become twisted and deformed, and in severe cases, the inflorescence may fail to develop entirely. Endoparasitic feeding by the nematodes on leaves, bracts, and petals leads to the formation of dark necrotic spots of varying shapes and

1 Department of Entomology, School of Agriculture, Lovely Professional University, Punjab, India-144411.
2 Department of Agronomy, School of Agriculture, Lovely Professional University, Punjab, India-144411.
3 College of Agriculture Food and Wine, The University of Adelaide, Urrbrae, South Australia, Australia.
* Corresponding author: shimpy610@gmail.com

DOI: 10.1201/9781003614869-12

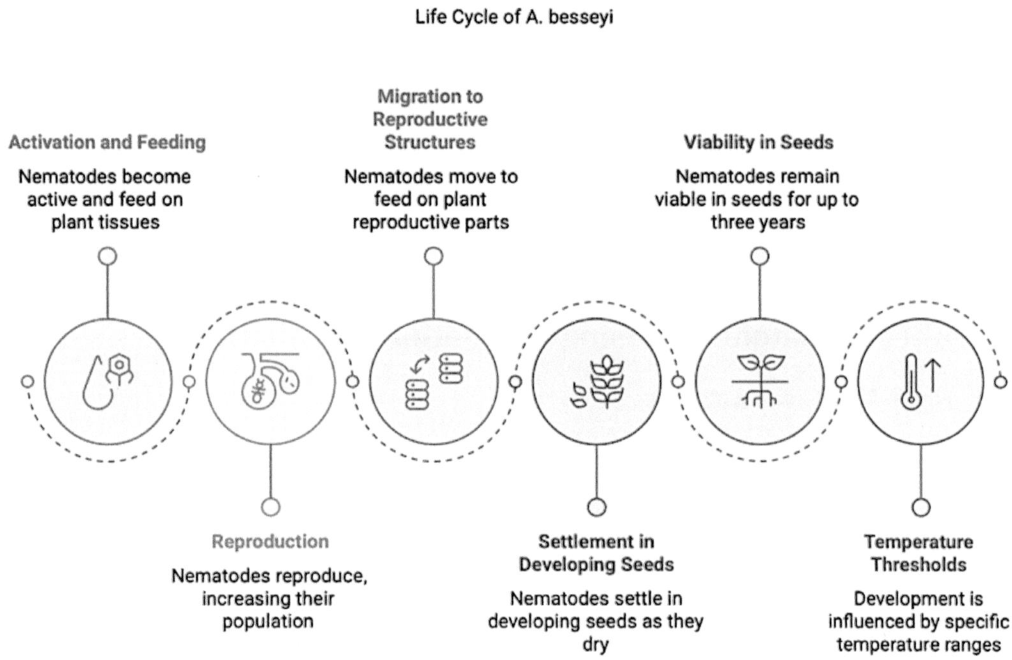

Fig. 1 Life cycle of foliar nematode, *Aphelenchoides besseyi* on tuberose.

sizes. These nematodes do not persist in the soil but can survive on plant parts such as bulbs, leaves, bracts, and anthers, making them a persistent challenge for growers (Roy et al., 2010, 2011). The impact of *A. besseyi* infestation is particularly severe in certain tuberose cultivars. The "Calcutta Single" variety has been observed to suffer an up to 59% yield loss, while the "Calcutta Double" variety experiences an annual reduction of approximately 30–40% (Khan & Pal, 2001; Pathak & Khan, 2009). Given the economic significance of tuberose, effective management strategies for nematode infestations are crucial. Integrated pest management (IPM) approaches, including low-input plant protection technologies, have gained prominence in recent years. These include practices such as summer ploughing, soil solarization, organic amendments, proper planting schedules, optimized spacing, and the selection of resistant varieties. Additionally, maintaining field sanitation, clean cultivation, and judicious use of environmentally safe pesticides are critical components of a sustainable pest management strategy. A combination of these methods can significantly reduce the prevalence and impact of *A. besseyi* on tuberose crops, ensuring higher yields and better flower quality. Apart from nematodes, several other insect pests also contribute to reduced yield and quality in tuberose cultivation (Gautam et al., 2020). Some of the major pests of tuberose include:

- **Bud Borer (*Helicoverpa armigera*):** Larvae bore into buds and flowers, causing direct damage. Control methods include light traps, manual collection, and spraying of Endosulfan 0.07% or Methyl Parathion 0.05% (Sridhar et al., 2022).
- **Aphids (*Aphis craccivora*):** These soft-bodied insects cluster around flower buds and young leaves, feeding on plant sap. Spraying Malathion @0.1% at 15-day intervals is an effective control method (Sridhar et al., 2022).
- **Grasshoppers (*Atractomorpha renulata*):** Feeding on leaves and flower buds, they cause defoliation and loss of floral elegance. Dusting with 5% Cythione/DDT/Folidol or spraying Quinalphos @0.05% helps in control (Sridhar et al., 2022).
- **Red Spider Mites (*Tetranychus urticae*):** These pests cause yellow streaks on leaves, leading to discoloration and leaf drop. Spraying with Kelthane @ 1.2% concentration is effective (Sridhar et al., 2022).

- **Rodents (*Rattus rattus* and *Apodemus* spp.):** Rodents create burrows in the field, damaging tuberose plants. Poison baits such as 'Roban' are commonly used for control (Sridhar et al., 2022).
- **Thrips (*Thrips flavus, Thrips palmi* and *Thrips hawaiiensis*):** Thrips suck sap from flower stalks and leaves, sometimes causing a disease known as "bunchy top". Spraying with 0.1% Malathion is an effective control measure (Sridhar et al., 2022).
- **Weevils (*Myllocerus* sp.):** These pests notch leaf edges and damage bulbs by tunnelling. Applying 10% BHC dust in the soil before planting helps manage infestations (Sridhar et al., 2022).

2. Other Nematodes Causing Infestation in Tuberose

2.1 Root-Knot Nematodes

Root-knot nematodes (*Meloidogyne incognita*) are among the most damaging pests affecting tuberose cultivation. These nematodes induce the formation of characteristic root galls, leading to compromised water and nutrient uptake. Infected plants exhibit symptoms such as chlorotic leaves, wilting, and stunted growth. Studies have shown that heavy infestations can result in a 10% reduction in flower yield. Additionally, these nematodes facilitate secondary infections by soil-borne pathogens, including *Fusarium oxysporum*.

Nematodes persist in tuberose bulbs, thereby facilitating their spread through planting material. Investigations have demonstrated that up to 58.35% of nematodes are found in the outer layers of the bulbs (Gaonkar, 2021). Control measures include hot water treatment of bulbs, biological control using *Paecilomyces lilacinus* and *Trichoderma harzianum*, and the application of neem-based products (Holajjer et al., 2020). Screening for resistant tuberose genotypes has also been explored as a management strategy.

2.2 Reniform Nematodes

Reniform nematodes (*Rotylenchulus reniformis*) attack the root system by feeding on cortical tissues, leading to root necrosis and discoloration. Symptoms of infestation include leaf chlorosis, defoliation, and malformed flowers. This nematode often interacts with other soil pathogens, exacerbating plant stress. Studies indicate that nematode populations vary across different tuberose cultivars, with some genotypes exhibiting a degree of tolerance. Chemical control measures such as the application of Carbofuran, as well as organic amendments like neem cake, have shown promise in managing reniform nematode populations (Chawla et al., 2005) (Figure 2).

3. Case Studies on *Aphelenchoides besseyi* Infestation and Management

3.1 Symptomatology and Yield Loss due to Foliar Nematodes

A field study conducted in West Bengal (Kadam et al., 2019) observed progressive symptom development in *Aphelenchoides besseyi*-infected tuberose stalks and flowers. Symptoms included prickle-like structures on the floral stalks, chlorotic leaves, malformed flowers, and reduced flower spike length. Infested flowers failed to bloom properly, leading to a 50% reduction in marketable yield.

The white tip nematode, *A. besseyi*, is a seed and vegetative propagule transmitted ectoparasitic nematode known for causing serious economic damage in various ornamental and agronomic crops (CABI, 2022; Raski, 1958). While it is most famously associated with white tip disease of rice, increasing reports have emerged of its impact on ornamental crops such as tuberose, chrysanthemum, and gladiolus in several parts of India (Chakraborty & Saha, 2016).

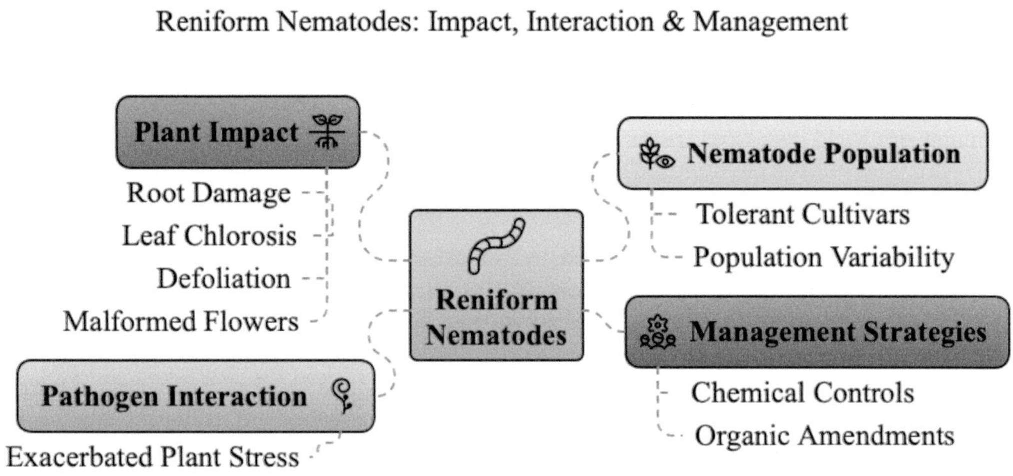

Fig. 2 Impact, interaction, and management of reniform nematode.

In tuberose, the nematodes predominantly infest bulb scales and floral tissues. Observations by Khan et al. (2018) in Uttar Pradesh corroborated the findings of Kadam et al. (2019), reporting stunted inflorescences, curled and thickened leaves, and a delay or failure in blooming. Infected bulbs serve as primary sources of inoculum, perpetuating the nematode in successive planting seasons (Patil & Shukla, 2017).

Scanning electron microscopy studies conducted by Mishra et al. (2020) revealed that nematode invasion causes disruption of epidermal and sub-epidermal tissues in the floral stalks and ovaries, which explains the malformed and non-functional floral structures. Yield losses as high as 40–60% have been recorded in severely infested fields, not only due to reduction in spike number and length but also due to lower aesthetic value of the flowers (Singh et al., 2021).

3.2 Impact of Nematode Management Practices on Yield

A study by Pathak and Khan (2009) reported that foliar nematodes, particularly *Aphelenchoides besseyi*, can lead to as much as 59% reduction in loose flower yield in tuberose (*Agave amica*) when left unmanaged. These microscopic nematodes, which typically enter plant tissues through stomata, migrate intercellularly, and feed on mesophyll and parenchyma, causing extensive damage. Infected plants exhibit chlorotic streaks on leaves, twisted or stunted buds, malformed petals, and in severe cases, spike abortion and floral sterility. The nematodes primarily spread through infected planting material, especially bulbs, and are favoured by warm, humid environments such as those prevalent during monsoons in eastern and northern India. Studies from other regions support these findings, with Kadam et al. (2019) and Singh et al. (2021) documenting 35–70% yield reductions and significant aesthetic losses in spike quality and marketability. Ravichandra (2014) also noted that foliar nematodes could cause up to 70% economic loss in bulbous ornamentals under favourable conditions. Managing foliar nematodes in tuberose requires an integrated approach. Hot water treatment of bulbs (at 52°C for 20 minutes) has proven effective in eliminating nematodes from planting material (Chakraborty & Saha, 2016). Chemical nematicides like carbofuran and oxamyl can reduce nematode populations in soil and foliage, though their use must be balanced with environmental safety considerations. Organic amendments such as neem cake, mustard cake, and vermicompost have shown suppressive effects on nematode activity and also improve soil health (Patil et al., 2019). Biological control agents like *Paecilomyces lilacinus* and *Pochonia chlamydosporia* have demonstrated efficacy in parasitizing nematode eggs and reducing infestation under both greenhouse and field conditions (Kumar & Mishra, 2020). Additionally, cultural practices such as using certified nematode-free bulbs, maintaining field hygiene, and rotating with non-host crops like marigold or maize are crucial for long-term management. Overall, combining these strategies into a comprehensive Integrated Nematode Management (INM) programme is essential for minimizing yield losses and ensuring sustainable tuberose production.

3.3 Integrated Pest Management Approach

Research conducted by Sarkar and Roy (2019) in the Gangetic Plains of West Bengal provided strong evidence supporting the effectiveness of IPM strategies against *Aphelenchoides besseyi*, the foliar nematode responsible for substantial yield and quality losses in tuberose (*Agave amica*). Their study involved multi-season field trials where plots were subjected to a comprehensive IPM approach, including pre-planting hot water treatment of bulbs (at 52°C for 20 minutes), rigorous field sanitation to remove infected residues, and strategic applications of granular nematicides such as phorate at recommended doses. This integrated strategy significantly suppressed nematode populations in both soil and plant tissues, delayed symptom development (e.g., floral malformation, spike abortion, and leaf chlorosis), and improved overall plant health. Compared to conventional practices, IPM-managed plots showed reduced nematode incidence throughout the season.

Moreover, the study reported that spikes from IPM-treated fields had longer stalks, more florets per spike, and significantly better marketability, resulting in a 42% increase in saleable yield. The quality of loose flowers was also enhanced, with reduced petal blemishes and better fragrance retention. Similar results were observed by Chakraborty and Saha (2016), who demonstrated that integrating hot water-treated bulbs with neem cake application and *Trichoderma harzianum* as a soil amendment reduced *A. besseyi* populations and improved tuberose spike quality in West Bengal's red and lateritic zone. Additionally, a study by Patil et al. (2019) showed that incorporating organic amendments such as mustard cake and vermicompost along with crop rotation significantly reduced nematode infestation while enriching soil microbial activity, thereby supporting plant vigour. These studies underscore that relying solely on chemical nematicides is insufficient for long-term management of foliar nematodes, especially in nematode-prone environments. Instead, a seasonally adapted, multi-pronged IPM approach combining physical, biological, organic, and chemical tools is both economically sustainable and ecologically sound for managing nematode-induced damage in tuberose cultivation.

3.4 Influence of Spacing and Variety on Nematode Infestation

A comparative study on varietal susceptibility of tuberose to foliar nematodes revealed that *Agave amica* cv. "Calcutta Single" exhibited significantly greater nematode-induced damage than "Calcutta Double". Plants of "Calcutta Single" showed severer symptoms such as chlorotic streaks, distorted flower spikes, malformed petals, and reduced flower yield under the same infestation conditions. The higher susceptibility of "Calcutta Single" has been attributed to its more open floral structure and softer tissues, which may facilitate easier entry and movement of *Aphelenchoides besseyi* (Khan et al., 2018). In contrast, the relatively compact floral architecture of "Calcutta Double" may act as a partial physical barrier to nematode colonization and feeding. Similar observations were made by Singh et al. (2021), who reported higher nematode population densities and symptom expression in "Calcutta Single" under identical agroecological conditions, confirming its greater vulnerability across locations.

In addition to varietal differences, the study also emphasized the role of planting density in the development and severity of foliar nematode infestations. Narrow plant spacing (e.g., 20 × 20 cm) resulted in significantly higher disease incidence and nematode populations than wider spacing (e.g., 30 × 25 cm). Denser canopy formation in closely spaced plants leads to increased humidity and reduced air circulation around the foliage, factors that promote the survival, movement, and reproduction of foliar nematodes (Ravichandra, 2014). Chakraborty and Saha (2016) also noted that narrow spacing facilitates quicker nematode dispersal through leaf contact, further exacerbating infestation. Thus, wider spacing not only reduces plant-to-plant nematode transmission but also creates a less favourable microclimate for nematode proliferation, serving as an effective cultural control method.

The findings underline the importance of integrating varietal resistance and agronomic practices such as appropriate planting density in nematode management strategies for tuberose cultivation. An integrated approach that includes the use of moderately resistant varieties like "Calcutta Double", wider spacing, and complementary practices such as hot water treatment of bulbs and organic soil amendments can significantly reduce nematode pressure and improve flower quality and yield. Kadam et al. (2019) and Patil et al. (2019) emphasized that cultural practices, especially planting geometry, when combined with biological and chemical interventions, enhance the efficacy of INM programmes. Therefore, optimizing plant spacing and cultivar selection should be prioritized as cost-effective, environmentally sustainable tools in the overall pest management framework for commercial tuberose cultivation.

3.5 Effectiveness of Biological Control Agents

A recent field trial evaluated the effectiveness of biological control agents, particularly *Paecilomyces lilacinus* and *Pseudomonas fluorescens*, in managing *Aphelenchoides besseyi* populations in tuberose (*Agave amica*). The application of these agents, either as bulb dips or soil drenches, resulted in a significant reduction in nematode density in both rhizospheric soil and plant tissues. Treated plots showed visibly healthier plants, with enhanced spike length, floret number, and overall vigour compared to untreated controls. These results are in line with earlier findings by Rao et al. (2003), who demonstrated the nematode-suppressing ability of *P. lilacinus* through egg parasitism and hyphal penetration. Similarly, Naznin et al. (2015) showed that *P. fluorescens* not only inhibits nematode development through the production of antimicrobial compounds and lytic enzymes, but also promotes plant growth by stimulating root development and nutrient uptake.

Such biological agents offer a promising, eco-friendly alternative to chemical nematicides, which are often associated with residual toxicity, environmental contamination, and resistance development. Keter et al. (2020) observed significant nematode suppression using *P. fluorescens* in African floricultural crops, further validating its wide applicability. Kavitha and Thirukumaran (2019) also reported that combining *P. lilacinus* with neem-based organic amendments enhanced its efficacy, likely due to improved microbial activity and rhizosphere colonization. These international studies reinforce the growing global consensus on the potential of biocontrol agents in nematode management. As sustainability and environmental safety become central to pest management programmes, integrating biological control into tuberose IPM strategies not only reduces chemical dependency but also supports long-term soil health and crop productivity.

3.6 Economic Losses in Commercial Tuberose Farms

An economic assessment of commercial tuberose (*Agave amica*) farms in Odisha revealed that infestations of the foliar nematode *Aphelenchoides besseyi* resulted in substantial financial losses. Infested plants exhibited characteristic symptoms such as chlorotic and twisted leaves, malformed flowers, aborted spikes, and general decline in plant vigour, all of which severely affected both the aesthetic value and quantity of the floral yield (Figure 3). On average, affected farms reported an annual revenue loss of approximately 40%, primarily due to a reduction in the number of marketable spikes and deterioration in flower quality. Similar economic impacts have been documented internationally; for instance, Cuenca et al. (2008) observed that nematode damage in ornamental flower production in Colombia resulted in an up to 50% loss in export quality flowers. Likewise, Bridge et al. (2005) emphasized that *A. besseyi* can significantly impair ornamental crops' commercial viability in tropical and subtropical climates, where warm and humid conditions promote nematode proliferation.

Farmers who implemented INM strategies were able to mitigate these losses effectively. One of the most successful approaches involved crop rotation with non-host plants, such as marigold (*Tagetes spp.*), which has been widely reported to suppress nematode populations due to its bioactive root exudates (Hooks et al., 2010). In addition, the use of resistant or moderately tolerant cultivars, such as "Calcutta Double", contributed to lower nematode infestation levels and better yield outcomes. Studies by Palomares-Rius et al. (2017) in Spain and Ntalli and Caboni (2017) in Greece confirmed that integrating resistant varieties with cultural practices like rotation and sanitation significantly reduces plant-parasitic nematode populations and supports crop sustainability. These approaches not only proved cost-effective, but also aligned with international goals of reducing synthetic nematicide dependency in ornamental crop production systems.

The Odisha study highlights the broader relevance of adopting preventive and sustainable nematode management practices. Economic modelling within the study showed that farms adopting INM protocols experienced a notable recovery in net returns, often within 1–2 seasons, due to improved flower quality, increased market acceptability, and reduced input costs related to chemical treatments. International evidence supports these findings: Seid et al. (2021) reported that smallholder floriculture growers in East Africa saw an up to 35% increase in profits after shifting to biocontrol-based and rotational nematode control strategies. The convergence of local and global research underscores the importance of early detection, varietal resistance, and ecological management in safeguarding the profitability and sustainability of commercial tuberose cultivation against A. besseyi infestations (Figure 3).

Fig. 3 Comparison between healthy flowers and infested flowers in tuberose.

3.7 Temperature and Seasonal Influence on Nematode Infestation

A study conducted in the subtropical regions of India by Amin et al. (2021) found that populations of the foliar nematode *Aphelenchoides besseyi* peaked during the monsoon and post-monsoon seasons. The nematode population dynamics closely correlated with environmental conditions-particularly high relative humidity (above 80%), moderate temperatures (25–30°C), and frequent leaf wetness, which favour their reproduction, dispersal, and infection potential. These climatic conditions, typical of the monsoon period in South Asia, create an ideal microenvironment for nematodes to move intercellularly through the stomata and colonize above-ground plant parts. Similar seasonal population surges have been reported in studies from other subtropical regions. For example, Li et al. (2019) observed peak *A. besseyi* activity in southern China during the rainy season in rice fields, while Sikora et al. (2005) noted that nematode outbreaks in tropical ornamentals tend to align with wet and moderately warm periods, which increase host susceptibility.

The timing of nematode population peaks has important implications for pest management in crops such as tuberose (*Agave amica*), which are highly sensitive to foliar nematode infestations. The study by Amin et al. (2021) emphasized that adjusting the planting schedule-either advancing or delaying it to avoid peak infestation windows-can significantly reduce the initial nematode load and subsequent disease severity. Additionally, timely implementation of control measures, such as hot water treatment of bulbs, application of bioagents like *Paecilomyces lilacinus*, and cultural practices including field sanitation and proper drainage, should be synchronized with environmental cues. These recommendations align with international findings: Youssef and El-Nagdi (2020) reported that pre-emptive nematode management based on weather monitoring reduced *Meloidogyne* spp. damage in floricultural crops in Egypt, while Trudgill et al. (2000) highlighted the importance of matching nematode life cycles with control timing in sustainable IPM programmes.

Given the clear seasonal trends in nematode abundance, incorporating climate-based forecasting tools into pest management strategies offers an opportunity to improve nematode control efficiency and reduce reliance on reactive chemical treatments. In a study from Brazil, Inomoto and Tavares (2018) successfully used climatic data to model and predict the risk of foliar nematode outbreaks in ornamental crops, enabling farmers to apply biocontrol and cultural interventions at critical periods. This approach has also been advocated in Southeast Asia, where climate-resilient nematode management is being promoted in high-value horticultural production systems (Gowen et al., 2005). Overall, the integration of seasonal nematode monitoring, weather-based forecasting and strategic timing of interventions represents a forward-looking and environmentally sustainable pathway to managing *A. besseyi* and other nematodes in tuberose and related crops.

Table 1 Incidence of *Aphelenchoides besseyi* infestation in tuberose across different regions.

Location	*Incidence/Impact*	*Reference*
Hawaii, USA	First recorded infestation on tuberose	Holtzmann (1968)
West Bengal, India	Severe loss due to malformed flowers	Chakraborti & Ghosh (1993)
Mekong Delta, Vietnam	Significant foliar disease in tuberose	Cuc & Pilon (2007)
Odisha, India	Widespread infestation in tuberose and rice	Roy et al. (2010)
Nadia, West Bengal, India	59% yield loss in "Calcutta Single" cultivar	Pathak & Khan (2009)

Table 2 Symptoms of nematode infestation in tuberose.

Plant Part Affected	*Observed Symptoms*	*Possible Cause*
Leaves	Chlorotic lesions, yellowing, necrotic streaks	*A. besseyi* feeding
Scape (Flower Stalk)	Twisting, curling, reduced spike length	Nematode infection
Flowers	Malformation, incomplete blooming, reduced fragrance	Infestation in floral tissues
Bulbs	Presence of nematodes in outer layers	Carryover infestation

Table 3 Effectiveness of different nematode management strategies.

Management Method	*Effectiveness (%)*	*Additional Benefits*
Hot water treatment of bulbs	85%	Reduces initial inoculum
Biocontrol (*Paecilomyces lilacinus*, *Pseudomonas fluorescens*)	70%	Eco-friendly alternative
Chemical control (Carbofuran, Monocrotophos)	90%	Quick suppression but environmental concerns
IPM approach (bulb treatment, field sanitation, nematicides)	95%	Sustainable and long-term control

Table 4 Yield losses due to *A. besseyi* in tuberose.

Variety	*Yield Loss (%)*	*Reference*
Calcutta Single	Up to 59%	Pathak & Khan (2009)
Calcutta Double	30–40%	Khan & Pal (2001)

Table 5 Seasonal influence on *A. besseyi* population.

Season	*Nematode Population Trend*	*Recommended Management Strategy*
Summer	Low	Preventive bulb treatment
Monsoon	High	Active monitoring and foliar sprays
Post-Monsoon	Peak	Intensive control measures
Winter	Decline	Crop rotation and sanitation

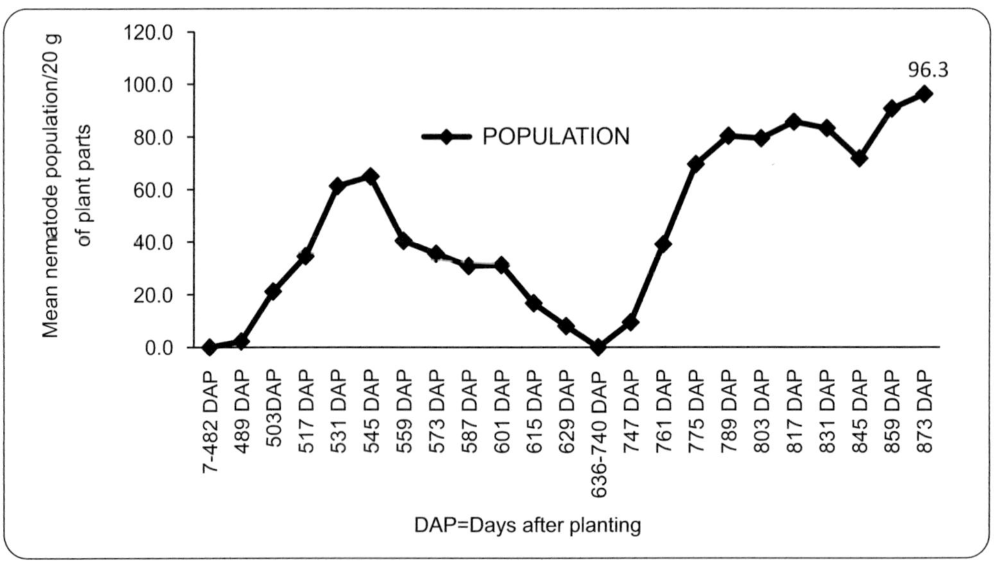

Fig. 4 Population of *Aphelenchoides besseyi* in different days after planting of tuberose.

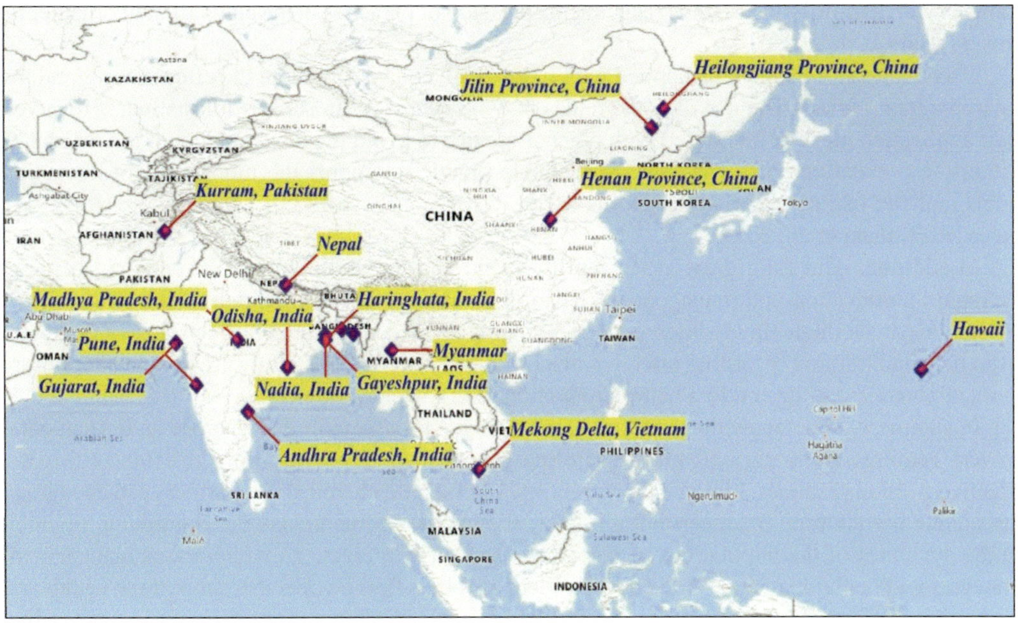

Fig. 5 World map representing *Aphelenchoides besseyi* nematode distribution.

4. Integrating Pest Management Strategies for Sustainable Control of Aphelenchoides besseyi Infestations in Tuberose and Ornamental Flowers

Effective management strategies for the control of the infestations of the *Aphelenchoids besseyi* in tuberose and other ornamental flowers are essential, given the harmful impact of the pathogen on the surrender and quality of the crops. The integrated approaches for the management of parasites (IPM), characterized by the integration of biological, cultural, physical and chemical methods, offer a holistic picture to mitigate these infestations of nematodes. The use of resistant cultivar is one of the main IPM strategies. According to Bala and Ravindranath (2018), the specific tuberose cultivars have levels of susceptibility to *A. besseyi*, with some varieties that demonstrate greater resilience against flowering malformations related to nematodes. Consequently, the selection of resistant cultivars is a critical tactic to reduce the populations of nematodes and mitigate the associated damage.

Cultural practices also play a fundamental role in the management of the infestations of *A. besseyi*. Khan et al. (2024) highlight the importance of maintaining soil health through the rotation of crops and organic amendments, which can improve the microbial soil activity and create a less favourable environment for the proliferation of nematodes. In addition, the maintenance of optimal irrigation practices is vital. Excessive humidity can exacerbate the problems of nematodes, indicating that water management is an essential aspect of sustainable horticultural practices (Reddy & Reddy, 2021). The implementation of adequate drainage can significantly reduce the survival of nematodes and reduce the risk of infestation in ornamental crops.

Biological control methods, in particular the use of natural antagonists, can also contribute to IPM strategies. The introduction of nematopathogenic mushrooms, for example, has shown promise in suppressing the populations of *A. besseyi* (Khan, 2020). In addition, the application of entomopathogenic nematodes (EPN) has an ecologically sustainable option that aligns with the principles of IPM. The various researches indicate that these beneficial nematodes can actually compete with harmful species, thus reducing the pressure of parasites in the cultivation of ornamental flowers (Bala et al., 2019).

Chemical control should be applied with judgement, considering both the effectiveness and the environmental impact. The use of nematicides may be necessary in serious infestations; However, their application must be integrated with other strategies to avoid the development of the resistance within the populations of nematodes. Sustainable practices support the monitoring of nematode levels and the use of chemical solutions as the last resource when the damage thresholds are exceeded (Kadam et al., 2019).

In addition, educating growers on the life cycle and the symptoms of *A. besseyi* is crucial for rapid identification and management (Khan et al., 2024). The recognition of the first signs of infestation can facilitate timely interventions, reducing the need for more drastic measures. A greater awareness and training regarding effective management strategies can allow growers to adopt more resilient practices, at the end to a better production of ornamental flowers.

In summary, the management of the infestations of *A. besseyi* in the tuberose and ornamental flowers requires an integrated multidisciplinary approach. By combining resistant cultivars, effective cultural practices, biological controls and responsible chemical use, farmers can encourage sustainable horticultural practices that minimize the impact of this nematode. Continuing research on the dynamics of the populations of nematodes and their interactions with various management strategies will be fundamental for the development of robust frameworks that support the health and productivity of ornamental crops (Reddy & Reddy, 2021). Collaborative efforts in understanding and facing nematodes' issues promise to improve the resilience of floral cultivation systems against the infestations of *A. besseyi*, adhering to sustainable agricultural principles. Ibrahim and Kurniawati (2020) explored an innovative and environmentally sound strategy to manage *A. besseyi* in rice by inducing plant resistance using chitosan, a natural biopolymer derived from chitin. Chitosan acts as an elicitor of plant defence responses and has been widely reported to activate systemic resistance against various pathogens and pests. In their study, rice seeds treated with chitosan showed significantly reduced nematode infestation, improved seedling vigour, and enhanced root and shoot biomass. This induced resistance approach provides an alternative to chemical nematicides, particularly relevant in sustainable agriculture systems where environmental and human health concerns are prioritized. Similar findings have been reported in other crops; for instance, El Hadrami et al. (2010) highlighted the role of chitosan in priming defence pathways in plants, enhancing tolerance to both biotic and abiotic stresses.

Beyond induced resistance, *A. besseyi* management is also entering a new frontier through the application of genetic technologies such as RNA interference (RNAi) and CRISPR/Cas9 gene editing. These methods rely on a detailed understanding of the molecular basis of host-parasite interactions, particularly the identification of nematode effectors-proteins secreted by the nematode to manipulate host cell functions. Recent advances in nematode genomics have enabled researchers to identify candidate effectors that could be targeted to disrupt nematode feeding or reproduction. For example, Vieira and Gleason (2019) discussed how transcriptomic studies in *A. besseyi* have uncovered several effector genes involved in plant parasitism, providing targets for RNAi-based suppression. When delivered via transgenic plants or topical applications, RNAi can silence essential nematode genes, reducing their viability and reproduction.

CRISPR/Cas9 gene editing further expands the possibilities for managing nematodes at the genetic level. Though still in early research phases for plant-parasitic nematodes, CRISPR can be used to either knock out susceptibility genes in host plants or potentially disable virulence genes in the nematodes themselves. Ali et al. (2022) emphasized that editing host plant genomes to eliminate nematode-compatible interactions could offer durable resistance. However, the use of CRISPR in nematode management poses several technical and regulatory challenges, including delivery methods, off-target effects, and the complexity of nematode life cycles. It also demands substantial international collaboration, genomic resources, and infrastructure to advance these tools from laboratory research to field application.

While induced resistance using compounds like chitosan presents a more immediate and practical solution for smallholder and resource-limited farming systems, gene-editing technologies offer long-term, potentially species-specific solutions to *A. besseyi* and other nematode pests. The integration of these approaches: biochemical, molecular, and genomic, could revolutionize nematode management in rice and other crops, such as tuberose, where *A. besseyi* also poses a serious threat. As highlighted by Haegeman et al. (2012), a deeper molecular understanding of nematode parasitism mechanisms is crucial for developing resistant cultivars and tailored biotechnological tools. Ultimately, success will depend on collaborative efforts from nematologists, plant biotechnologists, breeders, and policymakers to translate laboratory innovations into field-level, farmer-friendly solutions.

5. Conclusion

Nematode infestations, particularly by *Aphelenchoides besseyi* (foliar nematode), present a major constraint to commercial tuberose (*Agave amica*) cultivation, leading to malformed flowers, reduced spike length, and as much as 40–50% yield loss in severe cases (Kadam et al., 2019; Pathak & Khan, 2009). These nematodes colonize aerial plant parts, making them difficult to manage once established. Effective INM strategies combine hot water treatment of bulbs, crop rotation, optimal plant spacing, and use of biocontrol agents like *Paecilomyces lilacinus* and *Pseudomonas fluorescens*, which have shown to suppress nematode populations and enhance plant vigour (Rao et al., 2003; Naznin et al., 2015; Kavitha & Thirukumaran, 2019). Cultural practices such as field sanitation, avoidance of monoculture, and timely planting, especially to avoid nematode population peaks during humid post-monsoon periods, are also vital components of sustainable management (Amin et al., 2021; Inomoto & Tavares, 2018). Chemical nematicides, though effective, should be used judiciously and preferably integrated with biological and cultural methods to minimize environmental impact and resistance build up.

In addition to nematodes, tuberose is vulnerable to a range of insect pests such as stem borers (e.g., *Sphenoptera spp.*), thrips (*Thrips tabaci*), and red spider mites (*Tetranychus urticae*), which further compromise flower quality and marketability. IPM approaches are necessary, involving regular monitoring, sticky traps for thrips, neem-based formulations, and predatory mites or entomopathogenic fungi for mite control (Gupta & Sharma, 2020). Recent studies have highlighted the use of resistant cultivars like "Calcutta Double" that exhibit reduced susceptibility to foliar nematodes and insects compared to more susceptible types like "Calcutta Single" (Reddy, 2025). As pest pressure continues to evolve under changing climatic conditions, continued research into host resistance mechanisms, RNAi-based pest suppression, and climate-resilient IPM models is essential for ensuring stable and profitable tuberose cultivation (Haegeman et al., 2012; Ali et al., 2022). Collaborative efforts between breeders, nematologists, and extension services will be critical to developing and disseminating effective, locally adapted pest management packages for farmers.

References

Ali, M.A., Shah, K.H. and Khan, M.R. (2022). CRISPR/Cas9-mediated resistance against plant parasitic nematodes: Progress and prospects. *Frontiers in Genetics*, *13*, 856612.

Amin, M.R., Islam, M.A., Afroz, M. and Suh, S.J. (2021). Abundance of sucking insect pests on some flowering plants in relation to weather parameters. *Bangladesh Journal of Scientific and Industrial Research*, *56*(4), 241–248.

Amin, M., Roy, S. and Dey, P. (2021). Seasonal nematode dynamics in tuberose and implications for management. *Indian Journal of Plant Protection*, *49*(2), 112–118.

Bala, S.C. and Ravindranath, N. (2018). Field evaluation of tuberose cultivars and symptom manifestation caused by foliar nematode, *Aphelenchoides besseyi* in tuberose. *Int. J. Curr. Microbiol. Appl. Sci.*, *7*, 1364–1370.

Bala, S.C., Nihal, R. and Khan, M.R. (2019). Pathogenicity and population dynamics of foliar nematode, *Aphelenchoides besseyi* infecting tuberose in West Bengal, India. *Journal of Environmental Biology*, *40*(6), 1159–1163.

Bridge, J., Luc, M. and Plowright, R.A. (2005). Nematode parasites of ornamental and flowering plants. In: M. Luc, R.A. Sikora and J. Bridge (Eds.), *Plant Parasitic Nematodes in Subtropical and Tropical Agriculture*. CABI Publishing, pp. 705–737.

CABI (2022). *Aphelenchoides besseyi* (white tip nematode) datasheet. In: *Invasive Species Compendium*. https://www.cabi.org.

Chakraborti, H.S. and Ghosh, S.C. (1993). Studies on the floral malady of *Polianthes tuberosa* L. in West Bengal. *Journal of Botanical Society of Bengal*.

Chakraborty, G. and Saha, A. (2016). Management of *Aphelenchoides besseyi* in tuberose through hot water treatment and bioagents. *Indian Journal of Nematology*, *46*(1), 52–56.

Chawla, G., Singh, K.P. and Srivastava, A.N. (2005). Screening of tuberose (Polianthes tuberosa Linn.) cultivars against reniform nematode (*Rotylenchulus reniformis*). *Journal of Ornamental Horticulture*, *8*(4), 272–274.

Cuc, N.T.T.C. and Pilon, M. (2007). An *Aphelenchoides sp.* nematode parasitic of Polianthes tuberosa in the Mekong Delta. *Journal of Nematology*, *39*(3), 248–257.

Cuenca, J., Ospina, C.A. and Gómez, L.A. (2008). Economic impact of nematodes in flower production in Colombia. *Nematropica*, *38*(2), 211–219.

El Hadrami, A., Adam, L.R., El Hadrami, I. and Daayf, F. (2010). Chitosan in plant protection. *Marine Drugs*, *8*(4), 968–987.

Gaonkar, P.N. (2021). *Bio Management of Root-Knot Nematode, Meloidogyne incognita (Kofoid and White, 1919) Chitwood, 1949 in Tuberose (Polianthes tuberosa L.)* (Doctoral dissertation, Department of Nematology, OUAT, Bhubaneswar).

Gautam, R.D., Parihar, S.G. and Swaroop, K. (2020). Specialty Flower Production and Trade Affected by Insect Pests and Quarantine Issues. In: *Advances in Pest Management in Commercial Flowers*. Apple Academic Press, pp. 29–56.

Gowen, S.R., Quénéhervé, P. and Fogain, R. (2005). Nematode parasites of bananas and plantains. In: *Plant Parasitic Nematodes in Subtropical and Tropical Agriculture*. Wallingford UK: CABI Publishing, pp. 611–643.

Gupta, P. and Sharma, D. (2020). Integrated management of insect pests in tuberose. *International Journal of Entomology Research*, 5(2), 15–20.

Haegeman, A., Mantelin, S., Jones, J.T. and Gheysen, G. (2012). Functional roles of effectors of plant-parasitic nematodes. *Gene*, *492*(1), 19–31.

Holajjer, P., Saha, T.N., Girish, K.S., Deepak, E. and Prasad, K.V. (2020). Effectiveness of bio-formulations for management of root-knot nematode, *Meloidogyne incognita* in tuberose. *Indian Journal of Nematology*, *50*(1), 57–58.

Holtzmann, O.V. (1968). A foliar disease of tuberose caused by *Aphelenchoides besseyi. The Plant Disease Reporter*, *56*(52), 1.

Hooks, C.R.R., Wang, K.H. and Ploeg, A. (2010). Using marigold (*Tagetes spp.*) as a cover crop to protect crops from plant-parasitic nematodes. *Applied Soil Ecology*, *46*(3), 307–320.

Ibrahim, M. and Kurniawati, E. (2020). Induced resistance in rice using chitosan to manage *Aphelenchoides besseyi*. *Journal of Plant Protection Research*, *60*(4), 425–432.

Iftikhar, T., Majeed, H., Waheed, M., Zahra, S.S., Niaz, M., Bilal, B. and Riaz, M. (2023). Tuberose. In: *Essentials of Medicinal and Aromatic Crops*. Cham: Springer International Publishing, pp. 373–397.

Inomoto, M. M. and Tavares, C. D. (2018). Predicting foliar nematode outbreaks in ornamentals. *Nematologia Brasileira*, *42*(1), 12–18.

Kadam, A.R., Banerjee, A. and Sinha, A.K. (2019). Occurrence and pathogenicity of *Aphelenchoides besseyi* in tuberose (*Polianthes tuberosa*) in West Bengal. *Journal of Ornamental Horticulture*, *22*(2), 124–128.

Kadam, V., Chettri, D. and Mukhopadhyay, A.K. (2019). Study on the symptomatology of floral malady of tuberose associated with foliar nematode infestation on the plants. *Pharma Innov.*, *8*, 29–32.

Kavitha, K. and Thirukumaran, K. (2019). Integrated Management of Nematode Disease Complex in Tuberose (*Polianthes tuberose* L). *Journal of Krishi Vigyan*, *8*(1), 170–173.

Keter, S.J., Nyalala, S. and Ogweno, J. (2020). Biofumigation Efficacy of Spider Plant (*Cleome gynandra* L.) Accessions on Nematode Control in Tuberose (*Polianthes tuberosa* L.). *NASS Journal of Agricultural Sciences*, *2*(2), 15–15.

Khan, M.R. (2020). Nematode Pest Problems of Tuberose. In: *Advances in Pest Management in Commercial Flowers*. Apple Academic Press, pp. 119–136.

Khan, M.R. and Pal, A. (2001). Plant parasitic nematodes associated with tuberose (*Polianthes tuberosa* L.) in West Bengal.

Khan, M.R., Haque, Z., Akram, M. and Ansari, M.S.A. (2024). Nematode Problems in Ornamental Plants. In: *Diseases of Ornamental, Aromatic, and Medicinal Plants*. Bentham Science Publishers, pp. 182–206.

Khan, R., Prasad, J.S. and Pandey, S. (2018). White tip nematode-associated deformities in tuberose: A new report from Uttar Pradesh. *Indian Journal of Plant Protection*, *46*(4), 377–380.

Kumar, V. and Mishra, A. (2020). Evaluation of biological control agents against white tip nematode in tuberose under polyhouse conditions. *BioControl Science and Technology*, *30*(4), 373–380.

Li, Y., Liu, Y. and Chen, S. (2019). Seasonal dynamics of *Aphelenchoides besseyi* in rice fields of southern China. *Journal of Nematology*, *51*(1), 1–8.

Mishra, M., Bera, S. and Majumdar, S. (2020). Structural damage to floral organs of tuberose due to infestation by *Aphelenchoides besseyi*: A microscopic analysis. *Plant Pathology Journal*, *36*(1), 17–23.

Ntalli, N. and Caboni, P. (2017). A review of isothiocyanates biofumigation activity on plant parasitic nematodes. *Phytochemistry Reviews, 16*(5), 827–834.

Naznin, A., Hossain, M.M., Ara, K.A., Hoque, A., Islam, M. and Hasan, T. (2015). Influence of organic amendments and bio-control agent on yield and quality of tuberose. *J. Hort*, *2*(4), 1–8.

Palomares-Rius, J.E., Cantalapiedra-Navarrete, C. and Castillo, P. (2017). Strategies for control of plant-parasitic nematodes in ornamental plants. *Journal of Nematology*, *49*(4), 432–442.

Pathak, B. and Khan, M.R. (2009). Yield loss potential of foliar nematode, *Aphelenchoides besseyi* in tuberose. *Indian Journal of Nematology*, *39*(1), 38–40.

Pathak, K. and Khan, M.R. (2009). Evaluation of yield loss and nematode management in tuberose. *Nematologia Mediterranea*, *37*(2), 145–149.

Patil, R.R. and Shukla, Y.M. (2017). Bulb-transmitted nematodes in ornamental plants: A hidden menace. *Nematology Today*, *16*(2), 45–49.

Patil, R.R., Singh, V. and Joshi, K. (2019). Integration of organic amendments for sustainable nematode management in floricultural crops. *Journal of Eco-friendly Agriculture*, *14*(3), 55–60.

Rao, M.S., Shylaja, M. and Naik, D. (2003). Management of nematode induced wilt disease complex in tuberose (*Polianthus tuberosa* L.) cultivar prajwal using *Pochonia chlamydosporia* (*Verticillium chlamydosporium*) and *Trichoderma harzianum*. *Journal of Ornamental Horticulture*, *6*(4), 341–346.

Raski, D.J. (1958). Nematodes causing white tip of rice. *Plant Disease Reporter*, *42*(6), 563–566.

Ravichandra, N.G. (2014). *Plant Nematology*. I.K. International Publishing House.

Reddy, N. S., Sivakumar, T. and Balabaskar, P. (2021). Bio-Control Efficiency of *Trichoderma viride* against Stem Rot of Tuberose Caused by *Sclerotium rolfsii*. *International Journal of Current Microbiology and Applied Sciences*, *10*(11), 407–415.

Reddy, P.P. (2025). *Nematode Diseases of Vegetable Crops and their Management*. CRC Press.

Reddy, P.P. and Reddy, P.P. (2021). Protected Cultivation of Flower Crops. *Nematode Diseases of Crops and Their Management. Spring Nature, Singapore Pte Ltd.*, pp. 373–396.

Roy, K., Sarkar, H.K. and Pramanik, A. (2010). Morpho-anatomical changes in Gladiolus due to infection by *Aphelenchoides besseyi* from West Bengal, India. *Floriculture and Ornamental Biotechnology*, *4*, 96–98.

Roy, K., Sarkar, H.K. and Pramanik, A. (2011). Symptoms and morpho-anatomical changes in leaves of *Polianthes tuberosa* with infection of *Aphelenchoides besseyi*. *Annals of Plant Protection Sciences*, *19*(1), 249–250. Sarkar, S. and Roy, K. (2019). Integrated Management of *Aphelenchoides besseyi* infesting Tuberose in Gangetic Plains of West Bengal. *Int. J. Curr. Microbiol. App. Sci.*, *8*(2), 3299–3315.

Seid, A., Hailemariam, B. and Mugo, H. (2021). Economic benefits of biological control strategies in nematode management for floriculture. *African Journal of Horticultural Science*, *18*(1), 45–54.

Sikora, R.A., Bridge, J. and Starr, J.L. (2005). Management practices: An overview of integrated and biological control. In *Plant Parasitic Nematodes in Subtropical and Tropical Agriculture*. CABI Publishing, pp. 793–825.

Singh, A., Kaur, M. and Bansal, R.K. (2021). Yield loss estimation and varietal screening of tuberose against *Aphelenchoides besseyi*. *International Journal of Plant Protection*, *14*(2), 147–151.

Sridhar, V., Naik, S.O., Swathi, P. and Mani, M. (2022). Pests and Their Management in Ornamental Plants: (Rose, Jasmine, Chrysanthemum, Crossandra, Marigold, Tuberose, Carnation, China aster, Gerbera, Gladiolus, Hibiscus, etc.). *Trends in Horticultural Entomology*, 1189–1237.

Trudgill, D.L., Bala, G., Blok, V.C., Daudi, A., Davies, K.G., Gowen, S.R., ... and Voyoukallou, E. (2000). The importance of tropical root-knot nematodes (Meloidogyne spp.) and factors affecting the utility of Pasteuria penetrans as a biocontrol agent. *Nematology*, *2*(8), 823–845.

Vieira, P. and Gleason, C. (2019). Plant-parasitic nematode effectors: Insights into their diversity and new tools for their identification. *Current Opinion in Plant Biology*, *50*, 37–43.

Youssef, M.M.A. and El-Nagdi, W.M.A. (2020). Forecasting-based management of *Meloidogyne* spp. in floricultural crops in Egypt. *Egyptian Journal of Agronematology*, *19*(2), 102–112.

Chapter 13

Advanced Crop Protection Strategies for Management of Cryptic Species of *Bemisia tabaci*

Kalavathi K. Kambali[1*], G. R. Vishwas Gowda[2] and Belaghihalli N Gnanesh[3]

1. Introduction

The whitefly, *Bemisia tabaci* (Gennadius) (Hemiptera: Aleyrodidae), is recognized among the world's 100 most invasive alien species. Known for its extensive polyphagy, it infests 1,000 plant species and can transmit 400 plant viruses (Gilbertson et al., 2015). Yield losses caused by *B. tabaci* can range widely, from 10% to complete crop failure, depending on factors such as crop type, season, and pest density. Two populations, Sida race and Jatropha race, based on distinct host plant differences, were reported in Puerto Rico in 1957. Among over 1,000 whitefly species, *B. tabaci* stands out for its invasiveness and environmental adaptability. Its broad host range includes plants from Cucurbitaceae, Euphorbiaceae, Malvaceae, and Solanaceae families, such as cotton, tomato, eggplant, cucumber, squash, cassava, and okra. The pest drains nutrients by sucking phloem sap and indirectly damages plants by secreting honeydew, fostering fungal growth. It transmits over 400 plant viruses, with the Begomovirus *B. tabaci* complex particularly devastating to crops in developing countries, causing billions of dollars in economic losses annually. Differentiation of biotypes was initially based on biological traits like virus transmission capacity, host range variation, insecticide resistance, physiological changes in hosts, and endosymbionts. However, these traits of plasticity and variability made group separation challenging. Allozyme markers first resolved diversity, designating biotypes from A to T, later refined by mt16S ribosomal DNA, mitochondrial cytochrome oxidase I (mtCOI), and nuclear ribosomal intergenic spacer I (ITS1), using a 4.0% genetic divergence threshold for differentiation (Mahmood et al., 2022).

The first record of *B. tabaci* as a pest dates back to 1889, when it was observed on tobacco in Greece, described as *Aleyrodes tabaci* by Gennadius. Quaintance (1900) identified another whitefly species, *Aleyrodes inconspicua*, in the USA, which was later reclassified into the new genus, *Bemisia.* Due to numerous reports of whiteflies on various host plants from different countries, the species *'tabaci'* was synonymized and placed in the genus *Bemisia* (Takahashi, 1992). During this period, the host range of *B. tabaci* expanded, initially emerging from weeds like *Convolvulus arvensis* and *Brassica* spp. early in the year and spreading to cotton, cucumber, and hibiscus as temperatures rose.

[1] ICAR-KVK, Dharwad, University of Agricultural Sciences, Dharwad-580005.
[2] Department of Plant Pathology, University of Agricultural Sciences, Dharwad-580005.
[3] Genetics and Plant Breeding, ICAR JSS - Krishi Vigyan Kendra, Mysuru-571 129, Karnataka, India.
* Corresponding author: kambalikk@uasd.in

DOI: 10.1201/9781003614869-13

The population declines in October, with adults returning to *Brassica* spp., *Solanum* spp., and other weeds. By this time, *B. tabaci* was reported on 315 host plants across countries like Sudan, Egypt, Taiwan, the USA, and Israel, becoming a highly polyphagous pest that favors hosts from families such as Malvaceae, Solanaceae, Cucurbitaceae, Leguminosae, and Verbenaceae. Its distribution now extends to over 60 countries (De Barro et al., 2011).

In India, *B. tabaci* was first recorded in 1905, with outbreaks in cotton ecosystems in Rajasthan and Punjab during the 1920s–1930s. It later spread to other crops like brinjal, tomato, okra, tobacco, and ornamental plants. The associated virus diseases also gained significance, with differences in virus transmission modes. *B. tabaci* transmits viruses from genera such as Begomovirus, Crinivirus, Ipomovirus, Carlavirus, and Torradovirus. The genus Begomovirus includes cotton leaf curl virus (CLCV), a major cotton disease prevalent in Asia and Africa, and tomato yellow leaf curl virus (TYLCV), which affects tomato crops globally. Other significant viruses transmitted by *B. tabaci* include cucurbit yellow stunting disorder virus (CYSDV; Crinivirus), tomato chlorosis virus (ToCV; Crinivirus), cucumber vein yellowing virus (Ipomovirus), and tomato torrado virus (ToTV; Torradovirus) (Horowitz et al., 2020). Sriganganagar in Rajasthan and nearby districts in Punjab experienced a CLCV outbreak in 1994, which then spread to other cotton-growing regions. Subsequent outbreaks occurred during the 2004–2005, 2009–2010, 2013–2014, and 2015–2016 crop seasons, devastating cotton crops in North India (Datta et al., 2017). This review aims to provide a comprehensive overview of the *Bemisia tabaci* species complex and its insecticide resistance dynamics, factors influencing resistance development, and the geographic distribution of resistant populations. Additionally, we will discuss current and emerging strategies for managing resistance, including the integration of chemical, biological, and cultural control methods. Understanding these dynamics is crucial for developing sustainable and effective management programs to mitigate the impact of *B. tabaci* on global agriculture.

2. Host Races of *Bemisia tabaci*

Host associations were used in Bird's (1957) studies to identify different populations of *Bemisia tabaci* in Puerto Rico according to their host preferences. He had identified two races, Sida and Jatropha. The "Sida race" population was characterized by a wide host range that included species belonging to the Sida genus and also an effective vector of gemini viruses and "Jatropha race" exhibited effective Jatropha mosaic virus transmission as well as a preference for Jatropha as a host. The introduction of Brazilian cassava (*Manihot esculenta* Crantz) varieties into Africa led to a situation similar to this. Although *B. tabaci's* host range was wide, it did not colonize cassava in Brazil (Costa & Russell, 1975). On the other hand, *B. tabaci* quickly became practically monophagous on *M. esculenta* in Africa after adapting to cassava. Following their introduction to Africa, Brazilian cultivars faced significant risks of infection by the African cassava mosaic virus (Robertson, 1987). These instances emphasize the idea of phytophagic variants, also known as host races, in which populations of *B. tabaci* have similar physical characteristics but differing characteristics related to survival and development on various hosts (Walsh, 1864; Claridge & Den Hollander, 1983).

3. Biotypes of *Bemisia tabaci*

The term 'biotype' refers to insect individuals and populations of species that share certain biological characteristics (such as host survival and development, or host feeding and/or oviposition characteristics) but have little or no genetic basis (Claridge & Den Hollander 1983; Diehl & Bush, 1984). Initially, biotypes were classified according to biological properties such as viral transmission capability, host range fluctuation, pesticide resistance, physiological changes in host plants, and endosymbionts. Later assessments of biological features were insufficient to differentiate groups due to the flexibility and diversity of qualities within and across groups (Brown et al., 1995a; Costa

& Brown, 1991). Allozyme markers were the first molecular tools to resolve diversity and identify biotypes from A to T. Furthermore, mt16S ribosomal DNA, mitochondrial cytochrome oxidase I (mtCOI), and the nuclear ribosomal intergenic spacer I (ITS1) were investigated, and 4.0% pairwise genetic divergence was used to distinguish the isolated groups (Lee et al., 2013).

Considerable confusion has existed regarding the classification of whitefly species and biotypes within *Bemisia*. Notably, the silverleaf whitefly, once described as *B. argentifolii*, is now recognized as the 'B' biotype.

The accidental introduction of the B biotype into the United States and the Caribbean between 1986–1990 marked the first major recognition of *Bemisia tabaci*. During this period, populations escalated dramatically, causing severe outbreaks in irrigated cropping systems of the southwestern U.S. and widespread infestations across the southern states. By 1994, the B biotype had extended its range to the Caribbean and American Tropics, eventually reaching South America. Between 1990–1994, outbreaks of this biotype were also documented in Australia, China, Egypt, the Mediterranean region of Europe, Israel, Japan, Pakistan, and Turkey, establishing it as the first *B. tabaci* biotype with a truly cosmopolitan distribution and role as a major vector. In contrast, the whitefly commonly known as the sweet potato whitefly corresponds to the A biotype of *B. tabaci* (Bellows et al., 1994; Brown et al., 1995b; Brown, 2007).

The Q biotype was initially identified in Spain and, along with its related clades from Europe, North Africa, and the Middle East, is native to the Mediterranean region. It shows a preference for vegetable crops, particularly tomatoes. The biotypes that have been of particular concern to the U.S. in recent years have been *B. tabaci* biotype 'Q'. Specific resistance issues have included: pyriproxyfen (Distance), neonicotinoids imidacloprid (Marathon or Merit), acetamiprid (TriStar), thiamethoxam (Flagship), and reduced susceptibility to IGR buprofezin (Talus). The 'Q' biotype originates from the Mediterranean and has been reported in the following countries: China, Egypt, France, Israel, Japan, Morocco, the Netherlands and Spain. The first report of the 'Q' biotype in the continental U.S. was during December 2004 at a poinsettia nursery in Arizona (Bellows et al., 1994; Bethke et al., 2006; Brown, 2007).

Placement of global populations into various categories has proven useful for examining geographical relationships, yet the biological characteristics of these groups are paramount. However, research in this area remains limited. For instance, researchers have conducted comparative studies on host ranges. Thus far, studies have evaluated 13 populations across 10 biotypes (A, B, B2, D, E, H, J, K, L, and N) in host range assessments. Some authors have also assessed virus transmission characteristics, detailing relative efficiencies among 21 populations from 13 biotypes (including an unknown biotype from Uganda, and biotypes A, B, B2, D, E, G, H, J, K, L, M, and N). Furthermore, 46 populations from 20 biotypes (including an unknown biotype from Spain, and biotypes 0, A, AN, B, B2, C, Cassava, D, E, F, G, H, J, K, L, M, N, P, and Q) have been studied for their ability to induce squash silverleaf (SSL), a plant disorder notably linked with biotype B. This bioassay consistently highlights distinctions between B and non-B biotypes. Only eight populations representing four biotypes (A, AN, B, H) have been examined for fundamental biological parameters such as developmental and oviposition rates. Lastly, mating experiments have been conducted with 15 populations from eight biotypes (A, AN, B, B2, D, K, L, and M). The study of restricted gene flow is crucial as it aligns with the biological species concept (Perring et al., 2001).

Scientists have employed diverse methodologies to differentiate populations, emphasizing the inherent risks of relying on a single technique. Specifically cautioned against is the use of a sole esterase for classifying whiteflies. Proper identification of any given *B. tabaci* population necessitates a combination of these distinguishing techniques. This approach was exemplified by Perring et al. (1993), who utilized RAPD-PCR banding patterns, isoelectric focusing electrophoresis, mating behavior assessments and the absence of mating to differentiate between biotypes A and B within the United States (Table 1).

Table 1 Identification of Bemisia tabaci biotypes globally.

Biotypes	*Identification Method/Marker Used*	*Marker Type*	*Country*	*Year*	*Purpose/ Significance*	*References*
A and B	RAPD-PCR	Molecular (RAPD)	USA	1993	Initial identification of cryptic biotypes in the US population	Gawel & Bartlett (1993)
B	RAPD-PCR	Molecular (RAPD)	Australia	1997	Identification of introduced B biotype in Australia	De Barro & Driver (1997)
Australian AN	RAPD-PCR	Molecular (RAPD)	Australia	1998	Detection of novel Australian biotype (AN)	De Barro et al. (1998)
E, L, G, B, A	Cytochrome oxidase I (COI) gene sequencing	Molecular (mtDNA)	Benin, Sudan, India, Israel, Yemen, USA, Costa Rica, Mexico, Puerto Rico, USA	1999	Global diversity and invasion pattern study	Frohlich et al. (1999)
B	16S rRNA gene sequencing	Molecular (mtDNA)	Australia, Israel, Japan, Korea, USA	2000	Confirmation of B biotype identity and distribution	Lee & De Barro (2000)
A	16S rRNA gene sequencing	Molecular (mtDNA)	USA	2000	Detection of A biotype presence in native US populations	Lee & De Barro (2000)
B	RAPD-PCR (primers OpB20 and OpB11)	Molecular (RAPD)	India	2001	Identification of invasive B biotype in India	Banks et al. (2001)
B, Q, Cv	Morphology, RAPD-PCR, COI sequencing	Combined (Molecular + Morphological)	China	2009	Biotype differentiation and pest risk categorization	Qiu et al. (2009)
B, Q, Cv, ZHJ-1, ZHJ-2, ZHJ-3	Nymphal and adult morphological traits	Morphological	China	2013	Discovery of new morphological types (ZHJ series)	Lee et al. (2013)
B	mtCOI, mitochondrial 16S rRNA sequencing, RAPD-PCR	Molecular (multi gene)	Saudi Arabia	2014	Confirmation of B biotype in Middle East populations	Alhudaib et al. (2014)
B, Q, BT B1, BT B3 (Brazilian)	SCAR markers	Molecular (nuclear DNA)	Brazil	2016	Identification of native and invasive biotypes in Brazil	Queiroz et al. (2016)
B	RAPD-PCR	Molecular (RAPD)	India	2016	Continued validation of B biotype spread in India	Mandali et al. (2016)

4. Distinct Groups within the *Bemisia tabaci* Species Complex

The populations studied using multiple methodologies are the easiest to classify within the *Bemisia tabaci* species complex. Comprehensive data is gathered by employing various techniques consistently across different populations, which helps identify specific biotypes in particular regions. Several authors have contributed findings from diverse studies: Bedford et al. (1994) on mating experiments, Rosell et al. (1997) on morphology, Frohlich et al. (1999) on mitochondrial 16S and COI genes, Brown et al. (2000) on allozymes, and De Barro et al. (2000) on ribosomal ITS1 sequences. These collective studies provide extensive information for grouping populations and distinguishing them from others. Based on this data, seven distinct groups have been identified within the *B. tabaci* species complex (Table 2).

Table 2 Grouping of Bemisia tabaci biotypes based on morphological, physiological and reproductive characteristics.

Group	*Associated Biotypes*	*Key Diagnostic Characters / Traits*	*Biological or Ecological Significance*	*References*
Group 1 –*New World Biotypes*	A, C, N, R (primarily from the Americas)	Presence of 4 Anterior Sub-Marginal Setae (ASMS4)	Typical of New World populations; used for morphological differentiation	Rosell et al. (1997)
Group 2 – *Cosmopolitan / Invasive Biotypes*	B (= *B. argentifolii*), B2	Known to induce plant physiological disorders such as Squash Silverleaf (SSL) and Tomato Irregular Ripening (TIR)	Highly invasive; associated with economic crop damage	Perring (1996)
Group 3 –*Non-SSL Old World Biotypes*	E (Benin), S (Spain)	Lack ASMS4; smaller anterior wax margin; do not cause SSL	Differentiates from invasive B biotype; less phytotoxic	Rosell et al. (1997)
Group 4 –*Indian Non-SSL Biotype*	H (India)	Does not induce SSL	Geographically isolated non-virulent population	Frohlich et al. (1999)
Group 5 – *Hybridizable Non-SSL Biotypes*	L (Sudan), Q (Spain), J (Nigeria), ? (Egypt)	Do not produce SSL; capable of hybridizing with B biotype	Intermediate compatibility with invasive strains	Bedford et al. (1994)
Group 6 – *Reproductively Isolated Biotypes*	M (Turkey), ? (Hainan), ? (Korea)	Unique esterase profiles; reproductively isolated from B, K, and D biotypes	Likely distinct cryptic species; no interbreeding with invasive types	Bedford et al. (1994)
Group 7 – *Australian Biotype*	AN (Australia)	Esterase banding indicates hybrid formation with B biotype	Suggests introgression or recent divergence	Gunning et al. (1997)
Other Biotypes	D (Nicaragua), K (Pakistan)	- **D**: No SSL; no crossing with B, K, M - **K**: No SSL; lacks ASMS4; reproductively isolated from B, D, M	Serve as genetic outgroups; display distinct evolutionary trajectories	Bedford et al , (1994); Rosell et al. (1997)

5. Evolutionary Development of Cryptic Species of Bemisia tabaci

The term "cryptic species" refers to morphologically identical species that can only be distinguished using molecular data (Gill et al., 2016; Saez & Lozano, 2005) (Table 3). The term biotype was abandoned and genetic groups were described as cryptic species of the *B. tabaci* complex (Mahmood et al., 2022). *Bemisia tabaci* isn't a single species but a complex comprising 46 cryptic species, including Africa, Asia I, Asia II 1–12, Asia 1-India, Asia III, Asia IV, Asia V, Australia, Australia/Indonesia, China 1-7, Indian Ocean, Ru, Middle East Asia Minor I-II (MEAM), Mediterranean (MED), MEAM K, New World 1–2, Japan 1–2, Uganda, Italy 1, and Sub-Saharan Africa 1–5 (Rehman et al., 2021). MED has been split into four groups based on its intraspecies genetic variability and geographic range: Q1, Q2, Q3 and African silver-leafing (ASL), referring to the silver-leafing symptoms caused by feeding on squash (Gueguen et al., 2010).

Table 3 Recent Identification of Cryptic Species of *Bemisia tabaci* Worldwide.

Cryptic Species	*Method*	*Crop*	*Country*	*Reference*
MEAM1 or biotype B, and MED or biotype Q.	Mitochondrial cytochrome oxidase gene subunit-1 (*mtCOI*-LAMP)	MED (Poinsettia Plants) MEAM1 (Cotton)	USA	Dickey et al. (2013)
MEAM1, MED, China 2, China3, China 4, Asia I, Asia II 1, and Asia II 6	*mtCOI*	Vegetables, pulses, and ornamental plants	China	Hu et al. (2014)
Asia II 1, Asia II 5, Asia II 7, Asia 1, MEAM 1, and a new species "Pakistan"	*COI 3 and COI 5*	Cotton	Pakistan	Ashfaq et al. (2014)
Asia I, Asia II 1, and Asia II 7.	*mtCOI*	Soyabean	India	Prasanna et al. (2015)
Asia-I, Asia-II-7, Asia-II-8, MEAM-1, MEAM- K	*mtCOI*	More than 30 host plants from 71 places	India	Roopa et al. (2015)
MEAM1, MED, Asia II 3, Asia II 9 and China 6	*mtCOI* (RAPD-PCR)	Vegetables, ornamentals, and weedy plants	China	Jiu et al. (2017)
Asia I, Asia II 1, Asia II 5, Asia II 7, Asia II 8, Asia II 11, Asia IV, MEAM 1, China 3 and China 7	*mtCOI*	-	India	Rehman et al. (2021)
Mediterranean (MED) Q1 and Q2, Middle East–Asia Minor 1 (MEAM1)	*mtCOI*	Horticultural and ornamental plants	Italy	Bertin et al. (2021)
Asia II 1 and Asia II 6,	*mtCOI*	Cassava	Lao PDR	Leiva et al. (2022)
MED ASL, MED Q1, SSA 1 and SSA3	Microsatellite and *mtCOI*	Vegetables majorly tomato	Côte d'Ivoire	N'cho et al., 2022
Asia I, Asia I India, Asia II (1–13), Asia III, Asia IV, Asia V, China 1–6, MEAM (1, 2, K), MED, Australia/Indonesia, Japan (1 and 2).	*mtCOI*	-	Bangladesh, Indonesia, Japan, Korea, Myanmar, Nepal, Philippines, Singapore, Taiwan, and Vietnam	Lestari et al. (2022)
Asia I	*mtCOI*	Cucumber	India	Sindhura et al. (2023)
Asia 1, Asia II 1, Asia II 5 and Asia II 7	*mtCOI*	Cotton plants (different agroclimatic zones)	India	Ramesh et al. (2023)

The phylogenetic tree outlines the relationship among various groups (Figure. 1), highlighting *Bemisia tabaci* as a monophyletic group (supported with a probability of 1.00), with *B. tuberculata*, *B. afer*, *B. berbericola*, and *T. vaporariorum* as outgroups. The tree features 12 major clades, each strongly supported with a posterior probability of 0.95 or higher. These clades are broadly categorized along geographic lines: five groups pertain to the New World (including Asian groups Asia I, Asia II, China, Italy (Sicily), and one from Australia/Indonesia), while another clade represents the Mediterranean/Asia Minor/Africa region. Additionally, three clades are associated with Mediterranean/Asia Minor/Africa, specifically the Indian Ocean, Mediterranean, and sub-Saharan Africa silver leafing (Boykin et al., 2007).

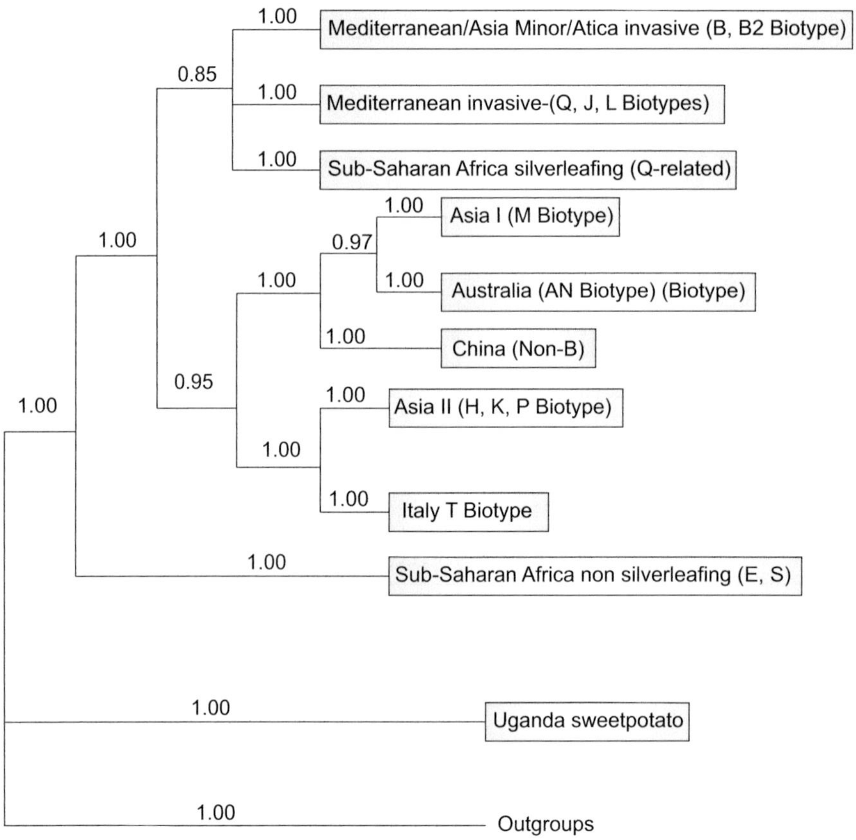

Fig. 1 Schematic representation showing relationships of major clades of *B. tabaci* generated using Bayesian analyses of 366 mitochondrial COI sequences (Boykin et al., 2007).

6. *Bemisia tabaci* and Its Association with Bacterial Endosymbionts

Candidatus Portiera aleyrodidarum is the primary endosymbiotic bacterium of the *Bemisia tabaci* species complex (Thano & Baumann, 2004; Sloan & Moran, 2012). This bacterium, which only lives in bacteriocyte cells in all whitefly individuals, is essential to the survival and growth of its host. It has undergone significant coevolution with members of the subfamily Aleyrodidae (Thano & Baumann, 2004; Baumann, 2005). In addition to *Portiera*, *B. tabaci* harbors seven secondary endosymbionts: *Rickettsia* (Gottlieb et al., 2006), *Wolbachia* (Nirgianaki et al., 2003), *Hamiltonella*, *Arsenophonus* (Sloan & Moran, 2012), *Cardinium* (Weeks et al., 2003), *Fritschea* (Everett et al., 2005), and *Hemipteriphilus* (Bing et al., 2013). These secondary symbionts vary widely in their prevalence within *B. tabaci* populations (Marubayashi et al., 2014), spatial localization throughout the developmental stages, and modes of transmission (Caspi-Fluger et al., 2011). The physiology, ecology, and reproduction of *B. tabaci* are significantly impacted by a number of secondary endosymbionts (Su et al., 2014). They impact host fitness (Shan et al., 2016), provide defense against pathogens (Hendry et al., 2014), improve thermotolerance (Brumin et al., 2011), confer resistance to insecticides (Kontsedalov et al., 2008), and influence the whitefly's ability to transmit viruses (Rana et al., 2012). They also contribute to rapid evolutionary changes. The complexity of interactions within the *B. tabaci* species complex is highlighted by these associations, which also have implications for approaches to disease and pest control.

Kanakala and Ghanim (2019) did a global investigation into the diversity of endosymbionts present in the *Bemisia tabaci* species complex. Their research utilized 16S/23S rRNA gene sequences from 1,226 reported endosymbiont accessions of *B. tabaci*, sourced from GenBank and spanning 32 different countries. Through Bayesian phylogenetic analysis, they described the worldwide distribution and genetic diversity of endosymbionts associated with *B. tabaci*, identifying multiple distinct groups of symbiotic bacteria within this pest species complex. A study by Tang et al. (2018) investigated the occurrence of secondary endosymbionts among various *B. tabaci* species in China. Their findings revealed that *Arsenophonus* was predominantly observed in native lineages like Asia II 1, Asia II 3, and China 1, while it was not present in the invasive MEAM1 population. In addition, Cardinium was commonly detected in MED, Asia II 1, and Asia II 3 species at moderate frequencies, but it was rarely present in MEAM1 and completely absent from China 1. These results suggest that secondary endosymbionts such as *Arsenophonus* and *Cardinium* are more frequently found in native whitefly species, which aligns with the distribution pattern of *Wolbachia*, but contrasts with *Hamiltonella*, an endosymbiont that is instead most abundant in invasive *B. tabaci* groups.

Rossitto De Marchi et al. (2020) found only two facultative or secondary endosymbionts like *Hamiltonella* and *Rickettsia,* association with the *B. tabaci* MEAM1 species populations collected from Florida. Li et al. (2023) found three facultative endosymbionts, *Cardinium* sp., *Hamiltonella* sp., and *Rickettsia* sp., were detected in all cryptic species obtained from 29 localities across China and the infection frequencies were significantly associated with their geographical distribution. In detail, *Cardinium* sp. was detected in a small number of MED (12.4%) and Asia II 2 (1.4%), whereas it was not detected in any of the specimens of MEAM1. *Rickettsia* sp. infected 77.5% of MED, 39.5% of MEAM1, and 10.8% of NCS individuals. The infection frequency of *Hamiltonella* sp. was 92.3% in MED, 71.6% in MEAM1, 0.13% in Asia I, and 3.57% in China 1.

7. Plant Viral Disease Transmitted by *Bemisia tabaci*

Plant viruses frequently require a biological vector for horizontal transmission in nature (i.e., plant-to-plant dissemination). Insects are the most widespread and significant biological vectors of plant viruses. Plant viruses are transmitted by insect vectors in three modes: nonpersistent, semipersistent, and persistent, depending on the minimum time period necessary for acquisition and the maximum length of virions retention (Ng & Falk, 2006). The most frequent plant virus vectors are hemipteroid insects, which include phloem-feeding aphids, whiteflies, planthoppers, and leafhoppers. It should be noted that the majority of the virus infections that have appeared in the last two decades are caused by viruses transmitted by whiteflies (Hemiptera: Aleyrodidae) (Navas-Castillo et al., 2011). These comprise ipomoviruses (Potyviridae), torradoviruses (Secoviridae), criniviruses (Closteroviridae), and begomoviruses (Geminiviridae). According to Zanardo and Carvalho (2017), whiteflies are the only vectors of the cowpea mild mottle virus (CPMMV) of the genus Carlavirus (Betaflexiviridae). Another surprising occurrence has surfaced recently: it has been demonstrated that the pepper whitefly-borne vein yellows virus (PeWBVYV), a polerovirus (Luteoviridae) family, is spread by whiteflies (Table 4). This is a variation from the family's general transmission pattern, which involves aphids (Ghosh et al., 2019).

Table 4 *Bemisia tabaci* cryptic complex species transmitted viruses worldwide.

Genera/Family	*Viral Species*	*B. tabaci Cryptic Species*	*Mode of Transmission*	*Country*	*Host*	*References*
Geminiviridae (Begomoviruses)	BGMV	NW 2, MEAM 1	Persistent, non propagative	Brazil	French bean	De Marchi et al. (2017)
	BYVMV	MEAM 1, Asia 1	Persistent, non propagative	India	Okra	Venkataravanappa et al. (2017)
	CLCuMuV	Asia II 7	Persistent, non propagative	China	Cotton	Chen et al. (2019)
	ToLCNDV	Asia II 5	Circulative persistent	India	Potato	Naga et al. (2021)
	EuYMV	NW 2	Persistent, non propagative	Brazil	French bean	De Marchi et al. (2017)
	SiGMV	Cryptic species B	Semi- persistent	US	Snap bean	Gautam et al. (2023)
Carlavirus (Betaflexiviridae)	CPMMV	MEAM 1, MED	Non-persistent	Brazil	Soyabean beans	Bello et al. (2021)
Crinivirus (Closteroviridae)	ToCV	MEAM 1, MED	Semi- persistent	China	Tomato	Shi et al. (2018)
	LCV	MEAM 1	Semi- persistent	Israel	Cannabis	Hadad et al. (2019)
	CYSDV	MEAM 1	Semi- persistent	Jamaica	Cucurbit	Pitter et al. (2024)
Ipomoviruses (Potyviridae)	SqVYV	MEAM 1	Semi- persistent	Florida	Watermelon	Webb et al. (2012)
Poleroviruses (Picornaviridae)	PeVYV-2 PeWBVYV	MEAM1	Semi- persistent	Israel	Pepper	Ghosh et al. (2021)

Cowpea mild mottle virus (CPMMV), tomato leaf curl New Delhi virus (ToLCNDV), Bean golden mosaic virus, cotton leaf curl multan virus (CLCuMuV), tomato chlorosis virus (ToCV), sida golden mosaic virus (SiGMV), lettuce chlorosis virus (LCV), squash vein yellowing virus (SqVYV), pepper vein yellows virus 2 (PeVYV-2), pepper whitefly-borne vein yellows virus (PeWBVYV), NW-2 –new world 2, MEAM 1- Middle East Asia Minor, MED- Mediterranean.

8. Insecticide Resistance

Resistance may be defined as a heritable change in the sensitivity of a pest population that is reflected in the repeated failure of a product to achieve the expected level of control when used according to the label recommendation for that pest species (Insecticide Resistance Action Committee (IRFC), 2013). The first documented resistance to an insecticide was in 1914 with the San Jose scale, *Comstockaspis perniciosus*, to lime sulfur (Melander, 1914). Insecticide resistance in *Bemisia tabaci* is a widespread issue, with resistance having developed to nearly all commonly used insecticides (Basit, 2019). Mota-Sanchez and Wise (2019) report that there are approximately 650 reported cases of insecticide resistance in the genus *Bemisia*, affecting over 60 different active ingredients. Resistance can be categorized into two types: cross-resistance and genetic resistance. Cross-resistance occurs when resistance to one insecticide or group of insecticides also provides resistance to another, while genetic resistance naturally arises within populations and is passed down through generations (IRFC, 2013).

8.1 Mechanisms of Insect Resistance

Insects exhibit four main mechanisms of resistance to insecticides: metabolic, altered target site, behavioral, and penetration resistance. Metabolic resistance occurs when insects can detoxify or neutralize the toxin more rapidly than their susceptible counterparts, or prevent the toxin from reaching its target sites by binding it to internal proteins. This is the most prevalent and challenging form of resistance, as it involves insects possessing higher levels or more effective forms of the enzymes responsible for breaking down insecticides. Altered target-site resistance happens when the binding site for the toxin on the insect is modified, diminishing the insecticide's effectiveness. Behavioral resistance involves changes in the insect's behavior to avoid exposure to the toxin, such as altering feeding habits or relocating to areas less affected by pesticides. For instance, some malaria-carrying mosquitoes in Africa have developed a preference for resting outdoors to avoid insecticides applied to indoor surfaces. Penetration resistance refers to the slower absorption of toxins by resistant insects, which can develop thicker cuticles or other barriers that impede the chemical's entry. This mechanism is often found alongside other resistance types (Buhler, 2014).

Bemisia tabaci develops resistance through several combined mechanisms, including increased detoxification and genetic changes in targets such as acetylcholinesterase (AChE), the GABA-gated chloride channel, and voltage-gated sodium channel (Figure 2). Enzymes that detoxify pesticides play a key part in diminishing the insect's sensitivity to chemical treatments. Since insecticides are usually hydrophobic, metabolic processes in *B. tabaci* convert these fat-soluble substances into more polar or less lipophilic forms, facilitating their removal and decreasing their effectiveness (Patra & Hath et al., 2022).

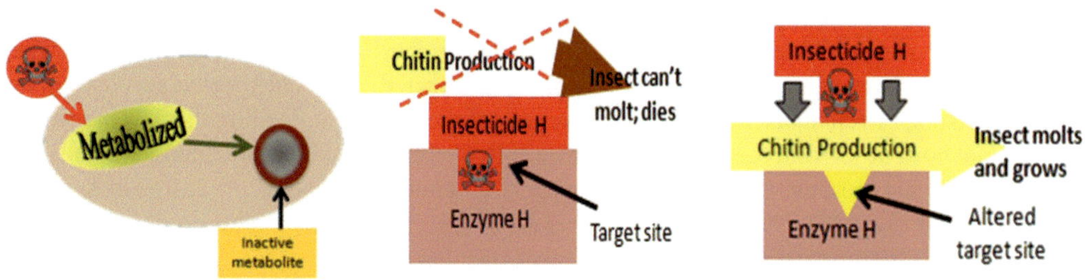

Fig. 2 Mechanisms through which *Bemisia tabaci* develops resistance: a) Metabolic resistance b) Alter site resistance

Table 5 Recent findings of resistance to insecticides in *Bemisia tabaci* cryptic species.

Species Type	*Insecticides*	*Country*	*References*
MEAM 1 (B)	Neonicotinoids, pyrethroids (bifenthrin), buprofezin,	USA	Caballero et al., 2013
MEAM 1 (B), MED (Q)	Pyriproxyfen, acetamiprid, thiamethoxam	Israel	Horowitz & Ishaaya, 2014
Asia I	Imidacloprid, thiamethoxam, monocrotophos, cypermethrin	India	Naveen et al., 2017
Asia II	Cypermethrin, deltamethrin, imidacloprid		
MEAM 1	Azadirachtin, spiromesifen, and lambda-cyhalothrin	Brazil	Dângelo et al. (2018)
Asia I, Asia II-7	Acephate, triazophos, Indoxacarb, dinotefuran, spiromesifen, pyriproxyfen, fonicamid	India	Roy et al. (2019)
MED (Q)	Spiromesifen, spirotetramat	Spain	Bielza et al. (2019)
MED, MEAM 1	Abamectin, Thiamethoxam, Acetamiprid, Nitenpyram, Imidacloprid, Pyriproxyfen, Pymetrozine, Chlorpyrifos	China	Wang et al. (2020)

Contd.

Table 5 *Contd.*

Species Type	*Insecticides*	*Country*	*References*
MEAM 1	Pyriproxyfen, bifenthrin	Australia	Hopkinson et al. (2020)
MED, MEAM 1	Spirotetramat, cyantraniliprole, pyriproxyfen	China	Zheng et al. (2021)
Asia I, Asia II 5, China 3, and Asia II 7	Thiamethoxam, imidacloprid, acetamiprid	India	Barman et al. (2022)
MED, MEAM 1	Pyriproxyfen, spirotetramat, cyantraniliprole	Türkiye	Yukselbaba & Ali (2022)
Asia II-1	Cypermethrin, imidacloprid, thiamethoxam	India	Rajna et al. (2024)
MEAM 1	Chlorpyrifos, imidacloprid, acetamiprid, deltamethrin, thiamethoxam	Türkiye	Erdogan et al. (2024)

8.2 Methods to Detect Insecticide Resistance

8.2.1 Field-Level Detection

When pest control methods don't work as expected in the field, people often think that pests have become resistant. Control problems can happen for many reasons like wrong ways of applying pesticides, bad weather, not enough pesticide used, or using poor quality products. But more often than not pesticide resistance is the main reason (IRAC, 2022; Sparks & Nauen, 2015). Resistance happens when pests become less sensitive to a pesticide that used to work well, usually because of repeated use of that pesticide (Whalon et al., 2008). At the field level, resistance is noticed when pests still survive even after the pesticide is applied as directed. This unexpected survival makes farmers, advisors, or scientists doubt the effectiveness of the treatment, leading them to investigate further. Right now, pests from the affected areas are collected and tested through bioassays and other tests to confirm and measure the level of resistance (Denholm & Rowland, 1992; Roush & Miller, 1986).

8.2.2 Laboratory Testing Methods

Specific area control failures require validation through laboratory testing procedures. Testing field activities is one way to identify resistance but is often influenced by other external factors such as weather, control method, control measure timing, pest count, and crop condition (Busvine, 1980; IRAC, 2022). Laboratory procedures provide a controlled environment where many of these factors are minimized, allowing for the accurate assessment of resistance and susceptibility of insect populations to specific insecticides.

Laboratory test aims at generating measurable of the pest population's response to some insecticide, Levels up to five dose levels may be tested for some precision, The accurate identification of important levels such as the lethal concentration, (LC_{50}, LC_{90}) or Lethal dose (LD_{50}, LD_{90}) is essential for determining resistance for different populations and for different time periods (Robertson et al., 2007). Adherence to a given protocol ensures the validity of inter-laboratory and inter-country comparisons.

Laboratory tests play a critical role in the finding of the resistance mechanisms. In bioassays, the phenotypic expression of resistance is captured (e.g., survival).

8.2.3 Leaf-Dip Bioassays

Leaf-dip bioassays are frequently used to assess resistance in whiteflies. The process involves preparing serial dilutions of insecticides from their commercial formulations, dipping excised leaves into the solutions, drying them, and then exposing the target insects under controlled conditions (IRAC, 2022; Nauen et al., 2015). This method is frequently used to assess resistance in whiteflies. The process involves preparing serial dilutions of insecticides from their commercial formulations. Leaves are collected from unsprayed fields, washed with tap water, and air-dried at room temperature. Leaf discs of suitable diameter are then prepared and dipped in the insecticidal solution, with slight agitation for 5–10 seconds to ensure complete coverage. Occasionally, entire leaves are used instead of discs. Treated leaves or discs are dried until the surface liquid evaporates. For control samples, leaves or discs are dipped in distilled water. The leaves are then placed in Petri plates or suitable containers with perforated lids or covered with muslin cloth. Approximately 20–30 adult whiteflies are introduced into each Petri plate. The tests are conducted at a temperature of 24–25°C, and mortality is assessed 48 to 72 hours post-treatment, depending on the insecticide used. Adults that show no movement are considered dead. Typically, three replicates are performed for each insecticide concentration and for controls. The mortality rate of nymphs at each insecticide concentration and in control samples is calculated using the following formula

$$\text{Percent mortality} = \frac{\text{Number of dead whiteflies}}{\text{Total number of whiteflies treated}} \times 100$$

Then, the corrected percent mortality was calculated using Abbott (1925) formula based on the control mortality. The corrected mortality data of each test insecticide of each location were subjected to probit analysis based on Finney (1971). The formula below is used to calculate resistance ratio (RR) or resistance factor (RF):

$$\text{Resistance Ratio (RR)} = \frac{\text{LC50 of a field population}}{\text{LC50 of the most susceptible population}}$$

8.2.4 Biochemical Assays

Biochemical assays have a central role in measuring detoxification enzymes in insect tissues to evaluate and explain the mechanisms of resistance to particular insecticides. These types of tests reveal the relationship between the insect's anatomical traits and the 'life processes' of biochemistry that lead to resistance. This enables the researcher to know whether the resistance is due to enhanced metabolism or other metabolic changes (Hemingway & Ranson, 2000).

The underlying principle is that populations of resistant insects have, most of the time, a higher concentration or altered form of detoxifying enzymes as compared to the susceptible populations. Analyzed enzyme activity in the homogenates of the insects' tissues enables one to make conclusions about the processes that are likely to be involved in resistance. Common biochemical assays include:

8.2.5 Esterase Tests

Identify elevated synthesis of carboxylesterases and cholinesterases that hydrolyze bond in organophosphorus compounds, carbamates and pyrethroids. Usually performed with α-naphthyl acetate or β-naphthyl acetate as substrates. The products are measured and quantified through colorimetric methods (Van Asperen, 1962).

8.2.6 Glutathione S-Transferase (GST) Test

Evaluates the identification of conjugates of reduced glutathione (GSH) with reactive metabolites of some insecticides which support detoxification. Evaluated with 1-chloro-2,4-dinitrobenzene (CDNB) as a substrate (Habig et al., 1974). Mixed-Function Oxidase (MFO) Tests/Cytochrome P450 Monooxygenases determine the insecticidal oxidative metabolism and the possible decrease of toxicity. Assessed with model substrates like p-nitroanisole or 7-ethoxycoumarin which were later analyzed by spectrophotometric or fluorometric detection (Rose et al., 1995).

8.2.7 Acetylcholinesterase (AChE) Enzyme

Organo-phosphates and carbamates target-site resistance was tested through evaluation of enzyme activity with specified inhibitors and non-inhibitory conditions.

8.3 Molecular Assays

Molecular assays are powerful tools for the detection and characterization of insecticide resistance mechanisms at the genetic level. Unlike bioassays or biochemical tests, which measure the phenotypic expression of resistance, molecular techniques identify specific genetic changes responsible for reduced insecticide sensitivity, often before resistance is detectable in mortality tests (Bass et al., 2014; Rinkevich et al., 2013).

These assays provide high sensitivity, specificity, and speed, making them invaluable for early detection, resistance monitoring, and population genetics studies. They are particularly useful in integrated pest management (IPM) programs because they can detect resistance alleles at low frequencies, enabling proactive interventions (Hemingway et al., 1998).

8.3.1 Molecular Approaches

1. **Polymerase Chain Reaction (PCR)-Based Detection**
   (a) Used to amplify specific DNA fragments containing resistance-associated mutations.
   (b) Examples include detection of *kdr* (knockdown resistance) mutations in the voltage-gated sodium channel gene, conferring pyrethroid resistance in many pest species (Martinez-Torres et al., 1999).
2. **Allele-Specific PCR (AS-PCR) and Quantitative PCR (qPCR)**
   (a) Allow for detection and quantification of known resistance alleles within populations.
   (b) qPCR can also measure changes in **gene** copy number or gene expression levels, such as overexpression of cytochrome P450 monooxygenases or esterases associated with metabolic resistance (Puinean et al., 2010).
3. **DNA Sequencing**
   (a) Both Sanger sequencing and next-generation sequencing (NGS) can identify known and novel mutations linked to resistance.
   (b) For example, mutations in the acetylcholinesterase (*ace-1*) gene confer organophosphate and carbamate resistance in mosquitoes (Weill et al., 2004).
4. **Microarray and RNA-Seq Analysis**
   (a) Allow genome-wide profiling of gene expression to identify candidate genes involved in resistance.
   (b) RNA-Seq provides higher resolution than microarrays and can detect novel transcripts, alternative splicing, and mutations simultaneously (David et al., 2014).
5. **High-Throughput SNP Genotyping**
   (a) Single nucleotide polymorphism (SNP) markers can be used to track resistance alleles across large populations quickly.
   (b) Particularly useful for resistance mapping and monitoring gene flow between pest populations (Bass et al., 2014).

9. Management of Insecticide Resistance

The major principles for successful IRM:

1. Right dosage according to label recommendation
2. Selection of insecticides based on efficacy and selectivity
3. Rotation of insecticides
4. Non-chemical control methods

9.1 Right Product and Right Dosage

Following the recommended rates on insecticide labels is a key practice for effective pest control. Applying products at rates higher or lower than instructed can accelerate the development of insect resistance, making control more difficult over time. This practice also protects non-target organisms and reduces potential negative impacts on the environment. Additionally, regularly inspect and maintain all spray equipment, checking for blocked nozzles or filters, to guarantee accurate application. Targeting the most susceptible life stages of the pest is another strategic step to maximize effectiveness and manage resistance

9.2 Selection of Insecticide Based on Efficacy and Selectivity

Cryptic whitefly species, while morphologically similar, exhibit distinct differences in insecticide susceptibility and resistance profiles, which can lead to divergent treatment outcomes. Successful management requires accurate species identification and careful selection of a product whose mode of action is effective against the target pest's specific life stage. Product labels provide essential guidance on dosage and application timing, and expert consultation is recommended when precise application is needed for optimal results (Nomikou et al., 2001).

9.3 Systemic Properties of Insecticide

Systemic activity may extend the residual efficacy and the length of the MoA spray window and needs to be considered to minimize resistance development (IRAC, 2020).

9.4 Insecticide Mixtures

It could delay the development of resistance. Insecticide mixture products should be used with careful consideration of the characteristics of the individual active substances, use pattern and pest complex targeted.

10. Management of *Bemisia tabaci*

Various methods are available to manage whitefly populations, including cultural, physical and mechanical methods, biological and chemical approaches.

10.1 Cultural Methods

Crop rotation patterns, specific crop types, spatial arrangements, and other cultural practices can be employed to extend host-free periods and minimize inter-crop migrations, thereby controlling whitefly populations (Hilje et al., 2001; Ellsworth & Martinez-Carrillo, 2001). For example, intercropping tomato with coriander and using overhead sprinkler irrigation and intercropping okra with coriander resulted in significantly lower *B. tabaci* colonization (Togni et al., 2018; Sharma et al., 2018). Similarly, timely sowing is crucial to manage white fly populations, *Bt* cotton planted at the normal sowing time (second week of May) experienced less whitefly infestation compared to crops sown later (fourth week of May) (Kataria et al., 2019) and Mahalakshmi et al. (2014) found that whitefly infestations in black gram were minimal in plots sown between June and December, regardless of the variety, compared to those sown between January and April. Moreover, planting soybean with border crops such as maize or moongbean (Abdallah, 2012), potato with mustard (Bhatnagar et al., 2018), and garlic with chickpea (Farhat et al., 2014) has proven to be effective in mitigating cotton whitefly infestations.

10.2 Physical and Mechanical Methods

Physical and mechanical strategies involve creating conditions that are unfavourable for pests, thus helping to manage sweet potato whitefly populations. These methods include removing breeding sites, alternate hosts, and crop residue, as well as using inert ground covers, healthy seedlings, and resistant crop varieties. Silver plastic mulching and row covers have been shown to effectively reduce whitefly populations in zucchini, leading to increased biomass accumulation and higher total yields (Rutz et al., 2023). In cotton, although yellow sticky traps (YSTs) did not significantly suppress the adult whitefly population compared to chemical control, they were useful for early detection of infestations and helped reduce population numbers when compared to untreated control plots (Kumar et al., 2019). In cotton, *Gossypium hirsutum* varieties such as Suchitra and Suraksha have shown resistance to whitefly (Anon, 2023). Similarly, in tomato, genotypes 2016/Res-1, 2015/Res-5, 2014/Res-1, and 2014/Res-4 have shown resistance to both whitefly and tomato leaf curl virus (Pal et al., 2021).

10.3 Biological Methods

To manage whitefly, the use of predators, parasitoids, and entomopathogenic organisms, has been found to be more effective.

10.3.1 Predators and Parasitoids

Predators of *Bemisia tabaci* primarily include beetles (Coccinellidae), true bugs (Miridae, Anthocoridae), lacewings (Chrysopidae, Coniopterygidae), mites (Phytoseiidae), and spiders (Araneae). *Amblyseius herbicolus*, a predatory mite well-suited to tomato plants, has proven more effective in controlling whitefly populations than *A. tamatavensis* (Cardoso et al., 2024). In cotton, Sanga et al. (2018) identified seven insect predators and seven spider species that target whitefly, with Chrysoperla as the most dominant species. Additionally, integrating *Chrysoperla carnea* with *Orius albidipennis* and *Phytoseiulus persimilis* significantly reduced whitefly numbers and improved the yield of greenhouse-grown cucumber (Adly, 2016). Parasitoids from the genera *Encarsia* and *Eretmocerus* (Hymenoptera: Aphelinidae) are among the most commonly used biological control agents for managing whitefly infestations (Perring et al., 2018; Horowitz et al., 2020). In a study conducted by Sanga et al. (2018), two species, *Encarsia lutea* (Masi) and *Encarsia sophia* (Girault & Dodd), were identified as emerging from whitefly pupae in cotton fields, highlighting their potential as effective parasitoids in cotton agroecosystems.

10.3.2 Entomopathogenic organisms

Entomophagous fungi, viruses, nematodes, and bacteria play a critical role in IPM of insect pests, as a potential substitute to inorganic insecticides, as they are harmless to farmers, non-target species, and the ecosystem (Shandhu et al., 2012). The sweet potato whitefly (*Bemisia tabaci*) is a significant pest of various crops. Entomopathogenic fungi (EPFs) like *Aschersonia aleyrodis* and *Isaria fumosorosea* have proven to be effective alternatives to chemical pesticides, with *A. aleyrodis* showing notable resilience under low humidity conditions (Bohatá et al., 2024). A study evaluated nine entomopathogenic nematode (EPN) species for their effectiveness against MEAM1 nymphs under lab conditions on Snap bean and tomato plants, with *Heterorhabditis floridensis* emerging as the most promising, achieving 99.25% mortality at 7 days post-inoculation and act as potential biocontrol agent for managing *B. tabaci* (Li et al., 2021). Three entomopathogenic fungi—*Isaria javanica* (Cjc-03, Cjw-01) and *Purpureocillium lilacinum* (TS-01)—were identified infecting *Bemisia tabaci* on chili and tomato plants. In laboratory and glasshouse trials, *I. javanica* exhibited the highest virulence, achieving significant mortality rates in *B. tabaci* nymphs and adults (Sani et al., 2023).

10.4 Synthetic Chemicals

Chemical pesticides are widely used to manage insect pests, including *Bemisia tabaci*, in open-field crops, with heavy reliance on several classes of insecticides. Commonly applied pesticides against *B. tabaci* include growth regulators like pyriproxyfen and buprofezin, ketoenols such as spiromesifen and spirotetramat, and diamides including anthranilic, cyantraniliprole, and chlorantraniliprole. Additionally, oils, detergents, and soaps are also frequently employed to control *B. tabaci* infestations (Perring et al., 2018; Horowitz et al., 2020). The neonicotinoid imidacloprid remains one of the most effective and widely used pesticides globally for controlling *B. tabaci* (Mullins, 1993). Flupyradifurone, a newer insecticide, shows promise in managing *B. tabaci* during vegetable production (Ran et al., 2022). In China, an integrated approach using systemic insecticides (such as thiamethoxam, sulfoxaflor, or cyantraniliprole) applied via soil drenching before transplanting, combined with anti-insect nets, has proven to be an effective and economically viable method for *B. tabaci* control throughout the tomato growth period (Mao et al., 2022). Eweis et al. (2022) show that sulfoxaflor and cyantraniliprole are particularly effective in controlling *Bemisia tabaci* on tomato, leading to a significant reduction in infestations and an increase in fruit yield when compared to azadirachtin. Kareh and Nemer (2023) compared the effectiveness of pyrifluquinazon and cyantraniliprole against *B tabaci* on eggplants under greenhouse conditions. Pyrifluquinazon proved to be highly effective, reducing whitefly eggs by over 99% and nymphs by over 97%, resulting in the highest yield (3.9 kg/plant). Cyantraniliprole also performed well, surpassing other conventional insecticides. A field study conducted at the Agricultural Research Station, Kota, found that two sprays of diafenthiuron 50WP (350 g a.i./ha) mixed with castor oil (0.05%) provided the highest whitefly mortality (82%) and tuber yield (24.7 t/ha) in potatoes, with an incremental cost-benefit ratio of 1:2.1. These results were comparable to diafenthiuron without castor oil, which achieved 78% mortality and a yield of 23.8 t/ha (Yadav et al., 2022). A significant drawback of relying solely on synthetic chemicals is the development of insecticide resistance. To prevent this, it's essential to adopt integrated approaches for managing whitefly populations. This includes combining cultural, physical, and biological methods with the careful and strategic use of chemical insecticides, including mixtures, to effectively control whitefly infestations.

References

Abbott, W.S., (1925). A method of computing the effectiveness of an insecticide. *Journal of Economic Entomology*, *18*(2), 265–267.

Adly, D. (2016). Use of predators for controlling the whitefly, *Bemisia tabaci* Genn. and the two spotted spider mite, *Tetranychus urticae* koch, in cucumber greenhouses in Egypt. *Egyptian Journal of Biological Pest Control*, *26*, 701–706.

Alhudaib, K.A., Rezk, A.A.S., Abdel-Banat, B.M.A. and Soliman, A.M. (2014). Molecular identification of the biotype of whitefly (*Bemisia tabaci*) inhabiting the eastern region of Saudi Arabia. *Journal of Biological Sciences*, *14* (8), 494–500.

Anonymous. (2023). *Annual Report on Varieties and Hybrids of Cotton*. Central Institute of Cotton Research (CICR), Nagpur, India.

Ashfaq, M., Hebert, P.D., Mirza, M.S., Khan, A.M., Mansoor, S., Shah, G.S. and Zafar, Y. (2014). DNA barcoding of *Bemisia tabaci* complex (Hemiptera: Aleyrodidae) reveals southerly expansion of the dominant whitefly species on cotton in Pakistan. *PloS One*, *9*(8), e104485.

Banks, G.K., Colvin, J., Reddy, R.C., Maruthi, M.N., Muniyappa, V., Venkatesh, H.M., Kumar, M.K., Padmaja, A.S., Beitia, F.J. and Seal, S.E. (2001). First report of the *Bemisia tabaci* B biotype in India and an associated tomato leaf curl virus disease epidemic. *Plant Disease*, *85*(2), 231–231.

Barman, M., Samanta, S., Upadhyaya, G., Thakur, H., Chakraborty, S., Samanta, A. and Tarafdar, J. (2022). Unraveling the basis of neonicotinoid resistance in whitefly species complex: Role of endosymbiotic bacteria and insecticide resistance genes. *Frontiers in Microbiology*, *13*, 901793.

Basit, M. (2019). Status of insecticide resistance in *Bemisia tabaci*: Resistance, cross-resistance, stability of resistance, genetics, and fitness costs. *Phytoparasitica*, *47*, 207–225.

Bass, C., Puinean, A.M., Andrews, M., Cutler, P., Daniels, M., Elias, J. and Field, L.M. (2014). Mutation of a nicotinic acetylcholine receptor β subunit is associated with resistance to neonicotinoid insecticides in the aphid *Myzus persicae*. *BMC Neuroscience*, *15*(1), 1–11.

Baumann, P. (2005). Biology of bacteriocyte-associated endosymbionts of plant sap-sucking insects. *Annual Review of Microbiology*, *59*, 155–189.

Bedford, I.D., Markham, P.G., Brown, J.K. and Rosell, R.C. (1994). Geminivirus transmission and biological characterization of whitefly (*Bemisia tabaci*) biotypes from different world regions. *Annals of Applied Biology*, *125*, 311–325.

Bello, V. H., da Silva, F.B., Watanabe, L.F.M., Vicentin, E., Muller, C., de Freitas Bueno, R.C.O., Santos, J.C., De Marchi, B.R., Nogueira, A.M., Yuki, V.A. and Marubayashi, J.M. (2021). Detection of *Bemisia tabaci* Mediterranean cryptic species on soybean in São Paulo and Paraná States (Brazil) and interaction of cowpea mild mottle virus with whiteflies. *Plant Pathology*, *70*(6), 1508–1520.

Bellows Jr, T.S., Perring, T.M., Gill, R.J. and Headrick, D.H. (1994). Description of a species of *Bemisia* (Homoptera: Aleyrodidae). *Annals of the Entomological Society of America*, *87*(2), 195–206.

Bertin, S., Parrella, G., Nannini, M., Guercio, G., Troiano, E. and Tomassoli, L. (2021). Distribution and genetic variability of *Bemisia tabaci* cryptic species (Hemiptera: Aleyrodidae) in Italy. *Insects*, *12*(6), 521.

Bethke, J., D. Gilrein and S. Ludwig. (2006). The Q-biotype Whitefly. Found at: http://etipm.tamu.edu/publications/pdf/Whitefly%20GrowerTalks.

Bhatnagar, A., Pant R.P. and Chakrabarti, S.K. (2018). Effect of border crops on population buildup of whitefly, aphids, and virus incidence in potato crop. *Potato Journal*, *45*(2).

Bielza, P., Moreno, I., Belando, A., Grávalos, C., Izquierdo, J. and Nauen, R. (2019). Spiromesifen and spirotetramat resistance in field populations of *Bemisia tabaci* Gennadius in Spain. *Pest Management Science*, *75*, 45–52.

Bing, X.L., Yang, J., Zchori-Fein, E., Wang, X.W. and Liu, S.S. (2013). Characterization of a newly discovered symbiont of the whitefly *Bemisia tabaci* (Hemiptera: Aleyrodidae). *Applied and Environmental Microbiology*, *79*, 569–575.

Bird, J. (1957). A whitefly transmitted mosaic of *Jatropha gossypifolia*. *University of Puerto Rico Agricultural Experimentation Station*, *22*, 1–35.

Bohatá, A., Folorunso, E.A., Lencová, J., Osborne, L.S. and Mraz, J. (2024). Control of sweet potato whitefly (*Bemisia tabaci*) using entomopathogenic fungi under optimal and suboptimal relative humidity conditions. *Pest Management Science*, *80*(3), 1065–1075.

Boykin, L.M., Shatters Jr, R.G., Rosell, R.C., McKenzie, C.L., Bagnall, R.A., De Barro, P. and Frohlich, D.R. (2007). Global relationships of Bemisia tabaci (Hemiptera: Aleyrodidae) revealed using Bayesian analysis of mitochondrial COI DNA sequences. *Molecular Phylogenetics and Evolution*, *44*(3), 1306–1319.

Brown, J.K. (2007). The *Bemisia tabaci* complex: Genetic and phenotypic variability drives begomovirus spread and virus diversification. *Plant Disease*, *1*, 25–56.

Brown, J.K., Coats, S.A., Bedford, I.D., Markham, P.G., Bird, J. and Frohlich, D.R., (1995a). Characterization and distribution of esterase electromorphs in the whitefly, *Bemisia tabaci* (Genn.) (Homoptera: Aleyrodidae). *Biochemical Genetics*, *33*, 205–214.

Brown, J.K., D.R. Frohlich and R.C. Rosell. (1995b). The Sweet potato or Silver leaf Whiteflies: Biotypes of *Bemisia tabaci* or a Species Complex? *Annual Reviews of Entomology, 40*, 511–534.

Brown, J.K., Perring, T.M., Cooper, A.D., Bedford, I.D. and Markham, P.G. (2000). Genetic analysis of *Bemisia* (Homoptera: Aleyrodidae) populations by isoelectric focusing electrophoresis. *Biochemical Genetics*, *38*, 13–25.

Brumin, M., Kontsedalov, S. and Ghanim, M. (2011). *Rickettsia* influences thermotolerance in the whitefly *Bemisia tabaci* B biotype. *Insect Science, 18*, 57–66.

Buhler. (2014). Insecticide resistance, Pesticide Stewardship project. (https://pesticidestewardship.org/resistance/insecticide-resistance/insecticide-resistance-mechanisms).

Busvine, J.R. (1980). *Recommended Methods for Measurement of Pest Resistance to Pesticides*. FAO Plant Production and Protection Paper No. 21.

Caballero, R., Cyman, S. and Schuster, D.J. (2013). Monitoring insecticide resistance in biotype B of *Bemisia tabaci* (Hemiptera: Aleyrodidae) in Florida. *Florida Entomologist*, *96*(4), 1243–1256.

Cardoso, A.C., Marcossi, Í., Fonseca, M.M., Kalile, M.O., Francesco, L.S., Pallini, A. and Janssen, A. (2024). A predatory mite as potential biological control agent of *Bemisia tabaci* on tomato plants. *Journal of Pest Science*, 1–13.

Caspi-Fluger, A., Inbar, M., Mozes-Daube, N., Katzir, N., Portnoy, V., Belausov, E. et al. (2011). Horizontal transmission of the insect symbiont *Rickettsia* is plant-mediated. *Proceedings of the Royal Society London. Series B Biological Sciences*, *279*, 1791–1796.

Chen, T., Saeed, Q., He, Z. and Lu, L. (2019). Transmission efficiency of Cotton leaf curl Multan virus by three cryptic species of *Bemisia tabaci* complex in cotton cultivars. *Peer Journal*, *7*, 7788.

Claridge, M.F. and Den Hollander, J. (1983). The biotype concept and its application to insect pests of agriculture. *Crop Protection*, *1* (19), 85–95.

Costa, A.S. and Russell, L.M. (1975). Failure of *Bemisia tabaci* to breed on cassava plants in Brazil (Homoptera: Aleyrodidae). *Ciencia e Cultura*, Sao Paulo, *27*, 388–390.

Costa, H. and Brown, J.K. (1991). Variation in biological characteristics and esterase patterns among populations of *Bemisia tabaci*, and the association of one population with silverleaf symptom induction. *Entomologia Experimentalis, 61*, 211–219.

Dângelo, R.A.C., Michereff-Filho, M., Campos, M.R., Da Silva, P.S. and Guedes, R.N.C. (2018). Insecticide resistance and control failure likelihood of the whitefly *Bemisia tabaci* (MEAM1; B biotype): A Neotropical scenario. *Annals of Applied Biology*, *172*(1), 88–99.

Datta, S., Budhauliya, R., Das, B., Gopalakrishnan, R., Sharma, S., Chatterjee, S. and Veer, V. (2017). Rebound of Cotton leaf curl Multan virus and its exclusive detection in cotton leaf curl disease outbreak, Punjab (India). *Scientific Reports*, *7*(1), 17361.

David, J.P., Faucon, F., Chandor-Proust, A. and Bonin, A. (2014). Comparative analysis of response to xenobiotics in insect pests and vectors. *Current Opinion in Insect Science*, *4*, 68–74.

De Barro, P.J. and Driver, F. (1997). Use of RAPD PCR to distinguish the B biotype from other biotypes of Bemisia tabaci (Gennadius) (Hemiptera: Aleyrodidae). *Australian Journal of Entomology, 36*: 149–152.

De Barro PJ, Liebregts W and Carver M. (1998). Distribution and identity of biotypes of Bemisia tabaci (Gennadius) (Hemiptera: Aleyrodidae) in member countries of the Secretariat of the Pacific Community. *Australian Journal of Entomology*. *37*, 214–218.

De Barro, P. J., Trueman, J.W.H. and Frohlich, D.R. (2011). *Bemisia argentifolii* is a race of *B. tabaci* (Hemiptera: Aleyrodidae): The molecular genetic differentiation of *B. tabaci* populations around the world. *Bulletin of Entomological Research*, *95*(3), 193–203.

De Marchi, B.R., Marubayashi, J.M., Favara, G.M., Yuki, V.A., Watanabe, L.F.M., Barbosa, L.F., Pavan, M.A. and Krause-Sakate, R. (2017). Comparative transmission of five viruses by *Bemisia tabaci* NW2 and MEAM1. *Tropical Plant Pathology*, *42*, 495–499.

Denholm, I. and Rowland, M.W. (1992). Tactics for managing pesticide resistance in arthropods: Theory and practice. *Annual Review of Entomology*, *37*, 91–112.

Dickey, A.M., Osborne, L.S., Shatters, R.G. and Mckenzie, C.L. (2013). Identification of the Meam1 Cryptic Species of *Bemisia tabaci* (Hemiptera: Aleyrodidae) by Loop-Mediated Isothermal Amplification. *Florida Entomologist*, *96*(3), 756–764.

Diehl, S.R. and Bush. G.L. (1984). An evolutionary and applied perspective of insect biotypes, *Annual Reviews of Entomology*, 471–504.

Ellsworth, P.C. and Martinez-Carrillo' J.L. (2001). IPM for *Bemisia tabaci*: A case study from North America. *Crop Protection*, *20*, 853–869.

Erdogan, C., Toprak, U. and Gurkan, M.O. (2024). Biochemical and molecular analyses of insecticide resistance in greenhouse populations of *Bemisia tabaci* (Hemiptera: Aleyrodidae) in Türkiye. *Phytoparasitica, 52*, 41.

Everett, K.D., Thao, M., Horn, M., Dyszynski, G.E. and Baumann, P. (2005). Novel chlamydiae in whiteflies and scale insects: Endosymbionts '*Candidatus Fritschea bemisiae*' strain Falk and '*Candidatus Fritschea eriococci*' strain Elm. *International Journal of Systematic and Evolutionary Microbiology*, *55*, 1581–1587.

Eweis, E., Ibrahim, E., Helmy, W., Jawad, A., Ibrahim, W. and Ibrahim, A. (2022). The efficacy of certain insecticides against whitefly, *Bemisia tabaci* (Genn.) on tomato and their effects on fruit quality. *Polish Journal of Entomology*, *91*(3), 137–148.

Farhat, T., Ali, M.R., Rahman, N.M.F., Kabir, M.M. and Kochi, M.N. (2014). Effect of Border Crops on the Incidence of Insect Pests in Chickpea. *International Journal of Innovative Science, Engineering, and Technology*, *1*(10), 453–456.

Finney, D.J. (1971). *Probit Analysis: A Statistical Treatment of the Sigmoid Curve*. London: Cambridge University Press, 333 pp.

Frohlich, D.R., Torres-Jerez, I., Bedford, I.D., Markham, P.G. and Brown, J.K. (1999). A phylogeographical analysis of the *Bemisia tabaci* species complex based on mitochondrial DNA markers. *Molecular Entomology*, *8*, 1683–1691.

Gautam, S., Buck, J.W., Dutta, B., Coolong, T., Sanchez, T., Smith, H.A., Adkins, S. and Srinivasan, R. (2023). Sida golden mosaic virus, an emerging pathogen of snap bean (*Phaseolus vulgaris* L.) in the southeastern United States. *Viruses*, *15*(2), 357.

Gawal, N.J. and Bartlett, A.C. (1993). Characterization of differences between whiteflies using RAPD-P CR. *Insect Molecular Biology*, *2*, 33–38.

Georghiou, G.P. and Taylor, C.E. (1977). Genetic and biological influences in the evolution of insecticide resistance. *Journal of Economic Entomology*, *70*(3), 319–323.

Ghosh, S., Bello, V.H. and Ghanim, M. (2021). Transmission parameters of pepper whitefly-borne vein yellows virus (PeWBVYV) by *Bemisia tabaci* and identification of an insect protein with a putative role in polerovirus transmission. *Virology*, *560*, 54–65.

Ghosh, S., Kanakala, S., Lebedev, G., Kontsedalov, S., Silverman, D., Alon, T., Mor, N., Sela, N., Luria, N., Dombrovsky, A., Mawassi, M., Haviv, S., Czosnek, H. and Ghanim, M. (2019). Transmission of a new polerovirus infecting pepper by the whitefly *Bemisia tabaci*. *Journal of Virology*, *93*, e00488–19.

Gilbertson, R.L., Batuman, O., Webster, C.G. and Adkins, S. (2015). Role of the insect supervectors *Bemisia tabaci* and *Frankliniella occidentalis* in the emergence and global spread of plant viruses. *Annual Review of Virology*, *2*, 67–93.

Gill, B.A., Kondratieff, B.C., Casner, K.L., Encalada, A.C., Flecker, A.S., Gannon, D.G., Ghalambor, C.K., Guayasamin, J.M., Poff, N.L., Simmons, M.P. and Thomas, S.A. (2016). Cryptic species diversity reveals biogeographic support for the 'mountain passes are higher in the tropics' hypothesis. *Proceedings of the Royal Society B: Biological Sciences*, *283*(1832), 20160553.

Gottlieb, Y., Ghanim, M., Chiel, E., Gerling, D., Portnoy, V., Steinberg, S., et al. (2006). Identification and localization of a *Rickettsia* sp. in *Bemisia tabaci* (Homoptera: Aleyrodidae). *Applied and Environmental Microbiology*, *72*, 3646–3652.

Gueguen, G., Vavre, F., Gnankiné, O., et al. (2010). Endosymbiont metacommunities, mtDNA diversity and the evolution of the *Bemisia tabaci* (Hemiptera: Aleyrodidae) species complex. *Molecular Ecology*, *19*, 4365–4376.

Gunning, R.V., Byrne, F.J. and Devonshire, A.L. (1997). Electrophoretic analysis of non-B and G-biotype *Bemisia tabaci* (Gennadius) (Hemiptera: Aleyrodidae) in Australia. *Australian Journal of Entomology*, *36*, 245–249.

Hadad, L., Luria, N., Smith, E., Sela, N., Lachman, O. and Dombrovsky, A. (2019). Lettuce chlorosis virus disease: A new threat to cannabis production. *Viruses*, *11*(9), 802.

Hemingway, J., C.J.K. Reid, C.M. Curtis and N.M. Thieme (1998). Prediction of mosquito control needs in West Africa using molecular markers. *Parasitology Today 14*: 128–132.

Hemingway, J. and Ranson, H. (2000). Insecticide resistance in insect vectors of human disease. *Annual Review of Entomology*, *45*, 371–391.

Hendry, T.A., Hunter, M.S. and Baltrus, D.A. (2014). The facultative symbiont *Rickettsia* protects an invasive whitefly against entomopathogenic *Pseudomonas syringae* strains. *Applied Environmental Microbiology*, 02447.

Hilje, L., Costa, H.S. and Stansly, P.A. (2001). Cultural practices for managing *Bemisa tabaci* and associated viral diseases. In: *Challenges and Opportunities for Pest Management of Bemisia tabaci in the New Century* (Eds.), S. Naranjo and P. Ellsworth. *Crop Protection*, *20*(9), 801–812.

Hopkinson, J., Pumpa, S., van Brunschot, S., Fang, C., Frese, M., Tay, W.T. and Walsh, T. (2020). Insecticide resistance status of *Bemisia tabaci* MEAM1 (Hemiptera: Aleyrodidae) in Australian cotton production valleys. *Austral Entomology*, *59*(1), 202–214.

Horowitz, A.R. and Ishaaya, I. (2014). Dynamics of biotypes B and Q of the whitefly *Bemisia tabaci* and its impact on insecticide resistance. *Pest Management Science*, *70*, 1568–1572.

Horowitz, A.R., Ghanim, M., Roditakis, E., Nauen, R. and Ishaaya, I. (2020). Insecticide resistance and its management in *Bemisia tabaci* species. *Journal of Pest Science*, 1–18.

Hu, J., Jiang, Z.L., Nardi, F., Liu, Y.Y., Luo, X.R., Li, H.X. and Zhang, Z.K. (2014). Members of *Bemisia tabaci* (Hemiptera: Aleyrodidae) cryptic species and the status of two invasive alien species in the Yunnan Province (China). *Journal of Insect Science*, *14*(1), 281.

Insecticide Resistance Action Committee (IRAC). (2020). *Insecticide Resistance Management*. http://www.irac-online.org/about/resistance management/.

Insecticide Resistance Action Committee (IRAC). (2013). *Resistance Definition*. http://www.irac-online.org/about/resistance/.

IRAC (Insecticide Resistance Action Committee). (2022). *Resistance Management Guidelines*. www.irac-online.org.

IRAC. (2022). *Test Methods*. Insecticide Resistance Action Committee. Retrieved from https://www.irac-online.org/methods/.

Jiu, M., Hu, J., Wang, L.J., Dong, J.F., Song, Y.Q. and Sun, H.Z. (2017). Cryptic species identification and composition of *Bemisia tabaci* (Hemiptera: Aleyrodidae) complex in Henan province, China. *Journal of Insect Science*, *17*(3), 78.

Kanakala, S. and Ghanim, M. (2019). Global genetic diversity and geographical distribution of *Bemisia tabaci* and its bacterial endosymbionts. *PloS One*, *14*(3), 0213946.

Kareh, C. and Nemer, N. (2023). Evaluation of Insecticides in the Management of Whiteflies (*Bemisia tabaci* Gennadius) and their Impacts on Yield of Eggplants. *University Journal of Agricultural Research*, *11*, 715–722.

Kataria, S.K., Pal, R.K., Kumar, V. and Singh, P. (2019)., Population dynamics of whitefly, *Bemisia tabaci* (Gennadius), as influenced by weather conditions infesting *Bt* cotton hybrid. *Journal of Agrometeorology*, *21*(4), 504–509.

Kontsedalov, S., Zchori-Fein, E., Chiel, E., Gottlieb, Y., Inbar, M. and Ghanim, M. (2008). The presence of *Rickettsia* is associated with increased susceptibility of *Bemisia tabaci* (Homoptera: Aleyrodidae) to insecticides. *Pest Management Science*, *64*, 789–792.

Kumar, R., Kranthi, S., Monga, D., Kumar, S., Sain, S.K. and Chaudhary, A. (2019). Evaluation of moving yellow sticky traps for monitoring and management of whitefly, *Bemisia tabaci* infesting cotton. *The Indian Journal of Agricultural Sciences*, *89*(8), 1245–1250.

Lee, M.L. and De Barro, P.J. (2000). Characterization of two biotypes of *Bemisia tabaci* (Gennadius) (Homoptera: Aleyrodidae) in South Korea based on 16S ribosomal RNA sequences. *Korean Journal of Entomology*, *7*, 384–390.

Lee, W., Park, J., Lee, G.S., Lee, S. and Akimoto, S.I. (2013). Taxonomic status of the *Bemisia tabaci* complex (Hemiptera: Aleyrodidae) and reassessment of the number of its constituent species. *PloS One*, *8*(5), 63817.

Leiva, A.M., Chittarath, K., Lopez-Alvarez, D., Vongphachanh, P., Gomez, M.I., Sengsay, S., Wang, X.W., Rodriguez, R., Newby, J. and Cuellar, W.J. (2022). Mitochondrial genetic diversity of *Bemisia tabaci* (Gennadius) (Hemiptera: Aleyrodidae) associated with cassava in Lao PDR. *Insects*, *13*(10), 861.

Lestari, S.M., Khatun, M.F., Acharya, R., Sharma, S.R., Shrestha, Y.K., Jahan, S.H., Aye, T.T., Lynn, O.M., Win, N.K.K., Hoat, T.X. and Thi Dao, H. (2023). Genetic diversity of cryptic species of *Bemisia tabaci* in Asia. *Archives of Insect Biochemistry and Physiology*, *112*(2), e21981.

Li, H., Jiang, Z., Zhou, J., Liu, X., Zhang, Y. and Chu, D. (2023). Ecological factors associated with the distribution of *Bemisia tabaci* cryptic species and their facultative endosymbionts. *Insects*, *14*(3), 252.

Li, Y., Mbata, G.N. and Shapiro-Ilan, D.I. (2021). Laboratory virulence of entomopathogenic nematodes to the sweet potato whitefly. *Journal of Nematology*, *53*(1), 1–8.

Mahalakshmi, M.S., Adinarayana, M., Reddy, A.S. and Rao, Y.K. (2014). Influence of sowing time on incidence of whiteflies and severity of yellow mosaic virus disease in blackgram (*Vigna mungo* (L.) Hepper), *Journal of Research ANGRAU*, *42* (3 & 4), 60–65.

Mahmood, M.A., Ahmed, N., Hussain, S., Muntaha, S.T., Amin, I. and Mansoor, S. (2022). Dominance of Asia II1 species of *Bemisia tabaci* in Pakistan and beyond. *Scientific Reports*, *12*(1), 1–13.

Mandali, R., Lakshmi, K.V. and Sujatha, M. (2016). Survey and biotype identification of whitefly, *Bemisia tabaci* transmitting tomato leaf curl virus in Andhra Pradesh, India. *Journal of Agricultural Technology*, *12*(7.1), 1687–1694.

Mao, L., Zhang, L., Wang, S., Zhang, Y., Zhu, L., Jiang, H. and Liu, X. (2022). Application of insecticides by soil drenching before seedling transplanting combined with anti-insect nets to control tobacco whitefly in tomato greenhouses. *Scientific Reports*, *12*(1), 15939.

Martinez-Torres, D., Foster, S.P., Field, L.M., Devonshire, A.L. and Williamson, M.S. (1999). A sodium channel point mutation is associated with resistance to DDT and pyrethroid insecticides in the peach–potato aphid, *Myzus persicae*. *Insect Molecular Biology*, *8*(3), 339–346.

Marubayashi, J.M., Kliot, A., Yuki, V.A., Rezende, J.A.M., Krause-Sakate, R., Pavan, M.A. (2014), Diversity and localization of bacterial endosymbionts from whitefly species collected in Brazil. *PLoS One, 9*, e108363.

Melander, A.L. (1914). Can Insects Become Resistant to Sprays? *Journal of Economic Entomology*, *7*(2), 167–173.

Mota-Sanchez, D. and Wise, J.C. (2019). The arthropod pesticide resistance database. Michigan State University. http://www.pesticideresistance.

Mullins, J.W., 1993. Imidacloprid: A new nitroguanidine insecticide. *American Chemical Society Symposium*, *524*, 183–198.

N'cho, A.J., Seka, K., Assiri, K.P., Simiand, C., Otron, D.H., Ochou, G., Konan, K.A.J., Kouadio, M.F., Fondio, L., Atta Diallo, H. and Martin, T. (2022). Genetic diversity of whitefly species of the *Bemisia tabaci* Gennadius (Hemiptera: Aleyrodidae) species complex, associated with vegetable crops in Côte d'Ivoire. *Plos One*, *17*(10), e0276993.

Naga, K.C., Siddappa, S. and Kumar, R. (2021). A new record of Asia II 5 genetic group of *Bemisia tabaci* (Gennadius) in the major potato growing areas of India and its relationship with tomato leaf curl New Delhi virus infecting potato. *3 Biotech*, *11*, 421.

Nauen, R. and Denholm, I. (2005). Resistance of insect pests to neonicotinoid insecticides: Current status and future prospects. *Archives of Insect Biochemistry and Physiology*, *58*(4), 200–215.

Navas-Castillo, J., Fiallo-Olivé, E. and Sánchez-Campos, S. (2011). Emerging virus diseases transmitted by whiteflies. *Annual Review of Phytopathology*, *49*, 219–248.

Naveen, N.C., Chaubey, R., Kumar, D., Rebijith, K.B., Rajagopal, R., Subrahmanyam, B. and Subramanian, S. (2017). Insecticide resistance status in the whitefly, *Bemisia tabaci* genetic groups Asia-I, Asia-II-1 and Asia-II-7 on the Indian subcontinent. *Scientific Reports*, *7*(1), 40634.

Ng, J.C.K. and Falk, B.W. (2006). Virus-vector interactions mediating nonpersistent and semipersistent transmission of plant viruses. *Annual Review of Phytopathology*, *44*, 183–212.

Nirgianaki, A., Banks, G.K., Frohlich, D.R., Veneti, Z., Braig, H.R., Miller, T.A., et al. (2003). Wolbachia infections of the whitefly *Bemisia tabaci*. *Current Microbiology*, *47*(2), pp. 0093–0101.

Nomikou, M., Janssen, A., Schraag, R. and Sabelis, M.W. (2001). Phytoseiid predators as potential biological control agents for *Bemisia tabaci*. *Experimental and Applied Acarology*, *25*, 271–291.

Pal, S., Karmakar, P., Chattopadhyay, A. and Ghosh, S.K. (2021). Evaluation of tomato genotypes for resistance to whitefly (*Bemisia tabaci* Gennadius) and tomato leaf curl virus in eastern India. *Journal of Asia-Pacific Entomology*, *24*(2), 68–76.

Patra, B. and Hath, T.K. (2022). Insecticide resistance in whiteflies *Bemisia tabaci* (Gennadius): Current global status. In: *Insecticides-Impact and Benefits of Its Use for Humanity. IntechOpen*, *4*, 50–56.

Perring, T.M. (2001). The *Bemisia tabaci* species complex. *Crop Protection*, *20*(9), 725–737.

Perring, T.M. (1996). Biological differences of two species of *Bemisia* that contribute to adaptive advantage. In: Gerling, D. and Mayer, D. (Eds.), *Bemisia*: *1995 Taxonomy, Biology, Damage, Control and Management*. Intercept Ltd., Andover, Hants, UK, pp. 3–16.

Perring, T.M., Cooper, A.D., Rodriguez, R.J., Farrar, C.A. and Bellows Jr., T.S. (1993). Identification of a whitefly species by genomic and behavioural studies. *Science*, *259*, 74–77.

Perring, T.M., Stansly, P.A., Liu, T.X., Smith, H.A. and Andreason, S.A. (2018). Whiteflies: Biology, ecology, and management. In: *Sustainable Management of Arthropod Pests of Tomato* (1st Edn.), Wakil, W., Brust, G.E., Perring, T.M. (Eds.), Academic Press: Cambridge, MA, USA, pp. 73–110.

Pitter, P.L., Mondal, S., Chang, P.G., Morgan, L.M., Aikman, S., Wintermantel, W.M. and Tennant, P.F. (2024). First report of cucurbit yellow stunting disorder virus infecting cucurbit crops in Jamaica. *Plant Disease*, *108*(4), 1120.

Prasanna, H.C., Kanakala, S., Archana, K., Jyothsna, P., Varma, R.K. and Malathi, V.G. (2015). Cryptic species composition and genetic diversity within Bemisia tabaci complex in soybean in India revealed by *mtCOI* DNA sequence. *Journal of Integrative Agriculture*, *14*(9), 1786–1795.

Puinean, A.M., Foster, S.P., Oliphant, L., Denholm, I., Field, L.M., Millar, N.S. and Bass, C. (2010). Amplification of a cytochrome P450 gene is associated with resistance to neonicotinoid insecticides in the aphid *Myzus persicae*. *PLoS Genetics*, *6*(6), e1000999.

Qiu, B., Chen, Y., Liu, L., Peng, W., Li, X., Ahmed, M.Z., Mathur, V., Du, Y. and Ren, S. (2009). Identification of three major *Bemisia tabaci* biotypes in China based on morphological and DNA polymorphisms. *Progress in Natural Science*, *19*(6), 713–718.

Quaintance, A. L. and Banks, N. (1900). Contributions toward a monograph of the American Aleurodidae. Technical Series. *Bureau of Entomology*, *8*, 9–64.

Queiroz, P.R., Martins, E.S., Klautau, N., Lima, L., Praça, L. and Monnerat, R.G. (2016). Identification of the B, Q, and native Brazilian biotypes of the *Bemisia tabaci* species complex using Scar markers. *Pesquisa Agropecuária Brasileira*, *51*(05), 555–562.

Rajna, S., Mahapatro, G., Subramanian, S. and Chander, S. (2024). Determination of insecticide resistance in cotton whitefly in north India. *The Indian Journal of Agricultural Sciences*, *94*(4): 404–409.

Ramesh, K.B., Mahendra, C., Kelageri, S.S., Rajna, S. and Subramanian, S. (2023). Distribution and mitotype diversity of *Bemisia tabaci*. *Indian Journal of Entomology*, 35–39.

Ran, W.A.N.G., Zhang, J.S., Che, W.N., Wang, J.D. and Chen, L.U.O. (2022). Genetics and fitness costs of resistance to flupyradifurone in *Bemisia tabaci* from China. *Journal of Integrative Agriculture*, *21*(5), 1436–1443.

Rana, V.S., Singh, S.T., Priya, N.G., Kumar, J. and Rajagopal, R. (2012). *Arsenophonus* GroEL interacts with CLCuV and is localized in midgut and salivary gland of whitefly *B. tabaci*. *PLoS One*, 7, e42168.

Rehman, M., Chakraborty, P., Tanti, B., Mandal, B. and Ghosh, A. (2021). Occurrence of a new cryptic species of *Bemisia tabaci* (Hemiptera: Aleyrodidae): An updated record of cryptic diversity in India. *Phytoparasitica, 49*, 869–882.

Rinkevich, F.D., Du, Y. and Dong, K. (2013). Diversity and convergence of sodium channel mutations involved in resistance to pyrethroids. *Pesticide Biochemistry and Physiology*, *106*(3), 93–100.

Robertson, I.A.D. (1987). The whitefly, *Bemisia tabaci* (Gennadius), as a vector of African cassava mosaic virus at the Kenya coast, and ways in which the yield losses in cassava, Manihot esculenta Crantz, caused by the virus can be reduced. *Insect Science and Its Application, 8*, 797–801.

Robertson, J.L., Russell, R.M., Preisler, H.K. and Savin, N.E. (2007). *Bioassays with Arthropods*. CRC Press.

Roopa, H.K., Asokan, R., Rebijith, K.B., Hande, R.H., Mahmood, R. and Kumar, N.K. (2015). Prevalence of a new genetic group, MEAM-K, of the whitefly *Bemisia tabaci* (Hemiptera: Aleyrodidae) in Karnataka, India, as evident from *mtCOI* sequences. *Florida Entomologist*, 1062–1071.

Rose, H., E. L. Bradley Jr, and C. M. Price. (1995). Investigation of the metabolism of a fungicide, using a liver microsomal preparation. In: *Mechanisms and models in toxicology*, (E8 Suppl., vol. 11, pp. 615.9, ISBN 0-387-17614.4 (U.S.).

Rosell, R.C., Bedford, I.D., Frohlich, D.R., Gill, R.J., Markham, P.G. and Brown, J.K. (1997). Analyses of morphological variation in distinct populations of *Bemisia tabaci*. *Annals of Entomological Society America*, *90*, 575–589.

Rossitto, De. Marchi, Bruno and Hugh, A. Smith. (2020). Bacterial endosymbiont diversity among *Bemisia tabaci* (Hemiptera: Aleyrodidae) populations in Florida. *Insects*,*179*, 13–19.

Roush, R.T. and Miller, G.L. (1986). Considerations for design of insecticide resistance monitoring programs. *Journal of Economic Entomology*, *79*(2), 293–298.

Roy, D., Bhattacharjee, T. and Biswas, A. (2019). Resistance monitoring for conventional and new chemistry insecticides on *Bemisia tabaci* genetic group Asia-I in major vegetable crops from India. *Phytoparasitica*, *47*, 55–66.

Rutz, T., Coolong, T., Srinivasan, R., Sparks, A., Dutta, B., Codod, C. and da Silva, A.L.B.R. (2023). Use of Insect Exclusion Row Cover and Reflective Silver Plastic Mulching to Manage Whitefly in Zucchini Production. *Insects*, *14*(11), 863.

Sáez, A.G. and Lozano, E. (2005). Body doubles. *Nature*, *433*(7022), 111–111.

Sanga, K.S., Shera, P.S., Sharma, S. and Kaur, R. (2018). *Natural Enemies of Whitefly, Bemisia tabaci (Gennadius) on Cotton in Punjab, India*. Society for Biocontrol Advancement.

Sani, I., Jamian, S., Saad, N., Abdullah, S., Mohd Hata, E., Jalinas, J. and Ismail, S.I. (2023). Identification and virulence of entomopathogenic fungi, *Isaria javanica* and *Purpureocillium lilacinum* isolated from the whitefly, *Bemisia tabaci* (Gennadius) (Hemiptera: Aleyrodidae) in Malaysia. *Egyptian Journal of Biological Pest Control*, *33*(1), 14.

Shan, H.W., Zhang, C.R., Yan, T.T., Tang, H.Q., Wang, X.W. and Liu, S.S. (2016). Temporal changes of symbiont density and host fitness after rifampicin treatment in a whitefly of the *Bemisia tabaci* species complex. *Insect Science*, *23*, 200–214.

Shandhu, S.S., Sharma, A.K., Beniwal, V., Goel, G., Batra, P., Kumar, A., Jaglan, S. and Malhotra, S. (2012). Myco-biocontrol of insect pests: Factors involved, mechanism, and regulation. *Journal of Pathology*, 1–10.

Sharma, A., Neupane, K.R., Regmi, R. and Neupane, R.C. (2018). Effect of intercropping on the incidence of jassid (*Amrasca biguttula biguttula* Ish.) and whitefly (*Bemesia tabaci* Guen.) in okra (*Abelmoschus esculentus* L. Moench). *Journal of Agriculture and Natural Resources*, *1*(1), 179–188.

Shi, X., Tang, X., Zhang, X., Zhang, D., Li, F., Yan, F., Zhang, Y., Zhou, X. and Liu, Y. (2018). Transmission efficiency, preference and behavior of *Bemisia tabaci* MEAM1 and MED under the influence of Tomato chlorosis virus. *Frontiers in Plant Science*, *8*, 2271.

Sindhura, K.A.V., Reddy, K.S., Pooja, P.S., Bharath, M., Khatib, S.M., Sumithramma, N., Reddy, G.K. and Reddy, C.L. (2023). Molecular Characterization of Cryptic Species of *Bemisia tabaci* Associated with Cucumber in Eastern Dry Zone of Karnataka. *Mysore Journal of Agricultural Sciences*, *57*(4), 24–32.

Sloan, D.B. and Moran, N.A. (2012). Endosymbiotic bacteria as a source of carotenoids in whiteflies. *Biology Letters*, *8*, 986–989.

Sparks, T.C. and Nauen, R. (2015). IRAC: Mode of action classification and insecticide resistance management. *Pesticide Biochemistry and Physiology*, *121*, 122–128.

Sparks, T.C., Crossthwaite, A.J., Nauen, R., Banba, S., Cordova, D., Earley, F. and Watson, G.B. (2020). Insecticides, biologics, and nematicides: Updates to IRAC's mode of action classification –a tool for resistance management. *Pesticide Biochemistry and Physiology*, *167*, 104587.

Su, Q., Xie, W., Wang, S., Wu, Q., Liu, B. and Fang, Y. (2014). The endosymbiont *Hamiltonella* increases the growth rate of its host *Bemisia tabaci* during periods of nutritional stress. *PLoS One*, 9, e89002.

Takahashi, H., Mitsui, J., Takakusa, N., Matsuda, M., Yoneda, H., Suzuki, J., Ishimitsu, K. and Kishimoto, T. (1992). NI-25, a new type of systemic and broad-spectrum insecticide. In: *Proceedings, Brighton Crop Protection Conference, Pests, and Diseases, Brighton, British Crop Protection Council*, *1*, 25–27.

Tang, X.T., Cai, L., Shen, Y. and Du, Y.Z. (2018). Diversity and evolution of the endosymbionts of *Bemisia tabaci* in China. *Peer Journal*, *6*, 5516.

Thao, M.L. and Baumann, P. (2004). Evolutionary relationships of primary prokaryotic endosymbionts of whiteflies and their hosts. *Applied and Environmental Microbiology*, *70*(6), 3401–3406.

Togni, P.H.B., Marouelli, W.A., Inoue-Nagata, A.K., Pires, C.S.S. and Sujii, E.R. (2018). Integrated cultural practices for whitefly management in organic tomato. *Journal of Applied Entomology*, *142*(10), 998–1007.

Van Asperen, K. (1962). A Study of House Fly Esterase by Means of a Sensitive Colorimetric Method. *Journal of In sect Physiology*, *8*, 401–416.

Venkataravanappa, V., Kodandaram, M.H., Lakshminarayana Reddy, C.N., Shankarappa, K.S. and Krishna Reddy, M. (2017). Comparative transmission of Bhendi yellow vein mosaic virus by two cryptic species of the whitefly, *Bemisia tabaci* (Hemiptera: Aleyrodidae). *3 Biotech*, *7*, 1–13.

Walsh, B.J. (1864). On phytophagic varieties and phytophagic species. In: *Proceedings of Entomological Society*, Philadelphia, *3*, 403–430.

Wang, F., Liu, J., Chen, P., Li, H.Y., Ma, J.J., Liu, Y.J. and Wang, K. (2020). *Bemisia tabaci* (Hemiptera: Aleyrodidae) Insecticide Resistance in Shandong Province, China. *Journal of Economic Entomology*, *113*(2), 911–917.

Webb, S.E., Adkins, S. and Reitz, S.R. (2012). Semipersistent whitefly transmission of Squash vein yellowing virus, causal agent of viral watermelon vine decline. *Plant Disease*, *96*(6), 839–844.

Weeks, A.R., Velten, R. and Stouthamer, R. (2003). Incidence of a new sex–ratio–distorting endosymbiotic bacterium among arthropods. In: *Proceedings of the Royal Society London. Series B Biological Sciences*, *270*, 1857–1865.

Weill, M., Malcolm, C., Chandre, F., Mogensen, K., Berthomieu, A., Marquine, M. and Raymond, M. (2004). The unique mutation in ace-1 giving high insecticide resistance is easily detectable in mosquito vectors. *Insect Molecular Biology*, *13*(1), 1–7.

Whalon, M.E., Mota-Sanchez, D. and Hollingworth, R.M. (2008). *Global Pesticide Resistance in Arthropods*. CABI.

Yadav, D.L., Patidar, B.K. and Nagar, B.L. (2022). Efficacy of Insecticides against Whitefly *Bemisia tabaci* (Genn.) in Potato. *Indian Journal of Entomology*, 900–902.

Yükselbaba, U. and Ali, I.H. (2022). Insecticide resistance status of *Bemisia tabaci* (Genn., 1889) (Hemiptera: Aleyrodidae) populations to cyantraniliprole, pyriproxyfen, and spirotetramat in Antalya (Türkiye). *Turkish Journal of Entomology*, *46*(3), 263–274.

Zanardo, L.G. and Carvalho, C.M. (2017). *Cowpea mild mottle virus* (*Carlavirus, Betaflexiviridae*): A review. *Tropical Plant Pathology*, *42*, 417–430.

Zheng, H., Xie, W., Fu, B., Xiao, S., Tan, X., Ji, Y. and Zhang, Y. (2021), Annual analysis of field-evolved insecticide resistance in *Bemisia tabaci* across China. *Pest Management Science*, *77*(6), 2990–3001.

Chapter 14

Ochratoxin A
Occurrence, Health Implications, and Management Strategies in Food and Feed Products

Pandareesh M. D.[1,2*], Bhumika H. S.[1,2], Dinesha R.[1] , Belaghihalli N Gnanesh[3], Shyam Sundar S.[4] and Abha Chauhan[5]

1. Introduction

Ochratoxin A (OTA) is a toxic compound produced mainly by certain filamentous fungi belonging to the *Aspergillus* and *Penicillium* genera. It can contaminate a wide range of agricultural products, such as cereals, coffee beans, dried fruits, grapes, cocoa, spices, nuts, as well as meat, milk, and dairy products. OTA was first identified in South Africa as a deadly byproduct of *Aspergillus ochraceus* found in cornmeal (Ben Miri et al., 2024). OTA is known for its harmful effects on human and animal health, posing serious risks such as neurotoxicity, nephrotoxicity, hepatotoxicity, teratogenicity, genotoxicity, embryotoxicity, immuno-suppression, and carcinogenic properties. The International Agency for Research on Cancer designated OTA as a potential human carcinogen (group 2B), based on evidence obtained from animal studies demonstrating its carcinogenic effects. Exposure to OTA has also been associated with the development of Balkan endemic nephropathy and chronic interstitial nephropathy. Recent research has demonstrated that OTA is a potent neurotoxin (Mateo et al., 2022b). OTA is listed among the top 10 most hazardous chemicals in food products in Europe and non-European Union countries. The presence of OTA in food and feed products has led to significant regulatory concerns worldwide, prompting efforts to mitigate its occurrence and exposure.

Due to its global prevalence and continuous low-level human exposure, regulatory restrictions for OTA levels in food products have been established. The European Food Safety Authority sets strict limits for Ochratoxin A: no more than 5 parts per billion (ppb) in cereals, 2 ppb in wine or grape

[1] Department of Biochemistry, Adichunchanagiri School of Natural Sciences, Adichunchanagiri University, B.G. Nagara, Mandya-571448, India.

[2] Center for Research and Innovation, Adichunchanagiri University, B.G. Nagara, Mandya-571448, India.

[3] Genetics and Plant Breeding, ICAR JSS - Krishi Vigyan Kendra, Mysuru-571 129, Karnataka, India

[4] Department of Oral and Maxillofacial Surgery, JSS Dental College and Hospital, JSS Academy of Higher Education and Research, SS Nagar, Mysuru-570015, India.

[5] New York State Institute for Basic Research in Developmental Disabilities, Staten Island, New York, USA.

* Corresponding author: pandareesh@acu.ac.in

DOI: 10.1201/9781003614869-14

juice, and 5 ppb in coffee products. The OTA levels in food items collected from European countries were found to range from 1.4 to 20,000 ng/g. The average daily intake of OTA ranged from 0.6 to 17.8 ng/kg body weight (Liu et al., 2020). In humans, OTA is quickly absorbed and binds strongly to albumin, but it has a low metabolic rate, which leads to a long serum half-life of around 35 days (Bui-Klimke et al., 2015). OTA can accumulate in the food chain, leading to its detection in the tissues of animals and humans after they consume contaminated feed and food (Ganesan et al., 2022). Studies on healthy individuals in Germany and Spain have shown the presence of nanomolar concentrations of OTA in biological fluids, including breast milk in mothers, as well as in the blood and urine of the Spanish population (Soto et al., 2016). OTA has also been found in the urine of infants and adults, and even in the brains of the German and Turkish populations. Worryingly, Ochratoxin A has been detected in human umbilical cord blood serum in Poland, with levels ranging from 130 to 5,420 ng/L. In Tunisia, patients suffering from chronic interstitial nephropathy showed an average blood level of 50.4 ng/mL, with some cases reaching as high as 171.2 ng/ml. These findings highlight that OTA contamination is a real health concern for people of all ages including infants, toddlers, and adults. Infants are especially at risk because, compared to their body weight, they consume more food than older children or adults. The acceptable daily intake of OTA for humans is set at 1.5 ng/kg of body weight. This chapter explores the occurrence of OTA in different food and feed matrices, its health implications, and various management strategies aimed at reducing OTA contamination and ensuring food and feed safety. Understanding the occurrence of OTA and its effects is essential for developing effective prevention and control measures to protect human and animal health.

2. Chemistry of OTA

The IUPAC name of OTA is (N-[[(3R)-5-chloro-8-hydroxy-3-methyl-1-oxo-7-isochromanyl] carbonyl]-3-phenyl-l-alanine). OTA has a molecular formula of $C_{20}H_{18}O_6ClN$ and consists of a para-chlorophenolic moiety with a dihydroiso-coumarin group amide-linked to L-phenylalanine. Several derivatives of OTA are known, including hydroxylated forms, derivatives lacking the phenylalanine moiety, and conjugated forms. Studies have identified new OTA derivatives such as DC-OTHQ, a decarboxylated hydroquinone derivative often linked to glutathione, and products of thermal degradation like 2′-DC-OTA and 2′R-OTA. These compounds are produced during the coffee roasting process, along with Ochratoxin α amide, which forms at high temperatures. These derivatives are crucial for assessing their potential health risks associated with OTA exposure (Malir et al., 2016).

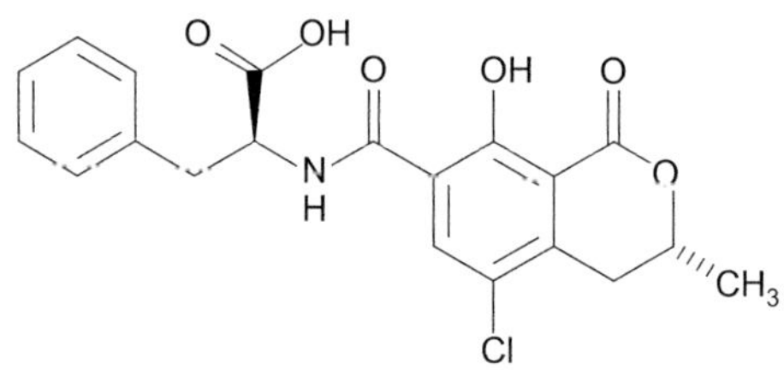

Fig. 1 Structure of Ochratoxin A.

The chemical structure of Ochratoxin A (Figure 1) features a dihydroisocoumarin unit connected to L-phenylalanine through an amide bond. OTA can undergo various transformations in biological systems, leading to the formation of various metabolites and derivatives (Table 1). For example, hydroxylation at different positions of the OTA molecule can lead to the formation of compounds like 4R-hydroxy Ochratoxin A (4R-OHOA) and 4S-hydroxy Ochratoxin A (4S-OHOA). Additionally, OTA can undergo methylation to form Ochratoxin A methyl ester (OTA-Me) or Ochratoxin A O-methyl (OM-OA). Decarboxylation of OTA results in Ochratoxin A decarboxylated

(DC-OA), while ethylation leads to the formation of ethylamine Ochratoxin A (OE-OA). Table 1 presents several Ochratoxin A derivatives and highlights the functional groups they contain. These derivatives may exhibit different toxicological properties compared to OTA, and understanding their formation and effects is essential for assessing the overall risk associated with OTA exposure (Malir et al., 2016).

Table 1 Structural Variants of Ochratoxins.

Abbreviation	*Metabolite Name*	*R1*	*R2*	*R3*	*R4*	*R5*	*R6*
10-OHOA	10-hydroxy Ochratoxin A	Phe	Cl	H	H	OH	OH
10-OHOA-Me	10-hydroxy OTA methyl ester	Phe (methyl ester)	Cl	H	H	OH	OH
4R-OHOA	4R-hydroxy Ochratoxin A	Phe	Cl	H	OH	H	OH
4R-OHOA-Me	4R-hydroxy OTA methyl ester	Phe (methyl ester)	Cl	H	OH	H	OH
4S-OHOA	4S-hydroxy Ochratoxin A	Phe	Cl	OH	H	H	OH
AcylHex-OTA	OTA–acyl hexose conjugate	Phe (acyl hexose)	Cl	H	H	H	OH
AcylPent-OTA	OTA–acyl pentose conjugate	Phe (acyl pentose)	Cl	H	H	H	OH
DC-OA	Decarboxylated OTA	Phe (decarboxylated)	Cl	H	H	H	OH
DC-OTHQ	Decarboxylated OTHQ	Decarboxylated Phe	OH	H	H	H	OH
d-OA	d-Ochratoxin A	d-Phe	Cl	H	H	H	OH
M-Oα	Methyl ester OTα	OCH_3	Cl	H	H	H	OH
OE-OA	Ethylamide OTA	Phe (ethyl amide)	Cl	H	H	H	OH
OM-OA	O-methyl OTA	Phe	Cl	H	H	H	OH
OP-OA	Ochratoxin A (open lactone)	Phe	Cl	H	H	–	OH
OP-OB	Ochratoxin B (open lactone)	Phe	H	H	H	–	OH
OP-OTα	Ochratoxin α (open lactone)	OH	Cl	H	H	–	OH
OP-OTβ	Ochratoxin β (open lactone)	OH	H	H	H	–	OH
OTA	Ochratoxin A	Phe	Cl	H	H	H	OH
OTA-Me	OTA methyl ester	Phe (methyl ester)	Cl	H	H	H	OH
OTA-Tyr	Tyrosine OTA	Tyrosine	Cl	H	H	H	OH
OTB	Ochratoxin B	Phe	H	H	H	H	OH
OTB-Et	OTB ethyl ester	Phe (ethyl ester)	H	H	H	H	OH
OTB-Me	OTB methyl ester	Phe (methyl ester)	H	H	H	H	OH
OTC	Ochratoxin C	Phe (ethyl ester)	Cl	H	H	H	OH
OTHQ	Ochratoxin A hydroquinone	Phe	OH	H	H	H	OH
OTQ	Ochratoxin A quinone	Phe	O	H	H	H	O
OTQ-GSH	OTA quinone–glutathione conjugate	Phe	O	H	H	H	O
OTα	Ochratoxin α	OH	Cl	H	H	H	OH
OTβ	Ochratoxin β	OH	H	H	H	H	OH

Ochratoxin A (OTA) consists of a phenylalanine group with chlorine at the R2 position, hydrogen at R3–R5, and a hydroxyl group at R6. Ochratoxin B (OTB) also contains phenylalanine but lacks chlorine at R2, having hydrogen instead, while maintaining hydrogen at R3–R5 and a hydroxyl group at R6. Ochratoxin C (OTC) is the ethyl ester form of phenylalanine, with chlorine at R2, hydrogen at R3–R5, and a hydroxyl group at R6.

Ochratoxin α (OTα) features a hydroxyl group at R1, chlorine at R2, hydrogen at R3–R5, and hydroxyl at R6. In contrast, Ochratoxin β (OTβ) has hydroxyl at R1, hydrogen at R2–R5, and hydroxyl at R6. The 4R-hydroxy OTA (4R-OHOA) derivative contains phenylalanine, chlorine at R2, hydrogen at R3, hydroxyl at R4, hydrogen at R5, and hydroxyl at R6. Its isomer, 4S-hydroxy OTA (4S-OHOA), has phenylalanine, chlorine at R2, hydroxyl group at R3, hydrogen at R4–R5, and hydroxyl at R6.

The 10-hydroxy OTA (10-OHOA) variant contains phenylalanine, chlorine at R2, hydrogen at R3–R4, hydroxyl at R5, and hydroxyl at R6. In open lactone forms, OTA (OP-OA) retains phenylalanine and chlorine at R2, hydrogen at R3–R4, no substituent at R5, and a hydroxyl at R6; OTB (OP-OB) is similar but with hydrogen at R2–R4. Open lactone OTα (OP-OTα) has a hydroxyl at R1, chlorine at R2, hydrogen at R3–R4, no substituent at R5, and hydroxyl at R6, while open lactone OTβ (OP-OTβ) has hydroxyl at R1, hydrogen at R2–R4, no substituent at R5, and hydroxyl at R6.

Ochratoxin A quinone (OTQ) carries phenylalanine, oxygen at R2, hydrogen at R3–R5, and oxygen at R6, whereas Ochratoxin A hydroquinone (OTHQ) has phenylalanine, hydroxyl at R2, hydrogen at R3–R5, and hydroxyl at R6. Decarboxylated OTHQ (DC-OTHQ) replaces phenylalanine with decarboxylated phenylalanine, retaining hydroxyl at R2 and R6. The OTA quinone–glutathione conjugate (OTQ-GSH) contains phenylalanine, oxygen at R2 and R6, and hydrogen at R3–R5.

Conjugated derivatives include OTA–acyl hexose (AcylHex-OTA) and OTA–acyl pentose (AcylPent-OTA), both with phenylalanine bound to an acyl sugar, chlorine at R2, hydrogen at R3–R5, and hydroxyl at R6. OTA methyl ester (OTA-Me) contains phenylalanine methyl ester, chlorine at R2, hydrogen at R3–R5, and hydroxyl at R6. OTB methyl ester (OTB-Me) and OTB ethyl ester (OTB-Et) both lack chlorine at R2, having hydrogen instead, while retaining hydroxyl at R6.

Hydroxylated methyl esters include 4R-hydroxy OTA methyl ester (4R-OHOA-Me), with phenylalanine methyl ester, chlorine at R2, hydrogen at R3, hydroxyl at R4, hydrogen at R5, and hydroxyl at R6, and 10-hydroxy OTA methyl ester (10-OHOA-Me), with chlorine at R2, hydrogen at R3–R4, hydroxyl at R5, and hydroxyl at R6. The ethylamide form of OTA (OE-OA) contains phenylalanine ethyl amide, chlorine at R2, hydrogen at R3–R5, and hydroxyl at R6.

Decarboxylated OTA (DC-OA) has decarboxylated phenylalanine, chlorine at R2, hydrogen at R3–R5, and hydroxyl at R6. O-methyl OTA (OM-OA) has phenylalanine with chlorine at R2, hydrogen at R3–R5, and hydroxyl at R6. The d-isomer of OTA (d-OA) features d-phenylalanine, chlorine at R2, hydrogen at R3–R5, and hydroxyl at R6. Methyl ester OTα (M-Oα) contains a methoxy group at R1, chlorine at R2, hydrogen at R3–R5, and hydroxyl at R6. Finally, tyrosine OTA (OTA-Tyr) contains a tyrosine group at R1, chlorine at R2, hydrogen at R3–R5, and hydroxyl at R6.

Table 1 summarizes naturally occurring and synthetic derivatives of ochratoxins, indicating structural modifications at substituent positions (R1–R6), corresponding molecular weights (MW), and literature references. Abbreviations denote the common names of each ochratoxin variant. Substituent groups are represented using standard chemical notations (e.g., Phe = phenylalanine, Cl = chlorine, OH = hydroxyl group, OCH_3 = methoxy group, — = position lacking a substituent). References correspond to the studies reporting the isolation, synthesis, or characterization of each compound (Malir et al., 2016).

3. Sources of OTA

OTA is a strong mycotoxin produced primarily by fungi from the *Aspergillus* and *Penicillium* genera, and it contaminates various agricultural products. Cereals such as wheat, barley, oats, and rye can be contaminated with OTA. Dietary exposure to OTA is a concern, especially in regions with high contamination levels in processed cereal foods. Contamination of cereals in Europe is primarily due to *Penicillium verrucosum*, while coffee and cocoa are potential sources of OTA exposure in tropical countries. When stored under conditions of high humidity and temperature (Kuruc et al., 2015). These grains are susceptible to fungal contamination during growth, harvesting, and storage, leading to OTA presence in food. Coffee beans are also susceptible to OTA contamination throughout different stages of growth, harvesting, processing, and storage. Similarly, oil seeds such as soybeans, sunflower seeds, and cotton seeds can be contaminated with OTA and other mycotoxins due to improper drying and storage. Dried fruits such as raisins, figs, and apricots are prone to OTA contamination, particularly if drying conditions are suboptimal or if they come into contact with contaminated surfaces. Grapes can become contaminated with OTA during cultivation, leading to OTA presence in wine and grape-derived products (Kuruc et al., 2015). In 1969, Shotwell and colleagues from the U.S. Department of Agriculture (USDA) published the first documented information about the presence of OTA in maize, with levels ranging from 110 to 150 ng/g. Scott and colleagues from Health Canada reported OTA levels in mouldy cereals, beans, and peanuts. Wheat, oats, barley, and rye showed OTA concentrations ranging from 30 to 27,000 ng/g, with 62.0% of samples testing positive. Over time, researchers have identified more than 90 different foods, both plant and animal-based, including milk, as contributors to dietary exposure to Ochratoxin A. In animal products such as pork blood products, edible offal, pork meat, chicken meat, and dry-cured ham, OTA concentrations typically range from 0.1 to 1 ng/g (Ostry et al., 2015). Similar levels of OTA are found in animal feed. In plant-based foods, OTA is found in cereal products, olives, beans, beer, wine, coffee, cocoa products, raisins, figs, licorice, pulses, pumpkin seeds, and tea (Ostry et al., 2015). Generally, OTA concentrations range from 0.1 to 100 ng/g. Black pepper, cayenne pepper, caraway, cardamom, coriander, chili powder, curcuma, and dried red pepper have OTA concentrations ranging from 1 to 100 ng/g. Food products derived from plants, such as wheat, oats, barley, rye, maize, rice, millet, sorghum, soybeans, horse beans, peas, beans, alfalfa, sunflower or pumpkin seeds, coconut, and peanut, also carry OTA concentrations between 1–100 ng/g (Ostry et al., 2015). Forage crops like hay and silage can also be contaminated, if harvested under conducive fungal growth conditions. Cocoa beans, spices like black pepper and paprika, nuts, meat products, milk, and dairy products can also be sources of OTA contamination, depending on factors such as environmental conditions, agricultural practices, storage conditions, and processing methods. OTA contamination spans a wide range of agricultural and food products, affecting both plant and animal-based products. The occurrence of this mycotoxin in cereals, coffee beans, oilseeds, dried fruits, and spices underscores the imperative of effective storage and handling to prevent risks of contamination. Ongoing surveillance and rigorous control strategies are indispensable for guaranteeing food safety and reducing OTA exposure in the food supply system. An understanding of the influencing factors on its occurrence is critical to fully comprehend the OTA contamination challenge. Thus, it is important to explore several factors that affect its occurrence. Environmental conditions like temperature, humidity, rainfall, and soil quality can contribute to the growth of fungi and OTA production.

4. Factors Influencing OTA Occurrence

Fruits and vegetables can become contaminated with OTA at various stages, including harvesting, processing, storage, and transportation. For example, commodities such as cereals, coffee beans, cocoa beans, and dried fruits get contaminated during storage, while grapes get contaminated in the field. The development of OTA is mostly post-harvest, and is affected by temperature, water activity, and seed type and quality. Biotic components such as lactose, urea, and amino acids may increase OTA biosynthesis, whereas glucose and sucrose may suppress it. Environmental factors significantly influence the occurrence of OTA contamination in agricultural products. Factors such as temperature, humidity, rainfall, and soil conditions directly impact fungal growth and OTA production (Wang et al., 2023). The growth of fungi, and consequently the production of OTA, is shaped by numerous physical, chemical, and biological influences. External conditions like light and temperature, along with internal factors such as pH, moisture levels, and the substrate's composition, all play a role. The presence of endogenous flora in food also adds to the biosynthesis of OTA. Aspergillus species, known for their ability to thrive in a wide range of climates and on various food products, are major contributors to the widespread presence of OTA. Cheong et al. (2016) investigated how different light wavelengths and day–night cycles affect the growth and OTA production of *Aspergillus carbonarius* and *Aspergillus westerdijkiae*. They found that light wavelength and photoperiod generally did not affect fungal growth, except that *A. westerdijkiae* showed reduced growth under UV-A light (366 nm) (Cheong et al., 2016). In another study, Lahouar et al. (2018) explored how water activity and temperature influence radial fungal growth and OTA production in three *Aspergillus tubingensis* isolates and three *Fusarium incarnatum* isolates grown on sorghum grain. These researchers established that 0.97 water activity and 37°C were the optimum conditions under which to produce OTA in *A. tubingensis* isolates, whereas 0.99 water activity and 25°C were optimum for *F. incarnatum* isolates. *A. tubingensis*, however, grew at only 15°C when water activity was greater (0.97 and 0.99) (Lahouar et al., 2018). Whereas warm and humid conditions provide conducive conditions for OTA-producing fungi, too much rainfall or over-irrigation can similarly raise the levels of moisture to enhance fungal growth. Physical and chemical properties of soil, such as pH and organic matter content, also affect fungal invasion. In addition to genetic makeup, the crop type of some varieties shows more resistance to fungal infection. Adequate crop rotation, cultural practices, and storage regimes are necessary to reduce OTA infection. It is important to understand these environmental factors to adopt appropriate prevention and control measures to minimize OTA levels in crops and maintain food safety.

5. Effect of OTA on Food and Feed

The European Food Safety Authority (EFSA) has noted OTA presence in animal tissues and products such as dairy, pork, and meat, which results from animals consuming OTA-contaminated feed (Kumar et al., 2020). OTA is quickly absorbed into the bloodstream and is slowly eliminated from the body (Kumar et al., 2020). OTA levels in animal feeds differ region to region, with increased levels of OTA reported in Northern Europe, North America, and Canada. OTA-polluted feeds significantly affect poultry, pigs, and monogastric animals, with pigs being especially sensitive to OTA-caused nephrotoxicity. Research has shown that the intake of feed containing 25 μg/kg of OTA in pigs led to a notable decrease in feed efficiency and final body weight loss. Poultry, particularly turkeys and chickens, can experience nephrotoxicity, weight loss, reduced egg production, and poor eggshell quality due to OTA toxicity (Stoev, 2023). The toxicity of OTA depends on the levels and duration of exposure, with contaminated feed resulting in reduced levels of albumin, cholesterol, globulin, potassium, total protein, and triglycerides in poultry (Ochieng et al., 2021). It also increases levels of uric acid and creatine, as well as the activity of serum alkaline phosphatase (ALP) and gamma-glutamyl transpeptidase (GGT). Ruminants, on the other hand, are more resistant to OTA toxicity due to rapid detoxification by rumen protozoa and bacterial enzymes into less harmful metabolites (Ochieng et al., 2021). OTA contamination is a serious threat to agricultural commodities and animal health, and mostly happens in the post-harvest phases. It can be controlled by several environmental and biotic factors. OTA contamination in animal feed causes detrimental effects on animals, especially pigs and poultry, which affect their productivity and health. Proper monitoring and control measures are imperative in preventing OTA contamination and maintaining the safety of agricultural commodities and animal-derived foodstuffs.

6. Importance of Understanding OTA Contamination

OTA contamination has far-reaching consequences on public health, food safety, economic stability, regulatory compliance, and sustainable agriculture practices. OTA is recognized as a potential human carcinogen with a broad spectrum of toxicity, including nephrotoxic, hepatotoxic, embryotoxic, teratogenic, neurotoxic, immunotoxic, and genotoxic activities. Through the understanding of OTA contamination routes, sources, and implications, various precautions can be taken to protect public health by identifying high-risk foods and preventing exposure. This knowledge will also contribute to food safety and quality through the ability to develop effective food safety management systems, curbing economic losses due to rejected shipments and ensuring adherence to regulatory standards. Additionally, it will promote sustainable agricultural practices such as crop rotation and integrated pest management to mitigate fungal development and mycotoxin production (Liu et al., 2022). In addition, OTA contamination research keeps scientific knowledge ahead by promoting awareness of fungal ecology, mycotoxin production mechanisms, and food safety technologies. Cooperation among stakeholders will effectively address OTA contamination comprehensively, finally contributing to food security, sustainable agriculture, and world health (Awuchi et al., 2022).

7. Metabolism of OTA

The health and toxic effects of OTA, which is widely present in cereals and cereal products for human and animal consumption, are well-documented. Previous studies indicated that most of the OTA does not change in the body, and it is presumed that its toxicity must be mediated through its metabolism. The different metabolic processes involving hydrolysis, hydroxylation, lactone opening, conjugation, and glucuronidation result in the production of OTA metabolites (Awuchi et al., 2022; Poor et al., 2014). In the case of rodents, hydrolysis activity is mainly due to anaerobic bacteria in the cecum and large intestine. While OTA hydrolysis also takes place in some rat tissues, including the duodenum, ileum, and pancreas, it is not significant in the liver and kidney. Hydrolysis of OTA in these tissues leads to the formation of OTα and phenylalanine, which are known to be non-toxic. Interestingly, it was previously believed that the resistance of ruminants to OTA's toxic effects was due to the ability of their rumen microflora to degrade the toxin. However, subsequent studies have shown that hydrolysis in ruminants is mainly associated with rumen protozoa (Cimbalo et al., 2020). Liver microsomes from various species, including humans, pigs, rats, chickens, cows, and goats, have been shown to metabolize OTA into different metabolites. These include 4-hydroxyochratoxin A (4-OH-OTA) and several novel hydroxyl metabolites (Yang et al., 2015). OTA metabolism incorporates several enzymes, such as cytochrome P450 (CYP450) enzymes, with phase I and phase II enzymes having important functions. Certain metabolites, like lactone-opened OTA (OP-OTA), have been reported to be more toxic than OTA. OTA detoxification by specific enzymes, like CYP2B1, provides an apparent mechanism for lowering OTA toxicity. Enzymes such as carboxypeptidases, which are secreted by the ruminant and rodent intestinal microflora, hydrolyze OTA's peptide bond to produce less toxic OTA. Healthy ruminants can break down and inactivate approximately 12 mg OTA/kg feed. Hydroxylation of OTA by enzymes such as peroxidases and carboxypeptidases produces metabolites such as 4-(R)-hydroxy ochratoxin A (4-OH-OTA) in human beings and rats, and 4-(S)-hydroxy ochratoxin A in pigs (Liu et al., 2022; Malir et al., 2016). OTA has a good binding affinity with proteins, particularly with the serum albumins, which facilitates its bioaccumulation in the organs of animals when they ingest contaminated feeds, thus resulting in the toxin being found in products such as eggs, milk, and liver. Unlike OTA, the contamination of food and animal feed with OTB is relatively low (Heussner et al., 2015). Conjugation with glutathione (to form glucuronides and sulphate conjugates, or GSH conjugates) is another route to inactivate OTA as a detoxification step. Yet the production of thiol radicals, which can result in peroxide radical formation, can cause bioactivation. Stable GSH conjugates like quinone-derived GSH conjugates may be produced in the liver and excreted from the body after N-acetylation. Identification of these GSH conjugates is difficult because they occur at low concentrations. In humans, the delayed excretion of OTA is caused by its reabsorption from urine. Its acidic character enables passive absorption in the intestine as its monoanionic or non-ionized form, and its plasma protein affinity enables absorption and transport to target organs through the bloodstream. One of the most researched effects of OTA is its nephrotoxicity, or its capacity to harm the kidneys. It has been found that OTA can increase in renal tissues and induce oxidative stress, inflammation, and cell death. Chronic exposure to low doses of OTA has been linked to several kidney ailments, including tubular degeneration, interstitial fibrosis, and even renal cancer (Imaoka et al., 2020). This renders OTA a critical kidney health concern, particularly among groups that are exposed to contaminated food for many years. OTA has also been known to be neurotoxic, which is an indicator of its potential to damage the nervous system. Studies have proven that OTA can penetrate the blood-brain barrier and accumulate in the brain, where it can cause neuroinflammation, oxidative stress, and death of neuronal cells (Pei et al., 2021). These effects can lead to neurodegenerative diseases, cognitive impairments, and behavioural changes. Thus, the neurotoxicity of OTA poses a significant risk to both the developing and mature nervous systems, with potential long-term consequences on neurological health.

8. Health Implications of OTA

OTA is a potent nephrotoxin, carcinogen, and neurotoxin, impacting various organs and systems. Chronic exposure primarily affects the kidneys, causing structural damage, impaired renal function, and increased risks of Balkan endemic nephropathy (BEN) and renal cell carcinoma. OTA also disrupts brain health, linked to neurodegenerative diseases like Alzheimer's and Parkinson's, and neurodevelopmental disorders such as autism. Additionally, OTA affects cardiovascular health by promoting oxidative stress and endothelial dysfunction, contributing to hypertension. It disrupts bone homeostasis, increasing osteoporosis risk. Preventive strategies, including reducing dietary OTA exposure, are crucial to mitigate its widespread toxic effects (Krogh et al., 1973). Figure 2 represents the effects of OTA on human health and its implications.

Figure 2 depicts the broad toxicological effects of a chlorinated compound found in various food sources, including cocoa beans, dry fruits, coffee beans, cereals, dairy products, meat, and fruits and vegetables. This compound exerts multiple toxic mechanisms, such as inducing oxidative stress, impairing mitochondrial function, disrupting calcium signaling, promoting inflammation, triggering apoptosis, and causing DNA damage. These mechanisms lead to a range of adverse effects across different organ systems. In the liver, it accumulates and disrupts protein synthesis, lipid metabolism, and detoxification pathways, contributing to hepatotoxicity and liver fibrosis. It also acts as a teratogen, interfering with fetal development, cell differentiation, and placental function, leading to malformations and growth retardation. The compound exhibits carcinogenic properties by forming DNA adducts, causing genomic instability, and triggering epigenetic changes that promote abnormal cell proliferation and immune suppression. Neurotoxic effects arise from its ability to cross the blood-brain barrier, impair neurotransmitter release, and induce cognitive and motor dysfunction. In the kidneys, it disrupts protein synthesis and causes tubular damage, fibrosis, and genotoxicity. Additionally, the compound affects bone homeostasis by inhibiting osteoblast function, increasing osteoclast activity, and altering mineral balance. These wide-ranging toxic effects underscore the potential health risks associated with exposure to this compound through contaminated food sources.

8.1 Nephrotoxicity

OTA is a powerful nephrotoxin whose major target organ in animals is the kidney. Long-term OTA exposure has been linked with numerous renal health issues in animals and humans. Even at low concentrations, OTA is nephrotoxic in mammals and related to human BEN, renal failure, and cancer. OTA is mostly aimed at the proximal tubules of the kidney and causes structural damage and functional impairment (Chernozemsky et al., 1977; Krogh et al., 1973). Repeated exposure can lead to tubular degeneration, tubular dilatation, and interstitial fibrosis, impairing renal function. OTA-induced nephrotoxicity is expressed as renal impairment by a fall in glomerular filtration rate, impaired urine concentration, and disturbances in electrolytes. Renal fibrosis, oxidative stress, and renal inflammation are other prevalent effects of OTA exposure, which add up to kidney injury

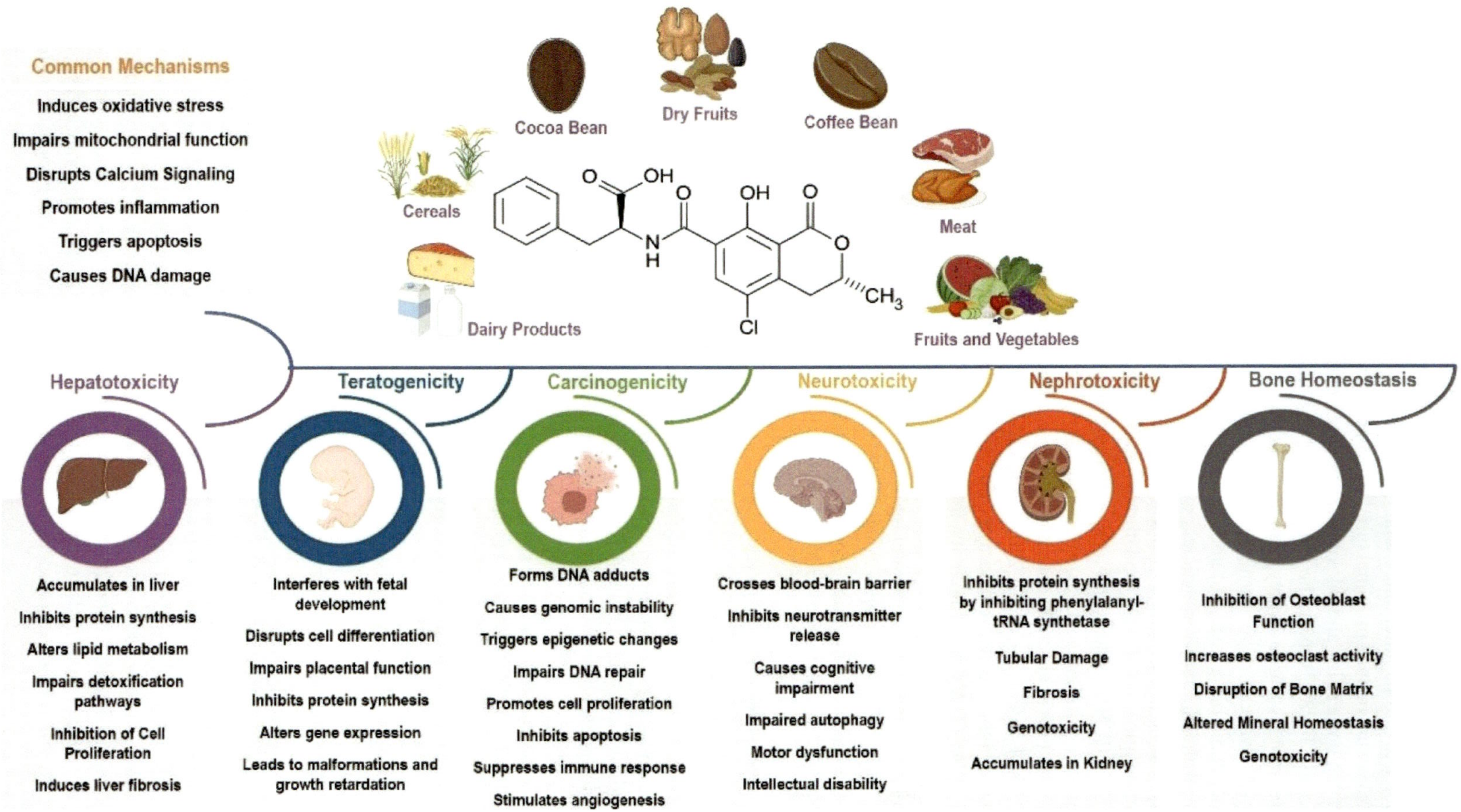

Fig. 2 Effects of OTA on human health and its implications.

and dysfunction. OTA has been linked to kidney health problems, including porcine and poultry nephropathy. It also plays a role in the development of certain kidney diseases, such as BEN in the Balkan region and chronic interstitial nephropathy in some North African countries. Kidney lesions due to OTA exposure include proximal tubular damage characterized by disruption of the membrane, decreased brush border size and thickness, condensed chromatin, and loss of the nuclear envelope (Kanisawa et al., 1978). BEN in the initial stages exhibits epithelial cell changes with no changes in organ size, progressing to kidney reduction and interstitial fibrosis on chronic exposure. Renal dysfunction at advanced stages results in enzymuria, polyuria with symptoms of polydipsia such as red tongue, thirst, bitter taste, but not edema or hypertension. Other symptoms can be headache, lumbar pain, asthenia, and iron deficiency anaemia (Luhe et al., 2003). Biochemical alterations comprise glycosuria, proteinuria, alkalinization of the urine, increased serum creatinine, and raised levels of immunoglobulin M (IgM) and immunoglobulin E (IgE). Additional studies revealed greater levels of OTA in BEN patients, yielding speculations for describing OTA's involvement in BEN and renal system tumors in Balkan areas. Subsequently, OTA's interference with intracellular metabolic mechanisms and proximal tubule activities, which lead to nephrotoxicity, was proved (Di Paolo et al., 1993; Pfohl-Leszkowicz et al., 2002). Interestingly, OTA-induced apoptosis was reported in human proximal tubule cells, reflecting its direct cytotoxic effect on renal cells. In the following years, focus is on OTA's nephrotoxicity following acute, sub-acute, and chronic exposures. It is essential to elucidate the mechanisms and health consequences of OTA-induced nephrotoxicity in order to devise preventive measures and interventions to eliminate kidney damage and ensure renal health. Reduction of OTA exposure by dietary management, application of food safety practices, and minimizing the levels of contamination in farm products are necessary interventions against OTA-caused nephrotoxicity and public health protection (Schwerdt et al., 1999).

8.2 Carcinogenicity of OTA

OTA is also classified as a human carcinogen and is categorized as such, and long-term exposure has been shown to increase the risk of renal cell carcinoma. OTA has been extensively studied for carcinogenicity. OTA's synergism with citrinin in promoting carcinogenicity was recorded. OTA was initially categorized under Group 3 as non-classifiable concerning its carcinogenicity to humans, but subsequently re-evaluated under Group 2B as possibly carcinogenic to humans. During 1989–2015, research on OTA's genotoxicity, DNA adduct formation, its role in oxidative stress and epigenetic processes, especially renal carcinogenesis, was carried out (Herman et al., 2021). The identification of OTA-DNA adducts in several tissues and the link between OTA exposure and renal tumours in animals as well as humans indicate OTA's potential to be carcinogenic. Though OTA's reclassification from Group 3 to Group 2B has shifted over the years, new findings suggest a potential upgrade to Group 2A or even Group 1, emphasizing the need for continued studies on OTA's mechanisms of carcinogenesis and human health implications.

8.3 Neurotoxicity of OTA

8.3.1 Implications of OTA in Neurodegenerative Disease

OTA exposure has been linked to neurodegenerative diseases such as Parkinson's disease (PD), Alzheimer's disease (AD), and multiple sclerosis (MS) (Izco et al., 2021). OTA exposure disturbs brain health, affecting crucial cellular mechanisms and causing neurodegeneration. Sub-chronic OTA exposure *in vivo* has been observed to cause motor dysfunction, dopaminergic cell dysfunction, and altered autophagy, all of which are characteristics of PD pathophysiology. Studies reveal that OTA can be harmful to the nervous system, with particularly damaging effects on the brain and spinal cord (Obafemi et al., 2023). OTA cytotoxicity in cerebellum brain tissue, where it induced high lactate dehydrogenase activity (LDH) and cell necrosis. Neurotoxicity of OTA through regional selectivity with specific emphasis on oxytocic cerebellum as a principal target and increased necrosis in ventral mesencephalon. OTA at low concentrations could also induce Parkinsonism in male mice, which affects striatal dopamine content, oxidative stress, and tyrosine hydroxylase immunoreactivity. It has been proven that proliferating neural stem cells were more sensitive to OTA compared to differentiated neurons, although there were robust DNA repair and antioxidant defenses. Further extrapolated these findings to pregnant mice to emphasize the impact of OTA on neurotransmitter content in various regions of the brain. Sub-chronic OTA treatment in young adult rats resulted in increased DNA damage, lipid peroxidation, and cellular cytotoxicity, particularly on amino acid levels and necrotic cell production in another study (Niaz et al., 2020). The brain distribution of OTA shows increased uptake in the cerebellum and changes in neurotransmitter synthesis, leading to signs such as tremor and muscle stiffness. All these findings together highlight the neurotoxicity potential of OTA, and its etiological role in neurodegenerative diseases, neurotransmitter disturbance, and cellular toxicity in various brain areas. OTA also evokes neuroinflammatory effects, especially through activation of pro-inflammatory microglia, which increases neurodegeneration. Human biomonitoring research has identified increased OTA concentrations in plasma samples of patients with neurodegenerative diseases and indicates that OTA might be involved in their pathogenesis (Niaz et al., 2020).

8.3.2 Implications of OTA in Neurodevelopmental Disorders

OTA exposure experiments have been linked with several neurodevelopmental disorders, especially in prenatal and early postnatal life. Neurodevelopmental disorders include a group of conditions impacting the development of the nervous system, resulting in cognitive, motor, social, and behavioural impairments (Tye et al., 2018). Although OTA exposure and specific neurodevelopmental disorders in humans have not been studied extensively, some studies on animal models and epidemiological studies indicate a probable connection defined by social interaction, communication, and repetitive behavior difficulties, which has been associated with prenatal exposure to environmental contaminants, such as mycotoxin OTA. Neurotoxic effects of OTA, involving oxidative stress , inflammation, and disruption of neuronal signalling pathways, may contribute to the pathogenesis of autism spectrum disorder (ASD) (De Santis et al., 2019; Tao et al., 2018). ADHD, which involves defective attention, hyperactivity, and impulsivity, has also been linked to environmental causes, such as prenatal toxin exposure. Exposure to OTA during brain developmental windows could interfere with neurotransmitter networks, neuronal relationships, and behavior control, and enhance vulnerability to ADHD (De Santis et al., 2019). Prenatal exposure to OTA has been linked to intellectual disability, characterized by significant limitations in thinking ability and adaptive behaviour. Such exposure can cause structural brain abnormalities, hinder cognitive function, and affect learning deficits in animal studies (Panerai et al., 2014). OTA-induced neurotoxicity, such as neuroinflammation, apoptosis of neurons, and synaptic dysfunction, has the potential to lead to intellectual disabilities. Delayed development, defined as delays or deviations from the attainment of developmental milestones, has also been associated with OTA exposure, possibly interfering with the normal development of the brain, resulting in delays in neuronal maturation, dendritic elongation, and synaptic elimination. OTA-induced disruption of neurotransmitter systems, neurogenesis, and neuronal migration may lead to developmental delay and deficits in cognitive and motor functions. Therefore, prenatal and early-life exposure to OTA can enhance the risk of neurodevelopmental disorders, such as intellectual disability and developmental delay by oxidative stress-mediated mechanisms, inflammation, neurotransmission disruption, and apoptotic cell death. Additional studies are required to determine the exact mechanisms and long-term implications of OTA exposure on neurodevelopment and to establish preventive measures to counteract its harmful effects on susceptible populations.

8.4 Impact of OTA on Vascular and Bone Balance

OTA presents an important risk to human health because it is nephrotoxic, hepatotoxic, carcinogenic, and immunosuppressive. Aside from its long-established toxic impact on the kidney and liver, new data indicate that OTA also has a tremendous impact on vascular and bone homeostasis. It is important to understand OTA's mechanism of action on such systems for the development of preventive strategies and therapeutic interventions to counteract its harmful effects (Jann et al., 2020).

8.4.1 Cardiovascular Effects of OTA

Vascular homeostasis, crucial for cardiovascular health, involves intricate mechanisms regulating blood flow, vascular tone, and endothelial integrity. Endothelial cells play a pivotal role, orchestrating vasodilation via NO and vasoconstriction via endothelin-1, ensuring proper blood distribution (Garoffolo et al., 2018). They also regulate platelet adhesion, leukocyte recruitment, and vascular permeability, vital for preventing thrombosis, inflammation, and vascular leakage. Studies on OTA's impact on vascular homeostasis imply its significance in cardiovascular health. Mitigating OTA's adverse effects necessitates strategies targeting its disruption of vascular function and integrity, emphasizing the importance of preserving vascular homeostasis for overall cardiovascular well-being. Endothelial damage caused by OTA plays a key role in the development of heart and blood vessel diseases. OTA exerts detrimental effects on endothelial cell function through multiple mechanisms, including oxidative stress, inflammation, and disruption of NO signalling pathways (Ramyaa et al., 2014). Oxidative stress serves as a central mediator in OTA-induced endothelial dysfunction, instigating the generation of ROS and subsequent oxidative damage to endothelial cells. The overproduction of ROS overwhelms the antioxidant defence mechanisms, leading to cellular injury, dysfunction, and impaired vascular homeostasis (Newsholme et al., 2019). In addition to oxidative stress, OTA promotes inflammation within the vascular endothelium, further exacerbating endothelial dysfunction. OTA exposure triggers the upregulation of pro-inflammatory cytokines and adhesion molecules, fostering a pro-inflammatory microenvironment. Investigators have assessed OTA effects on nasal epithelial cells in eosinophilic chronic rhinosinusitis and controls. OTA led to a significant increase in the levels of both IL-6 and IL-8 in both groups. Mycotoxins, like OTA, may induce inflammation in the nasal mucosa, suggesting a potential role in eosinophilic chronic rhinosinusitis pathophysiology. Furthermore, OTA disrupts NO signalling pathways, which are integral to the regulation of vascular tone and endothelial function. Nitric oxide (NO), generated by endothelial nitric oxide synthase (eNOS), is essential for vascular health. It helps blood vessels relax, prevents platelets from clumping together, and reduces inflammation. eNOS is pivotal in cardiovascular health, regulated by genetic polymorphisms and dimerization. Its impact on cardiac myocytes, vasculature, and autonomic nervous system (ANS) is crucial (Newsholme et al., 2019). Besides oxidative stress, OTA induces inflammation in the vascular endothelium and aggravates endothelial dysfunction. OTA exposure is responsible for the upregulation of pro-inflammatory cytokines and adhesion molecules, creating a pro-inflammatory microenvironment. Researchers have compared the effects of OTA on nasal epithelial cells in eosinophilic chronic rhinosinusitis with controls. OTA significantly raised levels of IL-6 and IL-8 in both groups. Mycotoxins, such as OTA, could trigger inflammation in the nasal mucosa, proposing a possible role in eosinophilic chronic rhinosinusitis pathogenesis. Additionally, OTA interferes with NO signalling pathways, which are central to vascular tone regulation and endothelial function. NO, released by endothelial nitric oxide synthase (eNOS), has the important function of preserving vascular homeostasis through vasodilation promotion, prevention of platelet aggregation, and anti-inflammatory effects. eNOS plays a central role in cardiovascular well-being, and its regulation through genetic polymorphisms and dimerization. Its influence on cardiac myocytes, vasculature, and ANS is vital. eNOS impacts both the sympathetic and parasympathetic ANS, which influences cardiovascular reflexes such as the baroreflex. Faulty eNOS is associated with hypertension, arteriosclerosis, and cardiovascular disease, affecting the parasympathetic and sympathetic ANS, influencing cardiovascular reflexes like the baroreflex. Dysfunctional eNOS correlates with hypertension, arteriosclerosis, and other cardiovascular diseases (Kietadisorn et al., 2012).

OTA was investigated for its association with hypertension in 50 freshly diagnosed hypertensive patients and 33 fit individuals. Most participants had detectable levels of OTA in their urine. High OTA levels were positively linked to hypertension in non-obese adolescents, suggesting OTA to be a potential risk factor (Dimkpa et al., 2010). By generating NO, the endothelium plays a key role in regulating the relaxation and constriction of blood vessels. One key mechanism through which OTA contributes to hypertension is endothelial dysfunction that initiates vascular inflammation, remodelling, and atherosclerosis, increasing cardiovascular risk. Mechanisms like oxidative stress and renin-angiotensin system (RAS) influence not only renal and cardiovascular systems but also the brain, remodelling, and exacerbate endothelial dysfunction through inflammation and oxidative stress. Inside cells, mitochondria, are vital for producing energy. Exposure to OTA reduces the expression of Lon, a key mitochondrial protein (Bota et al., 2016). When cells lack sufficient Lon, they become more vulnerable to OTA damage, which has been linked to kidney and heart diseases due to oxidative stress induced by angiotensin II. Overactivation of the renin-angiotensin system leads to mitochondrial damage, affecting cellular metabolism (Vajapey et al., 2014). Furthermore, OTA disrupts the renin-angiotensin-aldosterone system (RAAS), a fundamental regulatory pathway in blood pressure control. These alterations in RAAS dynamics result in enhanced vasoconstriction and sodium retention, contributing to hypertension and vascular dysfunction.

OTA induces Endoplasmic Reticulum (ER) stress, which in turn promotes atherosclerosis through various pathways. Previous research has verified that ER stress plays a role in dysregulating cholesterol metabolism in atherosclerosis, with upregulation of microRNA-33 and activation of C/EBP homologous protein contributing to lipid metabolism disorders in atherosclerotic macrophages (Christian et al., 2014). Overexpression of miR-384 is shown to mitigate angiotensin-II-induced apoptosis and ER stress in endothelial cells by downregulating HERP expression. Thus, in the context of OTA exposure, induction of ER stress may exacerbate atherosclerosis through mechanisms involving dysregulated lipid metabolism and endothelial dysfunction. Mononuclear white blood cells, especially monocytes and macrophages, produce tissue factor (TF) and plasminogen activator inhibitor-2 (PAI-2), which are important in regulating inflammation, immune responses, and blood vessel function (Gupta, 2013). Primary findings indicate that subtoxic levels of OTA have no direct influence on TF and PAI-2 production. In the presence of lipopolysaccharide or other inflammatory stimuli, OTA suppresses the induction of TF and PAI-2 in a dose-dependent fashion, with significant inhibition at the concentration of 1μg/ml. The inhibitory effect is reproducible in both mixed mononuclear cell cultures and in monocyte-enriched populations. OTA also suppresses the production of TF in response to various inflammatory triggers, such as phorbol myristate acetate, interleukin-1 beta (IL-1β), and tumour necrosis factor alpha (TNF-α). These observations indicate that OTA could have immunosuppressive activity by interfering with TF induction and PAI-2, which are central regulators of coagulation and fibrinolysis. Such an interference in the fine-tune of coagulation and fibrinolysis processes in the vasculature may result in disease processes, such as thrombosis and inflammation, thus compromising vascular homeostasis and cardiovascular health.

8.4.2 Effects of OTA on Bone Homeostasis

Bone homeostasis is vital for preserving skeletal integrity and mineral metabolism, relying on a delicate equilibrium between bone formation and resorption processes. OTA disrupts this balance through multifaceted mechanisms, compromising bone health and increasing the risk of skeletal disorders. OTA interferes with bone mineralization, further compromising bone integrity and strength. Chronic exposure to OTA has been correlated with reduced Bone mineral density (BMD) and an elevated risk of osteoporosis, a condition characterized by low BMD and increased susceptibility to fractures.

BMD serves as a critical indicator of bone strength and fracture risk, with decreased BMD being a hallmark of osteoporosis. Animal studies and epidemiological investigations have consistently demonstrated the detrimental impact of OTA on BMD and bone microarchitecture. OTA exposure, highlighting its detrimental effect on skeletal health, dysregulates lipid metabolism and decreases the availability of lipid-soluble vitamins. This puts stress on bone cells (osteocytes), prompting them to release calcium into the bloodstream, which over time can raise the risk of osteoporosis (Carina et al., 2020). In one of the studies, young chickens were subjected to varying levels of aflatoxin or ochratoxin in their diets, and the effects on bone strength were examined. Aflatoxin reduced bone-breaking strength at levels inhibitory to growth, while ochratoxin caused a decrease in breaking strength at slightly lower levels. The reduction in bone diameter occurred at higher toxin levels for both aflatoxin and ochratoxin. Birds with ochratoxicosis, even with bones of equal diameter as controls, exhibited lower breaking strength. Furthermore, prenatal or early-life exposure to OTA has been linked to skeletal deformities in animal models, suggesting vulnerability of developing skeletal issues to OTA toxicity. A study examined how injecting OTA into live highline layer eggs affected calcium metabolism and blood chemistry in both the embryos and the chicks after hatching. Throughout incubation and past-hatch, embryos and chicks from the OTA group exhibited lower serum calcium and inorganic phosphorus levels, along with elevated sodium, alkaline phosphatase, and triiodothyronine concentrations compared to controls. Allantoic fluid from OTA-injected eggs also displayed reduced calcium and phosphorus levels. Furthermore, OTA-exposed embryos showed decreased whole body and yolk calcium and phosphorus content compared to controls, indicating disrupted calcium metabolism. These findings are significant considering previous research linking prenatal or early-life exposure to OTA with skeletal deformities in animal models. Overall, the study reveals that OTA disrupts calcium metabolism both before and after hatching, highlighting the potential dangers of exposure during key stages of bone development (Ozawa et al., 2024). During parental and early postnatal stages, the skeletal system undergoes rapid growth and development, making it particularly susceptible to external influences. OTAs' interference with essential signalling pathways and cellular processes involved in skeletal development can disrupt the normal formation and patterning of bones, resulting in structural abnormalities (Jann et al., 2020; Terkawi et al., 2022).

9. Mechanisms of OTA Action

OTA damages the kidneys in many ways, working through multiple interconnected pathways. Studies show that OTA can block protein production, damage DNA, interrupt the cell cycle, and trigger cell death. When it builds up in the kidney's proximal tubule cells, it sets off damage by causing oxidative stress, harming DNA, triggering cell death, and sparking inflammation (Khoi et al., 2021). Oxidative stress is a major mechanism by which OTA induces nephrotoxicity. Studies have shown that OTA increases lipid peroxidation and suppresses antioxidant enzymes, leading to oxidative damage in rat kidneys and liver (Rašić et al., 2019). OTA also induces reactive oxygen species (ROS) formation in renal cells and increases DNA damage (Tao et al., 2018). Treatment with N-acetylcysteine, an antioxidant, can reduce ROS levels and enhance cell viability after OTA exposure (Yang et al., 2014). The NRF2 pathway is a critical cellular defence mechanism against oxidative stress and inflammation. NRF2 is a transcription factor that regulates the expression of genes encoding antioxidant proteins, detoxifying enzymes, and other cytoprotective proteins. OTA's impact on NRF2-mediated regulation has been found in human bone marrow-derived mesenchymal stem cells (MSCs). OTA hindered NRF2's nuclear import, leading to diminished self-renewal and osteogenic potential in early-passage MSCs (Hwang, 2014). Additionally, OTA may exacerbate bone loss by stimulating the activity of osteoclasts, cells responsible for bone resorption (Hwang, 2014; Terkawi et al., 2022). OTA also activates various signalling pathways, including apoptosis signal-regulated kinase 1 (ASK-1), aryl hydrocarbon receptor (AhR), pregnane X receptor, and nuclear factor erythroid 2-related factor 2 (Nrf2), which regulate ROS production and cell damage.

Another way OTA damages the kidneys is through nitrosative stress, which involves activating inducible nitric oxide synthase (iNOS) and causing protein nitration in kidney and liver cells. OTA has been found to raise nitric oxide (NO) levels in kidney cell lines, although some studies report mixed results on its impact on NO production (Longobardi et al., 2021). OTA in renal cells induces apoptosis through activation of caspase-3 and increased DNA fragmentation. OTA has also been shown to induce necrosis and apoptosis in human proximal tubular epithelial cells (HK-2) by activating various signalling pathways, including the mitogen-activated protein kinase/ extracellular signal-regulated protein kinase 1/2 and mesenchymal-epithelial transition factor/phosphoinositide 3-kinase /protein kinase B pathways (Darbuka et al., 2021). OTA can induce apoptosis through the p53 pathway and ER) stress. Heat shock proteins (HSPs), particularly glucose-regulated protein 75 (GRP-75), which is part of the HSP70 family, play a protective role in the kidney by enhancing cellular defences against apoptosis and necrosis. GRP-75 levels decrease with increasing OTA doses, suggesting that it may serve as a biomarker for renal tubular necrosis induced by OTA. OTA exposure has also shown varying effects on mitochondrial function by decreasing mitochondrial membrane potential, followed by hyperpolarization at later stages (Li et al., 2019). While some studies suggest that OTA induces hyperpolarization and increases ATP production, others have found that OTA reduces intracellular ATP levels (Li et al., 2019; Poor et al., 2014). OTA also inhibits mitochondrial oxygen consumption by suppressing complex I and II of the mitochondrial electron transport respiratory chain, indicating mitochondrial dysfunction (Wheaton et al., 2014). Other mechanisms of OTA-induced nephrotoxicity include inhibition of protein synthesis, disruption of the cell cycle, lipotoxicity, alterations in microRNA expression, and modulation of autophagy. These complex mechanisms highlight the diverse pathways in which OTA can exert its effects.

10. Management and Control Strategies for OTA

Managing OTA involves preventing its occurrence, detecting it early, enforcing safety regulations, and reducing its impact to limit exposure in people and animals. At the farm level, prevention focuses on following Good Agricultural, Manufacturing, and Storage Practices, along with proper handling, safe storage, and Integrated Pest Management to keep OTA contamination in check (Banahene et al., 2024). Management strategies for OTA encompass preventive measures, focusing on agricultural practices, harvesting, and post-harvest handling (Banahene et al., 2024). Such strategies play a vital role in keeping OTA contamination in agricultural goods to a minimum. At the agricultural level, practices such as crop rotation, selection of resistant varieties, proper irrigation management, and the use of biological control agents can disrupt the habitat of OTA-producing fungi and reduce contamination (Mondani et al., 2020). Timely harvesting, proper drying, and storage of crops are essential to minimize fungal growth and OTA production. Adhering to GAP, including sanitation, proper cleaning equipment, and maintaining hygienic conditions, can prevent contamination at various stages. Regular screening and monitoring of crops for fungal contamination and OTA levels, along with the use of mycotoxin binders in animal feed, can further reduce OTA exposure risks. Collaboration among stakeholders is one of the keys for effective implementation and enforcement of these strategies, while continuous research and innovation are essential for enhancing OTA management in agriculture.

Monitoring and surveillance efforts involve routine testing, sampling, and analysis, as well as establishing surveillance systems to track OTA contamination levels. Regulatory measures include setting maximum limits for OTA in food and feed products, implementing labelling and certification requirements, and enforcing regulations through inspections and audits. Risk assessment and management strategies encompass conducting risk assessments, communicating risks to stakeholders, and providing education and training programmes. Mitigation strategies involve using mycotoxin binders in animal feed, implementing post-harvest treatments, and exploring alternative control methods. International collaboration is crucial for harmonizing standards, supporting capacity-building initiatives, and ensuring global food safety (Casu et al., 2024). At harvest time, certain critical steps are essential to keep OTA contamination in check. These include keeping all farming apparatus clean and in good working order, and making sure crops are harvested at the exact moisture level. Only mature, contamination-free produce should be collected, while over-riped, fermented, or damaged items must be removed. Poor-quality produce should never be mixed with healthy crops. Finally, the harvest should be dried quickly and evenly in hygienic and controlled conditions. Although preventing OTA during crop growth and harvest is essential, any contamination that does occur is best handled after harvest. Post-harvest control focuses on critical points like keeping storage below 20°C with water activity under 0.70, and using safe wrapping materials such as metal containers instead of wood or clay (Casu et al., 2024).

10.1 Methods to Degrade and Absorb OTA

Various physical, chemical, enzymatic, and microbial approaches are used to degrade OTA in agricultural food and feed. Physical methods include heat treatment or irradiation, while physicochemical methods involve using substances like activated carbon, potassium, and wood fragments to absorb OTA (Barakat et al., 2022). Chemical methods for breaking down OTA depend on different compounds, such as solvents like dichloromethane, ethyl acetate, and methylene chloride; acids such as formic, propionic, and sorbic; and alkaline agents like sodium hydroxide, alkaline hydrogen peroxide, or monomethyl amine combined with calcium hydroxide. Sodium hypochlorite can also be used with these acids to degrade OTA. Fungicides like azoxystrobin effectively target OTA in wine, preservatives like potassium sorbate or calcium propionate work in bread, and antioxidants like vanillic acid can inhibit OTA synthesis (Dong et al., 2019). Depending on the commodity, these methods can be combined better to protect the safety and quality of food and feed. Enzymes, either crude or purified, are also used to break down OTA. For example, proteolytic enzymes such as trypsin, α-chymotrypsin, and carboxypeptidase A were first used in 1969 to degrade OTA, with carboxypeptidase A showing higher affinity for OTA than α-chymotrypsin (Xu et al., 2021). Additionally, crude metalloenzymes from *Aspergillus niger* have demonstrated significantly higher OTA hydrolytic activity than carboxypeptidase. Biological methods, which are considered the most cost-effective, involve using microorganisms such as bacteria, actinobacteria, filamentous fungi, and yeast to break down and/or absorb OTA. For instance, several bacterial strains like *Bacillus amyloliquefaciens* and *Sphingomonas paucimobilis* have successfully degraded OTA (Mwabulili et al., 2023). Actinobacteria strains can both degrade and absorb OTA, with some strains inhibiting OTA biosynthesis genes in fungi. Filamentous fungi like *Aspergillus niger* and yeasts such as *Yarrowia lipolytica* have also proven effective in breaking down OTA. Overall, comprehensive management strategies are essential to effectively reduce the risk of OTA contamination and protect human and animal health (Viegas et al., 2020).

10.2 Processing and Storage Techniques

Processing and storage techniques are crucial for minimizing OTA contamination in agricultural products during post-harvest handling. Implementing proper cleaning, sorting, drying, and storage practices can help mitigate the risk of OTA formation and accumulation (Schaarschmidt et al., 2018). Strategies such as removing damaged or mouldy grains, utilizing cleaning equipment, sorting grains by size and density, and employing colour sorting technologies can effectively remove contaminated grains (Matumba et al., 2015). Additionally, employing effective drying methods, maintaining optimal temperature and humidity levels, providing adequate aeration and ventilation, and using proper packaging materials can inhibit fungal development and OTA production during storage (Dallagnol et al., 2019). Regular monitoring and sampling protocols are essential for early detection of contamination and to enable timely intervention measures in a timely manner. By implementing these processing and storage techniques, producers can effectively reduce OTA contamination in agricultural products and ensure the safety and quality of food and feed products. Collaboration among stakeholders, adherence to regulatory standards, and ongoing research and innovation are crucial for improving OTA management and minimizing health risks associated with OTA exposure (Brandao et al., 2022).

10.3 Intervention and Mitigation Strategies for OTA

Intervention and mitigation strategies for OTA encompass dietary interventions, toxicity surveillance, and risk management practices that are aimed at reducing OTA exposure and mitigating health risks. Encouraging dietary diversity, implementing biomonitoring programmes, promoting good agricultural practices, and enforcing regulatory compliance are essential components of OTA management efforts. Continuous surveillance, research, and collaboration among stakeholders are critical for enhancing OTA mitigation strategies and safeguarding public health. Regulatory bodies like the Joint FAO/WHO Expert Committee on Food Additives (JECFA) have established provisional tolerable weekly intake (PTWI) limits for OTA based on risk assessments. Overall, effective intervention and mitigation strategies, coupled with robust surveillance and regulatory measures, are crucial for managing OTA contamination and ensuring food safety (Sharafi et al., 2023).

11. Regulatory Measures and Monitoring Programmes

Regulations and monitoring programmes play a vital role in managing OTA. They are key to protecting the safety and quality of both food and animal feed. Because of its toxic nature, OTA is regulated by law at both national and international levels. Its harmful effects became clear by the late 1970s. However, discussions about whether to control OTA in food and feed on a national or global scale didn't gain momentum until the 1990s. This was much later than for other mycotoxins, such as aflatoxins, which were regulated in the USA in the 1960s and the European Communities by the 1970s (Wei et al., 2021). Many countries and regions have set legal limits for OTA in food and animal feed to protect public health and support fair trade in agricultural products. These limits are not the same everywhere. They often differ between importing and exporting nations, and between developing and developed countries. For example, the maximum OTA levels allowed in raw cereal grains are 3–5 μg/kg in the European Union, Russia, and China. In contrast, countries like India and Brazil have looser limits, permitting up to 10–20 μg/kg. To control OTA risks, it is essential to set clear regulatory limits, carry out monitoring programmes, and enforce compliance with safety standards. These limits are based on risk assessments, scientific data, and international guidelines. While no large-scale global survey has been published in recent years, likely, the number of countries regulating OTA has not decreased, especially as new research continues to highlight its dangers and as global trade in food and feed expands (El Safadi et al., 2016).

Harmonizing regulatory limits across regions and countries facilitates international trade and ensures consistent protection of public health. Surveillance and monitoring programmes assess OTA levels in food and feed products at various steps of production, distribution, and consumption. Utilizing reliable analytical methods enables accurate detection and quantification of OTA. The Codex Alimentarius Commission, established by the Food and Agriculture Organization (FAO) and WHO, has debated the feasibility of establishing maximum limits on OTA. The Codex General Standard for Contaminants and Toxins in Food and Feed has set a maximum limit of 5 μg/kg for OTA in wheat, barley, and rye. Additionally, four codes of practice aiming at avoiding and decreasing OTA contamination in cereals, wine, coffee, and cocoa have been adopted between 2007–2014. The Regulation 466/2001, as amended by Regulation 472/2002, established the first maximum levels of OTA in various foodstuffs. For cereals (including rice and buckwheat) and derived cereal products, the maximum level was set at 5 ng/g. The same limit was applied to raw cereal grains (including uncooked rice and buckwheat) (El Safadi et al., 2016). For cereals, including both raw grains and processed products meant for direct human consumption, the maximum OTA limit is set at 3 ng/g.

Dried vine fruit such as currants, raisins, and sultanas were allowed a maximum level of 10 ng/g. Green and roasted coffee, as well as coffee products, wine, beer, grape juice, cocoa, cocoa products, and spices, did not have specified maximum levels under this regulation. In the European Union, Regulation 1881/2006 sets strict maximum limits for OTA in different food products. For animal feed, there is only a non-binding recommendation. Other countries, such as the USA, Australia, Canada, and Japan, have taken a different path. Instead of setting legal OTA limits, they rely on good agricultural and manufacturing practices. This approach has shaped international debates on mycotoxin regulation, including OTA. Under EU rules, unprocessed cereals are limited to 5.0 ng/g of OTA. All cereal-based products, including processed cereals and grains for direct human consumption, must not exceed 3.0 ng/g, with some exceptions. For coffee, roasted beans and ground roasted coffee (except instant coffee) are limited to 5.0 ng/g, while instant coffee can contain up to 10.0 ng/g. Wine, sparkling wine, fruit wine, aromatized wines, and grape juice products are all capped at 2.0 ng/g. Foods for infants and young children, including processed cereal-based products and dietary foods for special medical purposes, have the strictest limit of just 0.50 ng/g. Spices vary, with limits between 15–20 ng/g depending on the type. Liquorice and its extracts also have set limits, while wheat gluten, not sold directly to consumers, is capped at 8.0 ng/g. These maximum levels are designed to keep food products within safe OTA contamination limits and protect public health (Azam et al., 2021).

The European Commission's Recommendation 2006/576/EC sets guidance values for OTA in animal feed. For feed materials such as cereals and cereal-based products, the limit is 0.25 mg/kg, based on a feed moisture content of 12%. For complementary and complete feeds, the guidance values are lower, with 0.05 mg/kg for pigs and 0.1 mg/kg for poultry. These limits are designed to keep OTA levels in animal feed within safe ranges, helping to protect animal health. Risk assessments based on surveillance data allow regulatory agencies to evaluate exposure levels, assess health risks, and develop risk management strategies. Enforcing compliance with regulatory standards through inspections, audits, and enforcement actions ensures that food and feed producers adhere to established limits for OTA. Capacity building initiatives improve laboratory capabilities, data interpretation skills, and regulatory enforcement practices. Continuous monitoring, research, and collaboration among regulatory agencies, industry stakeholders, and international organizations are essential for enhancing OTA management strategies and protecting consumer safety (Bhatt et al., 2017).

12. Conclusion and Future Directions

The contamination of food and feed products with OTA is a significant concern due to its adverse health issues, including nephrotoxicity, neurotoxicity, carcinogenicity, hepatotoxicity, immunosuppression, teratogenicity, and its effects on vascular and bone homeostasis. Understanding the sources of OTA contamination and implementing effective mitigation strategies are essential steps toward ensuring food safety and minimizing human exposure to OTA. Collaboration among farmers, food processors, regulators, and researchers is crucial for developing and implementing comprehensive strategies to address OTA contamination in the food supply chain. Agricultural practices, environmental factors, and processing methods all play critical roles in influencing OTA occurrence in food and feed. By adopting sustainable farming practices, implementing proper storage and processing techniques, and adhering to stringent hygiene standards, stakeholders can minimize OTA contamination risks and ensure the safety and quality of agricultural products. Effective management practices, including crop rotation, proper storage, and climate-resilient agriculture, are essential for minimizing OTA contamination risks and safeguarding public health. Additionally, implementing preventive measures in animal feed production, such as identifying and addressing sources of contamination, is crucial for protecting animal health and food safety. Contamination of processed foods can arise from various sources, including raw ingredients, processing environments, packaging materials, and environmental pollutants. Implementing preventive measures, such as

Good Manufacturing Practices (GMPs) and hazard analysis and critical control points (HACCP), is essential for minimizing contamination risks and ensuring the safety and quality of processed foods. Regulatory measures and monitoring programs are essential for managing OTA contamination risks and safeguarding public health. Establishing regulatory limits, implementing surveillance programmes, and ensuring compliance with standards help minimize OTA exposure levels in food and feed products. Effective risk assessment, communication, and enforcement efforts contribute to the prevention and control of OTA-related health hazards (Bui-Klimke et al., 2015).

Future research directions should focus on innovative approaches to OTA control, including alternative control methods, molecular detection techniques, climate change resilience, gut microbiota interactions, and integrated management approaches. Overcoming challenges related to global harmonization, detection sensitivity, consumer awareness, sustainable agriculture, and multidisciplinary collaboration will require concerted efforts and strategic partnerships across sectors. By embracing a proactive and collaborative approach, stakeholders can work together to mitigate OTA risks, protect consumer health, and ensure the safety and sustainability of the food supply. Through ongoing research, innovation, and knowledge sharing, we can strive towards a future where OTA contamination is effectively managed and public health is safeguarded for future generations to come (Kumar et al., 2020).

References

Awuchi, C.G., Ondari, E.N., Nwozo, S., Odongo, G.A., Eseoghene, I.J., Twinomuhwezi, H., ... and Okpala, C.O.R. (2022). Mycotoxins' Toxicological Mechanisms Involving Humans, Livestock and Their Associated Health Concerns: A Review. *Toxins* (*Basel*), *14*(3). doi:10.3390/toxins14030167.

Azam, M.S., Ahmed, S., Islam, M.N., Maitra, P., Islam, M.M. and Yu, D. (2021). Critical Assessment of Mycotoxins in Beverages and Their Control Measures. *Toxins* (*Basel*), *13*(5). doi:10.3390/toxins13050323.

Banahene, J.C.M., Ofosu, I.W., Odai, B.T., Lutterodt, H.E., Agyemang, P.A. and Ellis, W.O. (2024). Ochratoxin A in food commodities: A review of occurrence, toxicity, and management strategies. *Heliyon*, *10*(20), e39313.

Barakat, N.A.M., Irfan, O.M. and Moustafa, H.M. (2022). H(3)PO(4)/KOH Activation Agent for High Performance Rice Husk Activated Carbon Electrode in Acidic Media Supercapacitors. *Molecules*, *28*(1). doi:10.3390/molecules28010296.

Ben Miri, Y., Benabdallah, A., Chentir, I., Djenane, D., Luvisi, A. and De Bellis, L. (2024). Comprehensive Insights into Ochratoxin A: Occurrence, Analysis, and Control Strategies. *Foods*, *13*(8). doi:10.3390/foods13081184.

Bhatt, T., Gooch, M., Dent, B. and Sylvia, G. (2017). Implementing Interoperability in the Seafood Industry: Learning from Experiences in Other Sectors. *J. Food Sci.*, *82*(S1), A22–A44.

Bota, D.A. and Davies, K.J. (2016). Mitochondrial Lon protease in human disease and aging: Including an etiologic classification of Lon-related diseases and disorders. *Free Radic. Biol. Med.*, *100*, 188–198.

Brandao, R.M., Batista, L.R., de Oliveira, J.E., Ferreira, V.R.F., Lunguinho, A.D.S., Nelson, D.L. and Cardoso, M.D.G. (2022). Active packaging of poly(lactic acid) nanofibers and essential oils with antifungal action on table grapes. *FEMS Microbiol. Lett.*, *369*(1). doi:10.1093/femsle/fnac116.

Bui-Klimke, T.R. and Wu, F. (2015). Ochratoxin A and human health risk: A review of the evidence. *Crit. Rev. Food Sci. Nutr.*, *55*(13), 1860–1869.

Carina, V., Della Bella, E., Costa, V., Bellavia, D., Veronesi, F., Cepollaro, S. ... and Giavaresi, G. (2020). Bone's Response to Mechanical Loading in Aging and Osteoporosis: Molecular Mechanisms. *Calcif. Tissue Int.*, *107*(4), 301–318.

Casu, A., Camardo Leggieri, M., Toscano, P. and Battilani, P. (2024). Changing climate, shifting mycotoxins: A comprehensive review of climate change impact on mycotoxin contamination. *Compr. Rev. Food Sci. Food Saf.*, *23*(2), e13323.

Cheong, K.K., Strub, C., Montet, D., Durand, N., Alter, P., Meile, J.C., ... and Fontana, A. (2016). Effect of different light wavelengths on the growth and ochratoxin A production in Aspergillus carbonarius and Aspergillus westerdijkiae. *Fungal Biol.*, *120*(5), 745–751.

Chernozemsky, I.N., Stoyanov, I.S., Petkova-Bocharova, T.K., Nicolov, I.G., Draganov, I.V., Stoichev, II, ... and Kalcheva, N.D. (1977). Geographic correlation between the occurrence of endemic nephropathy and urinary tract tumours in vratza district, Bulgaria. *Int. J. Cancer*, *19*(1), 1–11.

Christian, P. and Su, Q. (2014). MicroRNA regulation of mitochondrial and ER stress signaling pathways: Implications for lipoprotein metabolism in metabolic syndrome. *Am. J. Physiol. Endocrinol. Metab.*, *307*(9), E729–E737.

Cimbalo, A., Alonso-Garrido, M., Font, G. and Manyes, L. (2020). Toxicity of mycotoxins *in vivo* on vertebrate organisms: A review. *Food Chem. Toxicol.*, *137*, 111161.

Dallagnol, A.M., Bustos, A.Y., Martos, G.I., Valdez, G.F. and Gerez, C.L. (2019). Antifungal and antimycotoxigenic effect of Lactobacillus plantarum CRL 778 at different water activity values. *Rev. Argent. Microbiol.*, *51*(2), 164–169.

Darbuka, E., Gürkaşlar, C. and Yaman, I. (2021). Ochratoxin A induces ERK1/2 phosphorylation-dependent apoptosis through NF-κB/ERK axis in human proximal tubule HK-2 cell line. *Toxicon*, *199*, 79–86.

De Santis, B., Brera, C., Mezzelani, A., Soricelli, S., Ciceri, F., Moretti, G., … and Raggi, M.E. (2019). Role of mycotoxins in the pathobiology of autism: A first evidence. *Nutr. Neurosci.*, *22*(2), 132–144.

Di Paolo, N., Guarnieri, A., Loi, F., Sacchi, G., Mangiarotti, A.M. and Di Paolo, M. (1993). Acute renal failure from inhalation of mycotoxins. *Nephron.*, *64*(4), 621–625.

Dimkpa, U. and Oji, J.O. (2010). Relationship of body mass index with haemodynamic variables and abnormalities in young adults. *J. Hum. Hypertens.*, *24*(4), 230 236.

Dong, H., Xian, Y., Xiao, K., Wu, Y., Zhu, L. and He, J. (2019). Development and comparison of single-step solid phase extraction and QuEChERS clean-up for the analysis of 7 mycotoxins in fruits and vegetables during storage by UHPLC-MS/MS. *Food Chem.*, *274*, 471–479.

El Safadi, D., Cian, A., Nourrisson, C., Pereira, B., Morelle, C., Bastien, P., … and Viscogliosi, E. (2016). Prevalence, risk factors for infection, and subtype distribution of the intestinal parasite Blastocystis sp. from a large-scale multi-center study in France. *BMC Infect. Dis.*, *16*(1), 451.

Ganesan, A.R., Mohan, K., Karthick Rajan, D., Pillay, A.A., Palanisami, T., Sathishkumar, P. and Conterno, L. (2022). Distribution, toxicity, interactive effects, and detection of ochratoxin and deoxynivalenol in food: A review. *Food Chem.*, *378*, 131978.

Garoffolo, G., Madonna, R., de Caterina, R. and Pesce, M. (2018). Cell based mechanosensing in vascular patho-biology: More than a simple go-with the flow. *Vascul. Pharmacol.*, *111*, 7–14.

Gupta, G. (2013). Gingival crevicular fluid as a periodontal diagnostic indicator- II: Inflammatory mediators, host-response modifiers, and chair side diagnostic aids. *J. Med. Life*, *6*(1), 7–13.

Herman, D. and Mantle, P. (2021). Rat Tumour Histopathology Associated with Experimental Chronic Dietary Exposure to Ochratoxin A in Prediction of the Mycotoxin's Risk for Human Cancers. *Toxins* (*Basel*), *13*(3). doi:10.3390/toxins13030205.

Heussner, A.H. and Bingle, L.E. (2015). Comparative Ochratoxin Toxicity: A Review of the Available Data. *Toxins* (*Basel*), *7*(10), 4253–4282.

Hwang, E.S. (2014). Senescence suppressors: Their practical importance in replicative lifespan extension in stem cells. *Cell Mol. Life Sci.*, *71*(21), 4207–4219.

Imaoka, T., Yang, J., Wang, L., McDonald, M.G., Afsharinejad, Z., Bammler, T.K., … and Kelly, E.J. (2020). Microphysiological system modeling of ochratoxin A-associated nephrotoxicity. *Toxicology*, *444*, 152582.

Izco, M., Vettorazzi, A., Forcen, R., Blesa, J., de Toro, M., Alvarez-Herrera, N., … and Alvarez-Erviti, L. (2021). Oral subchronic exposure to the mycotoxin ochratoxin A induces key pathological features of Parkinson's disease in mice six months after the end of the treatment. *Food Chem Toxicol*, *152*, 112164.

Jann, J., Gascon, S., Roux, S. and Faucheux, N. (2020). Influence of the TGF-beta Superfamily on Osteoclasts/Osteoblasts Balance in Physiological and Pathological Bone Conditions. *Int. J. Mol. Sci.*, *21*(20). doi:10.3390/ijms21207597.

Kanisawa, M. and Suzuki, S. (1978). Induction of renal and hepatic tumors in mice by ochratoxin A, a mycotoxin. *Gan.*, *69*(4), 599–600.

Khoi, C.S., Chen, J.H., Lin, T.Y., Chiang, C.K. and Hung, K.Y. (2021). Ochratoxin A-Induced Nephrotoxicity: Up-to-Date Evidence. *Int. J. Mol. Sci.*, *22*(20). doi:10.3390/ijms222011237.

Kietadisorn, R., Juni, R.P. and Moens, A.L. (2012). Tackling endothelial dysfunction by modulating NOS uncoupling: New insights into its pathogenesis and therapeutic possibilities. *Am. J. Physiol. Endocrinol. Metab.*, *302*(5), E481–E495.

Krogh, P., Hald, B. and Pedersen, E.J. (1973). Occurrence of ochratoxin A and citrinin in cereals associated with mycotoxic porcine nephropathy. *Acta Pathol. Microbiol. Scand. B Microbiol. Immunol.*, *81*(6), 689–695.

Kumar, P., Mahato, D.K., Sharma, B., Borah, R., Haque, S., Mahmud, M.M.C., … and Bui, S. (2020). Ochratoxins in food and feed: Occurrence and its impact on human health and management strategies. *Toxicon*, *187*, 151–162.

Kuruc, J., Hegstad, J., Lee, H. J., Simons, K., Ryu, D. and Wolf-Hall, C. (2015). Infestation and Quantification of Ochratoxigenic Fungi in Barley and Wheat Naturally Contaminated with Ochratoxin A. *J. Food Prot.*, *78*(7), 1350–1356.

Lahouar, A., Jedidi, I., Sanchis, V. and Said, S. (2018). Aflatoxin B1, ochratoxin A, and zearalenone in sorghum grains marketed in Tunisia. *Food Addit. Contam. Part B Surveill.*, *11*(2), 103–110.

Li, Q., Dong, Z., Lian, W., Cui, J., Wang, J., Shen, H., … and Cui, H. (2019). Ochratoxin A causes mitochondrial dysfunction, apoptotic and autophagic cell death, and also induces mitochondrial biogenesis in human gastric epithelium cells. *Arch. Toxicol.*, *93*(4), 1141–1155.

Liu, K., Mao, X., Li, H., Liu, D., Le, G., Gan, F., … and Chen, X. (2020). Regulation of taurine in OTA-induced apoptosis and autophagy. *Toxicon*, *181*, 82–90.

Liu, W.C., Pushparaj, K., Meyyazhagan, A., Arumugam, V.A., Pappuswamy, M., Bhotla, H.K., … and Mousavi Khaneghah, A. (2022). Ochratoxin A as an alarming health threat for livestock and human: A review on molecular interactions, mechanism of toxicity, detection, detoxification, and dietary prophylaxis. *Toxicon, 213*, 59–75.

Longobardi, C., Damiano, S., Andretta, E., Prisco, F., Russo, V., Pagnini, F., … and Ciarcia, R. (2021). Curcumin Modulates Nitrosative Stress, Inflammation, and DNA Damage and Protects against Ochratoxin A-Induced Hepatotoxicity and Nephrotoxicity in Rats. *Antioxidants* (*Basel*), *10*(8) doi:10.3390/antiox10081239.

Luhe, A., Hildebrand, H., Bach, U., Dingermann, T. and Ahr, H.J. (2003). A new approach to studying ochratoxin A (OTA)-induced nephrotoxicity: Expression profiling *in vivo* and *in vitro* employing cDNA microarrays. *Toxicol. Sci.*, *73*(2), 315–328.

Malir, F., Ostry, V., Pfohl-Leszkowicz, A., Malir, J. and Toman, J. (2016). Ochratoxin A: 50 Years of Research. *Toxins* (*Basel*), *8*(7). doi:10.3390/toxins8070191.

Mateo, E., Tonino, R.P.B., Canto, A., Monroy Noyola, A., Miranda, M., Soria, J.M. and Garcia Esparza, M.A. (2022). The Neurotoxic Effect of Ochratoxin-A on the Hippocampal Neurogenic Niche of Adult Mouse Brain. *Toxins* (*Basel*), *14*(9). doi:10.3390/toxins14090624.

Matumba, L., Van Poucke, C., Njumbe Ediage, E., Jacobs, B. and De Saeger, S. (2015). Effectiveness of hand sorting, flotation/washing, dehulling and combinations thereof on the decontamination of mycotoxin-contaminated white maize. *Food Addit. Contam. Part A Chem. Anal Control Expo. Risk Assess.*, *32*(6), 960–969.

Mondani, L., Palumbo, R., Tsitsigiannis, D., Perdikis, D., Mazzoni, E. and Battilani, P. (2020). Pest Management and Ochratoxin A Contamination in Grapes: A Review. *Toxins*, *12*(5). doi:10.3390/toxins12050303.

Mwabulili, F., Xie, Y., Li, Q., Sun, S., Yang, Y. and Ma, W. (2023). Research progress of ochratoxin a bio-detoxification. *Toxicon*, *222*, 107005.

Newsholme, P., Keane, K.N., Carlessi, R. and Cruzat, V. (2019). Oxidative stress pathways in pancreatic beta-cells and insulin-sensitive cells and tissues: Importance to cell metabolism, function, and dysfunction. *Am. J. Physiol. Cell Physiol.*, *317*(3), C420–C433.

Niaz, K., Shah, S.Z.A., Khan, F. and Bule, M. (2020). Ochratoxin A-induced genotoxic and epigenetic mechanisms lead to Alzheimer disease: Its modulation with strategies. *Environ. Sci. Pollut. Res. Int.*, *27*(36), 44673–44700.

Obafemi, B.A., Adedara, I.A. and Rocha, J.B.T. (2023). Neurotoxicity of ochratoxin A: Molecular mechanisms and neurotherapeutic strategies. *Toxicology*, *497–498*, 153630.

Ochieng, P.E., Scippo, M.L., Kemboi, D.C., Croubels, S., Okoth, S., Kang'ethe, E.K., … and Antonissen, G. (2021). Mycotoxins in Poultry Feed and Feed Ingredients from Sub-Saharan Africa and Their Impact on the Production of Broiler and Layer Chickens: A Review. *Toxins* (*Basel*), *13*(9). doi:10.3390/toxins13090633.

Ostry, V., Malir, F., Dofkova, M., Skarkova, J., Pfohl-Leszkowicz, A. and Ruprich, J. (2015). Ochratoxin A Dietary Exposure of Ten Population Groups in the Czech Republic: Comparison with Data over the World. *Toxins* (*Basel*), *7*(9), 3608–3635.

Ozawa, S., Ojiro, R., Tang, Q., Zou, X., Jin, M., Yoshida, T. and Shibutani, M. (2024). *In vitro* and *in vivo* induction of ochratoxin A exposure-related micronucleus formation in rat proximal tubular epithelial cells and expression profiling of chromosomal instability-related genes. *Food Chem. Toxicol.*, *185*, 114486.

Panerai, S., Tasca, D., Ferri, R., Genitori D'Arrigo, V. and Elia, M. (2014). Executive Functions and Adaptive Behaviour in Autism Spectrum Disorders with and without Intellectual Disability. *Psychiatry J.*, *2014*, 941809.

Pei, X., Zhang, W., Jiang, H., Liu, D., Liu, X., Li, L., … and Li, D. (2021). Food-Origin Mycotoxin-Induced Neurotoxicity: Intend to Break the Rules of Neuroglia Cells. *Oxid. Med. Cell. Longev.*, *2021*, 9967334.

Pfohl-Leszkowicz, A., Petkova-Bocharova, T., Chernozemsky, I.N. and Castegnaro, M. (2002). Balkan endemic nephropathy and associated urinary tract tumours: A review on aetiological causes and the potential role of mycotoxins. *Food Addit. Contam.*, *19*(3), 282–302.

Poor, M., Veres, B., Jakus, P.B., Antus, C., Montsko, G., Zrinyi, Z., … and Koszegi, T. (2014). Flavonoid diosmetin increases ATP levels in kidney cells and relieves ATP depleting effect of ochratoxin A. *J. Photochem. Photobiol. B*, *132*, 1–9.

Ramyaa, P., Krishnaswamy, R. and Padma, V.V. (2014). Quercetin modulates OTA-induced oxidative stress and redox signalling in HepG2 cells - up regulation of Nrf2 expression and down regulation of NF-kappaB and COX-2. *Biochim. Biophys. Acta*, *1840*(1), 681–692.

Rašić, D., Micek, V., Klarić, M.S. and Peraica, M. (2019). Oxidative stress as a mechanism of combined OTA and CTN toxicity in rat plasma, liver, and kidney. *Human and Experimental Toxicology*, *38*(4), 434–445.

Schaarschmidt, S. and Fauhl-Hassek, C. (2018). The Fate of Mycotoxins During the Processing of Wheat for Human Consumption. *Compr. Rev. Food Sci. Food Saf.*, *17*(3), 556–593.

Schwerdt, G., Freudinger, R., Mildenberger, S., Silbernagl, S. and Gekle, M. (1999). The nephrotoxin ochratoxin A induces apoptosis in cultured human proximal tubule cells. *Cell Biol. Toxicol.*, *15*(6), 405–415.

Sharafi, H., Alizadeh, N., Moradi, M., Sharafi, K. and Mousavi Khaneghah, A. (2023). The prevalence and concentration of ochratoxin A in meat and edible offal: A global systematic review and meta-analysis. *Food Chem. Toxicol.*, *178*, 113921.

Soto, J.B., Ruiz, M.J., Manyes, L. and Juan-Garcia, A. (2016). Blood, breast milk and urine: potential biomarkers of exposure and estimated daily intake of ochratoxin A: A review. *Food Addit. Contam. Part A Chem. Anal Control Expo. Risk Assess*, *33*(2), 313–328.

Stoev, S.D. (2023). Foodborne Diseases Due to Underestimated Hazard of Joint Mycotoxin Exposure at Low Levels and Possible Risk Assessment. *Toxins* (*Basel*), *15*(7). doi:10.3390/toxins15070464.

Tao, Y., Xie, S., Xu, F., Liu, A., Wang, Y., Chen, D. … and Yuan, Z. (2018). Ochratoxin A: Toxicity, oxidative stress, and metabolism. *Food Chem. Toxicol.*, *112*, 320–331.

Terkawi, M.A., Matsumae, G., Shimizu, T., Takahashi, D., Kadoya, K. and Iwasaki, N. (2022). Interplay between Inflammation and Pathological Bone Resorption: Insights into Recent Mechanisms and Pathways in Related Diseases for Future Perspectives. *Int. J. Mol. Sci.*, *23*(3). doi:10.3390/ijms23031786.

Tye, C., Runicles, A.K., Whitehouse, A.J.O. and Alvares, G.A. (2018). Characterizing the Interplay Between Autism Spectrum Disorder and Comorbid Medical Conditions: An Integrative Review. *Front. Psychiatry*, *9*, 751.

Vajapey, R., Rini, D., Walston, J. and Abadir, P. (2014). The impact of age-related dysregulation of the angiotensin system on mitochondrial redox balance. *Front. Physiol.*, *5*, 439.

Viegas, S., Assuncao, R., Twaruzek, M., Kosicki, R., Grajewski, J. and Viegas, C. (2020). Mycotoxins feed contamination in a dairy farm: Potential implications for milk contamination and workers' exposure in a One Health approach. *J. Sci. Food Agric.*, *100*(3), 1118–1123.

Wang, G., Li, E., Gallo, A., Perrone, G., Varga, E., Ma, J., … and Xing, F. (2023). Impact of environmental factors on ochratoxin A: From natural occurrence to control strategy. *Environ. Pollut.*, *317*, 120767.

Wei, W., Liu, C., Ke, P., Chen, X., Zhou, T., Xu, J. and Zhou, Y. (2021). Toxicological and physiological effects of successive exposure to ochratoxin A at food regulatory limits. *Food Chem. Toxicol.*, *151*, 112128.

Wheaton, W.W., Weinberg, S.E., Hamanaka, R.B., Soberanes, S., Sullivan, L.B., Anso, E., … and Chandel, N.S. (2014). Metformin inhibits mitochondrial complex I of cancer cells to reduce tumorigenesis. *Elife*, *3*, e02242.

Xu, X., Pang, M., Liu, J., Wang, Y., Wu, X., Huang, K. and Liang, Z. (2021). Genome mining reveals the genes of carboxypeptidase for OTA-detoxification in Bacillus subtilis CW14. *Int. J. Biol. Macromol.*, *186*, 800–810.

Yang, Q., Shi, L., Huang, K. and Xu, W. (2014). Protective effect of N-acetylcysteine against DNA damage and S-phase arrest induced by ochratoxin A in human embryonic kidney cells (HEK-293). *Food Chem. Toxicol.*, *70*, 40–47.

Yang, S., Zhang, H., De Saeger, S., De Boevre, M., Sun, F., Zhang, S., … and Wang, Z. (2015). *In vitro* and *in vivo* metabolism of ochratoxin A: A comparative study using ultra-performance liquid chromatography-quadrupole/time-of-flight hybrid mass spectrometry. *Anal. Bioanal. Chem.*, *407*(13), 3579–3589.

CHAPTER 15

Farmer-led Initiatives and Participatory Approaches for Sustainable Crop Management

Devina Seram[1*], Haobijam James Watt[2], Habib Ali[3] and Megha N Parajulee[4]

1. Introduction

Farmers-led initiatives and participatory approach is a transformative move towards agricultural innovation. This shift places farmers at the center of the research process, which ultimately results in greater innovation. Farmers are empowered to take active and effective leadership roles in identifying challenges, co-developing solutions, and sharing knowledge. This contrasts with the traditional role of farmers, who are passive recipients of technologies developed by third parties such as scientists and researchers (Biggs, 2017). This approach places a high value on the profound and contextual understanding that farmers have regarding the local environments, crops, and farming systems that they are familiar with. Additionally, it seeks to integrate the understanding with the knowledge that researchers and development partners possess (PFO 2023). Participatory action research that is led by farmers recognizes the value of the involvement of farmer organizations in the governance of research, the formation of long-term partnerships, and the co-construction of traditional or innovative solutions, with a particular focus on the local context and the requirements of community development and upliftment of socio economic status. Farmer-centered agricultural research partnerships have the following characteristics:

i. The knowledge and experience of farmers are valued.
ii. Farmers are actively involved in the process of making decisions and participating in research activities, including the collection of data.
iii. It is recommended that farmers and researchers exchange their areas of expertise.
iv. Sustainable practices are encouraged and supported.
v. Farmers reap the benefits of the findings of research.
vi. The policies are inclusive and well-grounded in evidence.
vii. There is an emphasis placed on long-term commitment and scaling up.

[1] Department of Entomology, School of Agriculture, Lovely Professional University, Phagwara 144411, Punjab, India.
[2] Department of Agricultural Economics and Extension, School of Agriculture, Lovely Professional University, Phagwara 144411, Punjab, India.
[3] Khwaja Fareed University of Engineering and Information Technology, Rahim Yar Khan, Pakistan.
[4] Texas A&M University, Department of Entomology, Texas A&M AgriLife Research and Extension Center, Lubbock, Texas, USA.
* Corresponding author: devnah@gmail.com; devina.25125@lpu.co.in

DOI: 10.1201/9781003614869-15

It is possible for research led by farmers to improve rural livelihoods. The results of impact assessments indicate that there will be an increase in food and nutrition security as a result of improved production and increased crop diversity (Ensor & deBruin, 2022). Research conducted by farmers typically resulted in increased yields and household incomes in comparison to farming methods that were used in the past. Additionally, it enabled farmers to build up savings and make investments in assets. The majority of the research focused on reducing the number of chemical inputs that were used, which resulted in a positive impact on the environment (Waters-Bayer et al., 2015). At its core, farmer-led research is a participatory process that ensures the inclusion of all stakeholders throughout the entirety of the research cycle. These stakeholders include farmers, researchers, farmer organizations, and development partners. Continuous collaboration and mutual learning are essential to the success of the process, which encompasses everything from identifying baseline situations and establishing research agendas to designing, implementing, and evaluating effective solutions. Not only does it seek to generate innovations that are relevant and locally adapted, but it also seeks to build the capacities of all those involved, with the ultimate goal of maximizing the long-term impact of the things that are discovered through research (FundsforNGOs, nd). This collaborative model recognizes that both traditional knowledge and scientific inquiry have important roles to play in addressing contemporary agricultural challenges, and it encourages a complementarity between the two types of knowledge. Specifically, it encourages farmers to feel a sense of ownership over their land, which in turn motivates them to experiment, adapt, and share practices that have proven to be successful within their communities. The need for agricultural solutions that are resilient and driven by the community has become increasingly urgent as global concerns such as climate change, food insecurity, and the loss of biodiversity continue to emerge and become more severe. In order to achieve sustainable agriculture and to strengthen the resilience of smallholder farming communities all over the world, farmer-led research is becoming an increasingly important pathway (PFO, 2023). Figure 1 illustrates a simple step-by-step procedure for initiating solutions for an issue.

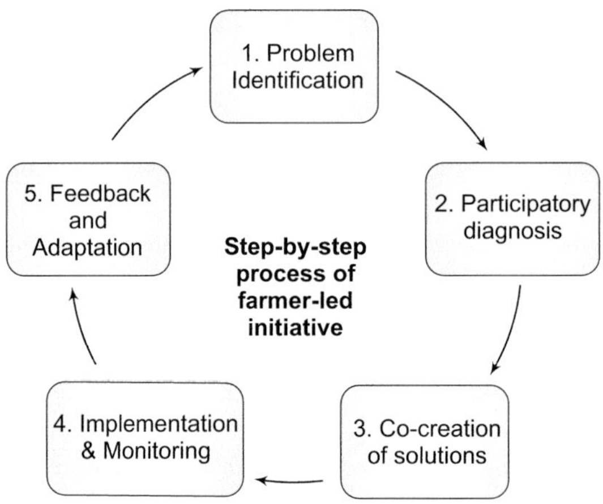

Fig. 1 A step-by-step process of launching a farmer-led initiative.

Traditional research and extension, which adheres to reductionist component research approaches, has been shown to be insufficient at times in order to address site-specific, complex, and value-laden challenges that are associated with environmentally sound and sustainable production. It has been reported that participating farmers in research projects results in the generation of locally relevant solutions and the capacity to adapt to changing conditions, both of which are essential for reducing the environmental damage caused by agriculture (Grabowski et al., 2018). Approaches to research that involve farmer participation are still relatively uncommon, in many parts because of the perceived difficulties and pitfalls that are typically associated with traditional agronomic research.

In particular, there are questions concerning who should be asked to participate, during which stages of the research and extension process, and in what manner (Lantz et al., 2006). These questions are among the most important ones. When compared to more conventional approaches, Farmer Participatory projects (FPR) typically only reach a relatively small number of farmers (Farrington & Martin, 1988). This is because FPR projects strive to be relevant to the local community. As an additional point of interest, FPR has been criticized for being a ' soft' science, appearing to be ad hoc, or lacking in rigor (Snapp & Heong, 2003). The necessity of documenting and systematizing FPR in order to provide responses to these questions and criticisms is something that we agree with (Carberry, 2001).

2. Importance of FIPA and a Shift from Conventional Agricultural Research

Farmer-led initiatives and participatory methodologies signify an important shift from traditional agricultural research, highlighting the essential role of farmers in formulating sustainable and context-specific agricultural solutions. This approach acknowledges that farmers have significant local knowledge, practical experience, and a profound comprehension of their unique environmental and socio economic contexts (Ensor & deBruin, 2022). By engaging farmers directly in the research process, from problem identification to solution implementation, these projects guarantee that agricultural innovations are more pertinent, adaptable, and likely to be implemented. This collaborative approach empowers farmers, improving their decision-making abilities and cultivating a sense of ownership about the study results (Grabowski et al., 2018). Moreover, farmer-led projects frequently result in more diversified and resilient agricultural systems, as they typically integrate traditional practices and local biodiversity with contemporary scientific understanding. This transition to participatory methods enhances the efficacy of agricultural research and fosters environment-friendly and balanced rural development. Some of the reasons for this shift include:

i. **Insufficient Farmer Engagement:** Conventional research frequently fails to involve farmers meaningfully, who are the principal stakeholders and ultimate beneficiaries of agricultural innovations (Bellon, 2001). In the absence of their active participation, research findings may not correspond with farmers' requirements, thereby diminishing the likelihood of adopting the suggested solutions. Research devoid of farmers' perspectives and practical contributions may fail to engage them, leading to sluggish adoption of suggested practices. If farmers do not recognize the significance or advantages of new technologies or methods, they may be reluctant to adopt them.

ii. **Limited Consideration of Local Context:** Conventional research often emphasizes generalized solutions, neglecting the varied local contexts in which agriculture functions. Agricultural practices effective in one region may be unsuitable in another due to differences in climate, soil, culture, and other variables.

iii. **Sustainability Issues:** Conventional research may neglect sustainable agricultural practices, resulting in possible environmental degradation, depletion of natural resources, and adverse effects on ecosystems. Environmental challenges may be specific to particular locations.

iv. **Disregarding Traditional Knowledge:** Conventional research may neglect the invaluable traditional knowledge that farmers have amassed over generations. Neglecting this knowledge may result in lost opportunities to utilize established practices and innovations.

v. **Insufficient Policy Influence:** Conventional research, when insufficiently connected with policymakers and farmers, may exert minimal impact on policy development. Policies may fail to address the genuine challenges encountered by farmers or may be impractical for implementation in the field.

vi. **Increasing Complexity of Agricultural Issues:** Contemporary agricultural challenges, including climate change, food security, and sustainability, necessitate a multidisciplinary and comprehensive approach. Conventional research, typically concentrating on isolated facets, may find it challenging to effectively tackle these intricate issues.

vii. **Emphasis on Immediate Outcomes:** Conventional research may prioritize short-term results, frequently influenced by donor funding cycles and publication pressures. Long-term sustainable solutions may necessitate additional time and investment, potentially conflicting with conventional research priorities.

There exists an escalating demand for initiatives and participatory approaches as farmers are increasingly acknowledged as knowledge holders and pivotal decision-makers in agriculture, emphasizing inclusiveness, relevance, and sustainability. In response to these challenges, there is a need for an increasing transition towards farmer-centered, participatory, and interdisciplinary methodologies in agricultural research, designed to effectively address the requirements and complexities of modern agriculture (Table 1).

Table 1 Comparative analysis of agricultural research methodologies between conventional research and farmer-led initiatives.

Aspect	*Traditional Agricultural Research*	*Farmer-led Initiatives and Participatory Approach*
Research Approach	Researcher-driven and Controlled	Farmer-driven and participatory
Decision-making	Researchers and policymakers decide	Collaborations among farmers, researchers, extension specialists
Farmers Involvement	Passive subjects of study	Active co-researchers and participants and disseminators of knowledge and innovation
Knowledge Contribution	Farmers' knowledge often overlooked	Farmers' traditional knowledge valued
Research Priorities	Set by researchers and Institutions	Identified based on local needs and issues
Data Collection and Analysis	Researchers collect and analyze data	Farmers involved in data collection
Research Outcomes	May not align with local realities	Context-specific and relevant outcomes
Sustainable Practices	May not prioritize sustainable methods	Emphasizes sustainable farming practices
Capacity Building	Limited focus on farmers' skills	Farmers are empowered through training. Farmers' capacity (confidence, knowledge, and skills) is strengthened to address challenges and manage experimentation and innovation.
Policy Influence	Limited farmer involvement in policymaking	Farmers actively contribute to policies ensuring policies address actual needs and issues
Impact and Adoption	Adoption rates can be low	High potential for widespread adoption
Long-Term Engagement	Limited ongoing engagement with farmers	Encourages long-term collaborations between farmers and other stakeholders
Equity and inclusiveness	May not address local disparities	Focuses on fair and equitable partnerships

Source: (PFO, 2023).

3. Importance of Fostering Research Initiatives and Approaches Led by Farmers

Participatory Action Research (PAR) is a methodology that involves farmers as active participants and co-researchers in the process of developing and implementing agricultural policies and practices, mainly led by farmers (Cornish et al., 2023). The following are five policy requirements that highlight the significance of this approach:

i. **Adopt Farmer-Centric Approaches:** These approaches should be prioritized by policymakers, and they should provide support for farmer-led initiatives and PAR projects. It is possible to foster solutions that are more sustainable and context-specific by recognizing farmers as key stakeholders in the agricultural sector and as holders of valuable knowledge.

ii. **Improve Agricultural Extension Services:** Governments should make investments in improving agricultural extension services to encourage participatory research and to make it easier for farmers to share their knowledge with one another (Bellon, 2001). In the context of farming communities, these services have the potential to facilitate the dissemination of valuable information and expertise. There is a win-win situation that occurs when farmers are involved in research, the generation of knowledge, and innovation (Ensor & deBruin, 2022). This helps to support the extension services provided by governments, which are frequently limited by funding and the challenges posed by the topography of many island countries.

iii. **Allocate Research Funding for Participatory Action Research:** Policymakers should allot a portion of the funding for agricultural research specifically to support action research projects that are led by farmers. The availability of specific funding guarantees that research efforts are driven by the needs of the local community and that farmers are actively involved in the decision-making process.

iv. **Encourage the formation of collaborative partnerships:** In order to advance participatory research, it is important to encourage collaborations between farmers, research institutions, non-governmental organizations (NGOs), and other relevant stakeholders. In order to increase the effectiveness of PAR initiatives, policymakers should enable the creation of platforms that promote communication, the exchange of information, and collaboration.

v. **Encourage Policy Co-creation with Farmers:** Policymakers have a responsibility to actively involve farmers in the process of formulating and evaluating agricultural policies. By involving farmers in the process of policy co-creation, it is possible to guarantee that policies are based on practical realities, that they are pertinent to the local context, and that they contribute to the sustainable development of the agricultural sector (PFO, 2023). All things considered, these messages highlight the importance of adopting a more inclusive and participatory approach in agricultural research and policymaking. This approach would ensure that the perspectives and experiences of farmers are valued, and that farmers themselves are active agents of change in the process of shaping the future of agriculture.

4. Farmer-led Initiatives and Approaches for Integrated Crop Protection and Farm Sustainability

Increasing pressures are being placed on the agricultural sectors as a result of social and economic challenges, as well as climate change and land degradation. To address these issues while also ensuring food security and environmental health, sustainable agriculture is an essential component. In order to promote sustainable land management practices, farmer-led initiatives and approaches are absolutely necessary (Figure 2). The implementation of sustainable practices and the contribution to the larger goals of sustainable agriculture are all made possible by these initiatives, which empower farmers of their agricultural enterprises (SAN, 2024).

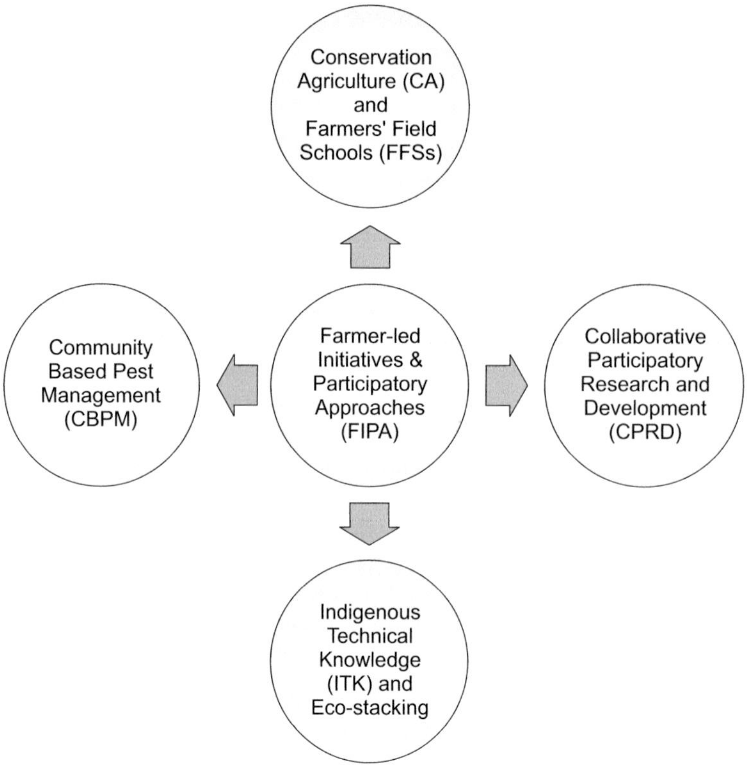

Fig. 2 Farmer-led initiatives and participatory approaches to integrate crop protection tactics, agriculture production and conservation practices, and research, outreach and knowledge sharing for farmscape sustainability.

4.1 Crop Protection Tactics and Crop Health

A multidimensional approach is paramount in good crop health and pest management by farmers. It is essential to adopt integrated pest management (IPM) by employing biologically integrated cultural and chemical strategies to lower the general equilibrium of pest populations and reduce the harmful effects of pests below economic threshold, while reducing pesticide load in the environment (Rezaei et al., 2020). Frequent field scouting or monitoring enables farmers to detect possible problems before they occur and employ appropriate precautionary/remedial measures. Insect pest and disease cycles can be overcome by crop rotation and habitat diversification that enhance crop health. Appropriate nutrient management based on soil testing ensures optimum plant growth and pest resistance. Disease susceptibility is minimized via effective irrigation and water management practices (Grabowski et al., 2018). Also, planting resistant crop cultivars at optimum planting date and recommended plant densities will maximize crop's ability to overcome pest-induced stresses. When farmers are continuously informed and engaged in matters involving mutual knowledge sharing, either at workshops or through the community programs, they will be better able to adopt the recommended methods to suit the changing conditions of the environment. On the other hand, agricultural cooperatives are powerful tools that can be used to promote environmentally responsible land management. Farmers can pool their resources, share their knowledge, and implement sustainable practices on a larger scale through the establishment of cooperative agreements (Bellon, 2001). Farmers can also gain access to better markets, obtain fair prices for their produce, and invest in environmentally friendly technologies when they work together to promote cooperative action. Cooperatives, which are formed through collaborative efforts, strengthen the resilience of farming communities and contribute to the development of sustainable agriculture.

4.2 Strategies for Implementation of Agroforestry

Agroforestry, which refers to the practice of incorporating trees and shrubs into agricultural landscapes, is a farmer-led initiative that offers a variety of advantages including habitat diversity and biodiversity conservation. In addition to enhancing carbon sequestration, this practice also improves soil health. Farmers who implement agroforestry practices could build farming systems that are more resistant to natural disasters, lessen their reliance on chemical inputs, and improve their standard of living. Agroforestry is an essential component of sustainable agriculture because it contributes to the preservation of ecological balance while simultaneously fostering agricultural productivity. The application of agroforestry captures the local knowledge and combines that wisdom with science as a way of initiating and conducting farmer-led integrated pest management, sustainable crop health management application, and participatory action. A well-known farmer-driven example with agroforestry projects is the Farmer Managed Natural Regeneration (FMNR) program in Niger (Suuk et al., 2015). This method incorporates farmers who choose and conserve the naturally growing tree seedlings on their farmlands, and this translates to better soil fertility, higher production of crops, and higher biodiversity. FMNR has increasingly become participatory and there has been large- scale adoption, whereby in Niger alone, more than 5 million hectares of land have been brought back to crop agriculture. In another instance, the Quesungual Slash and Mulch Agroforestry System (QSMAS) in Central America demonstrates how participatory methods can be effective in the management of crop health in a sustainable environment (Pauli et al., 2011). Created by small farmers in Honduras, QSMAS integrates knowledge used in the past with innovative concepts in agroforestry to create better soil moisture conservation, erosion mitigation, and increase crop yield. This innovative initiative by farmers has since grown to other countries like Nicaragua and El Salvador demonstrating that it is adaptable and effective. Moreover, the Participatory Domestication of Indigenous Fruit Trees program (PDIFTP) in Cameroon is the clear case of how farmers' input in agroforestry is used to enhance sustainable management of crop health (Leakey, 2017). The same program involves local communities in the process of choosing, cloning, and growing indigenous fruit trees together with their field crops. By including these trees in their farming systems, farmers are able to increase their source of income, provide better nutrition for the household and thus increase the overall health of the agro-ecosystem. The above examples help demonstrate the principles of more sustainable and effective crop health management practices when agroforestry is established at the initiative of farmers based on a participatory principle. While strengthening local resources and enhancing local expertise, placing leading farmers at the center of agroforestry initiative enables a stronger agroforestry system that is resilient, adaptable, and replicated by more farmers, leading to better livelihoods and a sustainable environment.

4.3 Agricultural Methods and Natural Resource Conservation

Initiatives that are led by farmers help to spread the practices of conservation agriculture throughout the communities in which they are implemented, which are essential for sustainability as they improve soil fertility, prevent soil erosion, and increase water retention (SAN, 2024). Farmers significantly contribute to the long-term health of their land and sustainability of their agricultural practices when they adopt conservation agriculture. Conservative farming practices in preserving natural resources are invaluable in assisting farmer-inspired efforts and participatory cropping health management. Agroecology, conservation tillage, IPM and other similar techniques allow farmers to operate on principles of ecology, thus ensuring increased production (and hence profits) through maintaining the ecology (Vikas & Ranjan, 2024). For example, Brazilian farmers use agroecology as an ecologically intensive practice that involves intercropping and crop rotation, which increase biodiversity and soil health (James et al., 2024). Likewise, in India, the Farmer Field Schools system is used as a way of promoting IPM, where farmers engage in identifying pests and beneficial insects, providing knowledgeable judgments in managing pests and reducing the use of chemicals (Mancini,

2006). These are participatory methods which encourage knowledge transfer and adapting to local circumstances, so that practices will remain effective and sustainable. In addition, practices such as organic farming, water conservation methods, and cover cropping exemplify the positive effects of resource-saving practices (Sharma 2023). Cooperative programs help smallholder farmers in Europe convert to organic growing, by offering training and resources, and boosting crop health and farming sustainability (Anderson et al., 2019). Successful implementation and wider adoption of the drip irrigation method in Israel was a direct result of the participatory programs that enabled farmers to learn and use these technologies. Also, cover crops such as clover and rye are being adopted in the Midwest region of the United States and local farmer networks are exchanging successful experiences to keep track of the positive impacts on soil health and productivity (Barnes et al., 2022). In general, such joint activities not only help to save natural resources but also increase resilience and sustainability of agricultural systems.

4.4 Collaborative Participatory Research and Development (CPRD)

Participatory research that is led by farmers encourages active participation from farmers in the process of developing and evaluating new agricultural practices. This strategy ensures that the innovations are applicable, suitable for the region, and adapted to meet the local needs of the farming community. Farmers are more likely to adopt and improve sustainable practices when they are given the opportunity to participate in research that fosters a sense of ownership. This method is essential for ensuring that new technologies and practices are effectively integrated into farming systems and for advancing sustainable agriculture (SAN, 2024). In addition, collaborative participatory research and development (CPRD) are essential in assisting farmer-led programs and participatory methods of maintaining sustainable crop health (Vaughn & Jacquez, 2020). CPRD involves local farmers in the research process, and therefore, knowledge created will be pertinent and applicable to the local situations. An example is the use of the Participatory Action Research (PAR) method in the Philippines aimed at getting the farmers to develop integrated pest management strategies (Pawera et al., 2024). In research and development, farmers partner with researchers to solve problems related to pests, and they carry out tests to arrive at more effective and locally appropriate pest control practices. Such collaborations do not only empower farmers but provide a sense of ownership towards the results of the research, which increases the rate of adoption of technology. CPRD will also enable the exchange of knowledge and experience among farmers since sharing of knowledge and experience is part and parcel of sharing of sustainable practices. In Kenya, the Farmer Research Networks (FRNs) successfully linked farmers with researchers and extension workers (Richardson et al., 2021). Because of such networks, farmers are able to collectively test several crop management procedures, including agroforestry practices and organic farming techniques. Farmers can share their experience and findings to get a better understanding of how to approach their crops' health enhancement to increase yields and make their environment sustainable. The Texas A&M University Cotton Entomology and Cropping Systems Program in the United States has successfully developed and implemented a network of partnerships amongst cotton farmers, agriculture industry partners, agriculture consultants, County Extension agents, and Texas A&M Research and Extension programs to cultivate/enhance synergy in multidisciplinary agricultural management practices including pest management, nutrient management, crop health improvement, and optimizing farm profitability and sustainability (Figure 3). Here, stakeholders, particularly producer clientele, rely heavily on participating partners for information on latest agriculture technology and research information. Producer organizations also support university research activities through grant funding and champion for additional funding from regional and national funding sources. Multidisciplinary research projects are deployed in farmers' fields with direct involvement of farmers in conducting such research, thereby having a true research partnership between university researchers and participating farmers. These research farms are used as demonstration plots and the outcome of the research is readily adopted by area farmers.

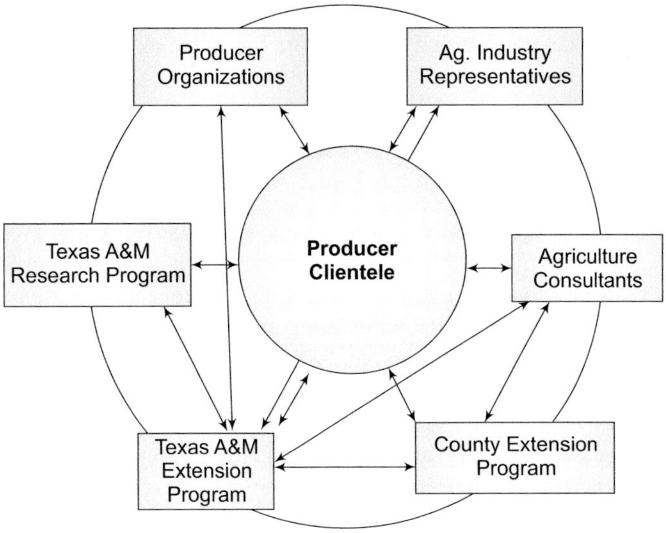

Fig. 3 Texas A&M University Cotton Entomology and Cropping System Program network for integrated pest and crop management.

4.5 *Indigenous Traditional Knowledge (ITK) in Crop Protection and Its Integration with Scientific Approaches*

Indigenous knowledge in crop protection includes traditional methods and wisdoms that have been passed down through generations (Khumalo et al., 2025). This information can prove useful in managing arthropod pests and diseases and is environmentally conducive. This expertise generally includes employing plant products that have pesticidal properties and are locally available and intercropping of such plants to utilize their natural ability to suppress pest populations and/or enhance natural enemies. For instance, farmers in some regions of Africa and India utilize neem tree (*Azadirachta indica*) extracts as a natural insecticide as they suppress several soft-bodied insect species and fungi (Dhakad et al., 2025). Combining this traditional knowledge with scientific methods can help develop better ways to protect crops that are more environmentally conducive. Scientists can evaluate the active ingredients in conventional plant-based insecticides to find the best ways to use them in ecologically intensive pest management programs. The use of marigolds as companion plants in vegetable gardens in India has been shown to suppress pests, especially nematodes. The traditional "Three Sisters" planting method of Native Americans, which uses maize, beans, and squash, has also shown to make soil more fertile and suppress crop pests (Kapayou et al., 2023). Researchers can create culturally appropriate, economically viable, and ecologically sustainable holistic integrated pest management systems by combining traditional knowledge with modern scientific methods. This will reduce the need for unilateral reliance on synthetic pesticides and enhance production sustainability. Moreover, Indigenous Traditional Knowledge (ITK) in crop protection has a pivotal role in bolstering measures such as farmer-led initiatives and participatory efforts toward sustainable crop health management by combining traditional practices passed down over the centuries and contemporary scientific approaches (Prinila et al., 2025). ITK constitutes a rich body of knowledge concerning local ecosystems, pest control, and crop types that have been established through generations. One example is that in India, farmers in some parts of the country apply different methods of traditional farming like planting certain kinds of companion plants to provide resistance to pests and increase the quality of the soil. The efforts of the International Crops Research Institute for the Semi-Arid Tropics (ICRISAT, India) which partners with the local farmers in order to integrate ITK with scientific research has resulted in better pest control measures that are both culturally acceptable and environmentally sound (ICRISAT, 2024). This partnership not only strengthens the farmers by acknowledging their wisdom but also makes the farming practice robust, flexible, and sustainable.

4.6 Farmers' Field Schools (FFSs) and Community-based Crop Health Management Programs

Farmers Field Schools (FFS) are practical, interactive learning bases, which provide farmers with the capacity to equally develop themselves as specialists in their areas of operation (Braun et al., 2000). Consequently, these schools strengthen the farmer-led approach, participatory methods of managing sustainable crop health through stimulating experiential learning, local knowledge, and encouragement of collaborative problem solving. FFS mostly engage communities of farmers, who regularly convene over a growing period to monitor, evaluate, and test different practices of crop management. This strategy will enable farmers to become active participants in the learning process and experience exchange and make relevant decisions regarding farming. Through farmer-to-farmer learning and experimentation, FFS implement ownership and promote adopting sustainable crop protection, and crop health management practices.

An important feature of FFS is the emphasis on analysis of agroecosystem wherein the farmers are taught precision in observation and comprehension of the intricate relationship between crops, pests, and natural enemies in the fields. This fact will help them make informed choices regarding the management of pests, lessening the desire to use chemical pesticides, and instead using more eco-friendly methods. The successful implementations of FFSs can be observed all over the world. For instance, in Indonesia, FFS played the role of minimizing the amount of pesticide used in rice production through the appreciation of use of IPM (Thorburn, 2015). They also taught farmers to recognize useful insects and natural ways to control pests and as such an overall large reduction in the usage of pesticides as well as improved crop health. In East Africa, the FFS was applied to address the problem of banana wilt disease (Tinzaara et al., 2019). In Uganda and Tanzania, farmers were enrolled in schools where they learned the identification, prevention, and management of diseases (Matowo et al., 2022). This participatory approach improved the use of locally adapted solutions and as a result, banana production was significantly improved. In Latin America, FFS have been used to enhance sustainability in cultivation of coffee. Through such schools, farmers in Colombia and Honduras have acquired skills on management of shade, soil conservation, and IPM (Gómez et al., 2024). This has increased the quality of coffee, addition of biodiversity on farms, and resilience towards climate change. FFS has also been triumphant in scaling the gender imbalance in the agricultural sector. The FFS targeting women have also empowered women farmers to participate as well as assume leadership positions in the management and decision-making of crops in Bangladesh and Nepal. Such schools have equipped women with knowledge and skills of practices in sustainable agriculture and as a result, there is improved household food security and economic empowerment.

Alternatively, the participatory aspect of FFS is not only in crops' management, but rather market access and value chain development. In West Africa, FFS have included modules on post-harvesting, quality control, and drafting crop marketing strategies such as cocoa and cashew. Such a holistic view has allowed farmers to not just enhance their production but also it has allowed them to be competitive in the market. FFS have an impact in promoting farmer-led and participatory strategy to establish locally appropriate, sustainable management of crop health (Khumalo et al., 2025). Through this approach, farmers are identified as proactive innovators as well as decision-makers, and not consumers of information. Consequently, FFS have turned out to be a powerful mechanism of promoting the sustainability of agriculture, enhancing food security, as well as empowering the rural population across the globe.

Community-based pest management (CBPM) is an approach by which a community works together with local residents, organizations, and authorities in controlling and preventing pest infestations (Aulia et al., 2024). This is a strategy that concentrates on education, prevention, and sustainable practices instead of focusing only on chemical treatment. And as an example, residents within a city may cooperate to remove potential sources of standing water that favor mosquitoes, e.g., emptied birdbaths on a regular basis and proper rain gutters. This means that in an agricultural environment, farmers can apply crop rotation and companion planting methods to naturally repel pests and share resources as well as ideas with surrounding farms. The second example involves the introduction of pheromone traps into community gardens as an alternative to non-invasive methods of managing the insects by tracking. CPBM can be used to help solve pest problems through the development of a feeling of community responsibility and the adoption of ecologically sound procedures, which can also assist in ensuring that the ecological and social aspects of the environment remain healthy. CBPM programs by farmers in India have become a powerful idea which is being used to solve agriculture problems (Chowdhury et al., 2023). Such programs allow local farmers to become their own masters of pest management strategies, relying on their knowledge of the region, and using it in conjunction with modern scientific methods. The most striking case study is that of the Community-Managed Sustainable Agriculture (CMSA) program in Andhra Pradesh that has managed to introduce non-pesticide management in thousands of villages (Kumar et al., 2009). This program has led farmers to adopt the use of pheromone traps, bio-pesticides, and crop diversification among others which has proved to create desirable reduction of pest infestation and improved crop production. Another such case is the IPM program in Maharashtra, which has been successful due to the incorporation of the farmer field school which has been relevant in transferring knowledge and expertise (FAO & DPPQS, 2021). In these schools, farmers undergo practical training so that they are updated on pests, their life cycles, and various types of control measures. The program also has significantly reduced crop chemical use and overall farm production costs. Also, the establishment of farmer groups and cooperatives has promoted knowledge-sharing and improved networking for increased effectiveness of pest management strategies at the community level.

4.7 Eco-stacking

Eco-stacking is a novel concept and strategy meant to maximize the advantages of ecosystem service providers in cropping systems to facilitate achievement of the objective of long-term sustainable agriculture and food production. By eco-stacking, we combine synergistically the useful services of functional biodiversity at all levels and types. It is an inclusive process with different ecosystem service providers linked together to enhance the overall cropping system enterprise (Hokkanen & Menzler-Hokkanen, 2024a). It is one that moves beyond the customary IPM practices and tries to exploit all of the functional biodiversity of a system. Eco-stacking has been concentrated primarily on maximizing biological control and pollination by beneficial arthropods and advocates the use of other invertebrates (e.g., earthworms) and beneficial vertebrates (e.g., bats, birds, and small mammals). Not the least of ecosystem services afforded by microbes are pest, disease and weed control, either directly (as part of so-called suppressive soils), or indirectly (as plant colonizers like endophytes or epiphytic microbial communities). Other nature-provided services should also be maximized using the eco-stacking approach whether it is maintaining their soil health or nutrient cycling (Hokkanen & Menzler-Hokkanen, 2024b). Eco-stacking greatly contributes to the farmer-led initiative and participatory mode of sustainable crop health management. The combination of numerous practical methods, including intercropping, relay cropping, agroforestry, and cover crops, will allow farmers to enhance biodiversity, optimize resource utilization, and improve ecosystem health. As another example, farmers in Brazil using eco-stacking were able to integrate coffee cultivation with shade trees, which is not only highly biodiverse, but also the solution to supporting beneficial insects that naturally stem off pests. Such a participatory process gives farmers the opportunity to exchange their experience and customize eco-stacking methodology to their circumstances, resulting in more crop resilience and vitality. Moreover, such programs as the "Agroecology in Action" in East Africa can promote and stimulate the design and application of eco-stacking practices in farming through community cooperation by sharing knowledge. Eco-stacking enables farmers to use the synergies among various modes of agricultural practices and network of stakeholder partners (Figure 3) to sustainably manage their crops and thereby increase services provided by the overall ecosystem (Wang et al., 2024).

5. Selected Case Studies

5.1 Africa RISING: The Malawian FPR Approach

Agriculture in Malawi is dependent on soils that are highly weathered and deficient in nutrients; this is a characteristic of farms throughout East and Southern Africa (Tittonell et al., 2005). The availability of economic resources and access to fertilizers is extremely restricted, particularly among farmers who are sole proprietors. This necessitates the implementation of an integrated strategy that makes use of both organic resources and traditional inorganic fertilizers (Kanyama-Phiri et al., 1998). The heterogeneity of soil parent material, household topology, and soil amendments can vary at the scale of 1, 10, or 100 meters, efforts to understand soil fertility and production potential have frequently failed (Tittonell et al., 2005). This is because soil parents can vary in composition. In addition to the fact that natural resources are not the same, the objectives of farmers and their orientation toward the agriculture market are also highly variable (Grabowski et al., 2018).

The Feed the Future initiative in Malawi, which is sponsored by the United States Agency for International Development and is called "Africa RISING", has made it easier to develop farmer participatory research (FPR) strategies in order to address the site-specific issues of environmental and food insecurity. The Africa RISING team seeks to fine-tune recommendations and understand the interaction of climate, soil type, and market goals to support the adoption of sustainable agriculture (SA) technologies (Snapp et al., 2018). This is in addition to on-farm monitoring and crop health management, which are used to understand yield gaps, profitability, and pests and disease incidences. Three administrative units in Malawi, which are referred to as Extension Planning Areas (EPAs), are currently conducting research that involves farmer participation. A stratified random approach was used to select the research sites to ensure that they were representative of three different agroecological zones, ranging from marginal to mesic. In light of this, it is reasonable to assume that the findings consider both low and high potential areas of smallholder agriculture in Central Malawi and beyond. Following is the list of the locations:

i. The Linthipe EPA experiences relatively high rainfall and high elevation (1000–1200 m asl) with high agricultural potential.
ii. The Kandeu EPA experiences medium rainfall and medium elevation (800–900 m asl) with intermediate agricultural potential.
iii. The Golomoti EPA experiences low effective rainfall because of high evapotranspiration and low elevation (550–650 m asl) as well as high temperature and marginal agricultural potential. The low-lying, flat terrain in this region is a contributing factor in the occurrence of flash flooding, and it is also characterized by the uneven distribution of rainfall.

At each of the three EPAs, the mother and baby trial (MBT) strategy was put into action in the year 2012. For evaluating a comprehensive set of SA technologies over a more extended period of time, a dozen mother trials were established and approximately 30 farmers were involved in each mother trial site. These farmers were provided with training and seed access to facilitate the testing of various technologies on their farms as part of the baby trial, which were later georeferenced. One of the technologies included organic growing of legume crops such as bean (*Phaseolus vulgaris*), cowpea (*Vigna unguiculata*), peanut (*Arachis hypogaea*), pigeon pea (*Cajanus cajan*), and soybean (*Glycine max*) (Snapp et al., 2018). Additionally, it included diversification of maize (*Zea mays*)-based systems through integrated nutrient management. This mother trial included the application of judicious doses of nitrogen and phosphorus fertilizer at the recommended rate of 69 kg of nitrogen per hectare and 9 kg of phosphorus per hectare. However, the rates were reduced based on the strict accounting for organic nutrient sources in the other treatments. In the case of baby trials, there was no fertilizer provided, which enabled the farmers to experiment with their own amendments. The MBTs, in general, served as a platform for collaborative learning regarding the technologies that performed exceptionally well. Over the course of five years, the MTBs engaged >1,000 farmers. Improvements such as plant- spacing arrangements, fertilizer combinations, and local management practices (such as ratooning) were made possible as a result of the research team's collaboration with Malawian extension staff (Rogé et al., 2016). The inference zone of the research was expanded through the use of additional tools such as crop modeling and socioeconomic surveys that were conducted as a panel series (repeated visits to the same farm households) (Wang et al., 2019). The preferences of farmers regarding technologies were documented using short semi-quantitative surveys that were carried out during field days and other demonstration events (Snapp et al., 2002). These surveys were conducted in addition to surveys of farmers who participated and farmers who did not. The doubled-up legume rotation was a novel South African application of technology that proved to be of interest to a great number of farmers. Specifically, this involved the cultivation of an early maturing grain legume (such as peanut or soybean) as an understory crop alongside a late maturing pigeon pea, which was then sequenced in rotation with maize. Other SA technologies included growing maize as an intercrop with a locally selected improved pigeon pea variety and growing maize in a rotation with peanut at a high population density and judicious fertilizer doses (Chimonyo et al., 2019). Both methods were used to increase the yield of maize and reduce incidence of pests on crops.

5.2 Farmer-led Research Program (FLRP): The Canadian Approach

A case study of Farmer-Led Research conducted in Ontario, Canada analyzes how the Ecological Farmers Association of Ontario (EFAO) has become an agent of change by supporting the development of farmer-led research programs (Nelson et al., 2023). Their study describes how farmers can embrace an active role in the research process and use their experience as a gateway to address particular agricultural issues in their local environments. In partnerships with others, farmers collaborate with researchers and extension services to construct and carry out research projects that are applicable to their farm operations, including sustainable farming practices and environmental protection. Some of the successful programs noted from this study include trials on cover cropping and organic pest control, which besides making crops healthier also led to enhancement and conservation of biodiversity and soil health. The EFAO helps initiate a participatory model of agricultural research as it identifies and institutionalizes farmer knowledge as an asset to the research process, supporting more sustainable and resilient systems of agriculture. The study highlights the need to incorporate field knowledge into the official farming methods, which will eventually lead to the emergence of new practices which will help both the environment and farmers in Ontario.

5.3 Farmer-led Innovations (FLIs): The Indian Approach

A case study was conducted to explore the unique features of several farmer-led innovations (FLIs) in maize developed across India (Kar et al., 2015). One of the innovations encompassed maize grain protection against insect pests during storage where a small farmer of Konachatra village in the Coochbehar District of West Bengal and an active member of a community-based pest management scheme, Mr. Rabindranath Burman, integrated the local preservation practices developed by local farmers in West Bengal. He also grows rice, maize, and tobacco with a relatively small landholding of 1.2 hectare, and believes in sustainable farming. Such an innovative approach to preservation would solve the general problem of storing cereal or food grains, which are exposed to insect pests and diseases after harvesting. Farmers can hang them in well-aerated places by leaving the grains in the cob and band a collection of 4–5 cobs together, which is more than enough to preserve the crop over a year, and some members of the community have cited even longer storage life. It is a cost-effective method that maintains crop quality in addition to letting the farmers have the advantage of retaining seeds till the next planting season and in case of scarcity may even sell their produce at prices higher than average. Based on the traditional preservation of onions and garlic, this community-based project becomes a clear indicator of a modest yet efficient solution concerning maize storage, evidencing merely how local expertise can help to improve crop maintenance and resistance to pests.

6. Outcomes and Impacts of Farmer-led Initiatives and Research Approaches

Research that is led by farmers results in a variety of innovations and adaptations that are suitable for the local environment (Rizzo et al., 2024). Most of the examples of such farmer-led initiatives were associated with land reclamation or improvement, the conservation of soil and water, the breeding of plants and the selection of varieties, crop husbandry, and the protection of crops from pests and diseases. These productive initiatives have improved production practices and improved farm profitability and agricultural sustainability across many regions of the world.

6.1 Dissemination of Research Findings and Methods Developed by Farmers

The results of the research led by farmers are primarily documented by the Civil Society Organizations (CSOs) that provided support. These CSOs also provide guidance to the farmers regarding data collection and documentation of their trials. However, this activity is neither a strength nor a focus of the CSOs, hence the data are frequently incomplete and rarely systematically analyzed. Despite this, the CSOs and farmer-researchers disseminate the results and information to other farmers, stakeholders and a wider audience through workshops, newsletters, and magazines (Wettasinha et al., 2014). The most common methods of sharing the information were directly from one farmer to another through informal networks and through opportunities that were purposefully created, such as innovation fairs, where farmer researchers shared their knowledge and experiences. These methods allowed for the rapid dissemination of innovations that required no or very few inputs from the outside world and brought about obvious benefits. With the intention of incorporating farmer-led research methodologies into government institutions, NGOs documented and disseminated the process as well as the lessons learned from utilizing the methodology. However, they did not include much information regarding the results of farmer-led research. Remarkably, external evaluators frequently failed to grasp the significance of the situation and instead focused on the dissemination of technologies rather than the dissemination of the approach (Waters-Bayer et al., 2015).

6.2 Impacts on Farmers' Livelihoods

It has been reported that farmer-led research has resulted in increased food and nutrition security (Rao & Swaminathan, 2017). This was achieved by increasing yields, making them more reliable and acceptable, improving storage, and increasing crop diversity in comparison to the practices that were used previously. Additionally, the increased agro-biodiversity contributed to a greater resilience to environmental risks, as well as to diseases and pests. When compared to conventional farming techniques that rely on external inputs, ecological farming techniques that are led by farmers typically resulted in higher household incomes, which was partly due to the lower cost of these locally devised techniques. Additionally, these techniques enabled farmers to increase savings and invest in economic assets. Such farming techniques that were developed locally also resulted in increased labour productivity. In contrast, other techniques required higher labor inputs, particularly in the beginning stages, such as the planting pits (zaï) (Carroll et al., 2025). These farmer-led research activities also resulted in reduced chemical inputs and had a positive impact on the environment. While several instances of stronger cohesion and joint action amongst households were observed within communities, the majority of the documented impacts were still at the household level. It was found that farmer-led experiments with newly introduced technologies were more likely to bring benefits to other farmers with medium and higher incomes. Alternatively, experiments based on endogenous innovation and the utilization of local resources were more likely to be relevant to those who were economically disadvantaged and for women. A strong correlation was observed between women being encouraged to become experimenters and their becoming more outspoken and active in the community, which extended beyond their involvement in farming activities.

6.3 Impacts on Local Capacity to Innovate

A description of both core and supportive capacity aspects that are required to strengthen the functionality of a system to innovate has been reported (Leeuwis et al., 2014). Core capabilities include the ability of individuals and communities to continuously identify and prioritize problems and opportunities in an environment that is constantly changing; the ability to take risks, experiment with social and technical options, and evaluate the trade-offs that result from these options; the ability to mobilize resources and form effective coalitions around promising options and visions for the future; and the ability to link with others in order to access, share, and process information and knowledge that is pertinent to the situation. One of the most significant areas of impact that was observed in the two case studies presented was the capacity of farmers to continue the process of innovation to address other and forthcoming challenges. The development of individual capabilities, including self-assurance, knowledge, and the ability to experiment and innovate, was a key feature that was mentioned. Peers in the farming industry and individuals from outside the agricultural research and development (ARD) recognized both male and female farmers as innovators. A great number of farmer-researchers went on to become skilled facilitators and continued to provide support to others. Farmers came together to work and learn in groups, initially for the purpose of experimentation and later for other activities, such as marketing, labor-sharing, savings and credit, and lobbying for farmers' rights. This resulted in the strengthening of the local organization's capacity within the communities. Farmers improved their ability to recognize and connect with relevant sources of information and partners in innovation, which enhanced their ability to innovate. In several instances, it was demonstrated that the establishment of social learning environments prompted innovation and, after the conclusion of the CSO intervention, the local capacity for innovation had been improved; however, there is very little documentation on how this occurred. After that, the processes of innovation at the local level continued.

6.4 Impacts on Formal and Informal ARD Institutions

Both formal and informal ARD institutions, such as government research and extension organizations at various levels, educational institutions, NGOs, community-based organizations (CBOs), and farmer organizations (FOs), underwent some changes because of the farmer-led initiatives and research approaches (Pound & Conroy, 2017). Formal ARD institutions were the primary route for institutionalization in six of the cases, while CBOs were the primary focus of the remaining five cases. Although there were some shifts in mentalities, skills, and knowledge among staff members, there were no cases that reported significant changes in the structures and working mechanisms of the organizations or in their budget allocations. Institutionalization through the formal sector was a slow process with limited success. NGOs, CBOs, FOs, and farmer movements have become more open and receptive to the incorporation of farmer-led participatory approaches.

6.5 Benefits and Limitations of Farmer-led Initiatives for Crop Protection

Initiatives developed through farmers' participation have various positive effects on farming communities and the food system as a whole. These are bottom-up strategies that create a sense of ownership among farmers in terms of innovation and problem-solving, which have higher adoption levels to new practices and technologies. The farmers are able to accept and uphold the changes over the period because, when developed and implemented, farmers played a part in coming up with the solutions. Also, farmers' community-level processes can tend to advance sustainable practices adapted to local conditions and ecosystems which lead to more environmentally friendly and financially profitable farming decisions. Such projects also help with greater resilience to pests and diseases because farmers can learn among themselves, conduct crop variety trials, and build on local expertise when it comes to managing pests. Farmer-led action can create a learning culture and increase the likelihood that everyone will learn constantly and enhance productivity through collective knowledge and experience to build the overall resilience of agricultural systems. For instance, the idea of sustainable crop protection through farmer-led initiatives against insects and diseases has gained popularity worldwide. Such methods frequently incorporate practices of IPM, including crop rotation, intercropping and natural predators. An example is the push-pull strategy involving planting of three crops, originated from Kenya (Pickett et al., 2014), which most of the farmers have adopted wherein beneficial insects and pests are attracted to the companion plants that subsequently avoid the main crop (Luttermoser et al., 2023). Some have resorted to the use of biological control in which predatory insects or even microorganisms are introduced as a way of controlling the population of pests. The local isolation of plant-derived insecticide solutions has also been created by farmers working together to share knowledge and resources, creating locally adapted insecticides such as neem-based pesticides, or the cultivation of disease resistant crops varieties. Also, farmer field schools and CBPM programmes have been found as an avenue to experiential learning whereby agriculturists have been able to carry out experiments and develop sustainable pest control methods specific to their locales.

Although farmed-led initiatives and approaches present several benefits, challenges, and limitations also exist in terms of crop protection. Scalability is a significant issue, where the methods applied in smaller pilot scales can be hard to implement in larger territories or diverse agricultural societies. Limited resources in terms of time, finances, and qualified personnel may therefore limit the scope and sustainability of participatory projects. The scattered educational skill sets of farmers, preservation of their locally adaptive technologies, and their willingness to absorb a new lifestyle can be set back to being applied dynamically. Furthermore, participatory projects often require long-term investment and follow-up that may be challenging to continue due to the changing priorities or funding schedule. The issues of engaging a wide range of stakeholders effectively might become even more problematic in terms of cultural and linguistic constraints. Despite these challenges, participatory techniques have remained a significant tool in developing locally accepted and farmer-accepted crop protection plans.

7. Conclusion

The increasing intricacy of global agricultural issues requires a transition from hierarchical, reductionist frameworks to collaborative, contextually relevant strategies. Farmer-led initiatives and the participatory approach have arisen as a credible and influential alternative that transforms the role of farmers from passive beneficiaries to empowered co-researchers and decision-makers. These approaches not only generate innovations that are locally pertinent and ecologically sustainable, but also fortify social capital, augment adaptive capacity, and enhance household-level outcomes, contributing to food security, income stability, and ecological resilience. A novel approach of eco-stacking has been included, which usually stems from the result of observations and experimentations done by farmers in their fields. Although scientific studies are required to confirm and promote development of this technique, its initiation and adaptation is quite frequently based on farmers' knowledge and experiences. Moreover, for all the farmer-led initiatives to achieve their maximum potential, continuous institutional support, comprehensive policy frameworks, and specific funding mechanisms are essential. Prioritizing capacity building, trust establishment, and local ownership is essential for the sustainability and efficacy of participatory methodologies. In the end, promoting farmer-led innovations will serve to achieve SA, and as a strategy for democratizing knowledge creation and enhancing rural development.

References

Anderson, C.R., Maughan, C. and Pimbert, M.P. (2019). Transformative agroecology learning in Europe: Building consciousness, skills, and collective capacity for food sovereignty. *Agriculture and Human Values, 36*, 531–547. https://doi.org/10.1007/s10460-018-9894-0.

Aulia, M.R., Eryanto, O, Siregar, M.P.A, Yuhendra, A. and Aminah, S. (2024). Empowering Communities with Integrated Pest Management for Sustainable Agriculture. *SPEKTA (Journal Pengabdian Kepada Masyarakat: Teknologi Dan Aplikasi)*, *5*(1), 51–59. https://doi.org/10.12928/spekta.v5i1.9326.

Barnes, P., Thompson, B. and Toma, L. (2022). Finding the ecological farmer: A farmer typology to understand ecological practice adoption within Europe. *Current Research in Environmental Sustainability, 4*, 100125. https://doi.org/10.1016/j.crsust.2022.100125.

Bellon, M.R. (2001). Participatory research methods for technology evaluation: A manual for scientists working with farmers. CIMMYT, Mexico DF, Mexico. (18)

Biggs, S. (2007). Building on the positive: An actor innovation systems approach to finding and promoting pro-poor natural resources, institutional, and technical innovations. *International Journal of Agricultural Resource Governance and Ecology, 6*(2), 144–64. https://ideas.repec.org/a/ids/ijarge/v6y2007i2p144-164.html.

Braun, A.R., Thiele, G. and Fernandez, M. (2000). *Farmer Field Schools and Local Agricultural Research Committees: Complimentary Platforms for Integrated Decision Making in Sustainable Agriculture*. Agricultural Research, and Extension Network. Network Paper No. 105, Overseas Development Institute.

Carberry, P.S. (2001). Are science rigour and industry relevance both achievable in participatory action research? In: *Proceedings of the 10th Australian Agronomy Conference*. Hobart, Tas. (Australian Society of Agronomy/The Regional Institute: Gosford, NSW). Available at: http://regional.org.au/au/asa/2001/plenary/5/carberry.htm.

Carroll, D.A., Wilson, K.R., Rogers, B.L., Coates, J. and Griffin, T.S. (2025). Appeals to environmental protection and farmer adoption of sustainable natural resource management: A case study of Kaya, Burkina Faso. *Journal of Environmental Management, 376*, 124415. https://doi.org/10.1016/j.jenvman.2025.124415.

Chimonyo, V., Snapp, S.S. and Chikowo, R. (2019). Grain legumes increase yield stability in maize based cropping systems. *Crop Science, 59*(3), 1222–1235. https://doi.org/10.2135/cropsci2018.09.0532.

Chowdhury, S., Sahoo, S.K., Sujatha, B. and Dubey, V K. (2023). Challenges of IPM Adoption in India. *Biotica Research Today, 5*(1), 116–118. 10.13140/RG.2.2.24775.78240.

Cornish, F., Breton, N., Moreno-Tabarez, U., Delgado, J., Rua, M., Aikins, A.G. and Hodgetts, D. (2023). Participatory action research. *Nature Reviews Methods Primers, 3*, 34. https://doi.org/10.1038/s43586-023-00214-1.

Dhakad, A.K., Kumar, R.., Choudhary, R., Singh, S., Khan, S., Poonia, P.K. (2025). Traditional to modern perspectives on Neem (*Azadirachta indica*): A gateway to bioactive compounds, sustainable agrochemicals, and industrial applications, *Industrial Crops and Products, 231*, 121155. https://doi.org/10.1016/j.indcrop.2025.121155.

Ensor, J. and deBruin, A. (2022). The role of learning in farmer-led innovation. *Agricultural Systems, 197*, 103356. https://doi.org/10.1016/j.agsy.2021.103356.

FAO & DPPQS. (2021). Integrated Pest Management (IPM) Farmer Field School (FFS): A *Guide for Facilitators of FFS on Maize with Special Emphasis on Fall Armyworm*. New Delhi. https://doi.org/10.4060/cb3963en.

Farrington, J. and Martin, A.M. (1988). Farmer participatory research: A review of concepts and recent fieldwork. *Agricultural Administration and Extension, 29*, 247–264. https://doi.org/10.1016/0269-7475(88)90107-9.

FundsforNGOs. (n.d.). A sample proposal on Encouraging farmer-led research in crop breeding and soil health. *FundsforNGOs*. https://www.fundsforngos.org/all-proposals/a-sample-proposal-on-encouraging-farmer-led-research-in-crop-breeding-and-soil-health/.

Gómez, M., Humphries, S., Kindsvater, S., Jiménez, J., Orellana, P., Wyngaarden, S., and Dodd, W. (2024). Advancing the Sustainable Development Goals through participatory research: Long-term impacts of farmer participation on sustainable land use and livelihoods in Honduras. *Agriculture and Food Security, 13*, 61. https://doi.org/10.1186/s40066-024-00513-0.

Grabowski, P., Musumba, M., Palm, C. and Snapp, S. (2018). Sustainable agricultural intensification and measuring the immeasurable: Do we have a choice? In: S. Bell and S. Morse, (Eds.), *Routledge Handbook of Sustainability Indicators and Indices*. Taylor and Francis Press, Abington, Oxfordshire, UK. https://www.taylorfrancis.com/chapters/edit/10.4324/9781315561103-29/sustainable-agricultural-intensification-measuring-immeasurable-philip-grabowski-mark-musumba-cheryl-palm-sieglinde-snapp.

Hokkanen, H.M.T. and Menzler-Hokkanen, I. (2024a). The concept of Eco-stacking. In: Jinjun Wang, Huai Liu, Ingeborg Menzler-Hokkanen, & Hongbo Jiang, *The Concept of Eco-stacking*, pp. 1–16. eISBN : 978-1-78924-871-5. https://doi.org/10.1079/9781789248715.000.

Hokkanen, H.M.T. and Menzler-Hokkanen, I. (2024b). Management of Agricultural Soils to Support Ecosystem Services for Pest Control. In: Jinjun Wang, Huai Liu, Ingeborg Menzler-Hokkanen, & Hongbo Jiang, *The Concept of Eco-stacking*, pp. 94–113. eISBN : 978-1-78924-871-5. https://doi.org/10.1079/9781789248715.0008.

ICRISAT. (2024). *ICRISAT and Plantix Celebrate a Decade of Collaboration*. International Crops Research Institute for the Semi-Arid Tropics. https://pressroom.icrisat.org/icrisat-and-plantix-celebrate-a-decade-of-collaboration.

James, D., Blesh, J., Liebert, J., Magnanti, N., Mehrabi, Z., Motter, C. and Wittman, H. (2024). Exploring the role of a social carrier in advancing agroecology and territorial autonomy. *Agroecology and Sustainable Food Systems*, *49*(4), 587–612. https://doi.org/10.1080/21683565.2024.2431863.

Kanyama-Phiri, G.Y., Snapp, S.S. and Minae, S. (1998). Partnership with Malawian farmers to develop organic matter technologies. *Outlook on Agriculture*, *27*, 167–175. http://doi:10.1177/003072709802700307.

Kapayou, D.G., Herrighty, E.M., Hill, C.G., Camacho, C.C., Winham, D.M. and McDamiel, M.D. (2023). Reuniting the Three Sisters: Collaborative science with Native growers to improve soil and community health. *Agriculture and Human Values*, *40*, 65–82. https://doi.org/10.1007/s10460-022-10336-z.

Kar, P., Sharma, R., Jat, S.L., Nigam, S., Debnath, S., Dar, Z.A., Sravani, D., Sen, D., Bhuyan, N. and Rakshit, S. (2025). Tracking the emergence and usage of farmer led innovations (FLIs) in maize ecosystem: A pan India exploration. *Indian Journal of Traditional Knowledge, 24*(2), 177–184. https://doi.org/10.56042/ijtk.v24i2.17209.

Khumalo, T.A., Chakale, M.V., Asong, J.A., Aremu, A.O. and Amoo, S.O. (2025). Indigenous farming methods and crop management practices used by local farmers in Madibeng local municipality, South Africa. *Scientific Reports, 15*, 8918. https://doi.org/10.1038/s41598-025-91210-w.

Kumar, V. T., Raidu, D.V., Killi, J., Pillai, M., Shah, P., Kalavakonda, V. and Lakhey, S. (2009). *Economically Sound, Economically Viable, Community Managed Sustainable Agriculture (CMSA) in Andhra Pradesh India*. 75961. The World Bank. https://documents1.worldbank.org/curated/en/805101468267916659/pdf/759610WP0P118800agriculture0AP02009.pdf.

Lantz, P.M., Israel, B.A., Schulz, A.J. and Reyes, A. (2006). Community based participatory research: Rationale and relevance for social epidemiology. In: J.M. Oakes and J.S. Kaufman (Eds.), *Methods for Social Epidemiology*. Jossey-Bass, San Francisco, CA, pp. 239–266.

Leakey, R.R.B. (2017). In: van Alfen, N. et al., (Eds.), *Encyclopedia of Agriculture and Food Systems* (Vol. 1). Elsevier, San Diego, pp. 253–269; Editor(s): Roger R.B. Leakey, *Multifunctional Agriculture*, pp. 297–314. ISBN: 9780128053560. https://doi.org/10.1016/B978-0-12-805356-0.00028-3.

Leeuwis, C., Schut, M., Waters-Bayer, A., Mur, R., Atta-Krah, K. and Douthwaite, B. (2014). *Capacity to Innovate from a System CGIAR Research Program Perspective*. Penang, Malaysia: CGIAR Research Program on Aquatic Agricultural Systems (AAS) Program Brief AAS. https://digitalarchive.worldfishcenter.org/server/api/core/bitstreams/19c0b499-fce2-4980-b8d6-33a9aec5932e/content.

Luttermoser, T., Khan, Z.R., Midega, C.A.O., Nyagol, D., Jonsson, M. and Poveda, K. (2023). Are pests adapting to the push-pull system? Ecologically intensified farms in Kenya maintain successful pest control over time. *Agriculture, Ecosystems, and Environment*, *347*, 108345. https://doi.org/10.1016/j.agee.2023.108345.

Mancini, F. (2006). *Impact of IPM Farmer Field Schools on the Environment, Health, and Livelihoods of Cotton Growers in Southern India*. Doctoral thesis, Biological Farming Systems Group, Wageningen University, The Netherlands. ISBN 90-8504-388-3.

Matowo, N.S., Tanner, M., Temba, B.A., Finda, M., Mlacha, Y.P., Utzinger, J., Okumu, F.O. (2022). Participatory approaches for raising awareness among subsistence farmers in Tanzania about the spread of insecticide resistance in malaria vectors and the possible link to improper agricultural pesticide use. *Malaria Journal*, *21*(1), 277. https://doi.org/10.1186/s12936-022-04289-1.

Nelson, E., Hargreaves, S. and Muldoon, D. (2023). Farmer knowledge as formal knowledge: A case study of farmer-led research in Ontario, Canada. *Journal of Agriculture, Food Systems, and Community Development*, *12*(4), 15–38. https://doi.org/10.5304/jafscd.2023.124.010.

Pauli, N., Barrios, E., Conacher, A.J. and Oberthür, T. (2011). Soil macrofauna in agricultural landscapes dominated by the Quesungual Slash-and-Mulch Agroforestry System (QSMAS), western Honduras. *Applied Soil Ecology*, *47*(2), 119–132. https://doi.org/10.1016/j.apsoil.2010.11.005.

Pawera, L., Manickam, R., Wangungu, C., Bonnarith, U., Schreinemachers, P. and Ramasamy, S. (2024). Guidance on farmer participation in the design, testing, and scaling of agricultural innovations. *Agricultural Systems, 218*, 104006. https://doi.org/10.1016/j.agsy.2024.104006.

PFO (2023). *Farmer-led, Participatory Action Research: A New Paradigm for Agricultural Research in the Pacific*. Pacific Farmers Organisations. https://pacificfarmers.com/resource/farmer-led-participatory-action-research-a-new-paradigm-for-agricultural-research-in-the-pacific/.

Pickett, J.A., Woodcock, C.M., Midega, C.A.O. and Khan, Z.R. (2014). Push–pull farming systems. *Current Opinion in Biotechnology, 26*, 125–132. https://doi.org/10.1016/j.copbio.2013.12.006.

Pound, B. and Conroy, C. (2017). Chapter 11 - The Innovation Systems Approach to Agricultural Research and Development, Editor(s): Sieglinde Snapp, & Barry Pound, *Agricultural Systems* (2nd Edn.), pp. 371–405. ISBN 9780128020708. https://doi.org/10.1016/B978-0-12-802070-8.00011-6.

Prinila, A.S., Ravi, G., Preetha, G., Anandham, R. and Manjunatha, C. (2025). Traditional insights into ITK and its delineation for sustainable crop protection. *Plant Science Today, 12*(1), 1–7. https://doi.org/10.14719/pst.6110.

Rao, N. and Swaminathan, M.S. (2017). A Farmer-Led Approach to Achieving a Malnutrition-Free India. *Agricultural Research, 6*, 1–7. https://doi.org/10.1007/s40003-016-0240-8.

Rezaei, R., Safa, L. and Ganjkhanloo, M.M. (2020). Understanding farmers' ecological conservation behavior regarding the use of integrated pest management: An application of the technology acceptance model. *Global Ecology and Conservation, 22*, e00941. https://doi.org/10.1016/j.gecco.2020.e00941.

Richardson, M., Coe, R., Descheemaeker, K., Haussmann, B., Wellard, K., Moore, M., Cady, J.M., Gubbels, P., Tchuwa, F., Paz, Y.R. and Nelson, R. (2021). Farmer research networks (FRN) in principle and practice. *International Journal of Agricultural Sustainability*, *20*(3), 247–264. https://doi.org/10.1080/14735903.2021.1930954.

Rizzo, G., Migliore, G., Schifani, G. and Vecchio, R. (2024). Key factors influencing farmers' adoption of sustainable innovations: A systematic literature review and research agenda. *Organic Agriculture, 14*, 57–84. https://doi.org/10.1007/s13165-023-00440-7.

Rogé, P., Snapp, S.S., Kakwera, M.K., Mungai, L., Jambo, I. and Peter, B. (2016). Ratooning and perennial staple crops in Malawi: A review. *Agronomy and Sustainable Development. 36*, 50. http://doi:10.1007/s13593-016-0384-8.

SAN. (2024). Farmer-Led Initiatives for Sustainable Land Management. *Sustainable Agricultural Network.* Assessed on 19 May 2025 from https://www.sustainableagriculture.eco/post/farmer-led-initiatives-for-sustainable-land-management#:~:text=Farmer%2Dled%20participatory%20research%20involves,needs%20of%20the%20farming%20community.

Sharma, S. (2023). Cultivating sustainable solutions: Integrated pest management (IPM) for safer and greener agronomy. *Corporate Sustainable Management Journal*, *1*(2), 103–108. http://doi.org/10.26480/csmj.02.2023.103.108.

Snapp, S. and K.L. Heong. (2003). Scaling up and out. In: B. Pound, S.S. Snapp, C. McDougal and A. Braun (Eds.), *Managing Natural Resources for Sustainable Livelihoods: Uniting Science and Participation*. Earthscan, London, UK.

Snapp, S.S., Grabowski, P., Chikowo, R., Smith, A. Anders, E., Sirrine, D., Chimonyo, V. and Bekunda, M. (2018). Maize yield and profitability trade-offs with social, human, and environmental performance: Is sustainable intensification feasible? *Agricultural Systems*, *162*, 77–88. http://doi:10.1016/j.agsy.2018.01.012.

Snapp, S.S., Kanyama-Phiri, G., Kamanga, B., Gilbert, R. and Wellard, K. (2002). Farmer and researcher partnerships in Malawi: Developing soil fertility technologies for the near-term and far-term. *Experimental Agriculture, 38*, 411–431. http://doi:10.1017/S0014479702000443.

Suuk, S.S., Laube, W., Seyni, A.A. and Cantrell, R.A. (2025). The Adoption of Farmer-managed Natural Regeneration in Dogonkiria, Niger. *Human Ecology, 53*, 127–136. https://doi.org/10.1007/s10745-025-00568-y.

Thorburn, C. (2015). The Rise and Demise of Integrated Pest Management in Rice in Indonesia. *Insects*, *6*(2), 381–408. PMCID: PMC4553486. https://doi.org/10.3390/insects6020381.

Tinzaara, W., Ssekiwoko, F., Kikulwe, E. and Karamura, E. (2019). Effectiveness of learning and experimentation approaches for farmers as a community based strategy for banana xanthomonas wilt management. *Journal of Agricultural Extension and Rural Development, 11*(7), 128–138. https://doi.org/10.5897/JAERD20198.1000.

Tittonell, P., Vanlauwe, B., Leffelaar, P.A., Rowe, E.C. and Giller, K.E. (2005). Exploring diversity in soil fertility management of smallholder farms in western Kenya: I. Heterogeneity at region and farm scale. *Agriculture, Ecosystems, and. Environment*, *110*, 149–165. http://doi:10.1016/j.agee.2005.04.001.

Vaughn, L.M. and Jacquez, F. (2020). Participatory Research Methods: Choice Points in the Research Process. *Journal of Participatory Research Methods, 1*(1), https://doi.org/10.35844/001c.13244.

Vikas and Ranjan, R. (2024). Agroecological approaches to sustainable development. *Frontiers in Sustainable Food Systems*, *8*, 1405409. https://doi.org/10.3389/fsufs.2024.1405409.

Wang, H., Snapp, S.S. Fisher, M. and Viens, F. (2019). A Bayesian analysis of longitudinal farm surveys in Central Malawi reveals yield determinants and site-specific management strategies. *PLoS One*, *14*(8), e0219296. https://doi.org/10.1371/journal.pone.0219296.

Wang, J., Liu, H, Menzler-Hokkanen, I. and Jiang, H. (2024). *The Concept of Eco-stacking*. eISBN:978-1-78924-871-5. https://www.cabidigitallibrary.org/doi/book/10.1079/9781789248715.0000.

Waters-Bayer, A., Kristjanson, P., Wettasinha, C., Veldhuizen, L. V., Quiroga, G., Swaans, K. and Douthwaite. (2015). Exploring the impact of farmer-led research supported by civil society organisations. *Agriculture and Food Security, 4,* 4. https://doi.org/10.1186/s40066-015-0023-7.

Wettasinha, C., Waters-Bayer, A., van Veldhuizen, L., Quiroga, G. and Swaans, K. (2014). *Study on Impacts of Farmer-led Research Supported by Civil Society Organizations*. Penang, Malaysia: CGIAR Research Program on Aquatic Agricultural Systems. Working Paper: AAS-2014-40. https://assets.publishing.service.gov.uk/media/57a089c5e5274a31e0000266/AAS-2014-40.pdf.

Index

About the Editors

Dr. Belaghihalli N Gnanesh is a Senior Scientist and Head at ICAR JSS Krishi Vigyan Kendra, Suttur, Mysuru, India. He previously held the position of Professor and Dean at the Sampoorna International Institute of Agricultural Science and Horticultural Technology, affiliated with the University of Mysore. He obtained a PhD in Genetics and Plant Breeding from the International Crops Research Institute for the Semi-Arid Tropics (ICRISAT), Hyderabad, and the University of Agricultural Sciences, Bangalore. With several years of research experience, Dr. Gnanesh has made significant contributions in the areas of germplasm resource management, crop genomics, and disease-resistance breeding. He is a recipient of the prestigious SERB Ramanujan Fellowship and was awarded a Visiting Fellowship by the Natural Sciences and Engineering Research Council of Canada. During his tenure at Agriculture and Agri-Food Canada (AAFC), Winnipeg, Canada he developed cost-effective, allele-specific diagnostic markers for oat crown rust resistance genes. Dr. Ganesh's current interests span plant genomics, genome survey, SNP discovery, and linkage mapping. He is also deeply involved in agricultural training and extension, focusing on field demonstrations, technology trials and precision farming that promote sustainable and climate-resilient agriculture.

Prof. Megha N Parajulee has a BSc (Agri. Econ. Hons.) from Y. S. Paramar University of Horticulture and Forestry, Solan, India, and an M.S. and PhD in Entomology and Quantitative Ecology from the University of Wisconsin-Madison, USA. She joined Texas A&M University in 1994 as a Cotton Entomologist and currently serves as Regents Fellow Professor and Cotton Entomology Program Leader. Prof. Parajulee is a leader in cotton entomology and cropping systems research, teaching, and service, with 135 refereed and over 400 non-refereed publications, 165 invited and 350 contributed presentations, and 35 M.S. /PhD students mentored during his 31-year tenure at Texas A&M University. Prof. Parajulee served as a Senior Fullbright Fellow to Nepal and Uzbekistan, USAID Advisor for agriculture curriculum development in Nepal, and has presented 45 invited plenary/keynote lectures in 16 countries in the areas of sustainable crop production and food security. Her scientific contributions and impact have been recognized with several awards, including the Texas A&M Vice Chancellor's Award in Research Excellence, Vice Chancellor's Award in Team Collaboration, AgriLife Faculty Fellow Award, Texas A&M Regents Fellow Award, Entomological Society of America IPM Excellence Award, US Department of Agriculture Southern Region IPM Hall of Fame Award, and US President's National and Community Service Award.

Dr. Ankush M Raut is an Associate Professor in the Department of Entomology at the School of Agriculture, Lovely Professional University, Punjab, India. He obtained a PhD in Agricultural Entomology from Bidhan Chandra Krishi Vishwavidyalaya, Mohanpur, India. He guided 1 PhD and 8 M.Sc. students during his professional career. He has published more than 20 research articles in international and national peer-reviewed journals, along with 5 book chapters and 4 patents/ copyrights during his career. He is also a Life Member of Indian Science Congress Association (West Bengal) and Association for Advancement in Plant Protection (West Bengal). He has organized 3 international conferences as organizing secretary/convener. Along with this, he has participated in the 106th Indian Congress (2019) as Local Sectional Secretary. His current research focuses on Biodiversity and conservation of predators and pollinators, Insect ecology, and Bee health.

Prof. Chittaranjan Kole is an internationally reputed scientist with an illustrious professional career spanning over forty years and original contributions in the fields of plant genomics, biotechnology, and molecular breeding leading to the publication of more than 170 impactful research articles and reviews. He has edited over 200 books for the globally leading publishers. His scientific contributions and editing acumen have been appreciated by seven Nobel Laureates including Prof. Norman Borlaug, Arthur Kornberg, Werner Arber, Phillip Sharp, Günter Blobel, Lee Hartwell and Roger Kornberg. He has been honored with several Fellowships, Honorary Fellowships, and national and international awards including the Outstanding Crop Scientist award conferred by the International Crop Science Society. He was also awarded the Raja Ramanna Fellow by the Department of Energy, Government of India. He has served in all prestigious positions in academia including as Vice-Chancellor of Bidhan Chandra Krishi Viswavidyalaya, Project Coordinator of Indo-Russian Center of Biotechnology in India, and Director of Research of Institute of Nutraceutical Research of Clemson University, in USA. He also served as a Visiting Professor at the Pennsylvania State University and Clemson University. He is also heading the International Climate-Resilient Crop Genomics Consortium, International consortium for Phytomedomics and Nutriomics and Genome India International as their Founder President.